Particle Physics

Particle Physics

Editor:

Olga Moreira, Ph.D.

ArclerPress

www.arclerpress.com

Particle Physics

Editor: Olga Moreira, Ph.D.

708 3rd Avenue, 6th Floor
New York
NY 10017
United States of America
www.arclerpress.com

ISBN: 978-1-68094-435-8

Library of Congress Control Number: 2016951127

About the Editor

Olga Moreira, Ph.D.

Olga Moreira obtained her Ph.D. in Astrophysics from the University of Liege (Belgium) in 2010, her BSc. in Physics and Applied Mathematics from the University of Porto (Portugal). Her post-graduate travels and international collaborations with the European Space Agency (ESA) and European Southern Observatory (ESO) led to great personal and professional growth as a scientist. Currently, she is working as an independent researcher, technical writer, and editor in the fields of Mathematics, Physics, Astronomy and Astrophysics.

Contents

List of Contributors

Abdelhak Djouadi
Laboratoire de Physique Théorique, U. Paris–Sud and CNRS, 91405 Orsay, France
Present address: TH Unit, CERN, Geneva, Switzerland

Rabindra N. Mohapatra
Maryland Center for Fundamental Physics and Department of Physics, University of Maryland, College Park, MD 20742, USA

Oleg Antipin
INFN — Sezione di Firenze, Via G. Sansone 1, I-50019 Sesto Fiorentino, Italy

Michele Redi
Dipartimento di Fisica dell'Universit`a di Pisa and INFN, Pisa, Italy
National Institute of Chemical Physics and Biophysics, Tallinn, Estonia

Alessandro Strumia
Dipartimento di Fisica dell'Universit`a di Pisa and INFN, Pisa, Italy
National Institute of Chemical Physics and Biophysics, Tallinn, Estonia

Tommi Alanne
Department of Physics, University of Jyväskylä, FI-40014 University of Jyväskylä, Finland
Helsinki Institute of Physics, FI-00014 University of Helsinki, Finland

Kimmo Tuominen
Department of Physics, University of Helsinki, FI-00014 University of Helsinki, Finland
Helsinki Institute of Physics, FI-00014 University of Helsinki, Finland

Ville Vaskonen
Department of Physics, University of Jyväskylä, FI-40014 University of Jyväskylä, Finland
Helsinki Institute of Physics, FI-00014 University of Helsinki, Finland

O. V. Teryaev
Bogoliubov Laboratory of Theoretical Physics, JINR, 141980 Dubna, Moscow Region, Russia

Michel Dubois-Violette
Laboratoire de Physique Théorique, CNRS, Université Paris-Sud, Université Paris-Saclay, Bâtiment 210, F-91 405 Orsay, France

J. L. Chkareuli
Center for Elementary Particle Physics, Ilia State University, 0162 Tbilisi, Georgia
E. Andronikashvili Institute of Physics, 0177 Tbilisi, Georgia

Morimitsu Tanimoto
Department of Physics, Niigata University, Niigata 950-2181, Japan

Tsutomu T. Yanagida
Kavli IPMU, TODIAS, University of Tokyo, Kashiwa 277-8583, Japan

Peter Lowdon
Physik-Institut, Universität Zürich, Winterthurerstrasse 190, 8057 Zürich, Switzerland

Wen-Ping Chen
Physics Department, National Taiwan University, Taipei 10617, Taiwan

Yu-Chih Chen
Physics Department, National Taiwan University, Taipei 10617, Taiwan

Ting-Wai Chiu
Physics Department, National Taiwan University, Taipei 10617, Taiwan
Center for Quantum Science and Engineering, National Taiwan University, Taipei 10617, Taiwan

Han-Yi Chou
Physics Department, National Taiwan University, Taipei 10617, Taiwan

Tian-Shin Guu
Center for Science Education, National Ilan University, I-Lan 260, Taiwan

Tung-Han Hsieh
Research Center for Applied Sciences, Academia Sinica, Taipei 115, Taiwan

L. I. Abou-Salem
Physics Department, Faculty of Science, Benha University, Benha 13518, Egypt

N. M. El Naggar
Physics Department, Faculty of Science, Benha University, Benha 13518, Egypt

I. A. Elmashad
Physics Department, Faculty of Science, Benha University, Benha 13518, Egypt

M. Monemzadeh
Department of Physics, University of Kashan, Ghotb Ravandi Boulevard, Kashan 87317-51167, Iran

N. Tazimi
Department of Physics, University of Kashan, Ghotb Ravandi Boulevard, Kashan 87317-51167, Iran

Sh. Babaghodrat
Department of Physics, University of Kashan, Ghotb Ravandi Boulevard, Kashan 87317-51167, Iran

Konstantin Ottnad
Institut für Strahlen- und Kernphysik (Theorie) and Bethe Center for Theoretical Physics, Universität Bonn, D-53115 Bonn, Germany

Carsten Urbach
Institut für Strahlen- und Kernphysik (Theorie) and Bethe Center for Theoretical Physics, Universität Bonn, D-53115 Bonn, Germany

Falk Zimmermann
Institut für Strahlen- und Kernphysik (Theorie) and Bethe Center for Theoretical Physics, Universität Bonn, D-53115 Bonn, Germany

S. J. Brodsky
SLAC National Accelerator Laboratory, Stanford University, Stanford, CA 94301, USA

A. Kusina
Laboratoire de Physique Subatomique et de Cosmologie, Université Grenoble-Alpes, CNRS/IN2P3, 53 avenue des Martyrs, 38026 Grenoble, France

F. Lyonnet
Southern Methodist University, Dallas, TX 75275, USA

I. Schienbein
Laboratoire de Physique Subatomique et de Cosmologie, Université Grenoble-Alpes, CNRS/IN2P3, 53 avenue des Martyrs, 38026 Grenoble, France

H. Spiesberger
PRISMA Cluster of Excellence, Institut für Physik, Johannes Gutenberg-Universität Mainz, 55099 Mainz, Germany
Centre for Theoretical and Mathematical Physics and Department of Physics, University of Cape Town, Rondebosch 7700, South Africa

R. Vogt
Nuclear and Chemical Sciences Division, Lawrence Livermore National Laboratory, Livermore, CA 94551, USA
Physics Department, University of California at Davis, Davis, CA 95616, USA

Md. Nasim
Department of Physics and Astronomy, UCLA, Los Angeles, CA 90095, USA

Vipul Bairathi
School of Physical Sciences, National Institute of Science Education and Research, Bhubaneswar 751005, India

Mukesh Kumar Sharma
Physics Department, University of Jammu, Jammu 180001, India

Bedangadas Mohanty
School of Physical Sciences, National Institute of Science Education and Research, Bhubaneswar 751005, India

Anju Bhasin
Physics Department, University of Jammu, Jammu 180001, India

Federico Alberto Ceccopieri
IFPA, Université de Liège, Allée du 6 Août, 4000 Liège, Belgium

A. B. Kurepin
Institute for Nuclear Research, RAS, 60th October Anniversary Prospect 7a, Moscow 117312, Russia

N. S. Topilskaya
Institute for Nuclear Research, RAS, 60th October Anniversary Prospect 7a, Moscow 117312, Russia

Abhay Deshpande
Department of Physics and Astronomy, Stony Brook University, Stony Brook, NY 11794, USA
RIKEN BNL Research Center, Brookhaven National Laboratory, Upton, NY 11973, USA

Jeppe R. Andersen
Department of Physics, Institute for Particle Physics Phenomenology, Durham University, Durham, UK

Michael Rauch
Institute for Theoretical Physics, Karlsruhe Institute of Technology (KIT), Karlsruhe, Germany

Michael Spannowsky
Department of Physics, Institute for Particle Physics Phenomenology, Durham University, Durham, UK

Preface

The recent discovery of the Higgs-Boson opened new horizons and instigated crucial debates about its consequences for the Standard Model (SM) of the electroweak and strong interactions and for new physics models beyond it such as supersymmetric theories (SUSY). This is leading to a next round of corresponding questions concerning the origin of elementary particle masses, the nature of dark matter and the role leptons, particularly, neutrinos which may play an important role in the matter-antimatter asymmetry of the Universe. This subject is discussed in detail in the first section of this book. The remaining content of this book focuses on the advances of specific theoretical modelling and experiments performed at the Large Hadronic Collider (LHC) and Relativistic Heavy Ion Collider (RHIC).

With regards to theoretical modelling discussed in this book, these include the use of Occam's Razor approach in the quark mass matrices; the study of the how boundary terms behave in Quantum Field theory (QFT) and how they influence the angular momentum operator in Quantum Chromodynamics (QCD); the study of the masses and decay constants of the pseudoscalar D mesons in two flavors lattice QCD with domain-wall fermion; the study of the quark-gluon plasma equation of state within the generalized Uncertainty Principle, the mass calculation of a pentaquark composed of baryons and mesons, the study η and η' mesons and their mixing angle. As for the experiments performed at the Large Hadronic Collider (LHC) and Relativistic Heavy Ion Collider (RHIC), these include a fixed-target experiment using the LHC beams for the study of backward particle production; relativistic heavy-ion experiments at RHIC to create extremely hot and dense matter to study the QCD phase structure; and a high luminosity fixed target experiment using the LHC beams to search for intrinsic charm of the nucleon. The energy scales relevant for the questions about the origin of the weak scale and of particle masses range from the TeV to the Planck scale. Experimental exploration of these questions requires advances in accelerator and detector technologies to an unprecedented energy reach, sensitivity and precision. An overview of the past and future experiments at LHC and RHIC as well as the potential use of proposed International Linear Collider (ILC) for the understanding of dark matter nature are also discussed.

Editor
Olga Moreira, Ph.D.

INTRODUCTION

Particle physics is the study of fundamental particles and the forces that guide them. Because many of the fundamental particles only appear during relativistic collisions in particle accelerators, colloquially called "atom smashers," particle physics is also known as "high-energy physics". Physicists have been slamming together particles at extreme speeds since 1929.

The best picture of particle physics we have today is called the Standard Model, developed painstakingly in the 1970s. It was a reaction to the "particle zoo", a huge proliferation of unusual fundamental particles discovered during high-energy physics experiments in the 1950s and 1960s. The final particle count ended up to be around 31, including 24 fermions (quarks, electrons, neutrinos, and their antiparticles), 6 bosons (one of which, the graviton, has yet to be observed), and one elusive particle responsible for the property of mass itself, which also has yet to observed, the Higgs boson. Basically, the fermions make up matter and the bosons mediate interactions between matter. The light coming from your computer screen is made up of photons which are bosons. They are interacting with the fermions making up your eyeball.

Most of the matter all around us is composed of only a few fundamental particles: up quarks, down quarks, and electrons. There are also 50 trillion low-mass neutrinos streaming through our bodies every second, passing through the entire Earth almost as if it wasn't even there. Neutrinos, whose name means "little neutral particle," are so elusive that it wasn't even known that they had a mass until 1998. One of the newest areas of astronomy is known as neutrino astronomy, whereby neutrino flux from the Sun and supernovae are observed using huge detectors.

Although only a few of particles in the particle zoo make up the matter we are familiar with, particle physics gives us great insight into the structure of reality by showing us the less common variants and how they fit together into a unified family. Particle physics can be said to be responsible for the existence of nuclear power, nuclear medicine, and nuclear bombs. Particle physics is considered one of the most respected areas of science because it ends up producing insights which are useful to other areas, sort of like math.

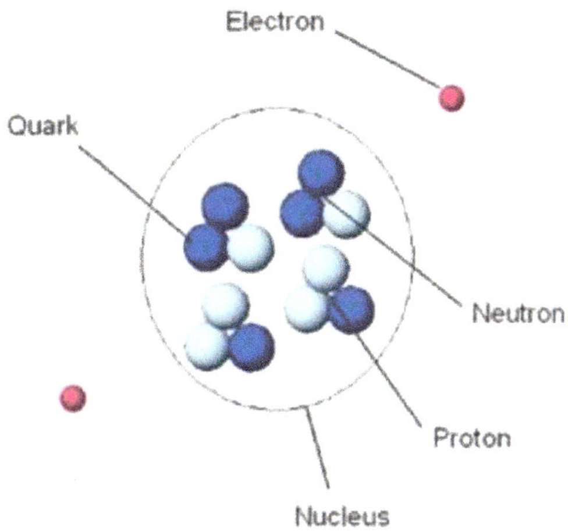

The Standard Model

Particle physicists now believe they can describe the behavior of all known subatomic particles within a single theoretical framework called the Standard Model, incorporating quarks and leptons and their interactions through the strong, weak and electromagnetic forces. Gravity is the one force not described by the Standard Model.

Quarks

The quark scheme was suggested by the symmetries in the way the many *mesons* and *baryons* seemed to be arranged in families. Theorists Gell-Mann and Zweig independently proposed in 1964 that just three fundamental "constituents" (and their anti-particles) combined in different ways according to the rules of mathematical symmetries could explain the whole zoo. Gell-Mann called these constituents *quarks*, and the three types were named *up*, *down* and *strange* quarks. Evidence for quark-like constituents of protons and neutrons became clear in the late 1960s and 1970s. In 1974, a new particle was unexpectedly discovered at *SLAC* (Stanford Linear Accelerator Center). It was given the unwieldy dual name *J/Psi*, because of its simultaneous discovery by two groups of experimenters! The *J/Psi* was later shown to be a bound state of a completely new quark-antiquark pair, which nevertheless had been predicted on the basis of a subtle phenomenon. The new fourth quark was named *charm*. (We do not wish to comment here on the choice of names!)

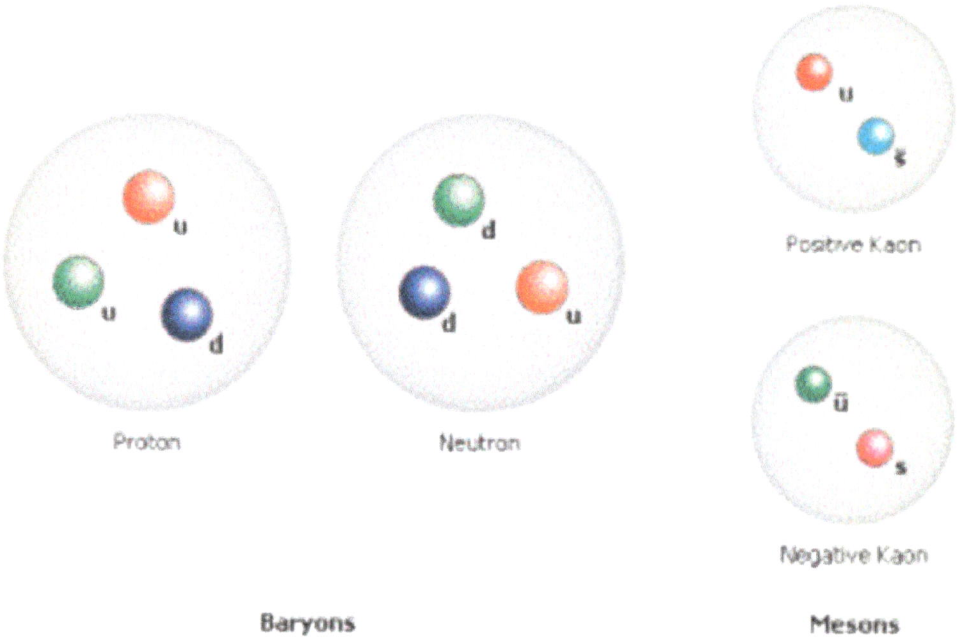

The four-quark scheme was extended to its present state of *six quarks* by the addition of a new pair, in a prediction by theorists Cabbibo and independently, Kobayashi and Maskawa (collectively known as CKM). So now we have the six quarks: *up*, *down*, *strange*, *charm*, *bottom* and *top* quarks and they each have their partner anti-quarks. The quarks are usually labeled by their first letters: *u, d, s, c, b* and ***t***. In various combinations they make up all the mesons and baryons that have been seen. The six-quark prediction was fulfilled when in 1977 a new heavy meson called the *Upsilon* was discovered at Fermilab and later shown to be the bound state of the *bottom* and *anti-bottom* quark pair. The *B* meson, containing an anti-*b* quark and a *u* or *d* quark was discovered by the CLEO experiment at Cornell in 1983. Finally, in 1998, conclusive evidence of the existence of the super heavy *top* quark was obtained at Fermilab.

Leptons

What about leptons? Only the electron, muon and neutrino were known before the 1960s. These behave differently from the mesons and baryons. First, they are much less massive. The mass of the electron is almost 2,000 times smaller than the mass of the proton, and the muon appears to be just a heavier version of the electron, its mass being nine times smaller than that of the proton. The neutrino has almost no mass at all, and up until recently, its mass was thought to be truly zero. Hence the name "leptons" or light particles. Second, the electron and muon interact with matter mainly through their electric charges; the neutrino being neutral, hardly at all. They all have a weak interaction with the matter in nuclei and, in high energy collisions, they do not produce the profusion of new mesons and baryons that protons and neutrons do when colliding with nuclei.

In 1962, the first experiment using a high-energy neutrino beam (the PhD thesis of this author) showed that the electron has its own electron-neutrino, and the muon its own distinct muon-neutrino. This was the very first evidence that there could be families or generations of pairs of fundamental particles. This notion was dramatically extended in 1974, when shortly after the discovery of the J/Psi, a new heavy lepton was discovered, called the tau, almost twice as massive as the proton, but behaving like the other leptons, sharing the weak interaction property! This was the first evidence that three pairs or families of leptons existed: the electron and electron-neutrino, the muon and muon-neutrino and the tau and tau-neutrino.

Quarks and leptons have an intrinsic angular momentum called spin, equal to a half-integer (1/2) of the basic unit and are labeled as fermions. Particles that have zero or integer spin are called bosons.

Table 1: The Quark and Lepton families

QUARKS	(*u*) up-quark mass = 0.005 (*d*) down-quark mass = 0.009	(*c*) charm-quark mass = 1.5 (*s*) strange-quark mass = 0.16	(*t*) top-quark mass = 186 (*b*) bottom-quark mass = 5.2	Charge = +2/3 Charge = -1/3
LEPTONS	(ν_e) elec-neutrino mass ~ 0 (**e**) electron mass = 0.00054	(ν_μ) muon-neutrino mass ~ 0 (μ) muon mass = 0.11	(ν_τ) tau-neutrino mass ~ 0 (τ) tau mass = 1.9	Charge = 0 Charge = -1

Forces and Interactions

Now we must tackle the fundamental forces or interactions among the quarks and leptons: Gravity, the Weak Force, Electromagnetism, and the Strong Force. Of these, our everyday world is controlled by gravity and electromagnetism. The strong force binds quarks together and holds nucleons (protons & neutrons) in nuclei. The weak force is responsible for the radioactive decay of unstable nuclei and for interactions of neutrinos and other leptons with matter.

The intrinsic strengths of the forces can be compared relative to the strong force, here considered to have unit strength (i.e., =1.) In these terms, the electromagnetic force has an intrinsic strength of (1/137). The weak force is a billion times weaker than the strong force. The weakest of them all is the gravitational force. This may seem strange, since it is strong enough to hold the massive Earth & planets in orbit around the Sun! But we know that that the gravitational force between two bodies a distance r apart is proportional to the product of the two masses (M & m) and inversely proportional to the distance r squared:

$$F_G = \frac{GMm}{r^2}$$

We see now what is meant by intrinsic strength. It is given by the magnitude of the universal force constant, in this case, G, independent of the masses or distances involved. In similar terms, the electromagnetic force between two particles is proportional to the product of the two charges (Q & q) and inversely to the distance r squared:

$$F_{em} = \frac{\alpha Qq}{r^2}$$

Here the universal constant alpha α , gives the intrinsic strength.

We can compare the relative strengths of the electromagnetic repulsion and the gravitational attraction between two protons of unit charge using the above equations. Independent of the distance, the ratio turns out to be 10^{36} Thus the two protons will repel each other and fly apart, easily overcoming the puny gravitational attraction.

As we noted before, forces can be represented in the theory as arising from the exchange of specific particles called gauge bosons, the quanta of the "force field". Just as photons are real (i.e., quanta of light!) and can be radiated (shaken off) when charged particles are accelerated or decelerated, the other gauge bosons (see below) can also be created and observed as real particles. All the bosons have 0 or integer spins.

Table 2: Forces and their quanta, the gauge bosons.Charge is in units of electron charge

FORCE	Relative Strength	Gauge Boson	Mass (rel. to proton)	Charge	Spin
Strong	1	Gluon (g)	0	0	1
Electromagnetic	1/137	Photon (γ)	0	0	1
Weak	10^{-9}	$W^{\pm}, Z$	86. 97	± 1. 0	1
Gravity	10^{-38}	Graviton (G)	0	0	2

The carriers of the strong force are called gluons, the glue that holds quarks together in protons and neutrons and also helps form nuclei. The carriers of the weak force come in three forms, and are called weak bosons: the $W^{\pm}$ and the Z^0. The carriers of the gravitational field are called gravitons and are unique in having a spin of 2.

Unification

For a universal theory, four forces are too many. Why is there not just one universal force? For decades physicists have been striving for the unification of the four forces into one universal force that existed at least in the primordial stage of the Universe. In such a picture, the four forces we observe today are just manifestations of the original single force. However, we must understand that our existence depends on having these different forces now. If gravity were not so weak, there might only have been one massive black hole instead of galaxies, stars and planets. If electromagnetic forces were not in delicate balance with the strong force, nuclei would disintegrate — no atoms or molecules, chemistry or biology! The weak force allows more subtle phenomena —the slow burning of stars like our Sun may not be possible without the weak interaction; supernova explosions which create all elements heavier than iron also depend on just the right strength of neutrino interactions; and radioactivity in its bowels allows the Earth to remain a warm hospitable body.

It is not quite satisfactory to have four different theories to account for these four forces. The electromagnetic interaction of particles is explained by a well established modern theory of Quantum Electrodynamics (QED). The weak interaction had its own theory but these two have now been combined as the Electroweak Theory in the Standard Model. The strong interaction between quarks and gluons has another theory called Quantum Chromodynamics (QCD), where the equivalent of electric charge is named "color". And Einstein's General Theory of Relativity explains how the gravity we know is a manifestation of the basic geometry of space-time.

Just as Maxwell showed that electricity and magnetism were manifestations of the same basic phenomenon of electromagnetism, the Electroweak theory, which in 1979 won the Nobel Prize for Glashow, Salam and Weinberg, succeeds in unifying the Weak and Electromagnetic interactions into what is called the Electroweak force. When we noted the intrinsic strengths of the four different interactions in Table 2, we omitted to say that these strengths could depend on the "temperature" or energy level of the interaction. Although these strengths are quite different at present temperatures (e.g., at 300K or equivalent energy of about 1/40 eV), the weak interaction depends strongly on the energy, and in collisions at near 1000 GeV, it gets just as strong as the electromagnetic interaction! The Electroweak theory of the Standard Model explains all this. The basic equations are symmetric in the way the two interactions occur and in fact the masses of all the quanta are zero. However, as the temperatures drops, the symmetry is broken and the quanta split up into four different gauge bosons of different masses: the W^+ and W^- (both 80 GeV), the Z^0 (91 GeV) and the photon γ with zero mass. At "room temperature" , the massive W and Z do not play an important part. But at very high energies of 300 GeV or more, the difference between the zero mass photon and the heavier W and Z bosons is erased, and they all act equally strongly. In 1983 the W boson and in 1984 the Z boson were observed at the CERN laboratory in Geneva, in high energy collisions of protons with antiprotons. They had the predicted masses. The Standard Model was on its way.

There is however one piece of evidence yet to be found. We mentioned above that the basic symmetry of the electroweak theory is broken as the temperature drops and the forces separate in strength as the bosons gain mass. The culprit that causes this is actually a new field called the Higgs field. It is possible to visualize how this works. Recall that mass is a manifestation of inertia or resistance to acceleration. If a Higgs field suddenly permeates all of space as the Universe cools, it can act as a drag on every particle moving in space, the drag depending on how well each interacts with the Higgs field. This drag shows up as inertia and thus a measurable mass of the particles that were originally massless. But now we have to look for the boson that carries this field — the Higgs boson. This is now the one feature of the Standard Model still needed to clinch the picture. It is expected to have a mass of about 100 GeV, within the reach of the largest accelerators planned for the immediate future.

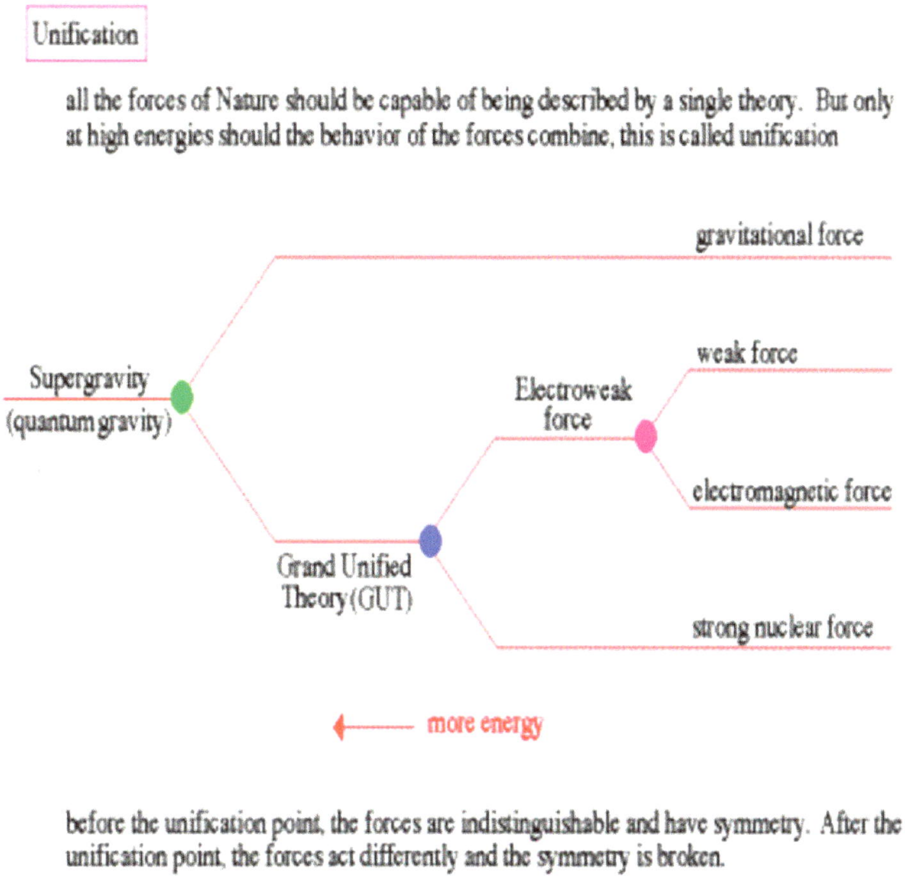

Figure 1: Unification.

Beyond the Standard Model

Theories, called "Grand Unification Theories" or GUTs, have been proposed to unify the electroweak force with the strong force. But so far no concrete evidence has been found for them. Beyond that, the holy grail of unification has long been the unification of gravity with all the other forces. Einstein himself labored in vain to fit gravity into a scheme where it could be compatible with quantum theory.

The theory of Supersymmetry requires a whole new set of particles beyond the Standard Model complement: a heavy partner for each quark, lepton and gauge boson of the old set, together all of them making up one great super-family of particles. The three forces strong, electromagnetic and weak all have exactly equal strengths in this theory at a very high energy. And of course, it gives experimentalists a whole new game of looking for new particles. It is just possible that one of these new super particles is a primordial relic of the Big Bang and makes up the Dark Matter in the Universe, a further incentive to discover these super-partners.

Meanwhile theoretical studies range far and wide in a search for the Theory of Everything (TOE). Most familiar is String Theory, which pictures particles as infinitesimal little vibrating loops of strings in 10 dimensions. Further refinements lead to Membrane Theory, with the entire Universe regarded as existing on multidimensional sheets or membranes, with particles as loops anchored on "our" sheet and gravitons ranging into the continuum between sheets. We await predictions that can be tested.

Particle Physics Experiments

Throughout the history of Physics, experimental discoveries and theoretical ideas and explanations have moved forward together, sometimes playing leap-frog, but always drawing inspiration one from the other. Modern versions of Rutherford's table-top experiment on the scattering of alpha particles occupy many square kilometers of land, with massive and costly apparatus in underground tunnels tens of kilometers long. These are the particle accelerators that speed protons, antiprotons, electrons, or positrons to near the speed of light and then make them collide head-on with each other or with stationary targets.

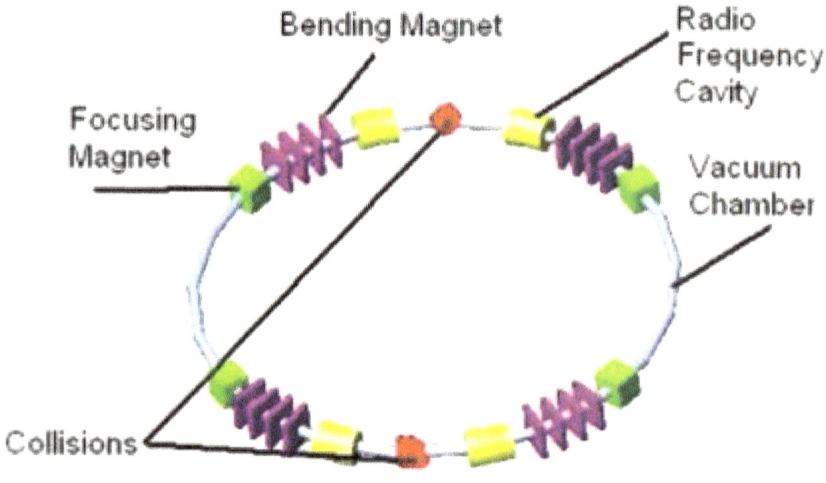

Figure 2: In an accelerator, focusing magnets and bending magnets guide the beam of particles around a ring.

The quest has mostly been for higher and higher collision energies. To make a pair of massive new particles and observe them flying apart, one has to generate excess energy over and above the equivalent of the mass ($2m_X$) of the pair: $E_{collision} > 2m_X c^2$. High energy is also needed to probe deeper and deeper to smaller length scales in studying the unknown — this is the equivalent of using X rays of shorter wave-lengths to probe smaller crystal structures. On the other hand, to look for rare phenomena, it is necessary to increase the intensity of particle beams and the collision rates. So accelerators have proceeded along parallel paths of ever higher energies and ever higher intensities.

To observe and interpret the results of collisions, particle detectors have to be developed that can track and analyze the particles that fly apart and disappear in nanoseconds. The detector consists of many different types of complex apparatus and electronics, requiring a cadre of experts in every conceivable technology. Collider experiments use large detectors completely surrounding the "interaction point" where high energy particles and antiparticles collide head-on.

Typical are electron-positron colliders, proton-antiproton colliders and massive detectors at the interaction points.

Other experiments study the collisions of intense beams with fixed (stationary) solid targets. Typical are several experiments with intense high energy neutrino beams and massive detectors in which neutrinos can interact. Many are studying the conversion of one type of neutrino (the muon-neutrino) into another (e.g., the tau-neutrino). Evidence for this is now pretty definite after decades of research, and precise measurements may pin down the non-zero mass of each neutrino. Relic neutrinos from the Big Bang populate the Universe, and even a tiny mass can explain some of the Dark Matter.

The art and science of particle accelerators and detectors has depended heavily on technology. The technology of solid state devices, superconducting magnets, electronics, computers and exotic materials, all have played leap frog with developments in experimental particle physics, sometimes driving and sometimes being driven by the inventions of particle physicists.

All these very complex detectors are built and operated by large numbers of physicists, in collaborations ranging from 100 to almost 1000 personnel. The collaborations extend across boundaries of countries and continents, in a typical illustration of science extending the hand of cooperation and friendship across national and political barriers.

CHAPTER 1 IMPLICATIONS OF THE HIGGS DISCOVERY FOR THE MSSM

Abdelhak Djouadi

INTRODUCTION

The historical discovery by ATLAS and CMS of a particle with a mass of approximately 125 GeV [1, 2] and properties that are compatible with those of a scalar Higgs boson [3–8] has far reaching consequences not only for the Standard Model (SM) of the electroweak and strong interactions, but also for new physics models beyond it. This is particularly true for supersymmetric theories (SUSY) [9–11] that are widely considered to be the most attractive extensions of the SM as they naturally protect the Higgs mass against large radiative corrections and stabilise the hierarchy between the electroweak and Planck scales, besides of allowing for the unification of the three gauge coupling constants and providing a good candidate for the dark matter in the universe, the lightest SUSY particle.

In the minimal supersymmetric extension of the SM (MSSM), two Higgs doublet fields H_u and H_d are required to break the electroweak symmetry, leading to a physical spectrum with five Higgs particles: two CP-even, h and H, a CP-odd, A, and two charged, $H^{\pm}$, states [6, 8]. Two parameters are needed to describe the MSSM Higgs sector at the tree level: one Higgs mass, which is generally taken to be that of the pseudoscalar boson M_A, and the ratio of vacuum expectation values of the two Higgs fields, $\tan\beta = v_d/v_u$, expected to lie in the range $1 \lesssim \tan\beta \lesssim 60$. The masses of the CP-even h,H and the charged $H^{\pm}$ states, as well as the mixing angle α in the CP-even sector are uniquely defined in terms of these two inputs at tree level, but this nice property is spoiled at higher orders [12–28].

At high M_A values, $M_A \gg M_Z$, one is in the so-called decoupling regime [29] in which the neutral CP-even state h is light and has almost exactly the properties of the SM Higgs boson, i.e. its couplings to fermions and gauge bosons are the same as the standard Higgs, while the other CP-even H and the charged $H^{\pm}$ bosons become heavy and mass degenerate with the A state, $M_H \approx M_{H^{\pm}} \approx M_A$, and they decouple from the massive gauge bosons. In this regime, the MSSM Higgs sector thus looks almost exactly like the one of the SM with its unique Higgs boson.

There is, however, one major difference between the two cases: while in the SM the Higgs mass is essentially a free parameter (and should simply be smaller than about 1 TeV in order to ensure unitarity in the high-energy scattering of massive gauge bosons), the lightest MSSM CP-even Higgs particle mass is bounded from above and, depending on the SUSY parameters that enter the important quantum corrections, is restricted to $M_h^{max} \approx 90$–130 GeV. The lower value comes from experimental constraints, in particular Higgs searches at LEP [30, 31], while the upper bound assumes a SUSY-breaking scale that is not too high, $M_S \lesssim \mathcal{O}$ (1 TeV), in order to avoid too much fine-tuning in the model. Hence, the requirement that the MSSM h boson coincides with the one observed at the LHC, i.e. with $M_h \approx 125$ GeV and almost SM-like couplings as the LHC data seem to indicate, would place very strong constraints on the MSSM parameters, in particular the SUSY scale M_S, through their contributions to the radiative corrections to the Higgs sector. This comes in addition to the limits that have been obtained from the search of the heavier Higgs states at the LHC, as well as from the negative search for supersymmetric particles.

In this review, we summarise the implications of the available LHC Higgs results for the MSSM Higgs sector. We first discuss the consequences of the M_h measured value for the

Citation: Abdelhak Djouadi, Implications of the Higgs discovery for the MSSM, DOI: 10.1140/epjc/s10052-013-2704-3.

various unconstrained (with the many free parameters defined at the weak scale) and constrained (with parameters obeying some universal boundary conditions at the high scale) versions of the MSSM. We then discuss the impact of the measured production and decay rates of the observed particle on the various Higgs couplings and, hence, the MSSM parameters. The impact of the negative search of the heavy H,A and $H^{\pm}$ states is summarised. An outlook is given in a concluding section.

IMPLICATIONS OF THE HIGGS MASS VALUE

The Higgs masses in the MSSM

In the MSSM, the tree-level masses of the CP-even h and H bosons depend only on M_A and tanβ. However, many parameters of the MSSM such as the masses of the third generation stop and sbottom squarks $m_{\tilde{t}_i}, m_{\tilde{b}_i}$ and their trilinear couplings A_t, A_b enter M_h and M_H through quantum corrections. In the basis (H_d,H_u), the CP-even Higgs mass matrix can be written in full generality as

$$\mathcal{M}^2 = M_Z^2 \begin{pmatrix} c_\beta^2 & -s_\beta c_\beta \\ -s_\beta c_\beta & s_\beta^2 \end{pmatrix} + M_A^2 \begin{pmatrix} s_\beta^2 & -s_\beta c_\beta \\ -s_\beta c_\beta & c_\beta^2 \end{pmatrix} + \begin{pmatrix} \Delta\mathcal{M}_{11}^2 & \Delta\mathcal{M}_{12}^2 \\ \Delta\mathcal{M}_{12}^2 & \Delta\mathcal{M}_{22}^2 \end{pmatrix} \quad (1)$$

where we use the short-hand notation sβ≡sinβ etc. and introduce the radiative corrections by a general 2×2 matrix $\Delta\mathcal{M}_{ij}^2$. One can then easily derive the neutral CP even Higgs boson masses and the mixing angle α that diagonalises the h and H states, $H = \cos\alpha H_d^0 + \sin\alpha H_u^0$ and $h = -\sin\alpha H_d^0 + \cos\alpha H_u^0$:

$$M_{h/H}^2 = \frac{1}{2}(M_A^2 + M_Z^2 + \Delta\mathcal{M}_{11}^2 + \Delta\mathcal{M}_{22}^2 \mp N) \quad (2)$$

$$\tan\alpha = \frac{2\Delta\mathcal{M}_{12}^2 - (M_A^2 + M_Z^2)s_\beta}{\Delta\mathcal{M}_{11}^2 - \Delta\mathcal{M}_{22}^2 + (M_Z^2 - M_A^2)c_{2\beta} + N} \quad (3)$$

$$N = \sqrt{M_A^4 + M_Z^4 - 2M_A^2 M_Z^2 c_{4\beta} + C}$$
$$C = 4\Delta\mathcal{M}_{12}^4 + (\Delta\mathcal{M}_{11}^2 - \Delta\mathcal{M}_{22}^2)^2 - 2(M_A^2 - M_Z^2) \times(\Delta\mathcal{M}_{11}^2 - \Delta\mathcal{M}_{22}^2)c_{2\beta} - 4(M_A^2 + M_Z^2)\Delta\mathcal{M}_{12}^2 s_{2\beta} \quad (4)$$

The by far leading one-loop radiative corrections to the mass matrix of Eq. (1) are controlled by the top Yukawa coupling, $\lambda_t = m_t / v\sin\beta$ with v=246 GeV, which appears with the fourth power. One obtains a very simple analytical expression for the radiative correction matrix $\Delta\mathcal{M}_{ij}^2$ if only this contribution is taken into account [12–14]

$$\Delta\mathcal{M}_{11}^2 \sim \Delta\mathcal{M}_{12}^2 \sim 0\,,$$

$$\Delta\mathcal{M}_{22}^2 \sim \epsilon = \frac{3\bar{m}_t^4}{2\pi^2 v^2 \sin^2\beta}\left[\log\frac{M_S^2}{\bar{m}_t^2} + \frac{X_t^2}{M_S^2}\left(1 - \frac{X_t^2}{12M_S^2}\right)\right] \quad (5)$$

where M_S is the geometric average of the two stop masses $M_S = \sqrt{m_{\tilde{t}_1} m_{\tilde{t}_2}}$ defined to be the SUSY-breaking scale and X_t is the stop mixing parameter given by $X_t = A_t - \mu/\tan\beta$ with μ the higgsino mass parameter; $\bar{m}_t$ is the running $\overline{\rm MS}$ top quark mass to account for the leading two-loop QCD corrections in a renormalisation-group improved approach (some refinements can be include as well).

Other soft SUSY-breaking parameters, in particular μ and Ab (and in general the corrections controlled by the bottom Yukawa coupling $\lambda_b = m_b / v\cos\beta$, which at large value of μtanβ become relevant) as well as the gaugino mass parameters $M_{1,2,3}$, provide a small but non-negligible correction to $\Delta\mathcal{M}_{ij}^2$ and can thus also have an impact on the loop corrections [15, 16, 25–28].

The maximal value of the h mass, $M_h^{\max}$ is given in the leading one–loop approximation above by

$$M_h^2 \overset{M_A \gg M_Z}{\rightarrow} M_Z^2 \cos^2 2\beta + \Delta\mathcal{M}_{22}^2 \quad (6)$$

and is obtained for the choice of parameters [25–28]:

- a decoupling regime with heavy A states, M_A~O(TeV);
- large values of the parameter tanβ, tanβ≳10;
- heavy stops, i.e. large M_S values and we choose in general M_S≲3 TeV to avoid a too large fine-tuning [32, 33];
- a stop trilinear coupling X_t=√6MS, the so-called maximal mixing scenario that maximises the stop loops [34].

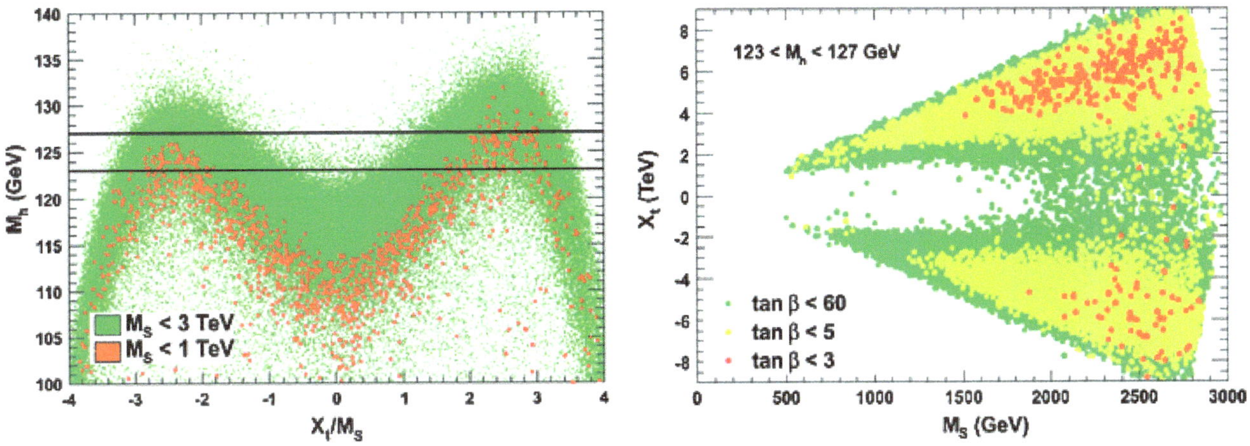

Figure 1: The maximal value of the h boson mass as a function of X_t/M_S in the pMSSM when all other soft SUSY-breaking parameters and tanβ are scanned (left) and the contours for the Higgs mass range 123 $<M_h<$ 127 GeV in the $[M_S,X_t]$ plane for some selected range of tanβ values (right); from Ref. [35].

If the parameters are optimised as above, the maximal h mass value can reach the level of $M_h^{\max}\approx$130 GeV.

An important aspect is that in the decoupling regime $M_A \gg M_Z$, the heavier CP-even and the charged Higgs states become almost degenerate in mass with the CP-odd state, $M_H\approx M_{H^\pm}\approx M_A$, while the mixing angle α becomes close to $\alpha\approx\pi 2-\beta$ making the couplings of the light h state to fermions and massive gauge bosons SM-like, and decoupling the H,$H^\pm$ from the weak bosons as is the case for the state A by virtue of CP invariance.

In this section, we discuss the implications of the measured mass value of the observed Higgs boson at the LHC [35–62] that we identify with the lightest state h of the MSSM. We consider the phenomenological MSSM [63] in which the relevant soft SUSY parameters are allowed to vary freely (but with some restrictions) and constrained MSSM scenarios such as the minimal supergravity (mSUGRA) [64–67], gauge mediated (GMSB) [68–72] and anomaly mediated (AMSB) [73–75] supersymmetry-breaking models (for a review, see again Ref. [8]). We also discuss the implications of such an M_h value for scenarios in which the supersymmetric spectrum is extremely heavy, the so-called split SUSY [76–78] or high-scale SUSY models [79, 80]. Finally, a new parametrisation of the Higgs sector, which uses the crucial information M_h=125 GeV, is discussed [81].

Implications for the phenomenological MSSM

In an unconstrained MSSM, there is a large number of soft SUSY-breaking parameters, O(100), but analyses can be performed in the so-called "phenomenological MSSM" (pMSSM) [63], in which CP conservation, flavour diagonal sfermion mass and coupling matrices and universality of the first and second sfermion generations are imposed. The pMSSM involves then 22 free parameters in addition to those of the SM: besides tanβ and M_A, these are the higgsino mass μ, the three gaugino masses $M_{1,2,3}$, the diagonal left- and right-handed sfermion mass parameters $m_{\tilde f_{L,R}}$ and the trilinear sfermion couplings A_f.

As discussed above, an estimate of the upper bound on M_h can be obtained by including the corrections that involve only the parameters M_S and X_t. However, to be more precise, one could scan the full pMSSM 22 parameter space in order to include the subleading corrections. To do so, one can use RGE programs such as Suspect [82] which calculate the Higgs and superparticle spectrum in the MSSM including the most up-to-date information [25].

To obtain the value $M_h^{\max}$ with the full radiative corrections, a large scan of the pMSSM parameters in an uncorrelated way was performed [35, 36] in the domains:

$$\begin{aligned} 1 \le \tan\beta \le 60,\ 50\,\mathrm{GeV} \le M_A \le 3\,\mathrm{TeV}, \\ -9\,\mathrm{TeV} \le A_t, A_b, A_\tau \le 9\,\mathrm{TeV}, \\ 50\,\mathrm{GeV} \le m_{\tilde f_L}, m_{\tilde f_R}, M_3 \le 3\,\mathrm{TeV}, \\ 50\,\mathrm{GeV} \le M_1, M_2, |\mu| \le 1.5\,\mathrm{TeV}. \end{aligned} \tag{7}$$

The results are shown in Fig. 1, where, in the left-hand side, the obtained maximal value $M_h^{\max}$ is displayed as a function of the ratio of parameters X_t/M_S. The resulting values are confronted to the mass range 123GeV$\le M_h \le$127GeV when the parametric uncertainties from the SM inputs such as the top quark mass and the theoretical uncertainties in the determination of M_h are included[1].

For $M_S \lesssim 1$ TeV, only the scenarios with X_t/M_S values close to maximal mixing $X_t/M_S\approx\sqrt{6}$survive. The no-mixing scenario $X_t\approx 0$ is ruled out for $M_S\lesssim 3$ TeV, while the typical mixing scenario, $X_t\approx M_S$, needs large M_S and moderate to large tanβ values. From the scan, one obtains a maximum $M_h^{\max}$=136, 126 and 123 GeV with maximal, typical and zero mixing, respectively.

What are the implications for the mass of the lightest stop state $\tilde t_1$? This is illustrated in the right-hand side of Fig. 1, where the contours are shown in the $[M_S,X_t]$ plane in which one obtains 123$<M_h<$127 GeV from the pMSSM scan; the regions in which tanβ≲3,5 and 60 are highlighted. One sees again that a large part of the parameter space is excluded if the Higgs mass constraint is imposed. In particular, large M_S values, in general corresponding to large $m_{\tilde t_1}$ 1are favoured. However, as $M_S = \sqrt{m_{\tilde t_1} m_{\tilde t_2}}$, the possibility that $m_{\tilde t_1}$ is of the order of a few 100 GeV is still allowed, provided that stop mixing (leading to a significant $m_{\tilde t_1}, m_{\tilde t_2}$ splitting) is large [36, 57–59].

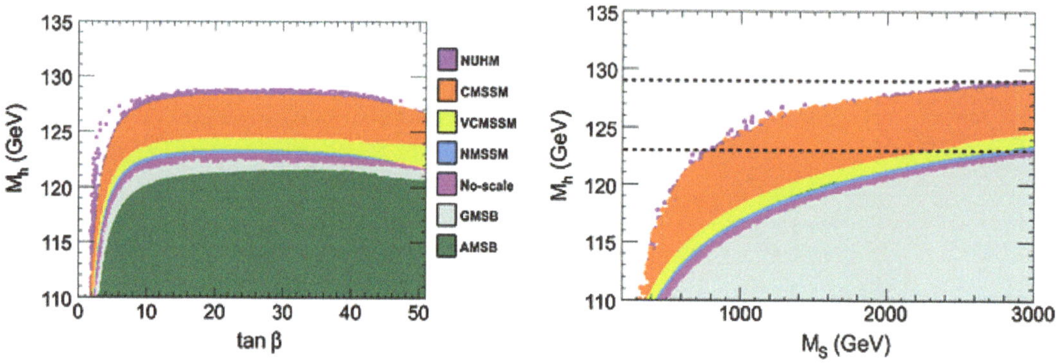

Figure 2: The maximal value of the h boson mass as a function of tanβ (left) and MS (right) with a scan of all other parameters in various constrained MSSM scenarios. The range $123<M_h<129$ GeV for the light h boson mass is highlighted. From Ref. [35].

Masses above 1 TeV for the scalar partners of light quarks and for the gluinos are also required by the direct searches of SUSY particles at the LHC [86, 87], confirming the need of high M_S values. Nevertheless, relatively light stops as well as electroweak sparticles such as sleptons, charginos and neutralinos are still possible allowing for a "natural SUSY" [33] in spite of the value $M_h \approx 125$ GeV. Nevertheless, the present LHC SUSY searches [86, 87] are constraining more and more this natural scenario.

Implications for constrained MSSM scenarios

In constrained MSSM scenarios (cMSSM), the various soft SUSY-breaking parameters obey a number of universal boundary conditions at a high energy scale, thus reducing the number of basic input parameters to a handful. The various soft SUSY-breaking parameters are evolved via the MSSM renormalisation group equations down to the low energy scale M_S where the conditions of proper electroweak symmetry breaking (EWSB) are imposed.

Three classes of such models have been widely discussed in the literature. There is first the minimal supergravity (mSUGRA) model [64–67] in which SUSY-breaking is assumed to occur in a hidden sector which communicates with the visible sector only via flavour-blind gravitational interactions, leading to universal soft breaking terms, namely a common $m_{1/2}, m_0, A_0$ values for the gaugino masses, sfermion masses and sfermion trilinear couplings. Then come the gauge mediated [68–72] and anomaly mediated [73–75] SUSY-breaking (GMSB and AMSB) scenarios in which SUSY-breaking is communicated to the visible sector via, respectively, gauge interactions and a super-Weyl anomaly.

These models are described by tanβ, the sign of μ and a few continuous parameters. Besides of allowing for both signs of μ, requiring $1 \leq \tan\beta \leq 60$ and, to avoid excessive fine-tuning in the EWSB conditions, imposing the bound $M_S = M_{EWSB} < 3$ TeV, we adopt the following ranges for the input parameters of these scenarios:

mSUGRA	50 GeV $\leq m_0 \leq$ 3 TeV	50 GeV $\leq m_{1/2} \leq$ 3 TeV	$\lvert A_0 \rvert \leq 9$ TeV
GMSB	10 TeV $\leq \Lambda \leq$ 1000 TeV	$1 \leq M_{\rm mess}/\Lambda \leq 10^{11}$	$N_{\rm mess} = 1$
AMSB	1 TeV $\leq m_{3/2} \leq$ 100 TeV	50 GeV $\leq m_0 \leq$ 3 TeV	

Hence, in contrast to the pMSSM, the various parameters which enter the radiative corrections to M_h are not all independent in these constrained scenarios, as a consequence of the relations between SUSY-breaking parameters that are set at the high-energy scale and the requirement that electroweak symmetry breaking is triggered radiatively for each set of input parameters. The additional constraints make that it is not possible to freely tune the parameters that enter the Higgs sector to obtain the pMSSM maximal value of M_h. In order to obtain even a rough determination of $M_h^{\rm max}$ in a given constrained SUSY scenario, it is necessary to scan through the allowed range of values for the basic input parameters.

Using again the program Suspect, a full scan of these scenarios has been performed in Ref. [35] and the results for $M_h^{\rm max}$ are shown in the left-hand side of Fig. 2 as a function of tanβ, the input parameter that is common to all models, and in the right-hand side of the figure as a function of M_S. In the adopted parameter space of the models and with the central values of the SM inputs, the obtained upper h mass value is $M_h^{\rm max} \approx 121$ GeV in the AMSB scenario, i.e. much less that 125 GeV, while in the GMSB scenario one has $M_h^{\rm max} \approx 122$ GeV (these values are obtained for $\tan\beta \approx 20$). Thus, clearly, these two scenarios are disfavoured if the lightest h particle has indeed a mass in the range 123–127 GeV and $M_S \lesssim 3$ TeV. In mSUGRA, one obtains $M_h^{\rm max} = 128$ GeV and, thus, some parameter space would still survive the M_h constraint.

The upper bound on M_h in these scenarios can be qualitatively understood by considering in each model the allowed values of the trilinear coupling At, which essentially determines the stop mixing parameter X_t and thus the value of M_h for a given scale M_S. In GMSB, one has $A_t \approx 0$ at relatively high scales and its magnitude does not significantly increase in the evolution down to the scale MS; this implies that we are almost in the no-mixing scenario which gives a low value of $M_h^{\rm max}$ as can be seen from Fig. 1. In AMSB, one has a non-zero At that is fully predicted at any renormalisation scale in terms of the Yukawa and gauge couplings; however, the ratio A_t/M_S with M_S determined from the overall SUSY breaking scale m3/2 turns out to be rather small, implying again that we are close to the no-mixing scenario. Finally, in the mSUGRA model, since we have allowed A_t to vary in a wide range as $|A0| \leq 9$ TeV, one can get a large A_t/M_S ratio, which le to a heavier Higgs particle. However, one cannot easily reach A_t values such that $X_t/M_S \approx \sqrt{6}$ so that we are not in the maximal-mixing scenario and the higher upper bound on M_h in the pMSSM cannot be reached.

In the case of mSUGRA, one can study several interesting special cases: the no-scale scenario with $m_0 \approx A_0 \approx 0$ [88, 89], the scenario $m_0 \approx 0$ and $A_0 \approx -1/4 m_{1/2}$, which approximately corresponds to the constrained next-to-MSSM (cNMSSM) [90, 91], $A_0 \approx -m_0$, which corresponds to a very constrained MSSM (VCMSSM) [92], and a non-universal Higgs mass model (NUHM) [93] in which the soft SUSY-breaking scalar mass terms are different for the sfermions and for the two Higgs doublet fields.

In two particular cases, namely the "no-scale" and the "approximate cNMSSM" scenarios, the upper bound on M_h is much lower than in the more general mSUGRA case and, in fact, barely reaches $M_h \approx 123$ GeV. The main reason is that these scenarios involve small values of A0 at the GUT scale, $A_0 \approx 0$ for no-scale and $A_0 \approx -1/4 m_{1/2}$ for the cNMSSM, which lead to A_t values at the weak scale that are too low to generate a significant stop mixing and, hence, one is again close to the no-mixing scenario. Thus, only a very small fraction of the parameter space of these two sub-classes of the mSUGRA model survive if we impose $123 < M_h < 127$ GeV. These models should thus have a very heavy sfermion spectrum as a value $M_S \gtrsim 3$ TeV is required to increase M_h^{max}. In the VCMSSM case, the value $M_h \simeq 125$ GeV can be reached as $|A_0|$ can be large for large m_0, $A_0 \approx -m_0$, allowing for typical mixing.

Finally, since the NUHM is more general than mSUGRA as we have two more free parameters, the [tanβ,M_h] area shown in Fig. 2 is larger than in mSUGRA. However, since we are in the decoupling regime and the value of MA does not matter much (as long as it a larger than a few hundred GeV) and the key weak-scale parameters entering the determination of M_h, i.e. tanβ,M_S and A_t are approximately the same in both models, one obtains a bound M_h^{max} that is only slightly higher in NUHM compared to the mSUGRA case.

In these constrained scenarios and, in particular in the general mSUGRA model, most of the scanned points giving the appropriate Higgs mass correspond to the decoupling regime of the MSSM Higgs sector and, hence, to an h boson with a SM–Higgs cross section and branching ratios. Furthermore, as the resulting SUSY spectrum for $M_h = 125 \pm 2$ GeV is rather heavy in these scenarios (easily evading the LHC limits from direct sparticle searches [86]), one obtains very small contributions to observables like the anomalous muon magnetic moment $(g-2)_\mu$ and to B-physics observables such as the rates $BR(B_s \to \mu^+\mu^-)$ or $BR(b \to s\gamma)$ [94]. Hence, the resulting spectrum complies with all currently available constraints. In addition, as will be discussed later, the correct cosmological density for the LSP neutralino required by recent measurements [95] can easily be satisfied. The M_h value provides thus a unique constraint in this decoupling regime.

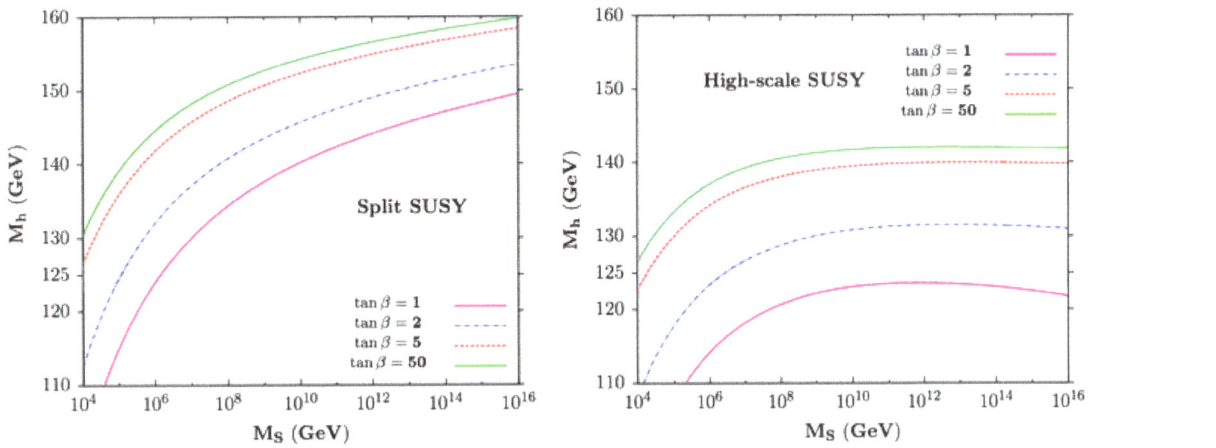

Figure 3: The value of h boson mass as a function of the SUSY scale M_S for several values of tanβ=1,2,5,50 in the split-SUSY (left) and high-scale SUSY (right) scenarios. From Ref. [35].

Split and high-scale SUSY models

In the preceding discussion, we have always assumed that the SUSY-breaking scale is relatively low, $M_S \lesssim 3$ TeV, which implies a natural SUSY scenario [33] with supersymmetric and heavier Higgs particles that could be observed at the LHC. However, as already mentioned, this choice is mainly dictated by fine-tuning considerations which are a rather subjective matter as there is no compelling criterion to quantify the acceptable amount of tuning. One could well abandon the SUSY solution to the hierarchy problem and have a very high M_S, which implies that, except for the lightest h boson, no other scalar particle is accessible at the LHC or at any foreseen collider.

This argument has been advocated to construct the so-called split SUSY scenario [76–78] in which the soft SUSY-breaking mass terms for all the scalars of the theory, except for one Higgs doublet, are extremely large, i.e. their common value M_S is such that $M_S \gg 1$ TeV (such a situation occurs e.g. in some string motivated models [96]). Instead, the mass parameters for the spin-12 particles, the gauginos and the higgsinos, are left in the vicinity of the EWSB scale, allowing for a solution to the dark matter problem and a successful gauge coupling unification, the two other SUSY virtues. The split SUSY models are much more predictive than the usual pMSSM as only a handful parameters are needed to describe the low-energy theory. Besides the common value M_S of the soft SUSY-breaking sfermion and one Higgs mass parameters, the basic inputs are essentially the three gaugino masses $M_{1,2,3}$ (which can be unified to a common value at M_{GUT} as in mSUGRA), the higgsino parameter μ and tanβ. The trilinear couplings A_f, which are expected to have values close to the EWSB scale set by the gaugino/higgsino masses that are much smaller than M_S, will play a negligible role.

Concerning the Higgs sector, the main feature of split SUSY is that at the high scale M_S, the boundary condition on the quartic Higgs coupling is determined by SUSY:

$$\lambda(M_S) = \frac{1}{4}\left[g^2(M_S) + g'^2(M_S)\right]\cos^2 2\beta. \tag{8}$$

where g and g′ are the SU(2) and U(1) gauge couplings. Here, tanβ is not a parameter of the low-energy effective theory as it enters only the boundary condition above and cannot be interpreted as the ratio of the two Higgs vevs.

If the scalars are very heavy, they will lead to radiative corrections in the Higgs sector that are significantly enhanced by large logarithms, $\log(M_S/M_{EWSB})$ where MEWSB≈|μ|,M_2. In order to have reliable predictions, one has to properly decouple the heavy states from the low-energy theory and resum the large logarithmic corrections; in addition, the radiative corrections due to the gauginos and the higgsinos have to be implemented. Following the early work of Ref. [76–78], a comprehensive study of the split SUSY spectrum has been performed in Ref. [97]. All the features of the model have been implemented in the code SuSpect [82] upon which the analysis presented in Ref. [35] and summarised here is based.

One can adopt an even more radical attitude than in split SUSY and assume that the gauginos and higgsinos are also very heavy, with a mass close to the scale M_S; this is the case in the so-called high-scale SUSY model [79, 80]. Here, one abandons not only the SUSY solution to the fine-tuning problem but also the solution to the dark matter problem by means of the LSP and the successful unification of the gauge couplings. However, there will still be a trace of SUSY at low energy: the matching of the SUSY and low-energy theories is indeed encoded in the Higgs quartic coupling λ of Eq. (8). Hence, even if broken at very high scales, SUSY would still lead to a "light" Higgs whose mass will give information on M_S and tanβ.

The treatment of the Higgs sector of the high-scale SUSY scenario is similar to that of split SUSY: one simply needs to decouple the gauginos and higgsinos from the low-energy spectrum (in particular remove their contributions to the renormalisation group evolution of the gauge and Yukawa couplings and to the radiative corrections to M_h) and set their masses to MS. The version of the program Suspect which handles the split SUSY case can be adapted to also cover the $M_1 \approx M_2 \approx M_3 \approx |\mu| \approx M_S$ case.

Using this tool, a scan in the [tanβ,M_S] plane has been performed to determine the value of M_h in the split SUSY and high-scale SUSY scenarios; in the former case, $M_{EWSB} \approx \sqrt{|M_2\mu|} \approx$246GeV was chosen for the low scale. The results are shown in Fig. 3, where M_h is displayed as a function of M_S for selected values of tanβ in both split (left plot) and high-scale (right plot) SUSY.

As expected, the maximal M_h values are obtained at high tanβ and M_S values and, at the scale M_S≈1016 GeV at which the couplings g and g′ approximately unify in the split SUSY scenario, one obtains M_h≈160 GeV for the higher tanβ=50 value. Not included is the error bands in the SM inputs that would lead to an uncertainty of about 2 GeV on M_h, which is now mainly due to the 1 GeV uncertainty on m_t. In addition, the zero-mixing scenario was assumed as the parameter A_t is expected to be much smaller than M_S; this approximation might not be valid for M_S values below 10 TeV and a maximal mixing $A_t/M_S=\sqrt{6}$ would increase the Higgs mass value by up to 10 GeV at M_S=O(1 TeV) as was discussed earlier for the pMSSM. In the high-scale SUSY scenario, one obtains a value M_h≈142 GeV (with again an uncertainty of approximately 2 GeV from the top mass) for high tanβ values and at the unification scale MS≈10^{14} GeV [79, 80]. Much smaller M_h values, in the 120 GeV range, can be obtained for lower scales and tanβ.

Hence, the requirement that the Higgs mass is in the range 123 ≲M_h≲127 GeV imposes strong constraints on the parameters of these two models. For this mass range, very large scales are needed for tanβ≈1 in the high-scale SUSY scenario, while scales not too far from M_S≈104 GeV are required at tanβ≫1 in both the split and the high-scale scenarios. In this case, SUSY should manifest itself at scales much below M_{GUT} if M_h≈125 GeV.

Splitting the Higgs and sfermion sectors

In the previous high-scale scenarios, the Higgs mass parameters were assumed to be related to the mass scale of the scalar fermions in such a way that the masses of the heavier Higgs particles are also of the order of the SUSY scale, $M_A \approx M_S$. However, this needs not to be true in general and one can, for instance, have a NUHM-like scenario where the Higgs masses are decoupled from those of the sfermions. If one is primarily concerned with the MSSM Higgs sector, one may be rather conservative and allow any value for M_A irrespective of the SUSY-breaking scale M_S. This is the quite "model-independent" approach that has been advocated in Refs. [98–101]: take M_A as a free parameter of the pMSSM, with values ranging from O(100 GeV) up to O(M_S), but make no restriction on M_S, which can be set to any value, even very high.

An important consequence of this possibility is that it reopens the low tanβ region, tanβ≲3, which was long thought to be forbidden if one requires a SUSY scale M_S≲1 TeV, as a result of the limit M_h≳114 GeV from the negative search of a SM-like Higgs boson at LEP [31]. If the SUSY scale is large enough, these small tanβ values would become viable again. To estimate the required magnitude of M_S, one can still use Suspect in which the possibility M_S≫1 TeV is implemented [97] with the full set of radiative corrections up to two loops included. In Fig. 4, displayed are the contours in the plane [tanβ,M_S] for fixed mass values M_h=120–132 GeV of the observed Higgs state (these include a 3 GeV theoretical uncertainty and also a 3 GeV uncertainty on the top quark mass [84] that is conservatively added linearly in the extreme cases). The maximal mixing $X_t=\sqrt{6}M_S$ scenario is assumed with 1 TeV gaugino/higgsino mass parameters.

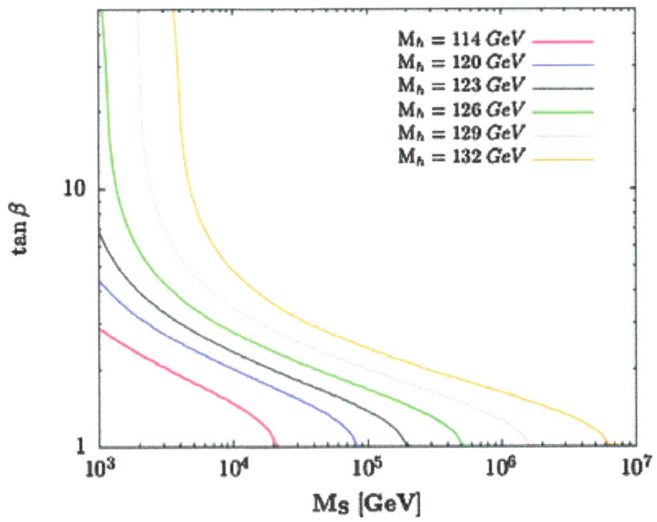

Figure 4: Contours for fixed values M_h=120,123,126,129 and 132 GeV in the [tanβ,M_S] plane in the decoupling limit $M_A \gg M_Z$; the "LEP2 contour" for M_h=114 GeV is also shown.

One observes that values of tanβ≈1 are possible and allow for an acceptable M_h value provided the scale M_S is large enough. For instance, while one can accommodate a scale M_S≈1 TeV with tanβ≈5, a large scale M_S≈20 TeV is required to obtain tanβ≈2; to reach the limit tanβ=1, an order of magnitude increase of M_S will be required. Outside the decoupling regime, the obtained M_S for a given M_h value will be of course larger. For completeness, also shown is the contour for the LEP2 limit M_h=114 GeV which illustrates the fact that tanβ≈1 is still allowed provided that $M_S \gtrsim$20 TeV.

A new parametrisation of the Higgs sector

It was pointed out in Refs. [98, 102–104] that when the measured value of the h boson mass M_h=125 GeV is taken into account, the MSSM Higgs sector with solely the dominant radiative corrections included can be again described with only two free parameters such as tanβ and M_A as was the case at tree level. In other words, the dominant radiative corrections that involve the SUSY parameters are fixed by the value of M_h. This observation le to a rather simple parametrisation of the MSSM Higgs sector.

More specifically, let us assume that in the 2×2 matrix for the radiative corrections to the CP-even Higgs mass matrix Eq. (1), only the leading $\Delta\mathcal{M}^2_{22}$ entry of Eq. (5) that involves the by far dominant stop–top sector contribution is taken into account; this is the so-called ☒approximation and its refinements [15, 16, 28]. In this $\Delta\mathcal{M}^2_{22} \gg \Delta\mathcal{M}^2_{11}, \Delta\mathcal{M}^2_{12}$ limit, one can simply trade $\Delta\mathcal{M}^2_{22}$ for the by now known h mass value M_h=125 GeV and obtain

$$M_H^2 = \frac{\left(M_A^2 + M_Z^2 - M_h^2\right)\left(M_Z^2 c_\beta^2 + M_A^2 s_\beta^2\right) - M_A^2 M_Z^2 c_{2\beta}^2}{M_Z^2 c_\beta^2 + M_A^2 s_\beta^2 - M_h^2}$$

$$\alpha = -\arctan\left(\frac{(M_Z^2 + M_A^2)c_\beta s_\beta}{M_Z^2 c_\beta^2 + M_A^2 s_\beta^2 - M_h^2}\right) \qquad (9)$$

This was called the habemus MSSM or hMSSM in Ref. [81].

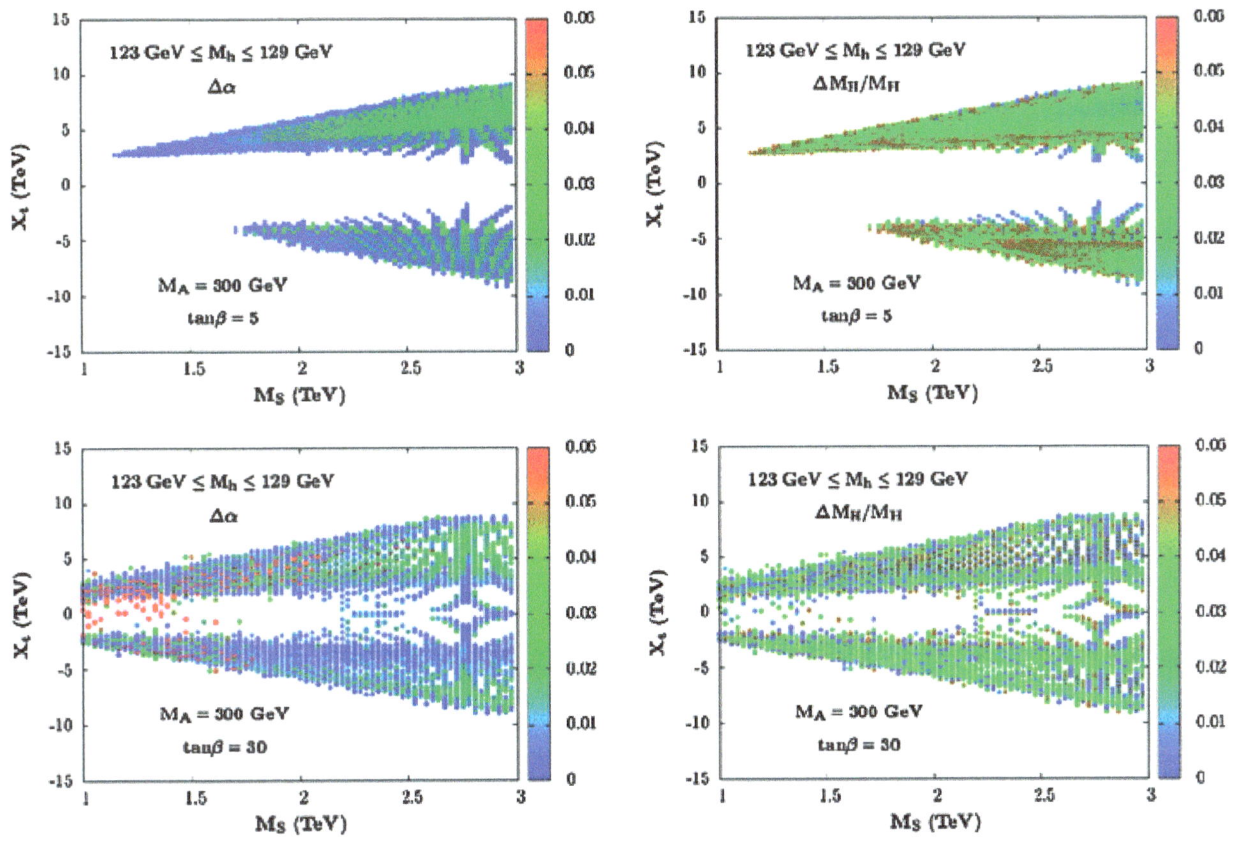

Figure 5: The variation of the mass M_H and the mixing angle α are shown as separate vertical colored scales, in the plane [M_S,X_t] when the full two loop corrections are included with and without the subleading matrix elements $\Delta\mathcal{M}^2_{11}$ and $\Delta\mathcal{M}^2_{12}$. . M_A=300 GeV, tanβ=5 or and 30 are taken and the other parameters are varied as described in the text [81].

However, this interesting and simplifying feature has to been demonstrated for all MSSM parameters and, in particular, one needs to prove that the impact of the subleading corrections $\Delta\mathcal{M}^2_{11}$ and $\Delta\mathcal{M}^2_{12}$ is small. To do so, a scan of the pMSSM parameter space using the program SuSpect, in which the full two-loop radiative corrections to the Higgs sector are implemented, has been performed [81]. For a chosen ($\tan\beta$,M_A) input set, the soft-SUSY parameters that play an important role in the Higgs sector are varied in the following ranges: $|\mu| \leq 3\,\text{TeV}, |A_t, A_b| \leq 3M_S, 1\,\text{TeV} \leq M_3 \leq 3\,\text{TeV}$ and 0.5 TeV $\leq M_S \leq 3$ TeV (≈3 TeV is the scale up to which programs such as SuSpect are expected to be reliable). The usual GUT relation between the weak scale gaugino masses $6M_1=3M_2=M_3$ has been assumed and A_u, A_d, $A_\tau=0$ has been set (these last parameters have little impact on the radiative corrections). The MSSM Higgs sector parameters have been computed all across the parameter space, selecting the points which satisfy the constraint $123 \leq M_h \leq 129$ GeV when uncertainties are included. For each of theses points, the Higgs parameters have been compared to those obtained in the simplified MSSM approximation, $\Delta\mathcal{M}^2_{11} = \Delta\mathcal{M}^2_{12} = 0$, with the lightest Higgs boson mass as input. While the requirement that M_h should lie in the range 123–129 GeV has been made, M_h was allowed to be different from the one obtained in the "exact" case $\Delta\mathcal{M}^2_{11}, \Delta\mathcal{M}^2_{12} \neq 0$.

Displayed in Fig. 5 are the differences between the values of the mass M_H and the mixing angle αthat are obtained when the two possibilities $\Delta\mathcal{M}^2_{11} = \Delta\mathcal{M}^2_{12} = 0$ and $\Delta\mathcal{M}^2_{11}, \Delta\mathcal{M}^2_{12} \neq 0$ are considered. This is shown in the plane [M_S,X_t] with $X_t=A_t-\mu\cot\beta$ when all other parameters are scanned as above. The A boson mass was fixed to M_A=300 GeV (a similar result was obtained for $M_A \approx 1$ TeV) and two representative values tanβ=5 and 30 are used. The conservative approach of plotting only points which maximise these differences has been adopted.

In all cases, the difference between the two M_H values is very small (in fact, much smaller than the H boson total decay width Γ_H), less than a few percent, while for α the difference does not exceed ≈0.025 for low values of tanβ, but at high tanβ values, one can reach the level of ≈0.05 in some rare situations (large values of μ, which enhance the μtanβ contributions). Nevertheless, at high enough tanβ, we are far in the decoupling regime already for $M_A \gtrsim 200$ GeV and such a difference does not significantly affect the couplings of the h and H bosons which, phenomenologically, are the main ingredients.

Hence, even when including the full set of radiative corrections, it remains a good approximation to use Eq. (9) to derive the parameters M_H and α in terms of the inputs tanβ,M_A and the measured M_h value.

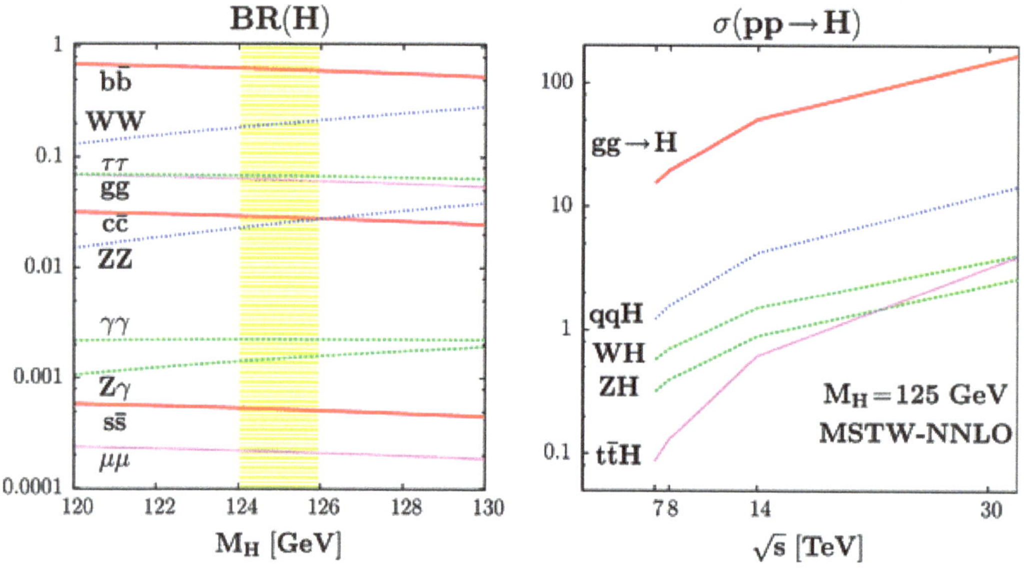

Figure 6: The SM-like Higgs boson branching ratios in the mass range 120–130 GeV (left) and its production cross sections at proton colliders as a function of the c.m. energy (right) [107].

In the case of the charged Higgs boson (whose physics is described by tanβ,$M_{H\pm}$ and eventually α), the radiative corrections to $M_{H\pm}$ are much smaller for large enough MA and one has, at the few percent level in most cases (which is again smaller than the total $H^\pm$ decay width),

$$M_{H^\pm} \simeq \sqrt{M_A^2 + M_W^2}. \tag{10}$$

In conclusion, this approximation allows to ignore the radiative corrections to the Higgs masses and their complicated dependence on the MSSM parameters and to use a simple formula to derive the other parameters of the Higgs sector, α, M_H as well as $M_{H\pm}$. This considerably simplifies phenomenological analyses in the MSSM Higgs sector which up to now rely either on large scans of the parameter space (as in the previous subsections) or resort to benchmark scenarios in which most of the MSSM parameters are fixed (as is the case of Ref. [105] for instance).

IMPLICATIONS OF THE HIGGS PRODUCTION RATES

Light Higgs decay and production at the LHC

In many respects, the Higgs particle was born under a very lucky star as the mass value of ≈125 GeV (although too high for a natural SUSY) allows to produce it at the LHC in many redundant channels and to detect it in a variety of decay modes. This allows detailed studies of the Higgs properties as will be discussed in this section.

We start by summarizing the production and decay at the LHC of a light SM-like Higgs particle, which should correspond to the lightest MSSM h boson in the decoupling regime. First, for Mh≈125 GeV, the Higgs mainly decays into $b\bar{b}$ pairs but the decays into WW* and ZZ* final states, before allowing the gauge bosons to decay leptonically W→ℓν and Z→ℓℓ (ℓ=e,μ), are also significant. The $h\to\tau^+\tau^-$ channel (as well as the gg and $c\bar{c}$ decays that are not detectable at the LHC) is also of significance, while the clean loop induced h→γγ mode can be easily detected albeit its small rates. The very rare h→Zγ and even $h\to\mu^+\mu^-$ channels should be accessible at the LHC but only with a much larger data sample. This is illustrated in the left-hand side of Fig. 6, where the decay branching fractions of a SM-like Higgs are displayed for the narrow mass range Mh=120–130 GeV

On the other hand, many Higgs production processes have significant cross sections as is shown in the right-hand side of Fig. 6, where they are displayed at a proton collider at various past, present and foreseen center of mass energies for a 125 GeV SM-like Higgs boson; the MSTW parton densities [106] have been used.

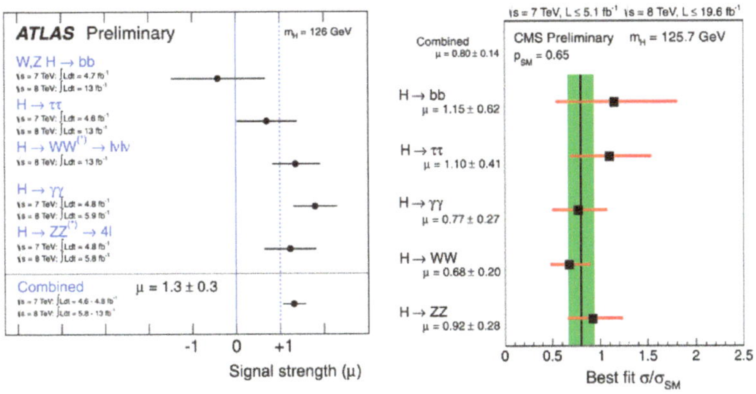

Figure 7: The signal strengths on the SM Higgs boson in the various search channels provided by ATLAS [108] and CMS [109] with the data collected so far at √s=7+8 TeV.

While the by far dominant gluon fusion mechanism gg→h (ggF) has extremely large rates (≈20 pb at √s=7–8 TeV), the subleading channels, i.e. the vector boson fusion (VBF) qq→hqqand the nHiggs-strahlung (HV) $q\bar{q}$→hV with V=W,Z mechanisms, have cross sections which should allow for a study of the Higgs particle already at √s≳8 TeV with the amount of integrated luminosity, ≈25 fb^{-1}, which has been collected by each experiment. The Higgs–top associated process pp→$t\bar{t}h$ (ttH) would require higher energy and luminosity.

This pattern already allows the ATLAS and CMS experiments to observe the Higgs boson in several channels and to measure some its couplings in a reasonably accurate way. The channels that have been searched are h→ZZ*→4ℓ±,h→WW*→2ℓ2ν,h→γγ where the Higgs is mainly produced in ggF with subleading contributions from hjj in the VBF process, h→ττ where the Higgs is produced in association with one (in ggF) and two (in VBF) jets, and finally h→$b\bar{b}$ with the Higgs produced in the HV process. One can ignore for the moment the additional search channels h→μμ and h→Zγ for which the sensitivity is still too low with the data collected so far.

A convenient way to scrutinise the couplings of the produced h boson is to consider their deviation from the SM expectation. One then considers for a given search channel the signal strength modifier μ, which, with some approximation, can be identified with the Higgs production cross section times decay branching fractions normalised to the SM value. For the h→XX decay channel, one would have in the narrow width approximation,

$$\mu_{XX}|_{\rm th} = \frac{\sigma(pp\to h\to XX)}{\sigma(pp\to h\to XX)|_{\rm SM}} = \frac{\sigma(pp\to h)\times {\rm BR}(h\to XX)}{\sigma(pp\to h)|_{\rm SM}\times {\rm BR}(h\to XX)|_{\rm SM}}. \tag{11}$$

which, from the experimental side would correspond to

$$\mu_{XX}|_{\rm exp} \simeq \frac{N^{\rm ev}_{XX}}{\epsilon\times\sigma(pp\to h)|_{\rm SM}\times {\rm BR}(h\to XX)|_{\rm SM}\times\mathcal{L}} \tag{12}$$

where $N^{\rm ev}_{XX}$ is the measured number of events in the XX channel, ϵ the selection efficiency and Lthe luminosity.

ATLAS and CMS have provided the signal strengths for the various final states with a luminosity of, respectively, ≈5 fb^{-1} for the 2011 run at √s=7 TeV and ≈20 fb^{-1} for the 2012 run at √s=8 TeV. The constraints given by the two collaborations are shown in Fig. 7.

When the various analysed Higgs search channels are combined, this le to a global signal strength [108, 109]

$$\begin{aligned} {\rm ATLAS}: \mu_{\rm tot} &= 1.30\pm 0.30 \\ {\rm CMS}: \mu_{\rm tot} &= 0.87\pm 0.23 \end{aligned} \tag{13}$$

which shows a good agreement with the SM expectation. In fact, when the ATLAS and CMS values are combined, one finds a global signal strength that is very close to unity, implying that the observed Higgs is rather SM-like.

Hence, already with the rather limited statistics at hand, the accuracy of the measurements in Eq. (13) is reaching the 20 % level for the ATLAS and CMS collaborations. This is at the same time impressive and worrisome. Indeed, as mentioned earlier the main Higgs production channel is the top and bottom quark loop mediated gluon fusion mechanism and, at √s=7 or 8 TeV, the three other mechanisms contribute at a level below 15 % when their rates are added and before kinematical cuts are applied. The majority of the signal events presently observed at the LHC, in particular in the main search channels $h\to\gamma\gamma, h\to ZZ^*\to 4\ell, h\to WW^*\to 2\ell 2\nu$and, to a lesser extent, $h\to\tau\tau$, thus come from the ggF mechanism which is known to be affected by large theoretical uncertainties.

As a matter of fact, although the cross section $\sigma(gg\to h)$ is known up next-to-next-to-leading order (NNLO) in perturbative QCD (and at least at NLO for the electroweak interaction) [110], there is a significant residual scale dependence which points to the possibility that still higher order contributions beyond NNLO cannot be totally excluded. In addition, as the process is of $\mathcal{O}(\alpha_s^2)$ at LO and is initiated by gluons, there are sizable uncertainties due to the gluon parton distribution function (PDF) and the value of the coupling α_s. A third source of theoretical uncertainties, the use of an effective field theory (EFT) approach to calculate the radiative corrections beyond the NLO approximation, should in principle also be considered [111, 112]. In addition, large uncertainties arise when the $gg\to h$ cross section is broken into the jet categories h+0j,h+1j and h+2j [113]. In total, the combined theoretical uncertainty has been estimated to be of order $\Delta th\approx\pm15\%$ by the LHC Higgs cross section working group [110] and it would increase up to $\Delta th\approx\pm20$ % if the EFT uncertainty is also included[2] [112].

Hence, the theoretical uncertainty is already at the level of the accuracy of the cross section measured by the ATLAS and CMS collaborations, Eq. (13). Another drawback of the analyses is that they involve strong theoretical assumptions on the total Higgs width since some contributing decay channels not accessible at the LHC are assumed to be SM-like and possible invisible Higgs decays in scenarios beyond the SM are supposed not to occur.

In Ref. [107], following earlier work [114–117], it has been suggested to consider the decay ratios D_{XX} defined as

$$D_{XX}^{\mathrm{p}} = \frac{\sigma^{\mathrm{p}}(pp\to h\to XX)}{\sigma^{\mathrm{p}}(pp\to h\to VV)} \tag{14}$$

$$= \frac{\sigma^{\mathrm{p}}(pp\to h)\times \mathrm{BR}(h\to XX)}{\sigma^{\mathrm{p}}(pp\to h)\times \mathrm{BR}(h\to VV)} \tag{15}$$

$$= \frac{\Gamma(h\to XX)}{\Gamma(h\to VV)} \propto \frac{c_X^2}{c_V^2} \tag{16}$$

for a specific production process p=ggF,VBF,VH or all (for inclusive production) and for a given decay channel $h\to XX$ when the reference channel $h\to VV$ is used. In these ratios, the cross sections $\sigma^{\mathrm{p}}(pp\to h)$ and hence, their significant theoretical uncertainties will cancel out, leaving out only the ratio of decay branching fractions and hence of partial decay widths. These can be obtained with the program HDECAY [118] which includes all higher order effects and are affected by much smaller uncertainties. Thus, the total decay width which includes contributions from channels not under control such as possible invisible Higgs decays, do not appear in the ratios D_{XX}^{p}. Some common experimental systematical uncertainties such as the one from the luminosity measurement and the small uncertainties in the Higgs decay branching ratios also cancel out. We are thus, in principle, left with only with the statistical uncertainty and some (non common) systematical errors. The ratios DXX involve, up to kinematical factors and known radiative corrections, only the ratios $|c_X|^2/|c_V|^2$ of the Higgs reduced couplings to the particles Xand V compared to the SM expectation, $c_X \equiv g_{hXX}/g_{hXX}^{\mathrm{SM}}$.

For the time being, three independent ratios can be considered: $D_{\gamma\gamma}$, $D_{\tau\tau}$ and D_{bb}. $D_{\gamma\gamma}$ is the ratio of the inclusive ATLAS and CMS di-photon and ZZ channels that are largely dominated by the ggF mechanism; $D_{\tau\tau}$ is the signal strength ratio in the $\tau\tau$ and WW searches where one selects Higgs production in ggF with an associated jet or in the VBF production mechanism; D_{bb} is the ratio of the $h\to b\bar{b}$ and $h\to WW$ decays in hV production for which the sensitivities are currently too low.

In order to test the compatibility of the couplings of the M_h=125 GeV Higgs state with the SM expectation, one can perform a fit based on the χ_R^2 function

$$\chi_R^2 = \frac{\left[D_{\gamma\gamma}^{ggF} - \frac{\mu_{\gamma\gamma}}{\mu_{ZZ}}\Big|_{\exp}^{ggF}\right]^2}{\left[\delta\left(\frac{\mu_{\gamma\gamma}}{\mu_{ZZ}}\right)_{ggF}\right]^2} + \frac{\left[D_{bb}^{VH} - \frac{\mu_{bb}}{\mu_{WW}}\Big|_{\exp}^{Vh}\right]^2}{\left[\delta\left(\frac{\mu_{bb}}{\mu_{WW}}\right)_{Vh}\right]^2} + \frac{\left[D_{\tau\tau}^{ggF} - \frac{\mu_{\tau\tau}}{\mu_{WW}}\Big|_{\exp}^{ggF}\right]^2}{\left[\delta\left(\frac{\mu_{\tau\tau}}{\mu_{WW}}\right)_{ggF}\right]^2} + \frac{\left[D_{\tau\tau}^{\mathrm{VBF}} - \frac{\mu_{\tau\tau}}{\mu_{WW}}\Big|_{\exp}^{\mathrm{VBF}}\right]^2}{\left[\delta\left(\frac{\mu_{\tau\tau}}{\mu_{WW}}\right)_{\mathrm{VBF}}\right]^2} \tag{17}$$

The errors $\delta(\mu_{XX}/\mu_{VV})$ are computed assuming no correlations between the different final state searches. The uncertainties on the ratios are derived from the individual errors that are dominated by the experimental uncertainties as one expects that the theoretical uncertainties largely cancel out in the ratios $D_{\gamma\gamma}$, D_{bb} and $D_{\tau\tau}$.

For the signal strengths above, the theoretical uncertainties have to be treated as a bias (and not as if they were associated with a statistical distribution) and the fit has to be performed for the two extremal values of the signal strengths: $\mu_i|_{\exp} \pm \delta\mu_i/\mu_i|_{\mathrm{th}}$ with the theoretical uncertainty $\delta\mu_i/\mu_i|_{\mathrm{th}}$ conservatively assumed to be ±20% for both the gluon and the vector boson fusion mechanisms (because of the contamination due to $gg\to h+2j$ in the latter case) and ≈5 % for hV associated production.

Fit of the Higgs couplings and their ratios

A large number of analyses of the Higgs couplings from the LHC data have been performed in the SM and its extensions and a partial list is given in Refs. [119–150].

In the MSSM, the couplings of the CP-even Higgs particles h and H to gauge bosons and fermions, compared to the SM Higgs couplings, are changed by factors that involve the sine and the cosine of the mixing angles β and α. Outside the decoupling regime where they reach unity, the reduced couplings (i.e. normalised to their SM values) of the lighter h state to third generation t,b,τ fermions and gauge bosons V=W/Z are for instance given by

$$c_V^0 = \sin(\beta - \alpha), c_t^0 = \cos\alpha / \sin\beta, c_b^0 = -\sin\alpha / \cos\beta \quad (18)$$

They thus depend not only on the two inputs [tanβ,MA], as occurs at tree level, but, a priori, on the entire MSSM spectrum as a result of the radiative corrections, in the same way as the Higgs masses. In principle, as discussed earlier, knowing tanβ and M_A and fixing M_h to its measured value, the couplings can be determined in general. However, this is true when only the radiative corrections to the Higgs masses are included. Outside the regime in which the pseudoscalar Aboson and the supersymmetric particles are very heavy, there are also direct radiative corrections to the Higgs couplings not contained in the mass matrix of Eq. (1) and which can alter this simple picture.

First, in the case of b-quarks, additional one-loop vertex corrections modify the tree–level $hb\bar{b}$ coupling: they grow as $m_b \mu \tan\beta$ and can be very large at high tanβ. The dominant component comes from the SUSY–QCD corrections with sbottom–gluino loops that can be approximated by $\Delta_b \simeq 2\alpha_s/(3\pi) \times \mu m_{\tilde{g}} \tan\beta / \max(m_{\tilde{g}}^2, m_{\tilde{b}_1}^2, m_{\tilde{b}_2}^2)$ [151]. Outside the decoupling regime the c_b coupling re

$$c_b \approx c_b^0 \times [1 - \Delta_b/(1+\Delta_b) \times (1 + \cot\alpha \cot\beta)] \quad (19)$$

with tanα→−1/tanβ for $M_A \gg M_Z$. A large Δb would significantly alter the dominant $h \to b\bar{b}$ partial width and affect the branching fractions of all other decay modes.

In addition, the $ht\bar{t}$ coupling is derived indirectly from the gg→h production cross section and the h→γγ decay branching ratio, two processes that are generated by triangular loops. In the MSSM, these loops involve not only the top quark (and the W boson in the decay h→γγ) but also contributions from supersymmetric particles, if not too heavy. In the case of gg→hproduction, only the contributions of stops is generally important. Including the later and working in the limit $M_h \ll m_t, m_{\tilde{t}_{1,2}}$, the coupling ct from the ggF process[3] is approximated by [152]

$$c_t \approx c_t^0 \left[1 + \frac{m_t^2}{4 m_{\tilde{t}_1}^2 m_{\tilde{t}_2}^2} \left(m_{\tilde{t}_1}^2 + m_{\tilde{t}_2}^2 - X_t^2 \right) \right] \quad (20)$$

which shows that indeed, $\tilde{t}$ contributions can be very large for light stops and for large stop mixing. In the h→γγ decay rate, because the t, $\tilde{t}$ electric charges are the same, the $ht\tilde{t}$ coupling is shifted by the same amount. If one ignores the usually small contributions of the other sparticles (to be discussed in the next subsection), the $ht\tilde{t}$ vertex can be simply parametrised by the effective coupling of Eq. (20).

We note that the h couplings to τ leptons and c quarks do not receive the direct corrections of Eqs. (19) and (20) and one should still have $c_c = c_t^0$ and $c_\tau = c_b^0$. However, using $c_{t,b}$ or $c_{t,b}^0$ in this case has almost no impact in practice as these couplings appear only in the branching ratios for the decays $h \to c\bar{c}$ and $\tau^+\tau^-$, which are small and the direct corrections should not be too large. One can thus, in a first approximation, assume that $c_c = c_t$ and $c_\tau = c_b$. Another caveat is due to the invisible Higgs decays which are assumed to be absent and which will be discussed later.

Hence, because of the direct corrections, the Higgs couplings cannot be described only by β and α as in Eq. (18). To characterise the Higgs particle at the LHC, it was advocated that at least three independent h couplings should be considered, namely c_t, c_b and $c_V = c_V^0$ [81]. One can thus define the following effective Lagrangian:

$$\mathcal{L}_h = c_V g_{hWW} h W_\mu^+ W^{-\mu} + c_V g_{hZZ} h Z_\mu^0 Z^{0\mu} - c_t y_t h \bar{t}_L t_R$$
$$- c_t y_c h \bar{c}_L c_R - c_b y_b h \bar{b}_L b_R - c_b y_\tau h \bar{\tau}_L \tau_R + \text{h.c.} \quad (21)$$

where $y_{t,c,b,\tau} = m_{t,c,b,\tau}/v$ are the Yukawa couplings of the heavy SM fermions, $g_h WW = 2M_W^2/v$ and $g_{hZZ} = M_Z^2/v$ the hWW and HZZ couplings and v the SM Higgs vev.

In Ref. [81], a three-dimensional fit of the h couplings was performed in the space $[c_t, c_b, c_V]$, assuming $c_c = c_t$ and $c_\tau = c_b$ as discussed above and of course the custodial symmetry relation $c_V = c_W = c_Z$, which holds in supersymmetric models. The results of this fit are presented in Fig. 8 for $c_t, c_b, c_V \geq 0$. The best-fit value for the couplings, with the √s=7+8 TeV ATLAS and CMS data turns out to be c_t=0.89,cb=1.01 and c_V=1.02.

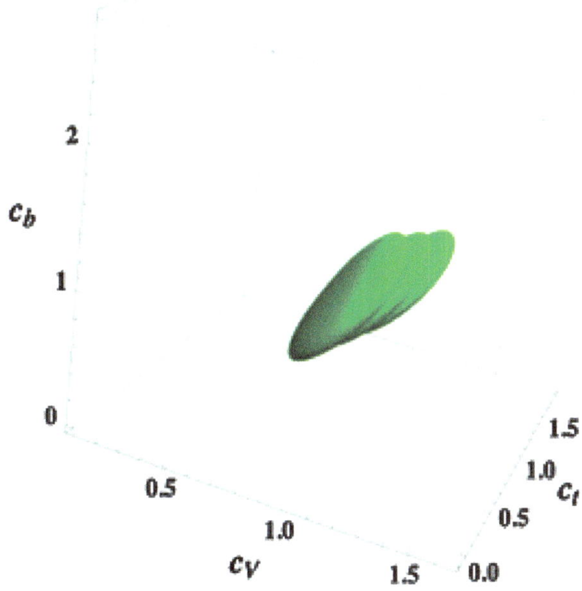

Figure 8: The best-fit region at 68%CL for the Higgs signal strengths in the $[c_t, c_b, c_V]$ space [81]. The three overlapping regions are for the central and extreme choices of the theoretical prediction for the Higgs rates including uncertainties.

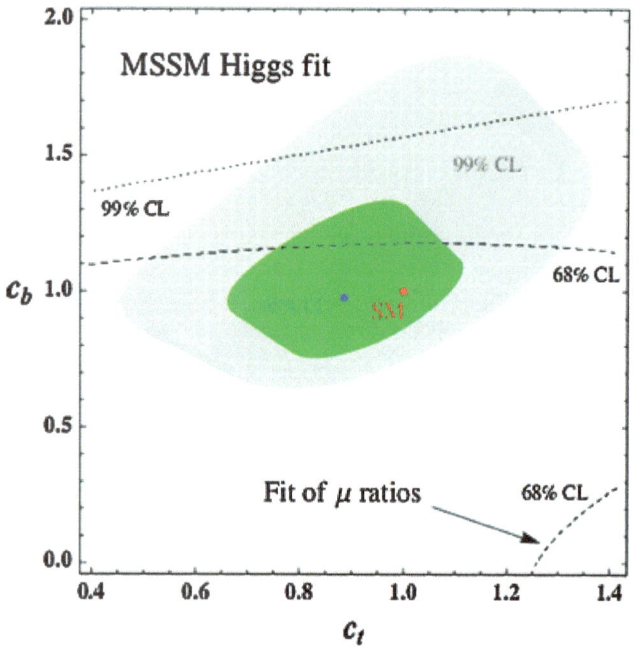

Figure 9: Best-fit regions at 68% and 99%CL for the Higgs signal strengths and their ratios in the plane $[c_t,c_b]$. The best-fit point is indicated in blue. From Ref. [81].

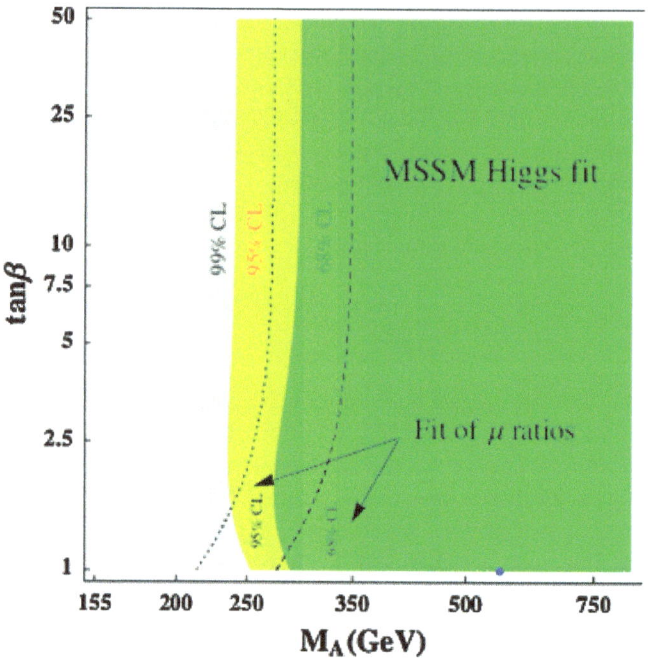

Figure 10: Best-fit regions for the signal strengths and their ratios in the plane [tanβ,M_A]; the best point is in blue [81].

In scenarios where the direct corrections in Eqs. (19) and (20) are not quantitatively significant (i.e. considering either not too large values of μtanβ or high sfermion masses), one can use the MSSM relations of Eq. (18) to reduce the number of effective parameters down to two. This allows to perform two-parameter fits in the planes $[c_V,c_t]$, $[c_V,c_b]$ and $[c_t,c_b]$. As an example, the fit of the signal strengths and their ratios in the $[c_t,c_b]$ plane is displayed in Fig. 9. In this two-dimensional case, the best-fit point is located at c_t=0.88 and c_b=0.97, while $c_V \simeq 1$. Note that although for the best-fit point one has $c_b \lesssim 1$, actually $c_b \gtrsim 1$ in most of the 1σ region.

Using the formulae Eq. (9) for the angle α and using the input $M_h \approx 125$ GeV, one can make a fit in the plane [tanβ,M_A]. This is shown in Fig. 10, where the 68, 95 and 99 % CL contours from the signal strengths and their ratios are displayed when the theory uncertainty is taken as a bias. The best-fit point when the latter uncertainty is set to zero is obtained for the values tanβ=1 and M_A=557GeV, which implies for the other parameters using M_h=125 GeV : M_H=580 GeV, $M_{H\pm}$=563 GeV and α=−0.837 rad, which le to cos(β−α)≃−0.05. Such a point with tanβ≈1 implies an extremely large value of the SUSY scale, M_S=O(100) TeV, for $M_h \approx 125$ GeV. One should note, however, that the χ^2 value is relatively stable all ovcr the 1σregion. Hence, larger values of tanβ (and lower values of M_A) could also be accommodated reasonably well by the fit, allowing thus for not too large M_S values. In all, cases one has $M_A \gtrsim 200$ GeV though.

An excess in the γγ channel?

In the early LHC data, a significant excess in the h→γγ detection channel was observed, raising the hope that it could be the first signal for physics beyond the SM. This excess has unfortunately faded away with more statistics and with the full 25 fb^{-1} data collected at $\sqrt{s}$=7+8 TeV, there is now only a ≈2σ excess in ATLAS which measures $\mu_{\gamma\gamma}$=1.6±0.3, while the signal strength measured by the CMS collaboration is $\mu_{\gamma\gamma}$=0.9±0.3, which is SM-like. Nevertheless, it would be interesting to briefly discuss this excess as, besides the fact that it has triggered a vast literature, the h→γγ channel is the one where new physics, and SUSY in particular, is most likely to manifest itself.

First, it has been realised early that this excess, if not due to a statistical fluctuation, can be easily explained or reduced in the context of the SM by invoking the large theoretical uncertainties that affect the production times decay rate in the dominant channel, gg→h→γγ. This is shown in Fig. 11, where the ATLAS and CMS ratios $R_{\gamma\gamma} \equiv \mu_{\gamma\gamma}$ and their combination, obtained with the ≈10 fb^{-1} data collected at $\sqrt{s}$=7+8 TeV, is compared to the theory uncertainty bands obtained by the LHC Higgs group [110] and in Ref. [112]. It is clear that including the theory uncertainty as a bias helps to reduce the discrepancy between measurement and expectation and e.g. the excess reduces to 1.3σ from the original ≳2σ value.

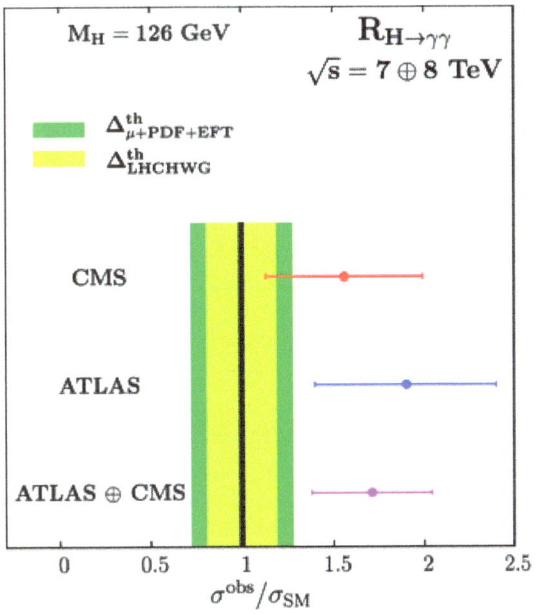

Figure 11: The value of $\mu_{\gamma\gamma}$ given by the ATLAS and CMS collaborations with the ≈10 fb^{-1} data collected at √s=7 and 8 TeV, as well as their combination, compared to two estimates of the theoretical uncertainty bands; from Ref. [153].

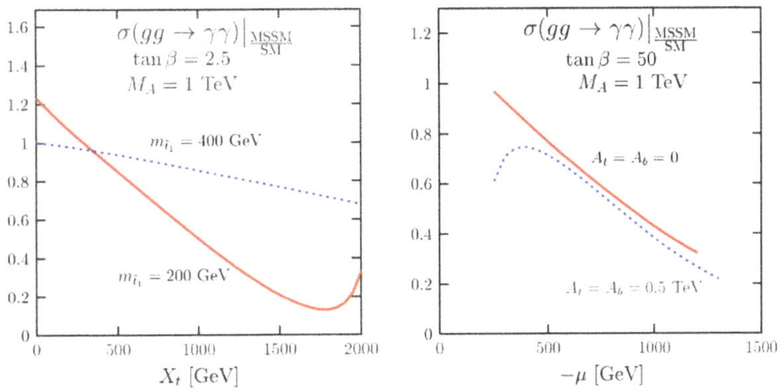

Figure 12: The deviation of $\mu_{\gamma\gamma}$ from its SM expectation from stop (left) and sbottom (right) contributions in various scenarios to the σ(gg→h)×BR(h→γγ) rate; from Ref. [152].

Ignoring this option, let us summarise the various possibilities that could explain this excess in the context of the MSSM. Deviations of $\mu_{\gamma\gamma}$ from the SM value may be due to modifications of either the production cross section or the decay branching fraction or to both. The h decay branching fractions may be modified by a change of the h total decay width. Since the dominant decay mode is h→$b\bar{b}$, a change of the effective h$b\bar{b}$ coupling by the direct vertex corrections of Eq. (19) outside the decoupling regime can change all other Higgs rates including h→γγ. The total width can also be modified by additional decay channels to SUSY particles and the only ones that are allowed by experimental constraints are invisible decays into the LSP that will be discussed later.

Nevertheless, these two possibilities would not only affect the h→γγ rate but also those of other channels such as h→ZZ where no excess has been observed. It is thus more appropriate to look at deviation in the h→γγ loop induced decay only. In the MSSM, the h→γγ process receives contributions from scalar top and bottom quarks, staus and charginos as briefly is summarised below.

- Light stops: as already discussed, the M_h=125 GeV constraint requires large $M_S = \sqrt{m_{\tilde{t}_1} m_{\tilde{t}_2}}$ and/or X_t values. If $m_{\tilde{t}_1} \lesssim 500$ GeV, one should have maximal mixing X_t≈√6MS and, in this case, the $h\tilde{t}_1\tilde{t}_1$ coupling of Eq. (20) is large and le to a sizeable change of the gg→h→γγ rate; cf. Fig. 12 (left). However, an enhancement of the h→γγ rate is over-compensated by a suppression of σ(gg→h) that seems not to occur. $\mu_{\gamma\gamma}$ is enhanced only in the no-mixing case, Xt≈0, which requires very heavy stops which decouple from the amplitude [57–59, 152].
- Light sbottoms: a light $\tilde{b}_R$ state does not conflict with the M_h value as its corrections to the mass are small. For $m_{\tilde{b}_1} \lesssim 500$ GeV, it contributes to the hgg loop but it reduces the gg→hproduction rate; Fig. 12 (right). In turn, it has little impact on the h→γγ rate because of the largely dominating W loop and the small ˜b1 electric charge. For $m_{\tilde{b}_1} \gtrsim 1$ TeV, as indicated by direct LHC searches, $\mu_{\gamma\gamma}$ is unaffected by sbottoms loops [152].
- Light staus: they lead to the largest contributions and have received most of the attention in the literature; see e.g. Ref. [154–156]. For low $m_{\tilde{\tau}_{L/R}}$ values,

a few 100 GeV, and large mixing $X_\tau = A_\tau - \mu\tan\beta$, with $\tan\beta \approx 60$ and $|\mu| = 0.5$–1 TeV, the lighter stau state has a mass close to the LEP2 bound, $m_{\tilde{\tau}_1} \approx 100$ GeV and its coupling to the h boson, $g_{h\tilde{\tau}\tilde{\tau}} \propto m_\tau X_\tau$, is huge. The $\tilde{\tau}_1$ contribution can hence significantly increase BR(h→γγ), up to 50 % [154–156], but this occurs only for extreme choices of the parameters.

- Light charginos: the $h\chi_i^+\chi_i^-$ couplings are in general small and are maximal when the $\chi_i^\pm$ istates are almost equal higgsino–wino mixtures. For a mass above 100 GeV and maximal couplings to the h boson, the $\chi\pm1$ contributions to the h→γγ rate do not exceed the 10–15 % level (with a sign being the same as the sign of μ) [157, 158].

Of course, different contributions can sum up resulting in more sizeable shifts. However, a 50 % deviation of the rate is unlikely and occurs only in extreme situations.

Invisible Higgs decays?

Invisible decays can also affect the properties of the observed h particle. In the MSSM, because of the LEP2 constraints, the only possible invisible channel for the h boson is into pairs of the LSP neutralinos, $h \to \chi_1^0\chi_1^0$. $\mathrm{BR_{inv}}$ can be important for $m_{\tilde{\chi}_1^0} < 60\mathrm{GeV}$ and for not too large M_1 and $|\mu|$ values which make the LSP a higgsino–gaugino mixture with significant couplings to the h state. Such an LSP would have the relic density Ωh^2 required by the WMAP results [95] since it will annihilate efficiently through the s-channel exchange of the h boson. However, BRinvshould be small in this case. This is exemplified in Fig. 13, where $\log_{10}(\Omega\chi h^2)$ is shown as a function of $m_{\chi_1^0}$ for the pMSSM points that satisfy the LHC Higgs constraints and $\mathrm{BR}(h \to \chi_1^0\chi_1^0) \geq 15\%$. Only a small area in the region $30 \lesssim m_{\chi_1^0} \lesssim 60$ GeV fulfils these conditions.

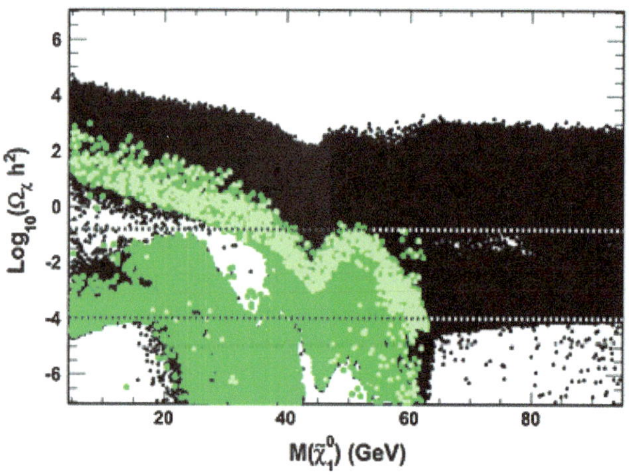

Figure 13: The neutralino relic density $\log10(\Omega\chi h^2)$ as a function of $m_{\chi_1^0}$ compatible with $\mathrm{BR}(h \to \chi_1^0\chi_1^0) \geq 15\%$ (green) and the LHC Higgs data at 90 % CL (light green). The horizontal lines show the WMAP constraint on $\Omega\chi h^2$. From Ref. [149].

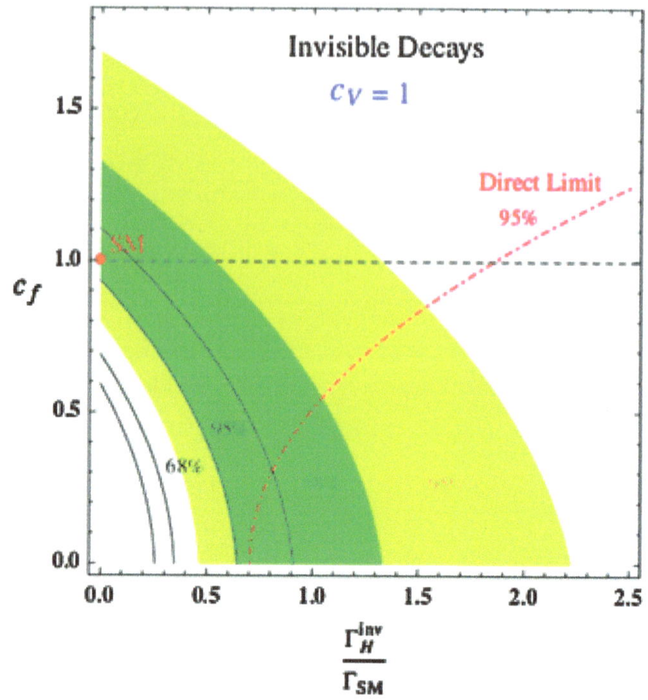

Figure 14: 1σ and 2σ domains from μ_{ZZ} for $c_V=1$ in the plane $[c_f, \Gamma_H^{\mathrm{inv}}/\Gamma_H^{\mathrm{tot}}]$ [150]. The dependence on the theory uncertainties are shown by the black curves that indicate the other possible extreme domains. The direct upper limit on Γ_H^{inv} from direct searches at LHC for $c_V=c_f=1$ [159] is also shown.

The invisible Higgs decay width can be constrained indirectly by a fit of the Higgs couplings and in particular with the signal strength μ_{ZZ}, which is the most accurate one and has the least theoretical ambiguities. Γ_H^{inv} enters in the signal strength through the total width Γ_H^{tot}, $\mu_{ZZ} \propto \Gamma(H \to ZZ)/\Gamma_H^{\mathrm{tot}}$ with $\Gamma_H^{\mathrm{tot}} = \Gamma_H^{\mathrm{inv}} + \Gamma_H^{\mathrm{SM}}$ and Γ_H^{SM} calculated with free coefficients c_f and c_V. The resulting 1σ or 2σ ranges are shown in Fig. 14, where c_f is freely varied while $c_V=1$ [150]. This gives $\Gamma_H^{\mathrm{inv}}/\Gamma_H^{\mathrm{SM}} \lesssim 50\%$ at the 95%CL if the assumption $c_f=c_V=1$ is made.

A more model independent approach would be to perform direct searches for missing transverse energy. These have been conducted by ATLAS [159] and CMS [163] in the pp→hV process with V→jj,ℓℓ and in the VBF channel, qq→qq E/T slash . As an example, we show in Fig. 15(left) the CMS results for the Higgs cross section times BRinv versus M_h when the two channels are combined. For Mh≈125 GeV a bound BRinv≲50% is obtained at the 95 % CL.

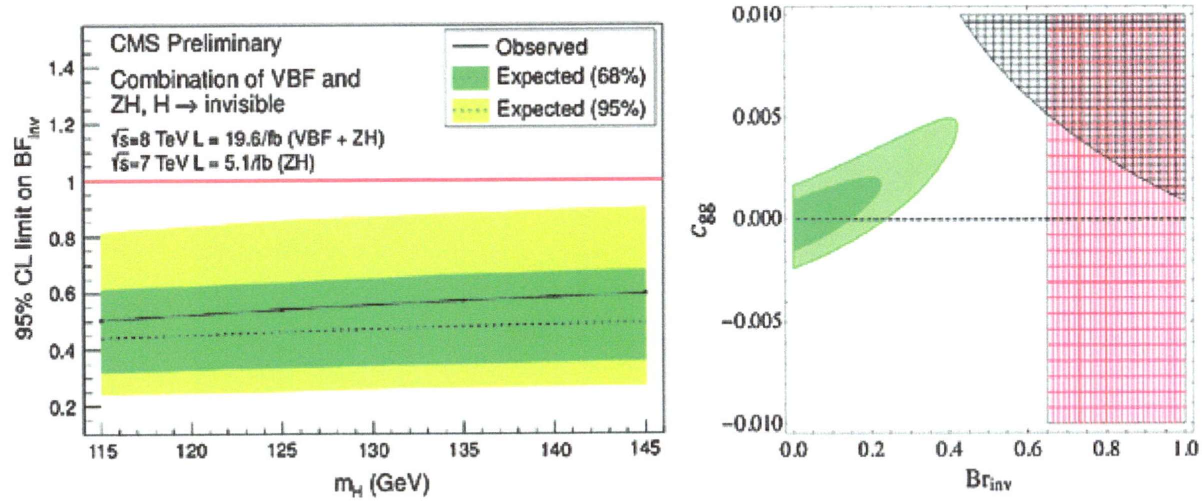

Figure 15: Left: the Higgs cross section times invisible Higgs decay branching ratio normalised to the total SM cross section in the combined hV and VBF channels from CMS with the ≈20fb^{-1} data at 8 TeV [163]. Right: 68% CL (light green) and 95% CL (dark green) best-fit regions to the combined LHC Higgs data. The black region is excluded by the monojet constraints while the red region is excluded by the ATLAS Z+E/T search [159]; from Ref. [160].

A more promising search for invisible decays is the monojet channel. In the ggF mode, an additional jet can be emitted at NLO leading to gg→hj final states and, because the QCD corrections are large, σ(H+1j) is not much smaller than σ(h+0j). The NNLO corrections besides significantly increasing the h+0j and h+1j rates, lead to h+2j events that also occur in VBF and VH. Hence, if the Higgs is coupled to invisible particles, it may recoil against hard QCD radiation, leading to monojet events.

In Refs. [160–162], it has been shown that the monojet signature carries a good potential to constrain the invisible decay width of a ≈125 GeV Higgs boson. In a model independent fashion, constraints can be placed on the rates

$$R_{\rm inv}^{\rm ggF} = \frac{\sigma(gg \to h) \times {\rm BR}(h \to {\rm inv.})}{\sigma(gg \to h)_{\rm SM}} \tag{22}$$

Recent monojet searches made by CMS and ATLAS [164, 165] are sensitive to R_{inv} close to unity. This is shown in Fig 15 (right) where the best-fit region to the LHC Higgs data is displayed in the Br_{inv}–c_{gg} plane, where c_{gg} is the deviation of σ(gg→h) from the SM expectation [160]. For the SM value c_{gg}=0, Brinv≳20% is disfavoured at 95% CL while for c_{gg}>0, a larger rate is allowed, up to Br_{inv}~50%.

The Higgs invisible rate and the dark matter detection rate in direct astrophysical searches are correlated in Higgs-portal models. Considering the generic cases of scalar, fermionic and vectorial dark matter particles χ that couple only to the Higgs, one can translate in each case the LHC constraint BR(h→inv.) into a constraint on the Higgs couplings to the χ particles. It turns out that these constraints are competitive with those derived from the bounds on the dark matter scattering cross section on nucleons from the best experiment so far, XENON [95].

This is shown in Fig. 16, where the maximum allowed values of the scattering cross sections are given in the three cases assuming ${\rm BR}_\chi^{\rm inv}$ ≲20%. The obtained spin-independent rates $\sigma_{\chi p}^{\rm SI}$ are stronger than the direct limit from the XENON100 experiment in the entire $M_\chi \ll \frac{1}{2} M_h$ range. In other words, the LHC is currently the most sensitive dark matter detection apparatus, at least in the context of simple Higgs-portal models.

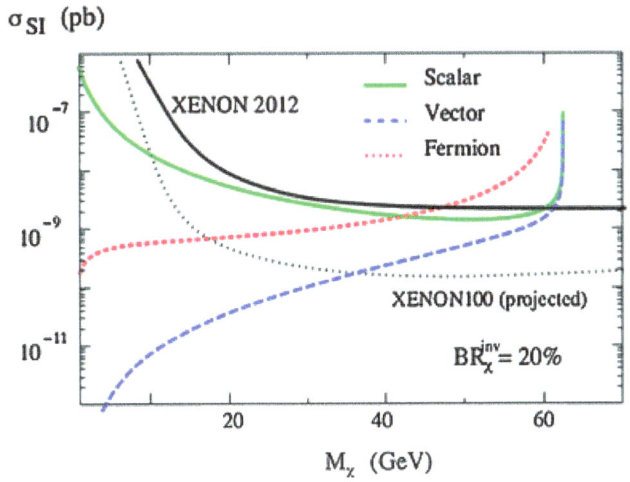

Figure 16: Bounds on the spin-independent direct detection cross section $\sigma_{\chi p}^{\rm SI}$ in Higgs-portal models derived for an invisible branching fraction of 20 % (colored lines) for a 125 GeV Higgs. These are compared to the current and future direct bounds from the XENON experiment (black lines). From Ref. [166].

Determination of the Higgs parity

Apart from the measurement of the couplings, one also needs in principle to establish that the observed Higgs state is indeed a CP even scalar particle and hence with $J^{PC}=0^{++}$ quantum numbers[4]. It is known that the Higgs to vector boson (hVV) coupling is a possible tool to probe these quantum numbers at the LHC [170, 171]. This can be done by studying various kinematical distributions in the Higgs decay and production processes. One example is the threshold behaviour of the M_{Z*} spectrum in the $h \to ZZ^* \to 4\ell$ decay channel and another is the azimuthal distribution between the decay planes of the two lepton pairs arising from the Z,Z* bosons from the Higgs decay. These are sensitive to both the spin and the parity of the Higgs.

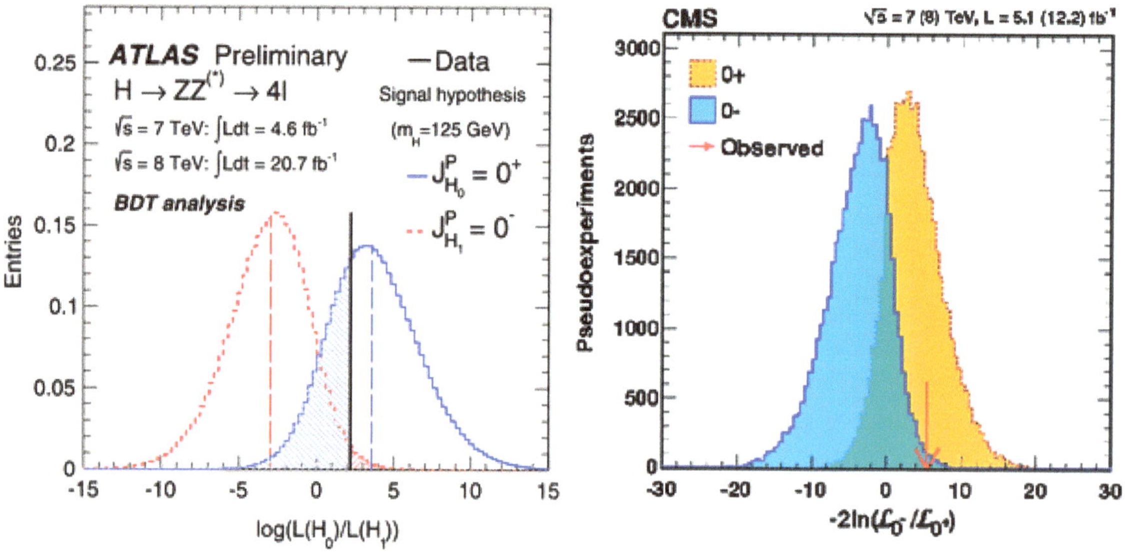

Figure 17: Discrimination between the 0^+ and 0^- parity hypotheses for the observed Higgs boson using the kinematics of the $h \to ZZ^* \to 4\ell$ channel by the ATLAS (left) and CMS (right) collaborations with the data collected at 7+8 TeV [172, 173].

With the 25 fb^{-1} data collected so far, the ATLAS and CMS collaborations performed a matrix-element likelihood analysis which exploits the kinematics and Lorenz structure of the $h \to ZZ^* \to 4\ell$ channel to see whether the angular distributions are more compatible with the 0^+ or 0^- hypothesis (as well as the spin-2 possibility) [172, 173]. Assuming that it has the same couplings as the SM Higgs boson and that it is produced mainly from the dominant ggF process, the observed particle is found to be compatible with a 0^+ state and the 0^- possibility is excluded at the 97.8% confidence level or higher; see Fig. 17.

Other useful diagnostics of the CP nature of the Higgs boson that also rely on the different tensorial structure of the hVV coupling can be made in the VBF process. It was known since a long time that in this channel, the distribution in the azimuthal angle between the two jets produced in association with the Higgs discriminates a CP-even from a CP-odd state [174]. This has been extended recently to other observables in this process, like the rapidity separation between the two jets [175–178].

Recently, the VBF channel $pp \to Hjj$ has been reanalysed in the presence of an anomalous hVVvertex that parametrises different spin and CP assignments of the produced Higgs boson [178]. The anomalous hVV coupling is introduced by allowing for an effective Lagrangian with higher dimensional operators, which include four momentum terms which are absent in the SM. It was shown that the kinematics of the forward tagging jets in this process is highly sensitive to the structure of the anomalous coupling and that it can effectively discriminate between different assignments for the spin (spin-0 versus spin-2) and the parity (CP-even versus CP-odd) of the produced particle. In particular, it was found that the correlation between the separation in rapidity and the transverse momenta of the scattered quarks, in addition to the already discussed distribution of the azimuthal jet separation, can be significantly altered compared to the SM expectation.

This is exemplified in Fig. 18, where the difference in rapidity between tagging jets (Δy_{jj}) for each of the higher dimensional operators in the hVV couplings is displayed.

These kinematical variables define new corners of the phase space that have not been explored by the experiments at the LHC to probe anomalous hVV couplings and to check the Higgs parity. In addition, some of these observables significantly depend on the c.m. energy and strong constraints on anomalous couplings can be obtained by performing measurements at the LHC with energies of $\sqrt{s}$=8 and 14 TeV. Finally, the associated hV production channel can be used as the invariant mass of the Vh system as well as the p_T and rapidities of the h and V bosons are also sensitive to anomalous hVV couplings.

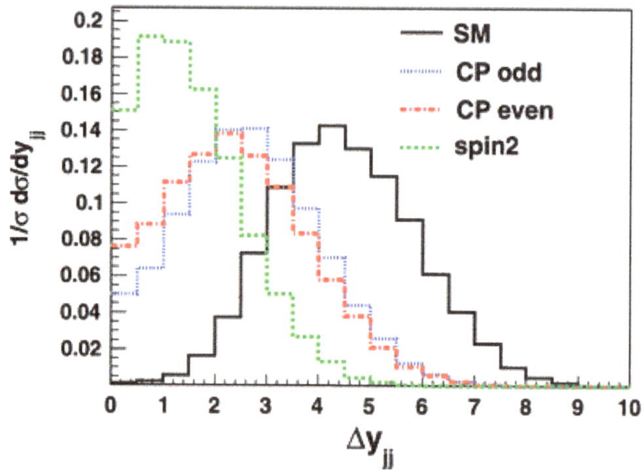

Figure 18: Normalised distribution of the difference in rapidity between the scattered jets in VBF for each of the SM and BSM operators (spin-2, CP-even and CP-odd) individually [178].

Nevertheless, there is a caveat in the analyses relying on the hVV couplings. Since a CP-odd state has no tree-level VV couplings, all the previous processes project out only the CP-even component of the hVV coupling [179–182] even if the state is a CP-even and -odd mixture. Thus, in the CP studies above, one is simply verifying a posteriori that indeed the CP-even component is projected out.

A better way to measure the parity of the Higgs boson is to study the signal strength in the $h \to VV$ channels [150, 183].

Indeed, the hVV coupling takes the general form $g^{\mu\nu}_{hVV} = -ic_V(M_V^2/v)\, g^{\mu\nu}$ where c_V measures the departure from the SM: cV=1 for a pure 0^+state with SM-like couplings and cV≈0 for a pure 0^- state. The measurement of c_V should allow to determine the CP composition of the Higgs if it is indeed a mixture of 0^+ and 0^- states.

However, having $c_V \neq 1$ does not automatically imply a CP-odd component: the Higgs sector can be enlarged to contain other states hi with squared h_iVV couplings $\Sigma_i c^2_{V_i}\, g^2_{h_iVV}$ that reduce to the SM coupling g^2_{hVV}. This is what occurs in the MSSM with complex soft parameters [170, 171]: one has three neutral states h_1, h_2 and h_3 with indefinite parity and their CP-even components share the SM hVV coupling, $c^2_{V_1} + c^2_{V_2} + c^2_{V_3} = 1$. But in all cases, the quantity $1 - c_V^2$ gives an upper bound on the CP-odd contribution to the hVV coupling.

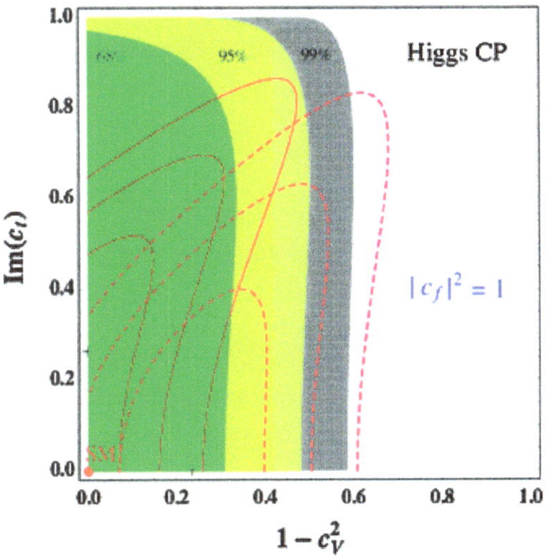

Figure 19: Best-fit regions at 68, 95 and 99%CL in the plane [$1-c_V^2$, $\mathrm{Im}(c_t)$] for $|c_t|^2 = |c_f|^2 = 1$. Superimposed are the best-fit regions when including a theory uncertainty of ±20% [150].

Using μVV and the ratios $\mu_{\gamma\gamma}/\mu VV$ and $\mu_{\tau\tau}/\mu VV$ as in Eq. (17), it was demonstrated that the particle has indeed a large CP component, ≳50 % at the 95 % CL, if the Higgs couplings to fermions are SM like. This is shown in Fig. 19, where one sees that the pure CP-odd possibility is excluded at the 3σ level, irrespective of the (mixed CP) Higgs couplings to fermions provided that $|cf|^2=1$.

IMPLICATIONS FROM HEAVY HIGGS SEARCHES

We turn now to the constraints on the MSSM Higgs sector that can be obtained from the search of the heavier H/A and H± states at the LHC and start with a brief summary of their production and decay properties.

H,A,H± decays and production at the LHC

The production and decay pattern of the MSSM Higgs bosons crucially depend on tanβ. In the decoupling regime that is indicated by the h properties, the heavier CP-even H boson has approximately the same mass as the A state and its interactions are similar. Hence, the MSSM Higgs spectrum will consist of a SM-like Higgs $h \equiv H_{SM}$ and two pseudoscalar-like particles, Φ=H/A. The H± boson will also be mass degenerate with the Φ states and the intensity of its couplings to fermions will be similar. In the high tanβ regime, the couplings of the non-SM like Higgs bosons to b quarks and to τ leptons are so strongly enhanced, and the couplings to top quarks and massive gauge bosons suppressed, that the pattern is rather simple.

This is first the case for the decays: the $\Phi \to t\bar{t}$ channel and all other decay modes are suppressed to a level where their branching ratios are negligible and the Φ states decay almost exclusively into $\tau^+\tau^-$ and $b\bar{b}$ pairs, with branching ratios of $\mathrm{BR}(\Phi \to \tau\tau) \approx 10\,\%$ and $\mathrm{BR}(\Phi \to b\bar{b}) \approx 90\,\%$. The H± boson decay into $\tau\nu_\tau$ final states with a branching fraction of almost 100 % for H± masses below the tb threshold, $M_{H^\pm} \lesssim m_t + m_b$, and a branching ratio of only ≈10 % for masses above this threshold while the rate for H±→tb will be at the ≈90 % level in most cases.

Concerning the production, the strong enhancement of the b-quark couplings at high tanβ makes that only two processes are relevant in this case: gg→Φ fusion with the b-loop included and associated production with b-quarks, $gg/q\bar{q} \to b\bar{b} + \Phi$, which is equivalent to the fusion process $b\bar{b} \to \Phi$ when no additional final b-quark is present. All other processes, in particular $V\Phi, t\bar{t}\Phi$ VBF have suppressed rates. In both the $b\bar{b}$ and the gg fusion cases, as $M_\Phi \gg m_b$, chiral symmetry holds and the rates are approximately the same for the CP-even H and CP-odd Abosons. While σ(gg→Φ) is known up to NLO in QCD [184], σ(bb→Φ) is instead known up to NNLO [185, 186].

The most powerful search channel for the heavier MSSM Higgs particles at the LHC is by far the process pp→gg+$b\bar{b} \to \Phi \to \tau^+\tau^-$. The precise values of the cross section times branching fraction for this process at the LHC have been updated in Refs. [110, 112] and an assessment of the associated theoretical uncertainties has been made. It turns out that, in the production cross section, the total uncertainty from scale variation, the PDFs and αs as well as from the b-quark mass are not that small: $\Delta^{\mathrm{TH}}\sigma(pp \to \Phi) \times \mathrm{BR}(\Phi \to \tau\tau) \approx \pm 25\%$ in the entire MΦ range probed at the LHC at √s=8 TeV; Fig. 20.

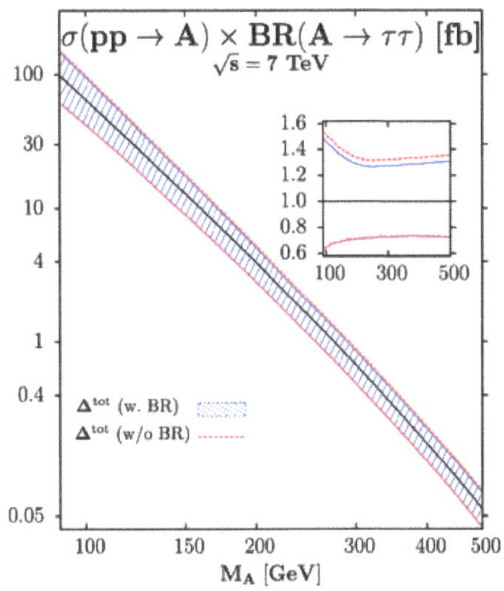

Figure 20: The combined σ(pp→A)×BR(A→ττ) rate with theoretical uncertainties with and without the branching ratio; in the inserts, shown are the uncertainties when the rates are normalised to the central values. From Ref. [112].

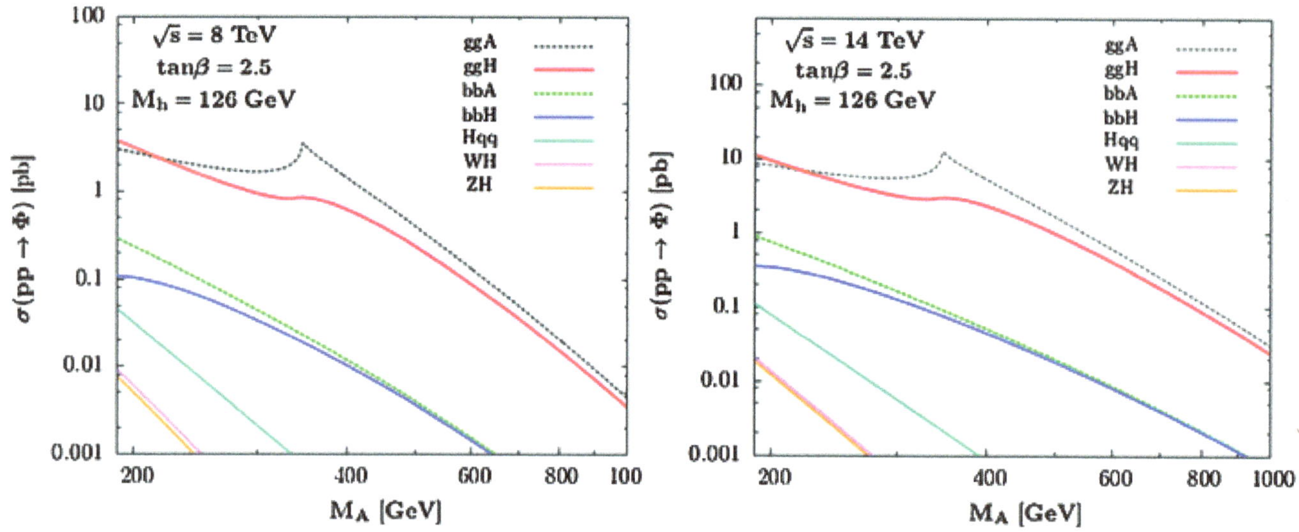

Figure 21: The production cross sections of the MSSM heavier neutral Higgs bosons at the LHC at √s=8 for tanβ=2.5; only the main production channels are considered [98].

Besides the QCD uncertainty, three other features could alter the rate σ(pp→Φ→ττ) in the MSSM and they are related to the impact of the SUSY particle contributions:

1. In the case of H (A does not couple to identical sfermions), there are squark (mainly stop) loops that contribute in addition in the gg→H process. But as they are damped by powers of $\tilde{m}_Q^2$ for $M_H \lesssim 2m_Q^2$, these contributions should be small for $\tilde{m}_Q \gtrsim$ 1TeV, in particular at high tanβ where the b-contribution is strongly enhanced.
2. The vertex correction to the $\Phi b\bar{b}$ couplings, Δ_b of Eq. (19), grows as μtanβ and can be very large in the high tanβ regime. However, in the full process $pp \to \Phi \to \tau^+\tau^-$, this correction appears in both the cross section and the branching fraction and largely cancels outs as one obtains, σ×BR×(1−Δ_b/5). A very large contribution Δ_b≈1changes the rate only by 20 %, i.e. less than the QCD uncertainty.
3. The possibility of light sparticles would lead to the opening of H/A decays into SUSY channels that would reduce BR(Φ→ττ). For M_Φ≲1 TeV, the only possibilities are decays into light neutralinos or charginos and sleptons. These are in general disfavoured at high tanβ as the Φ→ $b\bar{b}$ +ττ modes are strongly enhanced and dominant.

Thus, only in the unlikely cases where the decay H→ $\tilde{\tau}_1\tilde{\tau}_1$ has a branching rate of the order of 50 %, the squark loop contribution to the gg→H process is of the order 50 %, or the Δb SUSY correction is larger than 100 %, and one can change the pp→Φ→ττ rate by ≈25 %, which is the level of the QCD uncertainty. One thus expects σ(pp→Φ)×BR(Φ→ττ) to be extremely robust and to depend almost exclusively on M_A and tanβ.

Finally, for the charged Higgs boson, the dominant search channel is in $H^\pm$→τν final states with the $H^\pm$ bosons produced in top quark decays for $M_{H\pm}$≲m_t−m_b≈170 GeV, $pp \to t\bar{t}$ with $t \to H^+b \to \tau\nu b$. This is particularly true at high tanβ values when $\mathrm{BR}(t \to H^+b) \propto \bar{m}_b^2 \tan^2\beta$ is significant. For higher $H^\pm$ masses, one should rely on the three-body production process pp→tb$H^\pm$→tbτν, but the rates are presently rather small.

In the low tanβ regime, tanβ≲5, the phenomenology of the heavier A,H,$H^\pm$ bosons is richer [98, 187–190]. Starting with the cross sections, we display in Fig. 21 the rates for the relevant production processes at the LHC with √s=8 TeV assuming tanβ=2.5. For smaller tanβ values, the rates except for pp→H/A+$b\bar{b}$ are even larger as the H/A+tt and HVV couplings are less suppressed.

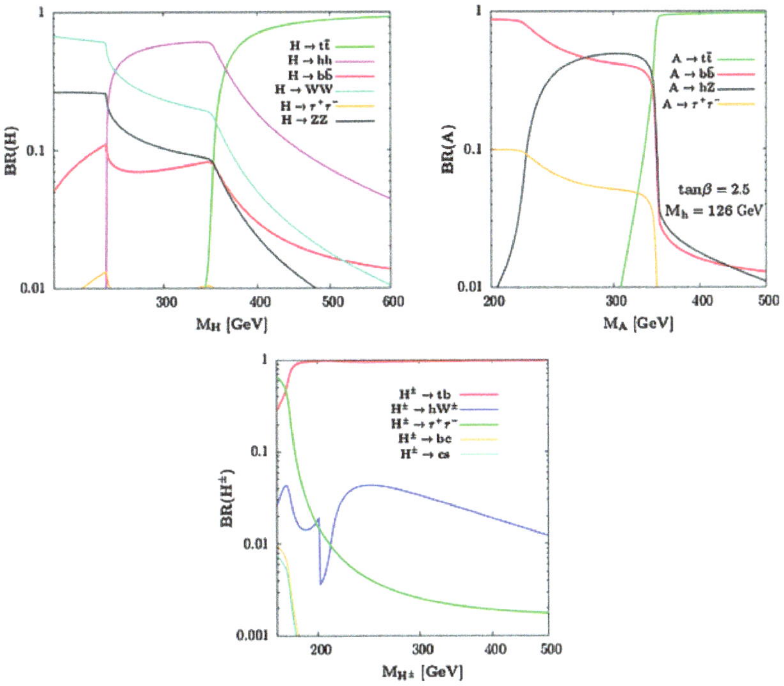

Figure 22: The H/A/$H^{\pm}$ decay branching ratios as functions of the Higgs masses for tanβ=2.5 [98].

Because of CP invariance which forbids AVV couplings, there is no AV and Aqq processes while the rates for associated $t\bar{t}A$ and $b\bar{b}A$ are small because the Att(Abb) couplings are suppressed (not sufficiently enhanced) compared to the SM Higgs. Only the gg→A process with the dominant t and sub-dominant b contributions included provides large rates. The situation is almost the same for H: only gg→H is significant at $M_H \gtrsim 300$ GeV and tanβ≲5; the VBF and HV modes give additional small contributions for tanβ≈1. For $H^{\pm}$, the dominant production channel is again top quark decays, t→H+b for $M_H \pm \lesssim 170$ GeV as for tanβ≲5, the mt/tanβpiece of the H^{+}tb coupling becomes large; for higher $H^{\pm}$ masses, the main process to be considered is gg/$q\bar{q}$ →H^{+}tb.

Turning to the H/A/$H^{\pm}$ decay pattern, it can be rather involved at low tanβ. A summary is as follows for tanβ≲3; see also Fig. 22, where the rates are shown for tanβ=2.5.

- Above the $t\bar{t}$ (tb) threshold for H/A($H^{\pm}$), the decay channels H/A→$t\bar{t}$ and H+→$t\bar{b}$ are by far dominant for tanβ≲3 and do not leave space for any other mode.
- Below the $t\bar{t}$ threshold, the H→WW,ZZ decay rates are still significant as gHVV is not completely suppressed.
- For $2M_h \lesssim M_H \lesssim 2m_t$, H→hh is the dominant H decay mode as the Hhh self-coupling is large at low tanβ.
- For $M_A \gtrsim M_h + M_Z$, A→hZ decays would occur but the A→ττ channel is still important with rates ≳5 %.
- In the case of H±, the channel H^{+}→Wh is important for $M_{H\pm} \lesssim 250$ GeV, similarly to the A→hZ case.

Hence, many decay and production channels need to be considered in this low tanβ regime.

Constraints from the LHC Higgs searches

The most efficient channel to probe the heavier MSSM Higgs bosons is by far pp→gg+bb→H/A→$\tau^{+}\tau^{-}$. Searches for this process have been performed by ATLAS with ≈5 fb^{-1} data at the 7 TeV run [191] and by CMS with ≈5+12 fb^{-1} data at the 7 and 8 TeV runs [192]. Upper limits on the production cross section times decay branching ratio have been set and they can be turned into constraints on the MSSM parameter space.

In Fig. 23, the sensitivity is displayed of the CMS pp→Φ→ττ analysis with 17 fb^{-1} of data in the [tanβ,M_A] plane. The excluded region, obtained from the observed limit at the 95 % CL, is drawn in blue. The dotted line represents the median expected limit which turns out to be weaker than the observed limit. As can be seen, this constraint is extremely restrictive and, for values $M_A \lesssim 250$ GeV, it excludes almost the entire intermediate and high tanβ regimes, tanβ≳5. The constraint is less effective for a heavier A boson, but even for MA≈400 GeV the high tanβ≳10 region is excluded and there is even sensitivity to large values $M_A \approx 800$ GeV for tanβ≳50.

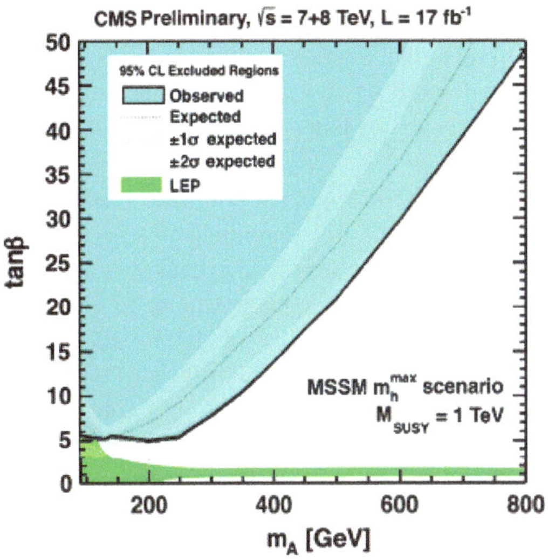

Figure 23: The expected and observed exclusion limits in the [tanβ,MA] plane in the CMS search of the MSSM neutral Higgs bosons in the channels pp→h/H/A→$\tau^{+}\tau^{-}$ with ≈17 fb^{-1} data collected at √s=7+8 TeV [192].

There are, however, some caveats to this exclusion limit as discussed previously. The first one is that there is a theoretical uncertainty of order of ±25% that affects the gg→Φ and $b\bar{b}$ →Φproduction cross sections which, when included, will make the constraint slightly weaker as one then needs to consider the lower value predicted for the production rate. A second caveat is that SUSY effects, direct corrections to the production and H/A decays into sparticles, could alter the rate. However, as previously argued, σ(pp→Φ)×BR(Φ→ττ) is robust against these SUSY effects and the latter will unlikely make a substantial change of the cross section times branching fraction. Finally, the constraint is specifically given in the maximal mixing scenario Xt/MS=√6 with MS=1 TeV. The robustness of σ×BR makes that the exclusion limit is actually almost model independent and is valid in far more situations than the "MSSM $M_h^{\max}$ scenario" quoted there, an assumption that can be removed without any loss.

In fact, the exclusion limit can also be extended to the low tanβ region which, in the chosen scenario with MS=1 TeV, is excluded by the LEP2 limit on the lighter h mass (the green area in the figure) but should resurrect if the SUSY scale is kept as a free parameter. Note also that H/Abosons have also been searched for in the channel gg→ $b\bar{b}$Φ with Φ→ $b\bar{b}$ (requiring more than three tagged b jets in the final state) but the constraints are much less severe than the ones derived from the ττ channel [193].

Turning to the H^+ boson [194, 195], the most recent result has been provided by the ATLAS collaboration using the full ≈20 fb^{-1} data collected at √s=8 TeV. The $H^\pm$ search as been performed using the τ plus jets channel with a hadronically decaying τ lepton in the final state. For $M_{H\pm}$≲160 GeV, the results are shown in Fig. 24. Here, the relevant process is top quark decays, t→H^+b with the decay H^+→τν having a branching ratio of almost 100 % at moderate to high tanβ. For these high values, the H^+tb coupling has a component ∝mbtanβ, which makes BR(t→!H^+b) rather large. Almost the entire tanβ≳10 region is excluded by the ATLAS analysis.

In addition, the branching fraction for the decay t→bH+ is also significant at low tanβ values, when the component of the coupling gtbH+∝¯mt/tanβ becomes dominant. On the other hand, the branching fraction for the decay $H^\pm$→τν does not become very small as it has competition only from H+→ $c\bar{s}$, which, even for tanβ≈1, does not dominate. Hence, the rates for pp→ $t\bar{t}$ with t→bH^+→bτν are comparable for tanβ≈3 and tanβ≈30 and the processes can also probe the low tanβ region. This is exemplified in Fig. 24 where one can see that the entire area below tanβ≈5 is also excluded. There remains then, for $H^\pm$ masses close to 90 GeV (where the detection efficiency is lower) and 160 GeV (where one is limited by the phase space), the intermediate area with tanβ≈5–10 where the $H^\pm$tb coupling is not strongly enhanced.

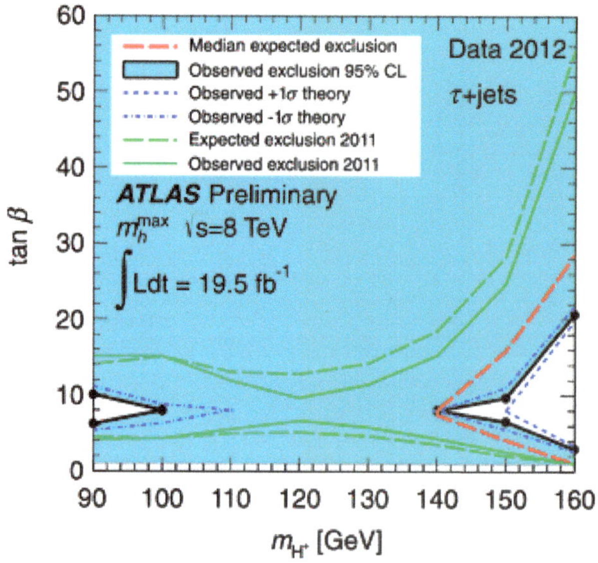

Figure 24: The $H^\pm$ limits from ATLAS with √s=8 TeV and ≈20 fb^{-1} data in the channel t→bH^+→bτν [194].

This ATLAS search has been extended to larger values of $M_{H\pm}$ where the charged Higgs is produced in association with top quarks, gb→tH^+, but the constraints are poor (only the region tanβ≳50 is excluded for $M_{H\pm}$=200–300 GeV) as the cross section for this process is low.

The reopening of the low tanβ region allows to consider a plethora of very interesting channels for the heavier Higgs bosons to be also investigated at the LHC: heavier CP-even H decays into massive gauge bosons H→WW,ZZ and Higgs bosons H→hh, CP-odd Higgs decays into a vector and a Higgs boson, A→hZ, CP-even and CP-odd Higgs decays into top quarks, H/A→ $t\bar{t}$, and even the charged Higgs decay $H^\pm$→Wh. These final states have been searched for in the context of a heavy SM Higgs boson or for new resonances in some non-SUSY beyond the SM scenarios and the analyses can be adapted to the case of the heavier MSSM Higgs bosons. They would then allow to cover a larger part of the parameter space of the MSSM Higgs sector in a model-independent way, i.e. without using the information on the scale MS and more generally on the SUSY particle spectrum that appear in the radiative corrections.

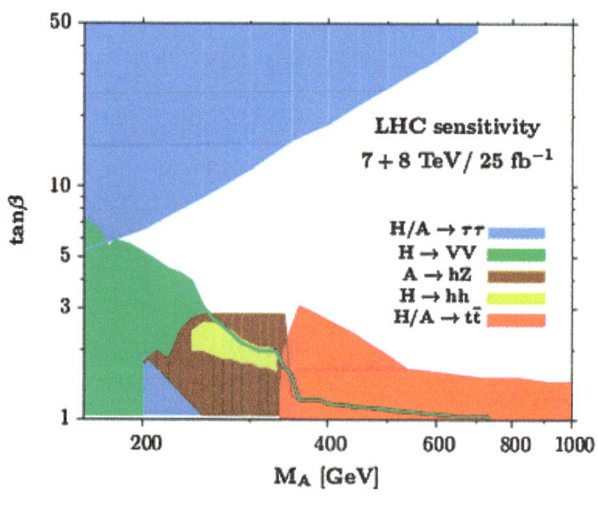

Figure 25: The estimated sensitivities in the various search channels for the heavier MSSM Higgs bosons in

the [tanβ,M_A] plane: H/A→ττ, H→WW+ZZ, H/A→ $t\bar{t}$, , A→hZ and H→hh[98]. The projection is made for the LHC with 7+8 TeV and the full 25 fb^{-1} of data collected so far. The radiative corrections are such that the h mass is M_h=126 GeV.

In Ref. [98] a preliminary analysis of these channels has been performed using current information given by the ATLAS and CMS collaborations in the context of searches for the SM Higgs boson or other heavy resonances (in particular new Z′ or Kaluza–Klein gauge bosons that decay into $t\bar{t}$ pairs). The results are shown in Fig. 25 with an extrapolation to the full 25 fb−1 data of the 7+8 TeV LHC run (it has been assumed that the sensitivity scales simply as the square root of the number of events). The sensitivities from the usual H/A→$\tau^+\tau^-$ and t→bH^+→bτν channels are also shown. The green and red areas correspond to the domains where the H→VV and H/A→ $t\bar{t}$ channels become constraining. The sensitivities in the H→hhand A→hZ modes are given by, respectively, the yellow and brown areas which peak in the mass range M^A=250–350 GeV that is visible at low tanβ values.

The outcome is impressive. These channels, in particular the H→VV and H/A→ $t\bar{t}$ processes, are very constraining as they cover the entire low tanβ area that was previously excluded by the LEP2 bound up to MA≈500 GeV. Even A→hZ and H→hh would be visible at the current LHC in small portions of the parameter space.

Could the observed state be the heavier H boson?

Let us briefly discuss the possibility, raised with the early LHC data, that the observed particle is the heavier MSSM H boson, as advocated for instance in Refs. [60–62, 105]. The possibility M_H≈125 GeV, with H couplings close to those of the SM Higgs, occurs for low values of M_A, ≈100–120 GeV, and moderate values of tanβ, ≈10. In this case, H has approximately SM-like properties, while h has a mass of order 100 GeV or below and suppressed couplings to vector bosons. A dedicated scan for this region of parameter space has been performed [36] and the results were confronted with the measured Higgs mass M_h=123–129 GeV and couplings that comply with the LHC ≈10 fb^{-1} data collected at √s= 7+8 TeV. Both the signal strengths in the various search channels of the observed Higgs boson and the limits from the pp→$\tau^+\tau^-$channel obtained by the CMS collaboration have been considered.

It was found that among the large flat scan with 10^8 points, only ≈2×10^{-5} of the generated points would remain after imposing these LHC constraints. These points were then excluded by applying the constraints from flavour physics [94] (see also Ref. [196]), mainly the radiative decay b→sγ and dark matter constraints [95] (as they do not satisfy the constraint of $10^{-4}<\Omega h^2<0.155$ when accounting for all uncertainties). The updated pp→$\tau^+\tau^-$ search performed by CMS with 17 fb^{-1} data, which excludes all values tanβ≳5 for M_A≲250 GeV as shown in Fig. 23, now definitely rules out this scenario.

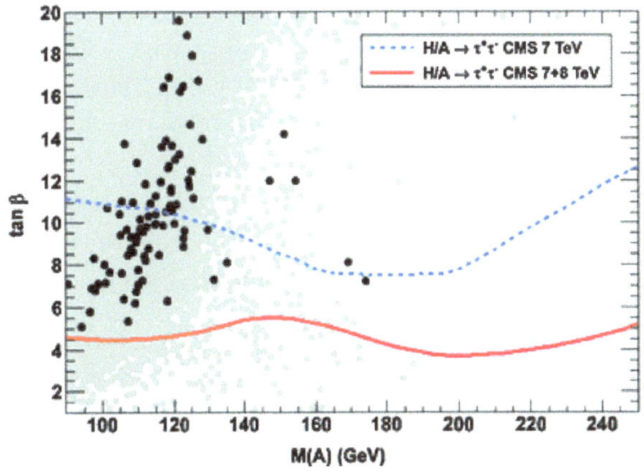

Figure 26: The parameter space [M_A,tanβ] with points for the heavier H boson to be observed with a mass in the range 123–129 GeV (light grey points) and after flavour and dark matter relic density constraints (black points) [36]. The CMS excluded regions from the 2011 and 2012 Φ→$\tau^+\tau^-$ searches are shown by the dashed blue and continuous red lines, respectively.

This is exemplified in Fig. 26, where we zoom in the [M^A,tanβ] plane for low values of the inputs and apply the constraints listed above. The small region in which the H boson was allowed to be the observed state (black points) by the previous H/A→$\tau^+\tau^-$ CMS search (dashed blue line) is excluded by the new data (in red). In fact, the latest ATLAS limits from $H^\pm$ searches given in Fig. 24 also exclude now the possibility M^A≈100–120 GeV and, hence, the scenario where H is the observed Higgs state[5].

Higgs production with SUSY particles

Finally, let us comment on the possibility of the Higgs bosons being produced in processes involving sparticles. First of all, there is the option of Higgs decays into SUSY particles. In the case of the lighter h boson, the only possibility when the LEP2 constraints are taken into account is the decay h→ $\chi_1^0\chi_1^0$, , which has been discussed in the context of invisible Higgs decays in Sect. 3.4. In view of the strong LHC limits on squark masses, the only SUSY channels of the heavier H/A/$H^\pm$ states that might be kinematically open would be the decays into chargino, neutralinos and sleptons. For H/A, these decays have been discussed in the context of the τsearches as they might reduce the H/A→ττ branching fractions but no specific search for these SUSY final states has been performed so far.

Turning to associated Higgs production with sparticles, the most important process was expected to be pp→ $\tilde{t}_1\tilde{t}_1$+ Higgs which could benefit from the possibly large Higgs–stop coupling [199–202]. The large value of MS and hence the lightest stop mass from the current constraint makes this process unlikely. Another possibility would be associated production with staus where the phase space could be more favourable but the rates are in general much smaller.

The only channel which could lead to a detectable signal with the data collected so far would be Higgs particles from decays of charginos and neutralinos. In particular the decays $\chi_2^0 \to \chi_1^0 h$, with χ_2^0 directly produced in association with $\chi_1^\pm$ in the process $pp \to \chi_2^0\chi_1^\pm$ leading a lepton, a Higgs (decaying either into $b\bar{b}$ or into multi-leptons via h→ZZ*,WW*) and missing energy [203–208].

The CMS collaboration has reported the results for searches of leptons and missing energy with a luminosity of ≈20 fb^{-1} data collected at √s=8 TeV [209]. They set a limit on the cross section times branching ratio for the possible SUSY process $pp \to \chi_2^0 \chi_1^\pm$ with $\chi_2^0 \to \chi_1^0 h \, \chi_1^\pm \to W \chi_1^0$. As can be observed from Fig. 27, where the cross section times branching ratio is displayed as a function of the masses $m_{\chi_1^\pm} = m_{\chi_2^0}$ (with the assumption that the LSP neutralino is very light, $m_{\chi_1^0}$=1 GeV), the data show no excess over the SM backgrounds.

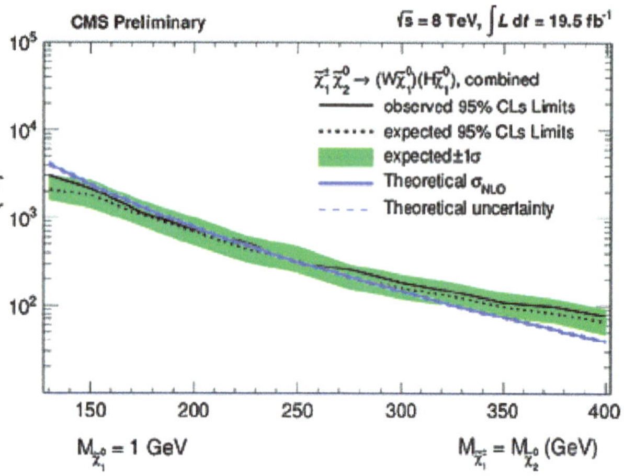

Figure 27: The interpretations of the CMS results from the combination of all lepton and E/Tsearches with ≈20 fb−1 data collected at √s=8 TeV. The expected and observed limits on the $pp \to \chi_2^0 \chi_1^\pm$ cross section times the $\chi_2^0 \chi_1^\pm \to W h \chi_1^0 \chi_1^0$ branching fraction (with the green band is for experimental uncertainties) is compared to the theoretical prediction [209].

5 What next?

The last 2 years were extremely rich and exciting for particle physics. With the historical discovery at the LHC of a Higgs boson by the ATLAS and CMS collaboration crowned by a Nobel price this fall, and the first probe of its basic properties, they witnessed a giant step in the unravelling of the mechanism that breaks the electroweak symmetry and generates the fundamental particle masses. They promoted the SM as the appropriate theory, up to at least the Fermi energy scale, to describe three of Nature's interactions, the electromagnetic, weak and strong forces,

However, it is clear that these 2 years have also led to some frustration, as no signal of physics beyond the SM has emerged from the LHC data. The hope of observing some signs of the new physics models that were put forward to address the hierarchy problem, which is deeply rooted in the Higgs mechanisms, with supersymmetric theories being the most attractive ones, did not materialise.

The discovery of the Higgs boson and the non-observation of new particles has nevertheless far reaching consequences for supersymmetric theories and, in particular, for their simplest low-energy formulation, the MSSM. The mass of approximately 125 GeV of the observed Higgs boson implies that the scale of SUSY-breaking is rather high, at least O(TeV). This is backed up by the limits on the masses of strongly interacting SUSY particles set by the ATLAS and CMS searches, which in most cases exceed the TeV range [86, 87]. This implies that if SUSY is indeed behind the stabilisation of the Higgs mass against very high scales that enter via quantum corrections, it is either fine-tuned at the permille level at least or its low-energy manifestation is more complicated than expected.

The production and decay rates of the observed Higgs particles, as well as its spin and parity quantum numbers, as measured by the ATLAS and CMS collaborations with the ≈ 25 fb^{-1}data collected at √s=7 and 8 TeV, indicate that its couplings to fermions and gauge bosons are approximately SM-like. In the context of the MSSM, this implies that we seem to be in the decoupling regime and this 125 GeV particle can be only identified with the lightest h boson, while the other H/A/$H^\pm$ states must be heavier than approximately the Fermi scale. This last feature is also backed up by the constraints from direct searches of these heavier Higgs states at the LHC.

This drives up to the question that is now very often asked in particle physics (and elsewhere): what to do next? The answer is, for me, obvious: we are only in the beginning of a new era.[6]Indeed, it was expected since a long time that the probing of the EWSB mechanism will be at least a two chapters story. The first one is the search and the observation of a Higgs-like particle that will confirm the scenario of the SM and most of its extensions, that is, a spontaneous symmetry breaking by a scalar field that develops a non-zero vacuum expectation value. This long chapter has just been closed by the ATLAS and CMS collaborations with the spectacular observation of a Higgs boson. This observation opens a second and equally important chapter: the precise determination of the Higgs profile and the unravelling of the EWSB mechanism itself.

A more accurate measurement of the Higgs couplings to fermions and gauge bosons will be mandatory to establish the exact nature of the mechanism and, eventually, to pin down effects of new physics if additional ingredients beyond those of the SM are involved. This is particularly true in weakly interacting theories such as SUSY in which the quantum effects are expected to be small. These measurements could be performed at the upgraded LHC with an energy close to √s=14 TeV, in particular if a very high luminosity, a few ab^{-1}, is achieved [210–212].

At this upgrade, besides improving the measurements performed so far, rare but important channels such as associated Higgs production with top quarks, pp→$t\bar{t}h$, and Higgs decays into $\mu^+\mu^-$ and Zγ states could be probed. Above all, a determination of the self-Higgs coupling could be made by searching for double Higgs production e.g. in the gluon fusion channel gg→hh[213–215]; this would be a first step towards the reconstruction of the scalar potential that is responsible of EWSB. A proton collider with an energy √s=30 to 100 TeV could do a similar job [212].

In a less near future, a high-energy lepton collider, which is nowadays discussed in various options (ILC, TLEP, CLIC, μ collider) would lead to a more accurate probing of the Higgs properties [216–224], promoting the scalar sector of the theory to the high-precision level of the gauge and fermionic sectors achieved by LEP and SLC [30].

Besides the high precision study of the already observed Higgs, one should also continue to search for the heavy states that are predicted by SUSY, not only the superparticles but also the heavier Higgs bosons. The energy upgrade to ≈14 TeV (and eventually beyond) and the planed order of magnitude (or more) increase in luminosity will allow to probe much higher mass scales than presently.

In conclusion, it is not yet time to give up on supersymmetry and on new physics in general but, rather, to work harder to be fully prepared for the more precise and larger data that will be delivered by the upgraded LHC. It will be soon enough to "philosophise" in 2 years from now, when the physics landscape will become clearer.

Footnotes

[1]This uncertainty is obtained by comparing the outputs of SuSpect and FeynHiggs [83] which use different schemes for the radiative corrections: while the former uses the $\overline{\rm DR}$ scheme, the latter uses the on-shell scheme; the difference in the obtained M_h amounts to ≈±2–3 GeV in general. To this, one has to add an uncertainty of ±1 GeV from the top quark mass measurement at the Tevatron, m_t=173±1 GeV [85]. Note that it is not entirely clear whether this mass is indeed the pole quark mass. A more rigorous determination of the pole mass from the measured top-pair cross section at the Tevatron gives a lower value with a larger uncertainty, $m_t^{\rm pole}$ =171±3 GeV [84].

[2]Note that also in the VBF process, despite the fact that the inclusive cross section has only a few percent combined scale and PDF uncertainty [110], the contamination by the gg→h+2j channel makes the total uncertainty in the h+jj final "VBF" sample rather large. Indeed O(30%)gg→h+2j events will remain even after the specific cuts that select the VBF configuration have been applied, and the rate is affected by a much larger uncertainty than the inclusive gg→h process, up to 50 % when one adds the scale and PDF uncertainties [113].

[3]In the case of the production process $gg/q\bar{q} \to ht\bar{t}$, it is still c_t^0, , which should describe the $ht\bar{t}$ coupling, but the constraints on the h properties from this process are presently very weak.

[4]To be more general, the spin of the particle needs also to be determined. The observation of the h→γγ decay channel rules out the spin-1 case by virtue of the Landau–Yang theorem [167, 168] and leaves only the spin 0 and ≥2 possibilities. The graviton-like spin-2 option is extremely unlikely and, already from the particle signal strengths in the various channels, it is ruled out in large classes of beyond the SM models; see e.g. Ref. [169].

[5]Note that the recent pp→ττ and H^+→τν limits also exclude the so-called "intense coupling regime" [197, 198], in which the three neutral Higgs bosons could be light and close in mass.

[6]One can rightfully use here the words of Winston Churchill in November 1942 after the battle of El Alamein (which in Arabic literally means "the two flags" but could also mean "the two worlds" or even "the two scientists"!): "Now, this is not the end; it is not even the beginning to the end; but it is, perhaps, the end of the beginning".

ACKNOWLEDGMENTS

This review relies heavily on work performed in the last two years in collaboration with S. Alekhin, A. Arbey, J. Baglio, M. Battaglia, A. Falkowski, R. Godbole, R. Grober, O. Lebedev, A. Lenz, N. Mahmoudi, L. Maiani, Y. Mambrini, B. Mellado, K. Mohan, G. Moreau, M. Muhlleitner, J. Quevillon, A. Polosa, V. Riquer and M. Spira. I thank them all for their input and for having made these two last years very fruitful and extremely exciting. Discussions with members of ATLAS and CMS (whom I congratulate in passing) are also acknowledged. I thank the CERN Theory Unit for its hospitality during this period. This work is supported by the ERC Advanced Grant Higgs@LHC.

REFERENCES

1. The ATLAS collaboration, Phys. Lett. B 716, 1 (2012)
2. The CMS collaboration, Phys. Lett. B 716, 30 (2012)
3. P. Higgs, Phys. Lett. 12, 132 (1964)
4. F. Englert, R. Brout, Phys. Rev. Lett. 13, 321 (1964)
5. G. Guralnik, C. Hagen, T. Kibble, Phys. Rev. Lett. 13, 585 (1964)
6. J. Gunion, H. Haber, G. Kane, S. Dawson, The Higgs Hunter's Guide (Reading, 1990)
7. A. Djouadi, Phys. Rep. 457, 1 (2008)
8. Djouadi, Phys. Rep. 459, 1 (2008)
9. M. Drees, R. Godbole, P. Roy, Theory and Phenomenology of Sparticles (World Sci., 2005)
10. H. Baer, X. Tata, Weak Scale Supersymmetry: From Superfields to Scattering Events (Cambridge U Press, Cambridge, 2006)
11. S. Martin, arXiv:hep-ph/9709356
12. Y. Okada, M. Yamaguchi, T. Yanagida, Prog. Theor. Phys. 85, 1 (1991)
13. J. Ellis, G. Ridolfi, F. Zwirner, Phys. Lett. B 257, 83 (1991)
14. H.E. Haber, R. Hempfling, Phys. Rev. Lett. 66, 1815 (1991)
15. M. Carena, J.R. Espinosa, M. Quiros, C.E. Wagner, Phys. Lett. B 355, 209 (1995)
16. H. Haber, R. Hempfling, A. Hoang, Z. Phys, C 75, 539 (1997)
17. S. Heinemeyer, W. Hollik, G. Weiglein, Phys. Rev. D 58, 091701 (1998)
18. S. Heinemeyer, W. Hollik, G. Weiglein, Eur. Phys. J. C 9, 343 (1999)
19. G. Degrassi, P. Slavich, F. Zwirner, Nucl. Phys. B 611, 403 (2001)
20. A. Brignole, G. Degrassi, P. Slavich, F. Zwirner, Nucl. Phys. B 631, 195 (2002)
21. Brignole, G. Degrassi, P. Slavich, F. Zwirner, Nucl. Phys. B 643, 79 (2002)
22. S. Martin, Phys. Rev. D 75, 055005 (2007)
23. P. Kant, R. Harlander, L. Mihaila, M. Steinhauser, JHEP 1008, 104 (2010)
24. J. Feng et al., Phys. Rev. Lett. 111, 131802 (2013)
25. Allanach et al., JHEP 0409, 044 (2004)
26. S. Heinemeyer, W. Hollik, G. Weiglein, Phys. Rep. 425, 265 (2006)
27. S. Heinemeyer, IJMPA 21, 2659 (2006)

28. M. Carena, H. Haber, Prog. Part. Nucl. Phys. 50, 63 (2003)
29. See for instance H.E. Haber, arXiv:hep-ph/9505240
30. J. Beringer,(PDG), et al., Phys. Rev. D 86, 010001 (2012)
31. LEP collaborations, Phys. Lett. B 565, 61 (2003)
32. R. Barbieri, G. Giudice, Nucl. Phys. B 306, 63 (1988)
33. M. Papucci, J. Ruderman, A. Weiler, JHEP 1209, 035 (2012)
34. M. Carena, S. Heinemeyer, C. Wagner, G. Weiglein, Eur. J. Phys. C 26, 601 (2003)
35. A. Arbey et al., Phys. Lett. B 708, 162 (2012)
36. Arbey et al., JHEP 1209, 107 (2012)
37. Among the vast literature on the subject, for the (mainly) early papers in the SUSY context, see e.g.: H. Baer, V. Barger, A. Mustafayev, Phys. Rev. D 85, 075010 (2012)
38. P. Draper, P. Meade, M. Reece, D. Shih, Phys. Rev. D 85, 095007 (2012)
39. O. Buchmueller et al., Eur. Phys. J. C 72, 2020 (2012)
40. S. Akula et al., Phys. Rev. D 85, 075001 (2012)
41. Strege et al., JCAP 1203, 030 (2012)
42. Beskidt et al., JHEP 1205, 094 (2012)
43. M. Carena et al., JHEP 1207, 175 (2012)
44. M. Carena, I. Low, C. Wagner, JHEP 1208, 060 (2012)
45. M. Cahill-Rowley et al., Phys. Rev. D 86, 075015 (2012)
46. P. Lodone, Int. J. Mod. Phys. A 27, 1230010 (2012)
47. M. Kadastik et al., JHEP 1205, 061 (2012)
48. U. Ellwanger, JHEP 1203, 044 (2012)
49. S. King, M. Muhlleitner, R. Nevzorov, Nucl. Phys. B 860, 207 (2012)
50. Ghilencea, Nucl. Phys. B 876, 16 (2013)
51. J. Cao et al., Phys. Lett. B 710, 665 (2012)
52. L. Aparicio, D. Cerdeno, L. Ibanez, JHEP 1204, 126 (2012)
53. J. Ellis, K. Olive, Eur. Phys. J. C 72, 2005 (2012)
54. J. Cao et al., JHEP 1203, 086 (2012)
55. Boudjema, G.D. La Rochelle, Phys. Rev. D 86, 015018 (2012)
56. Brummer, S. Kraml, S. Kulkarni, JHEP 1208, 089 (2012)
57. L. Hall, D. Pinner, J. Ruderman, JHEP 1204, 131 (2012)
58. A. Arvanitaki, G. Villadoro, JHEP 1202, 144 (2012)
59. Delgado et al., Eur. Phys. J. C 73, 2370 (2013)
60. S. Heinemeyer, O. Stal, G. Weiglein, Phys. Lett. B 710, 201 (2012)
61. P. Bechtle et al., Eur. Phys. J. C 73, 2354 (2013)
62. M. Drees, Phys. Rev. D 86, 115018 (2012)
63. A. Djouadi et al. (MSSM WG), arXiv:hep-ph/9901246
64. A.H. Chamseddine, R. Arnowitt, P. Nath, Phys. Rev. Lett. 49, 970 (1982)
65. R. Barbieri, S. Ferrara, C. Savoy, Phys. Lett. B 119, 343 (1982)
66. L. Hall, J. Lykken, S. Weinberg, Phys. Rev. D 27, 2359 (1983)
67. N. Ohta, Prog. Theor. Phys. 70, 542 (1983)
68. M. Dine, W. Fishler, Phys. Lett. B 110, 227 (1982)
69. Nappi, B. Ovrut, Phys. Lett. B 113, 1785 (1982)
70. L. Alvarez-Gaumé, M. Claudson, M. Wise, Nucl. Phys. B 207, 96 (1982)
71. M. Dine, A. Nelson, Y. Shirman, Phys. Rev. D 51, 1362 (1995)
72. G.F. Giudice, R. Rattazzi, Phys. Rep. 322, 419 (1999)
73. L. Randall, R. Sundrum, Nucl. Phys. B 557, 79 (1999)
74. Giudice, M. Luty, H. Murayama, R. Rattazzi, JHEP 9812, 027 (1998)
75. J. Bagger, T. Moroi, E. Poppitz, JHEP 0004, 009 (2000)
76. N. Arkani-Hamed, S. Dimopoulos, JHEP 0506, 073 (2005)
77. G.F. Giudice, A. Romanino, Nucl. Phys. B 699, 65 (2004)
78. J.D. Wells, Phys. Rev. D 71, 015013 (2005)
79. See, L. Hall, Y. Nomura, JHEP 1003, 076 (2010)
80. Giudice, A. Strumia, Nucl. Phys. B 858, 63 (2012)
81. A. Djouadi, L. Maiani, G. Moreau, A. Polosa, J. Quevillon, V. Riquer, arXiv:1307.5205 [hep-ph]
82. Djouadi, J.L. Kneur, G. Moultaka, Comput. Phys. Commun. 176, 426 (2007)
83. S. Heinemeyer, W. Hollik, G. Weiglein, Comput. Phys. Commun. 124, 76 (2000)
84. S. Alekhin et al., Phys. Lett. B 716, 214 (2012)
85. Tevatron, CDF+D0 EWWG, arXiv:1305.3929 [hep-ex]
86. See the talk given by O. Buchmuller for ATLAS and CMS at the EPS conference, Stockholm, July 2013
87. For a review, see N. Craig, arXiv:1309.0528 [hep-ph]
88. For recent studies, see: A. Benhenni et al., Phys. Rev. D 84, 075015 (2011)
89. T. Li et al., Phys. Lett. B 710, 207 (2012)
90. A. Djouadi, U. Ellwanger, A.M. Teixeira, Phys. Rev. Lett. 101, 101802 (2008)
91. Djouadi, U. Ellwanger, A.M. Teixeira, JHEP 0904, 031 (2009)
92. Ellis et al., Phys. Rev. D 70, 055005 (2004)
93. e.g. S.S. AbdusSalam et al., Eur. Phys. J. C 71, 1835 (2011)
94. For a recent review, see e.g. T. Hurth, F. Mahmoudi, Rev. Mod. Phys. 85, 795 (2013)
95. For a recent review, see e.g. R. Catena, L. Covi, SUSY dark matter(s). Eur. Phys. J. C 74, 2703 (2014). doi:10.1140/epjc/s10052-013-2703-4. arXiv:1310.4776
96. For a review, see e.g. G. Kane, R. Lu, B. Zheng, Int. J. Mod. Phys. A 28, 1330002 (2013)
97. N. Bernal, A. Djouadi, P. Slavich, JHEP 0707, 016 (2007)
98. A. Djouadi, J. Quevillon, arXiv:1304.1787 [hep-ph]
99. Delgado, G.F. Giudice, Phys. Lett. B 627, 155 (2005)
100. E. Arganda, J.L. Diaz-Cruz, A. Szynkman, Eur. Phys. J. C 73, 2384 (2013)

101. E. Arganda, J.L. Diaz-Cruz, A. Szynkman, Phys. Lett. B 722, 100 (2013)
102. L. Maiani, A.D. Polosa, V. Riquer, New J. Phys. 14, 073029 (2012)
103. L. Maiani, A.D. Polosa, V. Riquer, Phys. Lett. B 718, 465 (2012)
104. L. Maiani, A.D. Polosa, V. Riquer, Phys. Lett. B 724, 274 (2013)
105. M. Carena et al., Eur. Phys. J. C 73, 2552 (2013)
106. A. Martin et al., Eur. Phys. J. C 63, 189 (2009)
107. Djouadi, Eur. Phys. J. C 73, 2498 (2013)
108. The ATLAS collaboration, ATLAS-CONF-2013-034
109. The CMS collaboration, CMS-HIG-13-005
110. S. Dittmaier et al. (LHC Higgs WG), arXiv:1101.0593
111. Baglio, A. Djouadi, JHEP 1010, 064 (2010)
112. Baglio, A. Djouadi, JHEP 1103, 055 (2011)
113. S. Dittmaier et al. (LHC Higgs WG), arXiv:1201.3084
114. Zeppenfeld, R. Kinnunen, A. Nikitenko, E. Richter-Was, Phys. Rev. D 62, 013009 (2000)
115. A. Djouadi et al., arXiv:hep-ph/0002258
116. Dührssen et al., Phys. Rev. D 70, 113009 (2004)
117. K. Assamagan et al., arXiv:hep-ph/0406152
118. A. Djouadi, J. Kalinowski, M. Spira, Comput. Phys. Commun. 108, 56 (1998)
119. Carmi, A. Falkowski, E. Kuflik, T. Volansky, JHEP 1207, 136 (2012)
120. Azatov, R. Contino, J. Galloway, JHEP 1204, 127 (2012)
121. J. Espinosa, C. Grojean, M. Muhlleitner, M. Trott, JHEP 1205, 097 (2012)
122. J. Espinosa, C. Grojean, M. Muhlleitner, M. Trott, JHEP 1212, 045 (2012)
123. J. Espinosa, C. Grojean, M. Muhlleitner, M. Trott, JHEP 1209, 126 (2012)
124. P. Giardino, K. Kannike, M. Raidal, A. Strumia, JHEP 1206, 117 (2012)
125. P. Giardino, K. Kannike, M. Raidal, A. Strumia, Phys. Lett. B 718, 469 (2012)
126. T. Li, X. Wan, Y.-K. Wang, S.-H. Zhu, JHEP 1209, 086 (2012)
127. J. Ellis, T. You, JHEP 1206, 140 (2012)
128. A. Azatov et al., JHEP 1206, 134 (2012)
129. Klute, R. Lafaye, T. Plehn, M. Rauch, D. Zerwas, Phys. Rev. Lett. 109, 101801 (2012)
130. Djouadi, A. Lenz, Phys. Lett. B 715, 310 (2012)
131. J. Chang, K. Cheung, P.-Y. Tseng, T.-C. Yuan, JHEP 1212, 058 (2012)
132. S. Chang, C.A. Newby, N. Raj, C. Wanotayaroj, Phys. Rev. D 86, 095015 (2012)
133. Low, J. Lykken, G. Shaughnessy, Phys. Rev. D 86, 093012 (2012)
134. Montull, F. Riva, JHEP 1211, 018 (2012)
135. Carmi et al., JHEP 1210, 196 (2012)
136. S. Banerjee, S. Mukhopadhyay, B. Mukhopadhyaya, JHEP 1210, 062 (2012)
137. Bonnet, T. Ota, M. Rauch, W. Winter, Phys. Rev. D 86, 093014 (2012)
138. T. Plehn, M. Rauch, Europhys. Lett. 100, 11002 (2012)
139. Belanger, B. Dumont, U. Ellwanger, J. Gunion, S. Kraml, JHEP 1302, 053 (2013)
140. Belanger, B. Dumont, U. Ellwanger, J. Gunion, S. Kraml, arXiv:1306.2941
141. G. Altarelli, arXiv:1308.0545
142. T. Corbett et al., Phys. Rev. D 86, 075013 (2012)
143. A. Alves et al., Eur. Phys. J. C 73, 2288 (2013)
144. M. Peskin, arXiv:1207.2516v1 [hep-ph]
145. G. Cacciapaglia, A. Deandrea, G.D. La Rochelle, J.B. Flament, JHEP 1303, 029 (2013)
146. A. Falkowski, F. Riva, A. Urbano, arXiv:1303.1812 [hep-ph]
147. B. Cheung, S. McDermott, K. Zurek, JHEP 1304, 074 (2013)
148. K. Cheung, J.S. Lee, P. Tseng, JHEP 1305, 134 (2013)
149. A. Arbey et al., Phys. Lett. B 720, 153 (2013)
150. B. Djouadi, G. Moreau, arXiv:1303.6591
151. See e.g., M. Carena et al., Nucl. Phys. B 577, 88 (2000)
152. A. Djouadi, Phys. Lett. B 435, 101 (1998)
153. Baglio et al., Phys. Lett. B 716, 203 (2012)
154. See e.g. M. Carena, S. Gori, N. R. Shah, C.E.M. Wagner, JHEP 1207, 175 (2012) (and references therein)
155. G.F. Giudice, P. Paradisi, A. Strumia, JHEP 1210, 186 (2012)
156. U. Haisch, F. Mahmoudi, arXiv:1210.7806
157. A. Djouadi, V. Driesen, W. Hollik, J.I. Illana, Eur. Phys. J. C 1, 149 (1998)
158. J.A. Casas, J.M. Moreno, K. Rolbiecki, B. Zaldivar, JHEP 1309, 099 (2013)
159. The ATLAS collaboration, ATLAS-CONF-2013-011
160. A. Djouadi, A. Falkowski, Y. Mambrini, J. Quevillon, Eur. Phys. J. C 73, 2455 (2013)
161. Y. Bai, P. Draper, J. Shelton, JHEP 1207, 192 (2012)
162. A. Englert, J. Jaeckel, E. Re, M. Spannowsky, Phys. Rev. D 85, 035008 (2012)
163. The CMS collaboration, CMS-PAS-HIG-13-013
164. The CMS collaboration, arXiv:1206.5663 [hep-ex]
165. The ATLAS collaboration, ATLAS-CONF-2012-147
166. A. Djouadi, O. Lebedev, Y. Mambrini, J. Quevillon, Phys. Lett. B 709, 65 (2012)
167. L. Landau, Dokl. Akad. Nauk Ser. Fiz. 60, 207 (1948)
168. A. Yang, Phys. Rev. 77, 242 (1950)
169. Ellis, V. Sanz, T. You, Phys. Lett. B 726, 244 (2013)
170. For a review of the CP Higgs issue, see e.g. S. Kraml (ed.) et al., arXiv:hep-ph/0608079
171. a more recent study with references, see A. Alloul, B. Fuks, V. Sanz, arXiv:1310.5150
172. The ATLAS collaboration, Phys. Lett. B 726, 120 (2013)

173. The CMS collaboration, Phys. Rev. Lett. 110, 081803 (2013)
174. T. Plehn, D. Rainwater, D. Zeppenfeld, Phys. Rev. Lett. 88, 051801 (2002)
175. Hagiwara, Q. Li, K. Mawatari, JHEP 0907 (2009)
176. J. Frank, M. Rauch, D. Zeppenfeld, Phys. Rev. D 87, 055020 (2013)
177. A. Englert, D. Gonsalves-Netto, K. Mawatari, T. Plehn, JHEP 1301, 148 (2013)
178. B. Djouadi, R.M. Godbole, B. Mellado, K. Mohan, Phys. Lett. B 723, 307 (2013)
179. V. Barger et al., Phys. Rev. D 49, 79 (1994)
180. Grzadkowski, J. Gunion, X. He, Phys. Rev. Lett. 77, 5172 (1996)
181. J. Gunion, J. Pliszka, Phys. Lett. B 444, 136 (1998)
182. Bhupal Dev et al., Phys. Rev. Lett. 100, 051801 (2008)
183. A. Freitas, P. Schwaller, Phys. Rev. D 87, 055014 (2013)
184. Spira et al., Nucl. Phys. B 453, 17 (1995)
185. R. Harlander, W. Kilgore, Phys. Rev. D 68, 013001 (2003)
186. R. Harlander, S. Liebler, H. Mantler, Comp. Phys. Comm. 184, 1605 (2013)
187. A. Arbey, M. Battaglia, F. Mahmoudi, Phys. Rev. D 88, 015007 (2013)
188. Bechtle et al., arXiv:1305.1933
189. Craig, J. Galloway, S. Thomas, arXiv:1305.2424
190. Christensen, T. Han, S. Su, Phys. Rev. D 85, 115018 (2012)
191. The ATLAS collaboration, arXiv:1211.6956
192. The CMS collaboration, CMS-PAS-HIG-12-050
193. The CMS collaboration, CMS-PAS-HIG-12-033
194. The ATLAS collaboration, ATLAS-CONF-2013-090
195. The CMS collaboration, arXiv:1205.5736
196. G. Barenboim et al., arXiv:1307.5973 [hep-ph]
197. A. Boos et al., Phys. Rev. D 66, 055004 (2002)
198. B. Boos, A. Djouadi, A. Nikitenko, Phys. Lett. B 578, 384 (2004)
199. C. Djouadi, J.L. Kneur, G. Moultaka, Phys. Rev. Lett. 80, 1830 (1998)
200. Djouadi, J.L. Kneur, G. Moultaka, Nucl. Phys. B 569, 53 (2000)
201. A. Bélanger et al., Eur. Phys. J. C 9, 511 (1999)
202. Dedes, S. Moretti, Eur. Phys. J. C 10, 515 (1999)
203. Datta et al., Phys. Rev. D 65, 015007 (2002)
204. Datta et al., Nucl. Phys. B 681, 31 (2004)
205. Ghosh, M. Guchait, D. Sengupta, Eur. Phys. J. C 72, 2141 (2012)
206. K. Howe, P. Saraswat, JHEP 1210, 065 (2012)
207. A. Arbey, M. Battaglia, F. Mahmoudi, arXiv:1212.6865 [hep-ph]
208. T. Han, S. Padhi, S. Su, arXiv:1309.5966 [hep-ph]
209. The CMS collaboration, CMS-PAS-SUS-13-017
210. The ATLAS collaboration, arXiv:1307.7292 [hep-ex]
211. The CMS collaboration arXiv:1307.7135 [hep-ex]
212. S. Dawson et al., arXiv:1310.8361 [hep-ex]
213. See e.g., J. Baglio et al., JHEP 1304, 151 (2013)
214. M. Dolan, C. Englert, M. Spannowsky, JHEP 1210, 112 (2012)
215. W. Yao, arXiv:1308.6302 [hep-ph]
216. A. Abramowicz, arXiv:1307.5288 [hep-ex]
217. M. Bicer et al., arXiv:1308.6176 [hep-ex]
218. A. Baer et al., arXiv:1306.6352 [hep-ph]
219. B. Brau et al., arXiv:1210.0202 [hep-ph]
220. For earlier work, see e.g.: G. Arons et al., arXiv:0709.1893
221. A. Aguilar-Saavedra, arXiv:hep-ph/0106315
222. Accomando et al., Phys. Rep. 299, 1 (1998)
223. A. Peskin, H. Murayama, Ann. Rev. Nucl. Part. Sci. 46, 533 (1996)
224. B. Djouadi, Int. J. Mod. Phys. A 10, 1 (1995)

CHAPTER 2

NEUTRINO MASS AS A SIGNAL OF TEV SCALE PHYSICS

Rabindra N. Mohapatra

INTRODUCTION

The neutrino oscillation experiments, celebrated by this year's Nobel prize, have unambiguously established over the past decade and half that neutrinos have tiny but non-zero masses as well as nontrivial mixings between different flavors. They have provided the first unambiguous evidence for physics beyond the standard model (SM) which predicts that neutrino masses vanish. Since the origin of mass for all charged fermions of the standard model appears now to have been clarified by the discovery of the Higgs boson, an important question is whether there are new Higgs bosons and new fermions (and new forces) associated with understanding the origin of neutrino masses. As a starting point of this discussion, note that if we simply add three right handed neutrinos to the SM, one can write a Yukawa coupling of the form $\mathcal{L}_{\nu,Y} = h_\nu \bar{L}\tilde{H}\nu_R$ which via the SM Higgs vev can give a mass to neutrinos of magnitude $m_\nu = h_\nu <H^0>$. However, since $<H^0> \simeq 173$ GeV, to get sub-eV neutrino masses, we need to have $h_\nu \simeq 10^{-12}$, an "unnaturally" small number. So the strong suspicion among theorists is that there is some new physics beyond the standard model and new Higgs fields beyond the 125 GeV SM Higgs that are responsible for neutrino masses. The question then is at what scale this new physics manifests itself and can we test it experimentally. The goal of this article is to make a case that the origin of neutrino masses could be due to new physics at the multi-TeV scale and that there exists a natural class of TeV scale models for neutrino masses which can be probed at the LHC, the planned proton–proton collider at 100 TeV as well as low energy rare process search experiments such as $\mu \rightarrow e+\gamma$ and neutrinoless double beta decay.

Citation: Rabindra N. Mohapatra, Neutrino mass as a signal of TeV scale physics, Nuclear Physics B, Volume 908, July 2016, Pages 423-435, ISSN 0550-3213, http://dx.doi.org/10.1016/j.nuclphysb.2016.03.006.

The class of models, we focus on, is based on the simple paradigm of (type-I) seesaw mechanism [1] where the right handed neutrinos alluded to above have a heavy Majorana mass in addition to having a Yukawa coupling like all charged fermions. Neutrinos being electrically neutral allows for this possibility and it is not unreasonable to suspect that this also might be at the root of such diverse mass and mixing patterns for leptons compared to quarks. The crux of this physics is the seesaw mass matrix with the generic form in the (ν_L, N_R) space:

$$\begin{pmatrix} 0 & m_D \\ m_D^T & M_N \end{pmatrix} \quad (1)$$

where m_D mixes the ν and N states and is generated by the SM Higgs vev and M_N is the Majorana mass for N which embodies the new neutrino mass physics. For three generations of fermions, both m_D and M_N are 3×3 matrices. The mass of the light neutrinos is then given by the symbolic seesaw formula:

$$M_\nu \simeq -m_D M_N^{-1} m_D^T, \quad (2)$$

which also generates a mixing between the heavy (N) and the light neutrino (ν) again given by $\sim m_D/M_N \equiv V_{\ell N}$. The "unnaturalness" of the Yukawa couplings alluded to above are now considerably ameliorated due to two features of the formula above: first, it is the square of the Yukawa coupling h_ν that is present in the formula and secondly there is a heavy mass M_N in the denominator.

Clearly for a more fundamental understanding of the seesaw formula, one may ask the question: "Are the RH neutrinos N and their Majorana mass M_N, the seesaw scale put in by hand or is there a compelling theory that explains their origin in a natural manner?" Recall that these are the two key new ingredients in seesaw beyond the SM physics and their fundamental understanding would be a key step forward into the domain of Beyond the Standard Model (BSM) physics. Two immediate theories that present themselves and were discussed before the seesaw mechanism was suggested, are:

(i) an SO(10) grand unified theory that has a 16 dimensional spinor representation that, in addition to containing the fifteen standard model fermions, has room for just one more neutral fermion which can be identified with the right handed neutrino; and (ii) the left–right symmetric (LRSM) extension of the standard model which was proposed to explain the parity violation observed in weak processes to be truly a low energy phenomenon that will disappear at higher energies [2]. In the LRSM, the right handed neutrino appears as the parity partner of the left handed neutrino. As far as the seesaw scale is concerned, in SO(10) theories the seesaw scale is near $\sim 10^{14}$ GeV, which is close to and connected to the scale where all forces unify the GUT scale M_U. The high scale however has the disadvantage that the neutrino seesaw related new physics is well hidden from low energy and collider probes and the model is therefore virtually untestable. While this may well turn out to be the way nature operates, it behooves us to consider the second possibility where the seesaw scale which can be identified with the scale of the $SU(2)_R$gauge symmetry breaking. Since there are no a priori arguments which constrain this scale to be super high, it can be assumed to be in the TeV range and neutrino related new physics e.g. new gauge bosons, new Higgs bosons could all be in the TeV range making them accessible to collider and low energy probes and possibly testable in near future. This TeV scale theory is the one that we will explore in this article.

While this article focuses on type I TeV scale LR seesaw, there is another interesting alternative for TeV LR seesaw if small neutrino masses arise from the inverse seesaw [3] mechanism, which can be embedded into grand unified theories [4]. There have been extensive investigation of this class of models [5] and we do not discuss them here.

This paper is organized as follows: in sec. 2, I present then outline of the TeV scale model i.e. its particle content, symmetry breaking and Yukawa couplings in the lepton sector; in sec. 3, I discuss how type II contributions [6] are suppressed in the model; in sec. 4, I present the outline of the scenario that gives naturally small neutrino mass without necessarily using tiny Yukawa couplings; in sec. 5, I discuss the important limits (both lower and upper) from phenomenology as well as theory; in sec. 6, I discuss the experimental tests of the model.

THE BASIC INGREDIENTS OF LRSM

The left–right symmetric models are based on the gauge group $SU(2)_L \times SU(2)_R \times U(1)_{B-L}$[2], where the right handed fermions u_R, d_R, e_R, ν_R, which were $SU(2)_L$ singlets in the SM, are now assigned in a parity symmetric way to the right handed doublets of $SU(2)_R$ as follows:

$$Q_{L,R} = \begin{pmatrix} u \\ d \end{pmatrix}_{L,R}; \quad L, R = \begin{pmatrix} \nu \\ e \end{pmatrix}_{L,R}. \tag{3}$$

The right handed neutrinos are therefore a necessary part of the model and do not have to be added in an ad hoc manner just to implement the seesaw mechanism, as already noted. An important point is that the right handed neutrinos acquire a Majorana mass as soon as the $SU(2)_R$ symmetry is broken at scale v_R. This is quite analogous to the way the charged fermions get mass as soon as the standard model gauge symmetry $SU(2)_L$ is broken.

The $SU(2)_R$ breaking necessarily requires that there must be new Higgs bosons in addition to the 125 GeV Higgs boson discovered at the LHC. The minimal required Higgs sector of the model consists of the following multiplets [7]:

$$\Delta_R \equiv \begin{pmatrix} \Delta_R^+/\sqrt{2} & \Delta_R^{++} \\ \Delta_R^0 & -\Delta_R^+/\sqrt{2} \end{pmatrix}; \phi \equiv \begin{pmatrix} \phi_1^0 & \phi_2^+ \\ \phi_1^- & \phi_2^0 \end{pmatrix}. \tag{4}$$

The gauge symmetry $SU(2)_R \times U(1)_{B-L}$ is broken by the vacuum expectation value of $\langle \Delta_R^0 \rangle = v_R$ to the group $U(1)_Y$ of the standard model. There is also a left handed counter part to Δ_R, usually called Δ_L. We do not consider this field here for two reasons. There are versions of the model where parity and $SU(2)_R$ gauge symmetry scales are decoupled so that Δ_L fields are absent from the low energy theory [8], so that the Higgs sector is simpler. Another reason is that the presence of the Δ_L field in the TeV scale model generates a type II seesaw contribution to the neutrino masses which is large requiring fine tuning to understand neutrino masses; the decoupling of Δ_L fields avoids this problem and provides a natural way to realize the type I seesaw for neutrino masses. The vev of the ϕ field given by $\langle\phi\rangle = \mathrm{diag}(\kappa, \kappa')$ breaks the SM gauge group to $U(1)_{em}$.

To see how the fermions get their masses and how seesaw mechanism arises, we write down the Yukawa Lagrangian of the parity symmetric version of the model with one bi-doublet:

$$\mathcal{L}_Y = h_{\ell,ij} \bar{L}_i \phi R_j + \tilde{h}_{\ell,ij} \bar{L}_i \tilde{\phi} R_j + h_{q,ij} \bar{Q}_{Li} \phi Q_{Rj} + \tilde{h}_{q,ij} \bar{Q}_{Li} \tilde{\phi} Q_{R,j} + f_{\ell,ij}(L_i L_j \Delta_L + R_i R_j \Delta_R) + h.c. \tag{5}$$

where i, j stand for generations and a for labeling Higgs bi-doublets and $\tilde{\phi} = \tau_2 \phi^* \tau_2$and the coupling $RR\Delta_R$ is defined as $RR\Delta_R \equiv R_a^T 1 \tau_2 \Delta_R R_b$ (τ_2 is one of the Pauli matrices) and similarly for $LL\Delta_L$. We will drop the Δ_L coupling in subsequent sections since we will work in the version where it decouples in the low energy theory due to D-parity breaking.

After symmetry breaking, the quark and charged lepton masses are given by the generic formula $M_f = h\kappa + \tilde{h}\kappa'$ for up quarks and for down quarks and charged leptons, it is the same formula with κ and κ' interchanged. The above Yukawa Lagrangian leads to Dirac mass matrix for neutrinos $m_D = h^\ell \kappa + \tilde{h}^\ell \kappa'$ and the Majorana mass matrix or the heavy right handed neutrinos (N) fv_R which goes into Eq. (1) for calculating the neutrino mass and heavy–light neutrino mixing. It is also important to point out that due to the $U(1)_{B-L}$ part of the gauge symmetry [9], the Majorana mass of the right handed neutrino is connected to parity violation in nature, making connection with the original derivation of V-A theory using γ_5 invariance of leptonic weak interaction for massless neutrino [10].

NEUTRINO MASS AND TYPE I VS TYPE II SEE-SAW

It has been noted before [11] that the analysis of the minima of the most general Higgs potential with the above fields leads to $< \Delta_R^0 > = v_R$ and $< \Delta_L^0 > = v_L \equiv \frac{\gamma\kappa^2}{\lambda v_R}$ where $\gamma$and λ are Higgs self-couplings in the potential. The v_L arises from the terms in the potential of the form $\gamma\, \mathrm{Tr}(\phi \Delta_R \phi^\dagger \Delta_L^\dagger)$. The first vev v_R term contributes to the heavy Majorana mass for N whereas v_L contributes to the light Majorana mass an amount fv_L. Note that $v_L \leq 1$ GeV from electroweak precision measurements. The latter is the so-called type II seesaw [6]. Clearly for TeV v_R, unless $\gamma \sim 10^{-10}$ or so, the neutrino mass generated by the type II seesaw contribution is incompatible with oscillation data for TeV seesaw scale. Note of course that if $f \ll 1$ (say $\sim 10^{-10}$), in principle, one could have type II seesaw to dominate. We will not consider this possibility since, as eluded to in the introduction, our goal is to avoid tiny numbers like this from our model. It is therefore desirable to seek ways by which the type II seesaw contribution becomes naturally small without having to fine tune the scalar couplings. There are two ways that this can happen: (a) by decoupling parity and $SU(2)_R$ scales and (b) by making the model supersymmetric. These two strategies are described below.

Decoupling parity and $SU(2)_R$ scales

The left–right models have two new symmetry ingredients over the standard model: (i) there is a new $SU(2)_R$ group and (ii) there is a new discrete symmetry P which makes the left and right SU(2) gauge couplings equal. As it was pointed out in Ref. [8], the idea here is to introduce a real parity odd singlet scalar field into the Higgs sector and give a vev M_P to break parity while keeping the gauge symmetry $SU(2)_R$ unbroken. This can be done in such a way that the Δ_L field becomes much heavier and decouples from the low energy spectrum of the theory and thereby effectively suppressing v_L by an amount v_R^2/M_P^2, where M_P is the scale at which parity (also called D-parity for SO(10) models) is broken and $M_P \gg v_R$. This makes the type II contribution to neutrino masses negligible and allows the type I term to dominate.

Supersymmetric LRSM

An alternative way to suppress the type II contribution without giving up low energy parity symmetry is to consider the supersymmetric version of the minimal LRSM discussed above and keep only renormalizable terms in the superpotential. Since the superpotential must be holomorphic, terms such as $\gamma\,\mathrm{Tr}(\phi\Delta_R\phi^\dagger\Delta_L^\dagger)$ are absent and in fact there is no mutually coupled terms between ϕ and Δ's. As a result, at the tree level v_L is not induced. At the one loop level, the graph in Fig. 1, induces a term of the form $\mathrm{Tr}(\phi\Delta_R\phi^\dagger\Delta_L^\dagger)$ in the scalar potential with a coefficient $\gamma \sim \frac{h_\tau^2 f_\tau^2 M_S^2}{16\pi^2\Lambda^2}$, leading to a small induced v_L given by:

$$v_L \sim \frac{h_\tau^2 f_\tau^2 M_{SK}{}^2}{16\pi^2 v_R^2} \sim 10^{-9}\ \mathrm{GeV} \qquad (6)$$

where M_S is the SUSY breaking scale. This is of order eV scale and can be smaller than the type I contribution without arbitrary fine tuning, even though the $v_R \sim$few TeV.

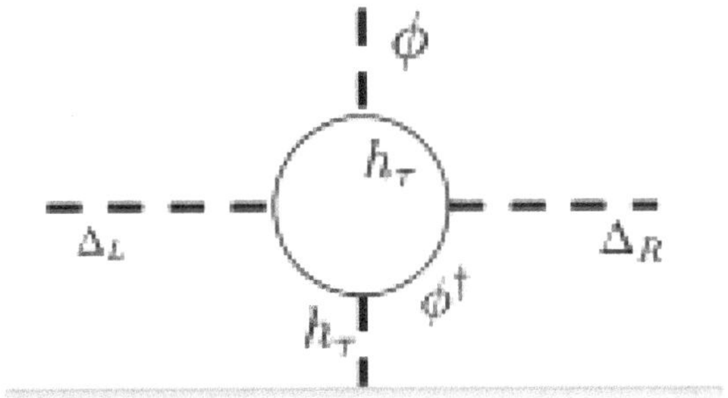

Figure 1: One loop induced type II seesaw contribution to neutrino masses in TeV SUSYLR model.

DETAILS OF MODEL FOR TYPE I FIT TO NEUTRINO OSCILLATION PARAMETERS

Having established that there are at least two reasonable ways to suppress the type II contribution to neutrino masses in the TeV scale model, we note that there are several ways to understand eV scale neutrino masses in TeV LR models from the type I contribution: (i) one can simply choose $h_v \sim h_e \sim 10^{-5.5}$ or (ii) one may use some symmetry so that in the symmetry limit the leading contribution from type I seesaw vanishes [12] i.e. $M_D^0 M_R^{-1} M_D^{0T} = 0$. Here we give an example that uses the second strategy. The mass matrices that use this strategy in the context of SM seesaw were discussed in [12]. Recently, we discuss the embedding of one such texture in the L–R model using an appropriate family symmetry [13]. The symmetry must not only guarantee the special leptonic textures but also must be free of light scalar bosons which can result if the effect of the discrete symmetry is to automatically lead to a U(1) symmetry. Moreover in LRSM, the charged-lepton mass matrix M_ℓ is related to m_D which puts additional constraints on phenomenological viability of the model. Therefore, we find it remarkable that the model presented below remains a viable TeV-scale L–R type-I seesaw model for neutrinos, and as a result, has interesting phenomenological implications beyond the minimal LRSM that could provide observable signals for TeV scale seesaw. The Dirac and Majorana mass matrices MD and MN considered here have the following form:

$$M_D = \begin{pmatrix} m_1 & \delta_1 & \epsilon_1 \\ m_2 & \delta_2 & \epsilon_2 \\ m_3 & \delta_3 & \epsilon_3 \end{pmatrix};\ M_N = \begin{pmatrix} \delta M & M & 0 \\ M & 0 & 0 \\ 0 & 0 & M_2 \end{pmatrix} \qquad (7)$$

with $\epsilon_i, \delta_i \ll m_i$ and $\delta M \ll M_i$. In the limit of $\epsilon_i, \delta_i, \delta M \to 0$, the neutrino masses vanish. The light neutrino masses given by the seesaw formula (1) become proportional to either ϵ_i, δ_i or their products. If the smallness of δ_i and ϵ_i can be guaranteed by some symmetry, then we have a "natural" TeV-scale seesaw model with tiny neutrino masses. As we show below, the mass textures in the above equation can be successfully embedded into the L–R framework, while satisfying all current experimental constraints [13]. Note that the choice of texture for the RH neutrino mass matrix in the above equation is also necessary to keep the model compatible with strong bounds from tree-level $\mu \to 3e$ decay as well as other rate charged lepton decays via TeV-scale Δ_R exchange.

In order to obtain the special Dirac and Majorana textures given in Eq. (7), we supplement the L–R gauge group $SU(2)_L\times SU(2)_R\times U(1)_{B-L}$ with a global discrete symmetry $D\equiv Z_4\times Z_4\times Z_4$. For the Higgs sector, we choose three bi-doublets ($\phi_{1,2,3}$) with B−L=0 and two RH triplets ($\Delta_{R,1,2}$) with B−L=2. The fermion and Higgs multiplets are assigned the D quantum numbers as shown in Table 1. The leptonic part of the Yukawa Lagrangian (3) invariant under the D-symmetry is given by

$$\mathcal{L}_{l,Y} = h_a\bar{L}_a\tilde{\phi}_1R_1 + h_{a2}\bar{L}_a\phi_2R_2 + h_{a3}\bar{L}_a\phi_3R_3 + f_{12}R_1R_2\Delta_{R,1} + f_{33}R_3R_3\Delta_{R,2} + H.c. \quad (8)$$

In the symmetry limit, the VEVs of $\phi_{1,2,3}$ have the form $<\phi_a>=diag(\kappa_a,0)$. This is because the terms of the form $Tr(\phi^\dagger_a\phi_b)$ which would change the ϕ VEV to the form $diag(\kappa_a,\kappa'_a)$ are forbidden by the D-symmetry from appearing in the scalar potential. Thus in the symmetry limit, the M_D elements in one column are big and non-zero, and the charged lepton mass matrix has one eigenvalue zero which can be identified as the electron flavor. The RH neutrino Majorana mass matrix has the form in Eq. (4) with $M_v=0$. To make the model realistic, we add small soft symmetry-breaking terms of the form $\delta V = \mu^2_{ab}\,Tr(\tilde{\phi}^\dagger_a\phi_b) + H.c.$ to the scalar potential. This will induce the ϕ_a VEVs of the form $<\phi_a> = diag(\kappa_a,\kappa'_a)$, where $\kappa'_a \sim \mu^2_{ab}/\kappa_b$. Choosing appropriately small μ^2_{ab}, we can get very small κ'_a as required to satisfy the neutrino oscillation data. Such symmetry breaking terms can be induced in UV complete theories where the quark sector has the right property, an example of which is given in Ref. [14].

Table 1: The discrete symmetry assignments for the fermion and Higgs fields in our L–R model that lead naturally to the special Dirac and Majorana textures given in Eq. (1)

Field	$Z_4\times Z_4\times Z_4$ transformation
$L_{1,2,3}$	(1,1,1)
R_1	(−i,1,1)
R_2	(1,−i,1)
R_3	(1,1,−i)
ϕ_1	(−i,1,1)
ϕ_2	(1,i,1)
ϕ_3	(1,1,i)
ΔR_1	(i,i,1)
ΔR_2	(1,1,−1)

CONSTRAINTS ON TEV SCALE LEFT–RIGHT MODEL FOR NEUTRINOS

In this section, we summarize constraints on the scale of left–right symmetry from low energy processes such as flavor changing neutral current effects, leptogenesis and supersymmetry embedding.

FCNC constraints on $SU(2)_R$ scale

In the TeV scale left–right model there are new contributions to FCNC processes such as K_L-K_S mass difference, $B\text{–}\bar{B}$ mixing, CP violation in kaon decay from both $W_R$exchange diagrams at one loop level as well as from the tree level neutral Higgs exchange. Most recent analysis of these constraints have been in Ref. [15] and we refer the reader to those papers for details. To summarize the results here in brief, the constraints on M_{WR} depend on the way parity is implemented in the model and the lowest lower bound from the above considerations is: $M_{WR}\geq 2.5$ TeV if $g_L=g_R$ and the bound scales with the value of g_R. In this connection, it may be worth noting that recent fits to the observed excesses at 2 TeV by both the ATLAS and CMS groups [16]require that $g_R<g_L$.

Constraints of supersymmetry on M_{WR}

When this minimal LR model is embedded into supersymmetry, the holomorphy of the superpotential imposes new upper bounds on the M_{WR}. The reason for this is that the global minimum of the SUSYLR model breaks electric charge due to the fact that the D-term always vanishes for the electric charge violating configuration $<\Delta^0_R> = \begin{pmatrix} 0 & v_R \\ v_R & 0 \end{pmatrix}$[17]. However, once one loop terms are taken into account, the charge conserving minimum becomes lower provided W_R mass has an upper limit depending on the value of the SUSY breaking scale. For instance, a one loop calculation [18] using SARAH package [19], this bound is found to be $M_{WR}\leq 7$ TeV for Msusy=5 TeV[19].

Leptogenesis constraints on W_R mass

One of the attractive consequences of seesaw mechanism is that it allows for right handed neutrino decays in combination with leptonic CP violation to explain the origin of matter–anti-matter asymmetry [20]. An important question then is whether in a TeV scale left–right seesaw model leptogenesis works. This was discussed in two recent papers [14] and [21]. It turns out that due to constraints of strong washout, there is a lower limit on the mass of the W_R boson above which lepton asymmetry generated by the decay of right handed neutrino gets washed out by W_R mediated scattering processes that violate lepton number. For a generic left–right model, this issue was investigated in Ref. [21] and it was shown the lower limit on W_R mass for leptogenesis to work is 18 TeV. This issue was subsequently reanalyzed in realistic models with neutrino fits and it was shown that this bound can be substantially lowered to ~10 TeV. All these results are important since it shows that the idea of type I seesaw leptogenesis can be ruled out if the W_R is discovered below 10 TeV. Note that, in the case of inverse seesaw TeV scale left–right models, these bounds get even lowered to TeV scale.

EXPERIMENTAL IMPLICATIONS

The TeV scale left–right seesaw model has a number experimentally testable implications. Here I give a brief glimpse of those consequences. This is by no means exhaustive and many important works have not been cited due to lack of space.

Collider signatures of TeV scale LR seesaw

There exist limits on the parameters of seesaw i.e. M_N and $V_{\ell N}$ from various low energy experiments and e^+e^- colliders [22]. It is clear from [22] that limits on these parameters for $M_N\geq 100$ GeV are very weak. Using the LHC Higgs data, these considerations have been extended to $M_N \leq 140$ GeV several years ago [23]. Broadly speaking in the post-100 GeV range for M_N, the limits on $V_{\ell N}$, the heavy–light mixing parameter are at the level of 10^{-2} or so. One can in principle search for the signals of seesaw at the LHC and future colliders [24].

As far as collider tests of SM-seesaw is concerned, we need combination of the Majorana mass M_N as well as $V_{\ell N}$ which gives the smoking gun signal of same-sign di-leptons plus two jets without missing energy ($\ell^\pm\ell^\pm jj$) [25], for TeV to sub-TeV M_N and $V_{\ell N} \geq 0.01$. This signal depends crucially on the heavy–light neutrino mixings and can effectively probe only the heavy neutrino masses M_N up to 300 GeV. It must be stressed that any positive signal would not only signify the Majorana character of the heavy sub-TeV neutrino N but also the detailed structure of m_D. To repeat, in generic ("vanilla") seesaw, we expect the heavy–light mixing $V_{\ell N} \sim \sqrt{\frac{m_\nu}{M_N}}$ to be rather tiny for TeV M_N and current upper limits on $m_\nu \leq 0.1$ eV from cosmological considerations e.g. Planck and WMAP. Only if the Dirac matrix m_D has specific forms [12] can $V_{\ell N}$ be significant enough to have observable collider signals..

In contrast with SM-seesaw case, in the LR symmetric embedding of TeV seesaw, the presence of right handed gauge interactions as well as the seesaw generated mixing between the heavy and light neutrinos lead via W_R production to significant signal strength for the $\ell^\pm\ell^\pm jj$ channel. The W_R exchange as was first pointed in [25] very early on after the seesaw mechanism was first proposed and more recently, the second mechanism which combines $V_{\ell R}$ and W_R was pointed out [26]. A crucial prediction of type I seesaw is that the number of same sign di-leptons must be equal to the number of opposite sign ones. If there is a difference between same and unlike sign di-leptons, it might signal an underlying inverse seesaw [27].

Important for the collider discussion is the relative values of the W_R and N masses. There are theoretical arguments based on vacuum stability which suggest that the heavy neutrinos in the minimal left–right seesaw models are lighter than the right handed gauge bosons [28] for a large range of parameters of the minimal LR model. We will therefore consider this mass ordering in this paper, although going beyond minimal version, one could avoid this restriction. It is also worth pointing out that for TeV scale left–right leptogenesis to work, we must have $M_N < M_{WR}$[29]. First implication of this is that for a right handed gauge boson (W_R) masses below 5–6 TeV, when it can be produced at the 14-TeV LHC with a decent cross section, its decay to the on-shell heavy RH neutrinos will allow a probe of its mixing with the light neutrinos for a wider mass range of up to a few TeVs from a study of di-leptons plus two jet final states. These two contributions can be distinguished from each other by the study of various end points in the invariant mass distributions, as has been noted recently [30]. For a sizable fraction of the allowed parameter space, the heavy–light mixing contribution can indeed dominate. Using this information, in combination with the existing experimental bounds, constraints on the mixing parameter that go into seesaw mechanism i.e. $V_{\ell N}$ can be substantially improved with 14 TeV LHC. This will further elucidate the details of seesaw formula. We close this subsection by noting that at LHC 14, the W_R reach is 5–6 TeV whereas at the future 100 TeV collider, the reach via the eejj channel can be up to 30 TeV [31].

On the collider front, another striking feature of the minimal seesaw left–right models is the existence of doubly charged Higgs fields Δ_R^{++}, which can be pair produced in pp colliders leading to four lepton final states. There have been extensive theoretical analysis of these processes [32] and there have also been LHC searches for this particle in specific lepton flavor channels. Current LHC limits on these particles are at the level of 400 GeV for particular channels. There are also a plethora of new neutral and singly charged Higgs bosons in the LR model that can be probed with 100 TeV collider [33].

Lepton Flavor Violation consequences

In the TeV scale LR seesaw models, lepton flavor violation receives new contributions not present in the SM-seesaw model. In the SM-seesaw model, the only class of graphs that lead to enhanced LFV signal (e.g. in $\mu\to e+\gamma$ process) arise from heavy–light mixing in second order and involve the W_L exchange. This contribution has been calculated most recently in [34] in cases where m_D has special forms and shown that lead to large $V_{\ell N}$ and it has been noted that $B(\mu\to e+\gamma)$ in this case can be as large as 10^{-13} for $M_N \leq 200$ GeV, whereas $B(\tau\to\mu+\gamma)$ can be as large 10^{-9}. All these are in the observable range of current and planned experiments. However, as the M_N increases, these effects go down as its fourth power and become unobservable.

In the generic left–right models, in addition to the heavy–light mixing contribution as in SM seesaw case, new contributions to $B(\mu\to e+\gamma)$ can arise from new sources depending on the details of the model. One contribution characteristic of LR models is that arising from the W_R–N virtual state in the loop as was noted in [35]. This scales like $M_{W_R}^{-8}$. The branching ratio from this source is given by [35]:

$$B_{\mu\to e+\gamma} \simeq \frac{3\alpha}{32\pi}\left(\frac{M_{W_L}}{M_{W_R}}\right)^8 \left(s_R c_R \frac{M_2^2 - M_1^2}{M_{W_L}^2}\right)^2 \quad (9)$$

where s_R, c_R are the mixings in the right handed charged leptonic current–W_R interaction. The interesting aspect of this contribution is that it only depends on the mixings in the right handed charged current interaction with W_R in a manner analogous to the well known GIM mechanism of standard model and can be used to constrain model parameters such as M_N and M_{WR}. In particular, for maximal θ_R, when $B_{\mu\to e\gamma}$ is searched down to 10^{-16} level, it can probe M_{WR} up to 30 TeV. Thus if there are no flavor symmetries endowing special structure to the right handed neutrino mass matrix, this can be a potent way to throw light on this class of TeV seesaw models.

Other contributions to LFV processes such as $\mu\to e+\gamma$ in LR models, come from the scalar sector of the model involving Δ_R^{++} fields in the loop [36].

$$B^{++}_{\mu\to e+\gamma} \simeq \frac{2\alpha m_W^4}{3\pi g^4}\left[\frac{(ff^\dagger)_{12}}{M^2_{\Delta^{++}}}\right]^2 \quad (10)$$

This has been calculated most recently under certain assumptions in [37] and current MEG bound implies a lower bound on $M_\Delta^{++} \geq 1.7$ TeV for right handed charged current mixing ~ 0.01. In the model described above, where all right handed neutrino masses and mixing angles are fixed by the neutrino mass fit, we estimate this contribution to $B_{\mu\to e+\gamma}$ to be $\sim 3\times 10^{-15}$ for $M_\Delta^{++} \sim 1$ TeV for $M_{WR} \sim 5$ TeV.

Turning now to another LFV process $\mu \to 3e$, again there are several other contributions e.g. (i) a photon mediated one loop graph that $\mu \to e+\gamma$ with a virtual γ , (ii) loop box graphs with W_R and N virtual states and (iii) a tree level graph involving the exchange of Δ_R^{++} states. The generic formula for the doubly charged boson mediated graph is given by [37]

$$B_{\mu\to 3e} \simeq \frac{1}{2}\left(\frac{m_W}{M_{W_R}}\right)^4 \left(\frac{M_{N,12}M_{N,11}}{M^2_{\Delta_R^{++}}}\right)^2 \quad (11)$$

Since in specific models such as the one discussed here, the neutrino mass fit fixes all the parameters of the model except M_{WR} and M_Δ^{++}, for M_{WR}=3 TeV and $M_{\Delta_R^{++}} = 1$ TeV, we predict $B(\mu \to 3e) \simeq 3\times 10^{-13}$, only a factor of 3 smaller than the current upper bound.

Implications for neutrino-less double beta decay

In this section, we discuss tests of the TeV LR seesaw model in neutrinoless double beta decay process. Since in LR-seesaw models both the light (ν_e) and heavy neutrinos (N) are Majorana fermions, they break lepton number by two units and lead to nuclear processes such as $(A,Z)\to(A,Z+2)+e^-+e^-$ where no neutrinos are emitted. The first contribution to this process comes from the well known light neutrino exchange and the amplitude $A_{\beta\beta0\nu}$ is proportional to $G_F^2 m_{\nu_e}$. The heavy neutrino contribution on the other hand, is given by $\sim G_F^2\left(\frac{m_W}{M_{W_R}}\right)^4 \frac{1}{M_N}$ [38]. There are also other contributions of similar order of magnitude coming from Δ_R^{++} exchange [39]. Current limits on this process already constrain the parameter space of this model and imply

$$M_N^{1/4} M_{W_R} \geq 1.1 \times 10^4 (\text{GeV})^{5/4} \quad (12)$$

Similarly Tello et al. [37] derive the following lower bound on the mass of Δ_R^{++} from $A_{\beta\beta0\nu}$ limits to be:

$$M_{\Delta_R^{++}} \geq 500\left(\frac{3.5\ \text{TeV}}{M_{W_R}}\right)^2 \times \sqrt{\frac{M_N}{3\ \text{TeV}}} \quad (13)$$

As the search for $\beta\beta_{0\nu}$ decay becomes more and more sensitive, it will probe a wider range parameters of the TeV LR seesaw model. Lack of observation of $\beta\beta_{0\nu}$ decay however cannot rule out TeV LR models since it turns out that for particular structure of the RH neutrino mass matrix, as in the model discussed here, $\beta\beta_{0\nu}$ process can be highly suppressed while the W_R mass is still in the TeV range.

Enhanced reach for W_R mass for sub-GeV right handed neutrino masses

It has been recently pointed out that [40] if the right handed neutrinos are in the sub-GeV mass range, they can be produced in beam dump experiments such as the planned SHIP experiment [41]. While this is theoretically a very improbable range in this class of models, it may be possible in other realizations of left–right symmetry and should be investigated.

There are also other tests for TeV left–right seesaw using displaced vertices at LHC [42] for lower mass range of the right handed neutrinos ($M_N \leq 100$ GeV) and for M_{WR} in the 4–4.5 TeV range.

SUMMARY

In this article, we have presented a brief overview of a class of TeV scale theories for neutrino masses based on the left–right extension of the standard model, originally proposed to understand observed parity violation in weak interactions as a low energy phenomenon. We discuss how type I seesaw can dominate the neutrino masses in this case and provide a glimpse of different low energy and collider tests of this model. Evidence for left–right symmetry at the TeV scale will revolutionize our thinking about the nature of unification of forces and matter as well as provide crucial insight into how the matter–anti-matter asymmetry in the universe emerged from a symmetric big bang early universe.

ACKNOWLEDGEMENTS

This work is supported by the US National Science Foundation Grant No. PHY-1315155. I would like to thank P.S. Bhupal Dev for discussion and collaboration on many of the topics discussed in this article.

REFERENCES

1. P. Minkowski, Phys. Lett. B 67 (1977) 421; T. Yanagida, in: A. Sawada, A. Sugamoto (Eds.), Workshop on Unified Theories and Baryon Number in the Universe, KEK, Tsukuba, 1979; S. Glashow, in: M. Lévy (Ed.), Quarks and Leptons, Cargèse, 1979, Plenum, NY, 1980; M. Gell-Mann, P. Ramond, R. Slansky, in: P. Van Niewenhuizen, D. Freeman (Eds.), Supergravity, North-Holland, Amsterdam, 1980; R.N. Mohapatra, G. Senjanovic,′ Phys. Rev. Lett. 44 (1980) 912.
2. J.C. Pati, A. Salam, Phys. Rev. D 10 (1974) 275; R.N. Mohapatra, J.C. Pati, Phys. Rev. D 11 (1975) 2558; G. Senjanovic,′ R.N. Mohapatra, Phys. Rev. D 12 (1975) 1502.
3. R.N. Mohapatra, Phys. Rev. Lett. 56 (1986) 561; R.N. Mohapatra, J.W.F. Valle, Phys. Rev. D 34 (1986) 1642.
4. P.S.B. Dev, R.N. Mohapatra, Phys. Rev. D 81 (2010) 013001, arXiv:0910.3924 [hep-ph].
5. For a recent discussion and references to literature, see A. Abada, M. Lucente, Nucl. Phys. B 885 (2014) 651, arXiv:1401.1507 [hep-ph]; A. Das, N. Okada, Phys. Rev. D 88 (2013) 113001, arXiv:1207.3734 [hep-ph]; C. Weiland, Ph.D. thesis, arXiv:1311.5860 [hep-ph].
6. G. Lazarides, Q. Shafi, C. Wetterich, Nucl. Phys. B 181 (1981) 287; J. Schechter, J.W.F. Valle, Phys. Rev. D 22 (1980) 2227; R.N. Mohapatra, G. Senjanovic, Phys. Rev. D 23 (1981) 165.
7. R.N. Mohapatra, G. Senjanovic,′ Phys. Rev. Lett. 44 (1980) 912.
8. D. Chang, R.N. Mohapatra, M.K. Parida, Phys. Rev. Lett. 52 (1984) 1072.
9. R.N. Mohapatra, R.E. Marshak, Phys. Rev. Lett. 44 (1980) 1316; R.N. Mohapatra, R.E. Marshak, Phys. Rev. Lett. 44 (1980) 1644 (Erratum).
10. E.C.G. Sudarshan, R.E. Marshak, Phys. Rev. 109 (1958) 1860.
11. R.N. Mohapatra, G. Senjanovic, Phys. Rev. D 23 (1981) 165.

12. A. Pilaftsis, T.E.J. Underwood, Nucl. Phys. B 692 (2004) 303, arXiv:hep-ph/0309342; J. Kersten, A.Yu. Smirnov, Phys. Rev. D 76 (2007) 073005.
13. C.H. Lee, P.S. Bhupal Dev, R.N. Mohapatra, Phys. Rev. D 88 (9) (2013) 093010.
14. P.S. Bhupal Dev, C.H. Lee, R.N. Mohapatra, J. Phys. Conf. Ser. 631 (1) (2015) 012007.
15. G. Beall, M. Bander, A. Soni, Phys. Rev. Lett. 48 (1982) 848; H. An, X. Ji, R.N. Mohapatra, Y. Zhang, Nucl. Phys. B 802 (2008) 247, arXiv:0712.4218 [hep-ph]; A. Maiezza, M. Nemevsek, F. Nesti, G. Senjanovic, Phys. Rev. D 82 (2010) 055022, arXiv:1005.5160 [hep-ph].
16. G. Aad, et al., ATLAS Collaboration, arXiv:1506.00962 [hep-ex]; See also G. Aad, et al., ATLAS Collaboration, Eur. Phys. J. C 75 (2015) 69, arXiv:1409.6190 [hep-ex]; G. Aad, et al., ATLAS Collaboration, Eur. Phys. J. C 75 (2015) 209, arXiv:1503.04677 [hep-ex]; V. Khachatryan, et al., CMS Collaboration, J. High Energy Phys. 1408 (2014) 173, arXiv:1405.1994 [hep-ex]; V. Khachatryan, et al., CMS Collaboration, J. High Energy Phys. 1408 (2014) 174, arXiv:1405.3447 [hep-ex].
17. R. Kuchimanchi, R.N. Mohapatra, Phys. Rev. D 48 (1993) 4352; R. Kuchimanchi, R.N. Mohapatra, Phys. Rev. Lett. 75 (1995) 3989.
18. K.S. Babu, R.N. Mohapatra, Phys. Lett. B 668 (2008) 404, arXiv:0807.0481 [hep-ph]; K.S. Babu, A. Patra, arXiv:1412.8714 [hep-ph].
19. L. Basso, B. Fuks, M.E. Krauss, W. Porod, J. High Energy Phys. 1507 (2015) 147, arXiv:1503.08211 [hep-ph]; W. Porod, private communication, 2015.
20. M. Fukugita, T. Yanagida, Phys. Lett. B 174 (1986) 45.
21. J.M. Frere, T. Hambye, G. Vertongen, J. High Energy Phys. 0901 (2009) 051, arXiv:0806.0841 [hep-ph].
22. A. Atre, T. Han, S. Pascoli, B. Zhang, J. High Energy Phys. 0905 (2009) 030, arXiv:0901.3589 [hep-ph].
23. P.S. Bhupal Dev, R. Franceschini, R.N. Mohapatra, Phys. Rev. D 86 (2012) 093010, arXiv:1207.2756 [hep-ph]; A.M. Gago, P. Hernández, J. Jones-Pérez, M. Losada, A. Moreno Briceño, Eur. Phys. J. C 75 (10) (2015) 470, http://dx.doi.org/10.1140/epjc/s10052-015-3693-1, arXiv:1505.05880 [hep-ph].
24. F.F. Deppisch, P.S. Bhupal Dev, A. Pilaftsis, New J. Phys. 17 (7) (2015) 075019, arXiv:1502.06541 [hep-ph]; P.S.B. Dev, A. Ibarra, arXiv:1601.01658 [hep-ph].
25. W.Y. Keung, G. Senjanovic, Phys. Rev. Lett. 50 (1983) 1427.
26. C.Y. Chen, P.S.B. Dev, R.N. Mohapatra, Phys. Rev. D 88 (2013) 033014, arXiv:1306.2342 [hep-ph]; M. Nemevsek, G. Senjanovic, V. Tello, Phys. Rev. Lett. 110 (15) (2013) 151802, arXiv:1211.2837 [hep-ph].
27. P.S. Bhupal Dev, R.N. Mohapatra, Phys. Rev. Lett. 115 (18) (2015) 181803, arXiv:1508.02277 [hep-ph].
28. R.N. Mohapatra, Phys. Rev. D 34 (1986) 909.
29. F.F. Deppisch, J. Harz, M. Hirsch, Phys. Rev. Lett. 112 (2014) 221601, arXiv:1312.4447 [hep-ph].
30. P.S.B. Dev, D. Kim, R.N. Mohapatra, J. High Energy Phys. 1601 (2016) 118, arXiv:1510.04328 [hep-ph].
31. J.N. Ng, A. de la Puente, B.W.P. Pan, J. High Energy Phys. 1512 (2015) 172, arXiv:1505.01934 [hep-ph].
32. K. Huitu, J. Maalampi, A. Pietila, M. Raidal, Nucl. Phys. B 487 (1997) 27, arXiv:hep-ph/9606311; For recent works and references to earlier literature, see G. Bambhaniya, J. Chakrabortty, J. Gluza, M. Kordiaczynska, ´ R. Szafron, J. High Energy Phys. 1405 (2014) 033, arXiv:1311.4144 [hep-ph]; G. Bambhaniya, J. Chakrabortty, J. Gluza, T. Jelinski, R. Szafron, Phys. Rev. D 92 (1) (2015) 015016, arXiv: 1504.03999 [hep-ph].
33. P.S. Bhupal Dev, R.N. Mohapatra, Yongchao Zhang, arXiv:1602.05947.
34. A. de Gouvea, J. Jenkins, Phys. Rev. D 77 (2008) 013008, arXiv:0708.1344 [hep-ph]; M.B. Gavela, T. Hambye, D. Hernandez, P. Hernandez, J. High Energy Phys. 0909 (2009) 038, http://dx.doi.org/ 10.1088/1126-6708/2009/09/038, arXiv:0906.1461 [hep-ph]; R. Alonso, M. Dhen, M.B. Gavela, T. Hambye, J. High Energy Phys. 1301 (2013) 118, arXiv:1209.2679 [hep-ph].
35. Riazuddin, R.E. Marshak, R.N. Mohapatra, Phys. Rev. D 24 (1981) 1310.
36. R.N. Mohapatra, Phys. Rev. D 46 (1992) 2990.
37. V. Tello, Ph.D. thesis, 2012, and references therein.
38. R.N. Mohapatra, Phys. Rev. D 34 (1986) 909; M. Hirsch, H.V. Klapdor-Kleingrothaus, O. Panella, Phys. Lett. B 374 (1996) 7, arXiv:hep-ph/9602306; V. Tello, M. Nemevsek, F. Nesti, G. Senjanovic, F. Vissani, Phys. Rev. Lett. 106 (2011) 151801, arXiv:1011.3522 [hep-ph]; F.F. Deppisch, T.E. Gonzalo, S. Patra, N. Sahu, U. Sarkar, Phys. Rev. D 91 (1) (2015) 015018, arXiv:1410.6427 [hep-ph]; P.S. Bhupal Dev, S. Goswami, M. Mitra, Phys. Rev. D 91 (11) (2015) 113004, arXiv:1405.1399 [hep-ph].
39. R.N. Mohapatra, J.D. Vergados, Phys. Rev. Lett. 47 (1981) 1713; C.E. Picciotto, M.S. Zahir, Phys. Rev. D 26 (1982) 2320.
40. O. Castillo-Felisola, C.O. Dib, J.C. Helo, S.G. Kovalenko, S.E. Ortiz, Phys. Rev. D 92 (1) (2015) 013001, arXiv: 1504.02489 [hep-ph].
41. S. Alekhin, et al., arXiv:1504.04855 [hep-ph].
42. J.C. Helo, M. Hirsch, S. Kovalenko, Phys. Rev. D 89 (2014) 073005, http://dx.doi.org/10.1103/PhysRevD.89.073005, arXiv:1312.2900 [hep-ph].

CHAPTER 3

DYNAMICAL GENERATION OF THE WEAK AND DARK MATTER SCALES FROM STRONG INTERACTIONS

Oleg Antipin, Michele Redi and Alessandro Strumia

INTRODUCTION

The idea that the weak scale could be dynamically generated from strong interactions has a long history. Originally, techni-color models were developed as an alternative to the Higgs: the weak interactions of the techni-quarks Q were chosen so that their condensates would break the SM electro-weak group and the weak scale was the techni-color scale. This scenario was disfavoured by flavour and precision data even before the first LHC run, where the Higgs and no new physics was observed.

Later, strong dynamics was invoked to generate a composite or partially-composite Higgs, although realising complete models is so complicated that model-building is usually substituted by postulating effective Lagrangians with the needed properties. Recently, models where new strong dynamics does not break the electro-weak symmetry nor provide a composite Higgs have been considered in the literature, just because they are simple, phenomenologically viable and lead to interesting LHC phenomenology [1].

With abuse of language we use the old name 'techni-color'. In this paper we show that these models

1. provide Dark Matter candidates;
2. provide a dynamical origin for the electro-weak scale, if we adopt the scenario of 'finite naturalness' [2–7].

Citation: Oleg Antipin, Michele Redi, Alessandro Strumia, Dynamical generation of the weak and Dark Matter scales from strong interactions, DOI: 10.1007/JHEP01(2015)157.

Point 2 amounts to assuming that quadratically divergent corrections to the Higgs mass have no physical meaning and can be ignored, possibly because the fundamental theory does not contain any mass term [7]. In this context, dynamical generation of the weak scale via dimensional transmutation has been realised with weakly-coupled dynamics, in models where an extra scalar S has interactions that drive its quartic $\lambda_S|S|^4$ negative around or above the weak scale: S acquires a vev at this scale, and its interaction $\lambda_{HS}|H|^2|S|^2$ effectively becomes a Higgs mass term, $m^2 = \lambda_{HS}\langle S\rangle^2$ [8–28]. A related possibility is that the scalar S is interacting with techni-quarks [29, 30] or charged under a techni-color gauge group ([31, 32]; see also [52], in the past other authors considered a similar idea, but without an elementary Higgs [53, 54]) and again S acquires a vev or forms a condensate. In all these models $\langle S\rangle$ can be pushed arbitrarily above the weak scale by making λ_{HS} arbitrarily small, leaving no observable signals.

We here consider simple models without any extra scalar S beside the Higgs doublet H. The SM is extended by adding a gauge group G_{TC} (for example SU(N)) and techni-quarks Q_L charged under the SM, as well as the corresponding Q_R in the conjugated representations of the gauge group $G_{SM} \otimes G_{TC}$, so that $Q_L \oplus Q_R$ is vectorial. As a consequence the condensate $\langle \mathcal{Q}_L \mathcal{Q}_R \rangle$ transforms as a singlet of G_{SM} and does not break it.

The techni-quarks have no mass terms because of our assumption that only dimensionless couplings exist;1 for certain assignments of their gauge quantum numbers, techniquarks can have Yukawa interactions y with the elementary SM Higgs doublet H. The scenario that we consider is described by the renormalizable Lagrangian

$$\mathscr{L} = \mathscr{L}_{\rm SM}^{m=0} - \frac{1}{4}\mathcal{G}_{\mu\nu}^{A2} + \bar{\mathcal{Q}}_L^i i\not{D}\mathcal{Q}_L^i + \bar{\mathcal{Q}}_R^j i\not{D}\mathcal{Q}_R^j + (y_{ij} H \mathcal{Q}_L^i \mathcal{Q}_R^j + \text{h.c.}) \tag{1.1}$$

where

$\mathscr{L}_{\rm SM}^{m=0}$ describes the SM without the Higgs mass term, and $\mathcal{G}_{\mu\nu}^A$ is the techni-color field strength. In models where Yukawa couplings y are not allowed (for example techniquarks in the 3 of $SU(2)_L$) the number of free parameters is the same as in the Standard Model: all new physics is univocally predicted. This new physics manifests as:

Strong dynamics generates a dynamical scale ΛTC that can be identified with the mass of the lightest vector meson resonance, the techni- ρ, and spontaneously breaks accidental chiral symmetries conserved by the techni-strong interactions producing light pseudo-Goldstone bosons (GB). Using large N counting $m_\rho = g_\rho f$ where f is the decay constant of the techni-pions and $g_\rho \approx 4\pi/\sqrt{N}$.

- In absence of techni-quark masses, the techni-pions $\pi \sim \mathcal{Q}_L \mathcal{Q}_R$ acquire mass $m_\pi^2 \approx \alpha_2 m_\rho^2/4\pi$ from the electro-weak gauge interactions that explicitly break the global techni-flavour accidental symmetries. Yukawa couplings also contribute to their masses; in absence of Yukawa couplings the lightest techni-pions could be a stable $SU(2)_L$ triplet providing a viable DM candidate.
- The heaviest new particles are techni-baryons with mass $m_B \approx N_{m\rho}$. The lightest techni-baryon is stable and is a natural DM candidate; if it is a thermal relic, the observed DM abundance is reproduced for $m_B \approx 100$ TeV [34, 35].

[1]Relaxing this hypothesis allows other interesting possibilities for Dark Matter that will be discussed in a separate publication [33].

The LHC phenomenology of techni-strong dynamics was discussed in [1]. The main new point of our work is the possible connection with the weak scale and implications for dark matter. Assuming that power divergences vanish [2, 3, 7], the techni-strong interactions give a finite negative contribution to the Higgs squared mass term, such that the weak scale is dynamically generated. The Higgs physical mass arises as

$$M_h^2 \approx +\alpha_2^2 f^2 + y^2 \frac{m_\rho^2 f^2}{m_\pi^2} \tag{1.2}$$

so that the techni-color scale is predicted to be $f \approx M_h/\alpha_2 \approx$ few TeV, or smaller in models where y is present and dominant in eq. (1.2). Unlike ordinary techni-color as a solution to the usual hierarchy problem, where the natural scale for new physics is the weak scale itself, in this scenario the natural mass scales are

$$m_\pi \sim 2\,\text{TeV}, \qquad m_\rho \sim 20\,\text{TeV}, \qquad m_B \sim 50\,\text{TeV}. \tag{1.3}$$

New physics effects in accelerator searches and precision experiments are well below the present sensitivity. In particular no new effects are generated in flavor physics. Technipions [37] and techni-baryons [38], stable due to accidental symmetries of the renormalizable Lagrangian, can provide a thermal Dark Matter candidate.

This work is organised as follows. In section 2 we consider the Higgs mass generated by the SM electro-weak gauge couplings, by the SM strong coupling, and by the Yukawa couplings of the Higgs with the techni-quarks, allowed in some models. Dark Matter is discussed in section 3. We conclude in section 4. In the appendix we present the technical details of the computation of the potential induced by Yukawa interactions.

HIGGS MASS

We write the tree-level potential of the SM Higgs doublet H as

$$V = m^2|H|^2 + \lambda|H|^4. \tag{2.1}$$

If $m^2 \equiv -M_h^2/2$ is negative, the Higgs doublet H develops the vacuum expectation value $v = M_h/\sqrt{2\lambda} \approx 246.2\,\text{GeV}$: expanding the potential V around its minimum as $H = (0, (v+h)/\sqrt{2})$ shows that $M_h \approx 125\,\text{GeV}$ is the tree-level mass of the physical Higgs boson h.

Under our assumptions, the only mass scale of the theory is set by the dynamical scale of the techni-color sector. Through loop corrections it induces other scales and in particular the Higgs mass parameter. Electro-weak interactions of the techni-quarks induce a 2-loop contribution, computed in section 2.1, and color charges give a 3 loop contribution to the Higgs mass, computed in section 2.2. If the Higgs couples to the techni-quarks through Yukawa interactions (for example if techni-quarks contain doublets and singlets under the electro-weak interactions) a contribution to the Higgs mass is also generated at 1-loop, computed in section 2.3.

Electro-weak interactions

Electro-weak gauge interactions give a minimal, quasi-model-independent, contribution to the Higgs mass, described by the non-perturbative techni-color multi-loop dressing of the two-loop Feynman diagram in figure 1a (plus the associated seagull diagram): the Higgs interacts with the electro-weak vectors that interact with the techni-quarks.

To leading order in the SM interaction, and to all orders in the techni-strong interactions, the techni-strong dynamics corrects the SM electro-weak gauge bosons propagator as

$$D^{YY}_{\mu\nu}(q) = -i\frac{\eta_{\mu\nu}}{q^2}(1 + g_Y^2\Pi_{YY}(q^2)) + i\xi_Y\frac{q_\mu q_\nu}{q^2} \quad (2.2)$$

$$D^{ab}_{\mu\nu}(q) = -i\frac{\eta_{\mu\nu}}{q^2}(1 + g_2^2\Pi_{WW}(q^2))\delta^{ab} + i\xi_W\frac{q_\mu q_\nu}{q^2}\delta^{ab} \quad (2.3)$$

where ξ_V are gauge-fixing parameters. Techni-strong dynamics is encoded in the $\Pi_{VV}(q^2)$ functions. From the point of view of the techni-strong dynamics, they are the renormalised two-point functions of the currents $J^a_\mu = \sum_i \bar{\mathcal{Q}}^i\gamma_\mu T^a_{\mathcal{Q}^i}\mathcal{Q}^i$ (where $\mathcal{Q}^i = (\mathcal{Q}^i_L, \bar{\mathcal{Q}}^i_R)$ is a Dirac spinor and T^a are the SM gauge generators) corresponding to the unbroken part of the accidental global techni-flavour symmetry, partially gauged by electro-weak interactions:

$$i\int d^4x\, e^{iq\cdot x}\langle 0|\mathrm{T}J^V_\mu(x)J^{V'}_\nu(0)|0\rangle \equiv \delta^{VV'}(q^2 g_{\mu\nu} - q_\mu q_\nu)\Pi_{VV}(q^2). \quad (2.4)$$

The correction to the Higgs mass is

$$\Delta m^2 = -\frac{3}{4i}\int\frac{d^4q}{(2\pi)^4}\frac{3g_2^4\Pi_{WW}(q^2) + g_Y^4\Pi_{YY}(q^2)}{q^2}, \quad (2.5)$$

and, performing the Wick rotation to the Eucliedan $Q^2 = -q^2 > 0$,

$$\Delta m^2 = \frac{3}{4(4\pi)^2}\int dQ^2\left[3g_2^4\Pi_{WW}(-Q^2) + g_Y^4\Pi_{YY}(-Q^2)\right]. \quad (2.6)$$

In general the integral above is UV-divergent, quadratically and logarithmically. In the case at hand, the unphysical power divergences are ignored because of our assumption of finite naturalness, and logarithmic divergences (that describe the RGE running of m^2) are absent, because of our assumption that the only mass scale, Λ_{TC}, is generated dynamically. Thereby the generated squared Higgs mass term is finite and scheme independent.

We next show that the electro-weak interactions induce a calculable negative Higgs mass so that the electro-weak symmetry is spontaneously broken. We proceed in 3 steps: dispersion relations in section 2.1.1 show in general that $\Delta m^2 < 0$, Operator Product Expansion in section 2.1.2 shows that Δm^2 is ultra-violet finite, vector meson dominance and/or large N in section 2.1.3 allow to give the estimate $\Delta m^2 \approx -\alpha_2^2 f^2$.

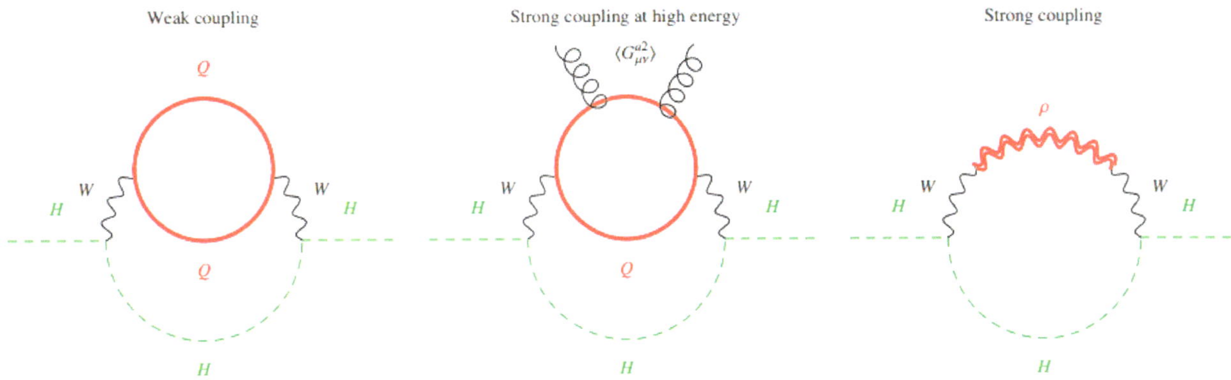

Figure 1: The two loop contribution to the Higgs mass coming from the electro-weak gauge interactions of: a) a techni-quark, to be dressed with non-perturbative techni-interactions, approximated as: b) the techni-gluon condensate; c) the techni- ρ. The extra seagull diagram is not explicitly plotted.

Dispersion relation

Under our assumptions, quadratically divergent terms are zero and we are interested in the dependence on the physical scales of the theory. To extract this we consider the variation of the Higgs mass with respect to the dynamical scale of the theory Λ_{TC},

$$\frac{\partial \Delta m^2}{\partial \Lambda_{\rm TC}^2} = \frac{3g_2^4}{4(4\pi)^2}\int dQ^2\left[3g_2^4\frac{\partial \Pi_{WW}}{\partial \Lambda_{\rm TC}^2} + g_Y^4\frac{\partial \Pi_{YY}}{\partial \Lambda_{\rm TC}^2}\right]. \quad (2.7)$$

The sign of the gauge correction Δm^2 can be determined using the dispersion relation [42]

$$\frac{\partial \Pi_{VV}(q^2)}{\partial q^2} = \frac{1}{\pi}\int_0^\infty ds\frac{\text{Im}\,\Pi_{VV}(s)}{(s-q^2-i\epsilon)^2}. \quad (2.8)$$

where we use the conventions of [43] (our Π_{VV} is defined with the same sign as Π in this book). The optical theorem relates the cross-sections $\sigma(s)$ to Im Π_{VV} (s), allowing to show in general that Im $\Pi_{VV} \le 0$.[2] For dimensional reasons, the dimension-less Π_{VV} can only depend on $Q^2/\Lambda_{\rm TC}^2$. Thereby

[2]As a check, replacing techni-color with a perturbative one-loop correction of fermions with explicit mass m_Q, one would obtain

$$\frac{\partial \Pi_{VV}(-Q^2)}{\partial m_Q^2} = -\frac{g^2}{2\pi^2}\frac{Q^2}{m_Q^2}\int_0^1 \frac{x^2(1-x)^2}{m_Q^2+x(1-x)Q^2}. \quad (2.9)$$

Inserting this into eq. (2.7) the integrand is negative definite but the integral is logarithmically divergent. This corresponds to a contribution proportional to $g^2m_Q^2$ in the RG equation for the Higgs mass m^2 . No such UV-divergent RGE effect is present in a techni-color theory that generates dynamically a mass scale $\Lambda_{\rm TC}$ from a dimension-less coupling $g_{\rm TC}$, given that, in any mass-independent scheme such as Minimal Subtraction, only $g_{\rm TC}$ can appear in the RGE.

$$\frac{\partial \Pi_{VV}}{\partial \Lambda_{\rm TC}^2} = -\frac{Q^2}{\Lambda_{\rm TC}^2}\frac{\partial \Pi_{VV}}{\partial Q^2} = \frac{Q^2}{\Lambda_{\rm TC}^2}\frac{1}{\pi}\int\frac{\text{Im}\Pi_{VV}(s)}{(s+Q^2)^2}ds < 0 \quad (2.10)$$

where in the last step we used the dispersion relation. A similar relation holds for the hypercharge contribution. The integrand in (2.7) is negative definite corresponding to a negative Δm^2 given the boundary condition $\Delta m^2 = 0$ for $\Lambda_{\rm TC} = 0$.

The ultra-violet tail

In a theory with a dynamical scale $\Lambda_{\rm TC}$, arguments based on Operator Product Expansion allow to show that $\partial\Delta m^2/\partial\Lambda_{\rm TC}^2$ is ultra-violet convergent as expected and to compute the high-energy tail of $\Pi_{VV}(q^2)$. Π_{VV} can be expanded as

$$\Pi_{VV}(q^2) \overset{q^2\gg\Lambda_{\rm TC}^2}{\simeq} c_1(q^2) + c_2(q^2)\langle 0|m_Q\mathcal{Q}_L\mathcal{Q}_R|0\rangle + c_3(q^2)\langle 0|\frac{\alpha_{\rm TC}}{4\pi}\mathcal{G}_{\mu\nu}^{A2}|0\rangle + \cdots. \quad (2.11)$$

The first term (unity operator) does not contribute to (2.7). Indeed, at leading order it describes the diagram in figure 1a with techni-quarks but neglecting their techni-color interactions, such that

$$c_1 = \frac{C}{12\pi^2}\ln(-q^2) + \cdots \quad (2.12)$$

where $C > 0$ is a model-dependent group theory factor given by $C = \text{Tr}T^aT^a$ in terms of the $SU(2)_L$ techni-quark generators (with a similar expression factor for the $U(1)_Y$ generators). This high energy tail does not contain any mass scale, so that the associated quadratically divergent no-scale integral in eq. (2.6) vanishes, under our assumptions. The second term also vanishes, because it is proportional to the techni-quark masses m_Q that vanish under our assumption that the theory does not contain any mass scale.

The third term in eq. (2.11) is represented by the Feynman diagram in figure 1b, which gives $c_3 = -C'/q^4$ [42], where $C' > 0$ is another order one model-dependent group theory factor. The techni-gluons form a positive condensate (the condensate is positive-defined in the Eucliedian path-integral [42], in agreement with QCD lattice computations)

$$\langle 0|\frac{\alpha_{\rm TC}}{4\pi}\mathcal{G}_{\mu\nu}^{A2}|0\rangle = \kappa\,\Lambda_{\rm TC}^4. \quad (2.13)$$

where $\kappa > 0$ is an order-one coefficient. This allows to show that the UV contribution to the squared Higgs mass term is negative as expected:

$$\Delta m^2|_{\rm UV} \simeq -\frac{3C'g_2^4}{4(4\pi)^2}\kappa\Lambda_{\rm TC}^4\int_{Q_{\rm min}^2}^\infty\frac{dQ^2}{Q^4} \approx -\alpha_2^2\frac{\kappa\Lambda_{\rm TC}^4}{Q_{\rm min}^2}. \quad (2.14)$$

The $1/Q_{\rm min}^2$ dependence on the artificial infra-red cut-off $Q_{\rm min} \sim \Lambda_{\rm TC}$ shows that the dominant effects comes from virtual momenta Q^2 around the techni-meson masses.

The infra-red and resonance region

The dominant contribution to the Higgs mass comes from the Q^2 region densely populated by the techni-meson resonances. A variety of methods have been proposed to approximatively describe such region: vector meson dominance, Weinberg sum rules, large N, holographic models. . . As long as the techni-quarks are charged under the electro-weak group, they form, among the various mesons, spin-1 resonances that mix with the SM electro-weak vectors V_μ. This is described by the effective Lagrangian

$$\mathscr{L}_{\text{eff}} = -\frac{1}{4g_0^2}V^a_{\mu\nu}V^{a\,\mu\nu} - \frac{1}{4g_\rho^2}\rho^a_{\mu\nu}\rho^{a\,\mu\nu} + \frac{f^2}{2}(V^a_\mu - \rho^a_\mu)^2 \tag{2.15}$$

such that the massless eigenstate has gauge coupling $1/g_2^2 = 1/g_0^2 + 1/g_\rho^2$ and the orthogonal heavy state has mass $m_\rho^2 = f^2(g_0^2 + g_\rho^2)$. Integrating out the ρ at tree-level one finds:

$$\Pi_{VV}(q^2) = \frac{m_\rho^2}{g_\rho^2(q^2 - m_\rho^2 + i\epsilon)}. \tag{2.16}$$

Plugging eq. (2.16) into eq. (2.6) we obtain a logarithmically divergent infra-red correction to the squared Higgs mass term:

$$\Delta m^2 \approx -\frac{9g_2^4}{4(4\pi)^2}\int dQ^2 \frac{m_\rho^2}{g_\rho^2(Q^2 + m_\rho^2)} \sim -\frac{g_2^4 m_\rho^2}{(4\pi)^2 g_\rho^2}\log\frac{\Lambda^2}{m_\rho^2} \sim -\alpha_2^2 f^2 . \tag{2.17}$$

The integrand is negative definite and its size agrees with the naive expectation based on the Feynman diagram plotted in figure 1c, including the $1/g_\rho^2$ suppression of vector mixing. The logarithmic UV divergence here arises because this is only an approximate description, where an explicit mass term m_ρ substitutes the dynamical mechanism of mass generation. An infinite number of states would be needed to properly describe the nonperturbative dynamics.

In theories with large N this can be made more rigorous: Π_{VV} can be represented exactly as an infinite sum of poles corresponding to the physical quasi-stable techni-mesons of the theory:

$$\Pi_{VV}(q^2) = \frac{N}{16\pi^2}m_\rho^2\sum_i \frac{c_i^2}{q^2 - m_i^2 + i\epsilon}. \tag{2.18}$$

where c_i are adimensional coefficients. The infinite number of resonances allows to reproduce the logarithmic divergence, that does not contribute to the Higgs mass zero under our assumption of finite naturalness.

These considerations offer an intuitive argument to understand the sign of Δm^2. The net effect of non-perturbative dynamics is creating a mass gap that stops the techni-quark contribution to the RGE running of g_2, g_Y below Λ_{TC}, effectively making g_2, g_Y smaller with respect to the perturbative case. As a consequence the unphysical power divergence present in the SM, $\Delta m^2 \sim +g_{2,Y}^2\Lambda^2$, gets replaced by a finite physical effect $\Delta m^2 \sim -g_{2,Y}^4\Lambda_{TC}^2$.

Color interactions

We next consider techni-color models where the techni-quarks have SM color interactions. For example, techni-quarks could be a color octet of $SU(3)_c$, charged also under the technicolor gauge group. Then techni-quarks cannot have any Yukawa coupling to the SM Higgs: both the Yukawa contribution of section 2.3 and the electro-weak contribution of section 2.1 are absent.

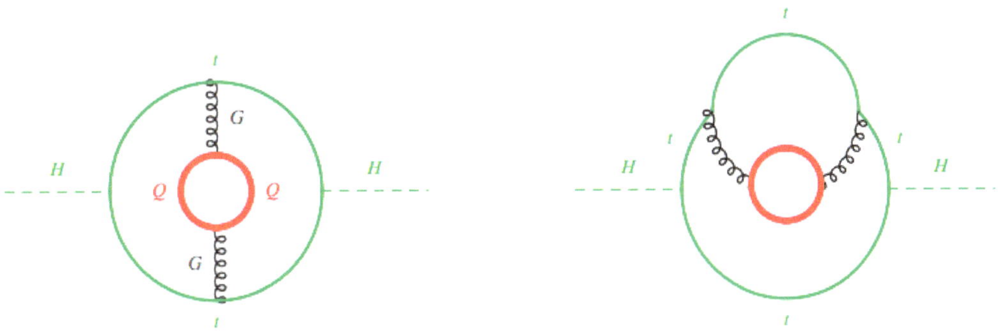

Figure 2: The three loop contribution to the Higgs mass coming from techni-quarks Q that only have color interactions. Similar diagrams can be drawn for graviton contributions.

In these models, the Higgs mass is dominantly generated at three loops: the Higgs interacts with the top quark, that interacts with the gluons, that interact with the techni-quarks, as plotted in figure 2. The computation can be performed along the lines of section 2.1 by defining $\Pi_{GG}(q^2)$, the techni-color correction to the gluon propagator. Summing the two diagrams of figure 2, the result is ultraviolet-convergent:

$$\Delta m^2 = -\frac{64 y_t^2 g_3^4}{(4\pi)^4}\int dQ^2 \Pi_{GG}(-Q^2) \sim \frac{y_t^2 g_3^4}{4\pi^4}f^2. \tag{2.19}$$

The computation of the sign is analogous to what described in the previous section (with Π_{WW} replaced by Π_{GG}): in the present case we find a positive Δm^2, such that this contribution does not induce electro-weak symmetry breaking. The sign of the effect also corresponds to the intuitive reasoning presented at the end of the previous section: the sign is opposite to the known negative sign of the naive quadratic divergence associated with y_t, because g_3 and thereby y_t are reduced by techni-strong dynamics.

We mention a final possibility. The techni-quarks could be completely neutral under the whole SM gauge group. In this situation only gravity mediates a contribution to the Higgs mass, proportional to the two-point function of the energy momentum tensor. Furthermore, a super-Planckian techni-color condensate would dynamically generate the Planck mass itself, within a dimensionless extension of Einstein gravity such as agravity [7]. The problem is that techni-color dynamics, dominated by a single non-perturbative coupling, has no free parameters and would also generate a large negative cosmological constant, which is at odd with observations.

Yukawa interactions

Finally, we consider the case where the gauge quantum numbers of the techni-quarks allow for Yukawa couplings to the elementary Higgs. This choice implies the existence of a techni-pion π^2 with the same quantum numbers of the Higgs doublet H, that can then mix with H.

The left panel of figure 3 shows the one-loop corrections to the squared Higgs mass generated by a weakly coupled techni-quark with Yukawa interactions to the Higgs. At strong coupling the physical degrees of freedom become bound state techni-hadrons that can be described using effective Lagrangian techniques. The techni-quark loop can be matched to an effective chiral Lagrangian, so that such diagrams collapses to a tree level diagram (right-handed panel of figure 3) dominated by the lightest techni-mesons, the techni-pions $\pi \approx Q_L Q_R$.

Figure 3: Correction to the Higgs mass coming from the Yukawa coupling with: a) a weakly coupled massive fermion; b) a massless strongly interacting fermion.

For simplicity we here consider Yukawa couplings that preserve the $Q_L \leftrightarrow Q_R$ parity of the techni-strong interactions; a more general discussion can be found in the appendix. Similarly to quark masses in QCD, the Yukawa interactions produce the following term in the chiral Lagrangian,

$$y\, m_\rho f^2 \,\mathrm{Tr}[HU] + \mathrm{h.c.} \tag{2.20}$$

where $U = \exp(i\pi^{\hat{a}} T^{\hat{a}}/f)$ is the Goldstone boson matrix. As we discuss in detail in the appendix, upon minimisation of the potential this term induces a mass mixing $\approx y m_\rho f H \pi^*$ between the techni-pion and the elementary Higgs. This term also explicitly breaks accidental symmetries respected by gauge interactions.

What emerges is a two-Higgs doublet system where the extra Higgs doublet π_2 is a heavy composite doublet with negligible vev. In order to compute the mass eigenstates, we need to compute the mass matrix. Including effects at tree and one-loop level in the SM couplings g_2 and y, the mass matrix has the structure

$$\begin{array}{c} \\ \pi_2 \\ H \end{array} \begin{array}{c} \begin{array}{cc} \qquad\pi_2^* \qquad\qquad & \qquad H^* \end{array} \\ \begin{pmatrix} (\mathcal{O}(g_2^2) \pm \mathcal{O}(y^2))/(4\pi)^2 & \mathcal{O}(y)\sqrt{N}/(4\pi) \\ \mathcal{O}(y)\sqrt{N}/(4\pi) & -\mathcal{O}(y^2)N/(4\pi)^2 \end{pmatrix} m_\rho^2 \end{array} \tag{2.21}$$

where we used the fact that the one-loop contribution of weak gauge interactions to $m_\pi^2 \approx g_2^2 m_\rho^2/(4\pi)^2$ is positive (as known from the SM analogous computation of the π^+/π^0 mass difference [39]), and added the one-loop Yukawa contribution (absent in the SM[3]). The HH^* entry describes the contribution of composite scalar resonances that can also mix with the Higgs giving a negative sub-leading contribution to its mass squared, see appendix for more details.

[3]The literature on composite Higgs models explored linear couplings of SM quarks to composite fermionic states, finding that they can give a negative contribution to the Higgs mass term. Simple UV completions require extra scalars as in the supersymmetric realisation of [40]. Here instead we compute the technipion potential induced by a bi-linear HQ_LQ_R Yukawa coupling, involving techni-quarks Q and a scalar H without techni-strong interactions.

We see that the phenomenologically acceptable regime is the one where the Yukawa coupling is small, y << g, such that: 1) the loop contribution coming from the Yukawa coupling can be ignored; 2) the heaviest eigenstate is the techni-pion with squared mass $m_\pi^2 > 0$; 3) the determinant of the mass matrix is dominated by the off-diagonal terms and is negative: the lightest eigenstate is the elementary Higgs, that acquires a negative squared mass term dominated by the mass mixing term in eq.(2.21) and given by a seesaw-like formula:

$$\Delta m^2 \sim -\frac{y^2}{(4\pi)^2}\frac{m_\rho^4 N}{m_\pi^2} \sim -y^2 \frac{m_\rho^2 f^2}{m_\pi^2}. \quad (2.22)$$

DARK MATTER

The models described in this paper contain two Dark Matter (DM) candidates: technibaryons and techni-pions. Their stability is guaranteed by accidental symmetries of the renormalizable Lagrangian, techni-baryon number and (possibly) G-parity [37].

In fact the presence of stable states is a generic prediction of the framework that implies restrictions on the representations of the techni-quarks under the SM gauge group, such that the stable states are viable DM candidates. In table 1 we summarise the simplest allowed charge assignments under the electroweak group and the resulting DM candidates. Introducing techni-quark masses allows several other possibilities [33].

The new matter modifies the running of SM gauge couplings. Adding n_2 weak doublets and n_3 weak triplets in the $N \oplus \bar{N}$ of $\mathrm{SU}(N)_{\mathrm{TC}}$ the beta-function of $\mathrm{SU}(2)_L$ becomes

$$b_2 = -\frac{19}{6} + \frac{2N}{3}(n_2 + 4n_3) \quad (3.1)$$

such that the $\mathrm{SU}(2)_L$ gauge coupling does not develop a Landau pole below the Planck scale ($b_2 < 0$) for small enough n_2, n_3, N. Higher $\mathrm{SU}(2)_L$ lead to Landau poles instead. The trans-Planckian Landau pole for hypercharge can be naturally avoided in models where hypercharge is embedded in $\mathrm{SU}(2)_R$ below a few TeV [36]; a technicolor sector could be used to dynamically break the extended gauge group.

Techni-pions

If techni-quarks fill N_F fundamentals and anti-fundamentals of the $\mathrm{SU(N)}_{\mathrm{TC}}$ gauge group with $N \geq 3$, the spontaneous symmetry breaking $\mathrm{SU}(N_F)_L \otimes \mathrm{SU}(N_F)_R / \mathrm{SU}(N_F)$ of the accidental global techni-flavor symmetry produces $N_F^2 - 1$ Goldstone bosons in the adjoint of the unbroken $\mathrm{SU}(N_F)$. These scalars acquire mass from effects that explicitly break the global symmetries. Within finite naturalness the only contribution to their masses is due to SM gauge interactions, and possibly to the techni-quark Yukawa couplings.

If Yukawa couplings are forbidden by the fermions quantum numbers, then the model is extremely predictive: it only has one free parameter — the techni-color scale — which is fixed by the Higgs mass under the hypothesis of finite naturalness. All the rest is univocally predicted: techni-pion masses, Dark Matter and its thermal relic abundance.

Table 1: Dimension-less techni-color models that give viable techni-baryon (TCb) and/or technipion (TC_π) Dark Matter candidates with Q = Y = 0. We consider models with SU(N) gauge group for N = {3, 4} and N_F = {2, 3, 4, 5} flavours of techni-quarks in its fundamental plus antifundamental. The darker rows give the techni-flavour content of the lightest TCb and TC_π considering only masses induced by techni-color interactions. The lighter rows consider models with viable assignments of electro-weak interactions and show, after including the mass splitting due to unbroken electro-weak interactions, the $\mathrm{SU(2)}_L$ content of the lighter DM candidates

number of techni-flavors	Yukawa	$N=3$ TCb	$N=3$ TCπ	$N=4$ TCb	$N=4$ TCπ	
$N_F = 2$		2	3	1	3	under TC-flavor SU(2)
model 1: $Q = 2_{Y=0}$	0	charged	3	1	3	DM, under $\mathrm{SU}(2)_L$
$N_F = 3$		8	8	$\bar{6}$	8	under TC-flavor SU(3)
model 1: $Q = 1_Y + 2_{Y'}$	1	1	no	1	no	DM, under $\mathrm{SU}(2)_L$
model 2: $Q = 3_{Y=0}$	0	3	3	1	3	DM, under $\mathrm{SU}(2)_L$
$N_F = 4$		$\overline{20}$	15	$20'$	15	under TC-flavor SU(4)
model 1: $Q = 4_{Y=0}$	0	charged	3	1	3	DM, under $\mathrm{SU}(2)_L$
$N_F = 5$		$\overline{40}$	24	$\overline{50}$	24	under TC-flavor SU(5)
model 1: $Q = 2_Y + 3_{Y'}$	1	1	no	charged	no	DM, under $\mathrm{SU}(2)_L$
model 2: $Q = 5_{Y=0}$	0	3	3	1	3	DM, under $\mathrm{SU}(2)_L$

The SM gauge interactions give positive squared masses to the gauge-charged technipions, while SM singlets remain exact massless Goldstone bosons. If the N_F techni-quark flavors are composed of k irreducible (real or pseudo-real) representations of G_{SM}, then the techni-pions decompose under G_{SM} as

$$\text{Adj}_{SU(N_F)} = \left[\sum_{i=1}^{k} r_i\right] \otimes \left[\sum_{i=1}^{k} \bar{r}_i\right] \ominus 1 \tag{3.2}$$

so that k − 1 techni-pions are neutral singlets (the extra scalar singlet analog of the η' in QCD acquires mass from anomalies with techni-interactions and will not play a role in what follows).

One combination of singlets corresponds to a global symmetry anomalous under $SU(2)_L$, so that the corresponding Goldstone boson acquires an axion-like coupling to SM vectors: an almost massless axion with a decay constant f ~ TeV would be grossly excluded by star cooling and other bounds. In absence of techni-quark Yukawa interactions, these bounds significantly reduce the space of models favouring the simplest models with k = 1. The techni-quarks should belong to a single irreducible representation $j = (N_F - 1)/2$ of $SU(2)_L$ and, in order to obtain a neutral lightest techni-baryon, the techni-quark hypercharge should vanish. Then the $N_F^2 - 1$ techni-pions lie in the following irreducible representations J of $SU(2)_L$:

$$\text{Adj}_{SU(N_F)} = \sum_{J=1}^{N_F-1} J. \tag{3.3}$$

Models of this kind were studied in [37], where it was pointed out that a discrete symmetry, "G-parity" exists in these theories (for zero hypercharge) due to the fact that SU(2)L representations are real or pseudo-real. G parity acts on techni-quarks as $Q \to \exp(i\pi T^2)Q^c$, replacing any $SU(2)_L$ representation with its conjugate representation, which is equivalent to the original representation. SM fields are neutral. On techni-pions G parity becomes the $(-1)^J$ Z_2 symmetry, so that techni-pions with even (odd) isospin (J) are even (odd). Summarizing:

- Techni-pion singlets under $SU(2)_L$ are G-even, do not acquire masses from SM gauge interactions and can have anomalous couplings to $SU(2)_L$ vectors: they are excluded in our framework unless Yukawa couplings make them massive. They are absent if techni-quarks fill a single irreducible representation of $SU(2)_L$.
- Techni-pions in the 3 of $SU(2)_L$ are G-odd and could be the lightest stable DM candidates. The simplest models are listed in table 1.
- Techni-pions in the 5 of SU(2)L are G-even and are heavier, $m\pi 5 \approx \sqrt{3} m\pi 3$: they undergo anomalous decays into electro-weak vectors, $\pi 5 \to WW$.
- Techni-pions in higher representations of $SU(2)_L$, if present, decay into lighter technipions respecting G-parity by emitting two $SU(2)_L$ vectors, e.g. $\pi_7 \to \pi_3 WW$.

The situation is different in models where Yukawa couplings y of techni-quarks to the elementary Higgs are present. The Yukawa couplings break explicitly G-parity and accidental global symmetries so that the SM singlet techni-pions η receive non-zero masses given by eq. (A.20), $M_\eta \sim |y| v m_\rho / m_{\pi 2}$ and star cooling bounds are easily avoided. Furthermore, techni-pions can now decay through the Higgs, so that only techni-baryons remain as dark matter candidates.

Models with techni-color gauge group SU(2) ~ Sp(2) are special: its fundamental representation is pseudo-real, $2 \sim 2^*$, so that the techni-flavour symmetry is enhanced becoming $SU(2N_F)/Sp(2N_F)$. The extra techni-pions are QQ scalars and there are no stable techni-baryons. Dangerous light techni-pions neutral under $SU(2)_L$ are again absent if techni-quarks lie in a single representation of $SU(2)_L$ with dimension $2N_F$. Within our assumptions however these models do not provide DM candidates because techni-pions are G-even.

Techni-baryons

Techni-baryons are techni-color singlet states constructed with N techni-quarks. The stability of the lightest techni-baryon follows from the accidental techni-baryon number global symmetry.

Using the non-relativistic quark model, group theory allows to compute the electroweak quantum numbers of the techni-baryons: their wave-function must be anti-symmetric in the techni-quarks. The wave function is assumed to be antisymmetric in techni-color, and so must be symmetric in spin and flavour for the lightest techni-baryons that have no orbital angular momentum. Different techni-baryons are split by spin-spin interactions that prefer, as lightest techni-baryon, the one with smallest spin. As a consequence, the lightest techni-baryons have spin 0 (1/2) for even (odd) N ≥ 2.

In general the $SU(N_F)$ techni-flavour representation of the lightest techni-baryon corresponds to a Young diagram with 2 rows having N/2 boxes each (for N even) and to a Young diagram with 2 rows having (N + 1)/2 and (N − 1)/2 boxes respectively (for N odd). In particular, they are

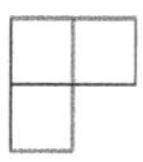 for $N = 3$ and 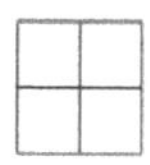 for $N = 4$. (3.4)

This is the end of the story, as long as techni-color interactions are involved.

Next, the components of a techni-baryon multiplet are split by SM gauge interactions, and possibly by techni-quark Yukawa interactions. The lightest components are those with smallest G_{SM} charge.

Furthermore, electro-weak symmetry breaking induces extra splitting within the components of any electro-weak multiplet, with the result that the component with smallest electric charge is the lightest stable state [44, 45]. Since DM direct detection constraints demand that DM does not couple at tree level to the Z, the DM hypercharge should be zero, which is possible for integer isospin.

Direct detection of Dark Matter

The previous discussion is summarised in table 1, which tells that the simplest TC models lead to the following viable stable DM candidates:

- Techni-baryons, fermions for odd N and scalars for even N. Their annihilation cross section is estimated to be $\sigma v \sim g_{TC}^4 / 4\pi M^2$, around the unitarity bound

[34, 35]. By performing a naive rescaling of the QCD non-relativistic $p\bar{p}$ cross section, $\sigma_{p\bar{p}}v \sim 100/m_p^2$, we estimate that the cosmological thermal relic abundance of a technibaryon equals the total DM abundance if its mass is loosely around $m_B \sim 200$ TeV. A cosmological techni-baryon asymmetry can leave a higher abundance, allowing for a lighter m_B.

- Scalar techni-pions, that fill a $SU(2)_L$ triplet with hypercharge Y = 0. Techni-pions have small residual techni-color interactions (as well as small quartic couplings) and thereby behave as Minimal Dark Matter [44, 45]. Their cosmological thermal relic abundance equals the total DM abundance if their mass is around 2.5 TeV [44, 45]. Their spin-independent cross section for direct detection is $\sigma_{SI} \approx 0.12\ 10^{-46}$ cm^2 [2, 3, 46, 47], as plotted in figure 4.

As already discussed, both mass scales suggested by the DM cosmological abundance arise naturally within the context of finite naturalness. Techni-baryons have distinctive features in direct detection experiments: if DM is a neutral composite particle made of charged techni-quarks, direct detection can be mediated by the photon [49]. Any such DM particle can have a non trivial form factor, dominated at low energy by the 'charge radius' interaction. For a scalar DM S this is the only interaction and can be written as

$$\frac{e}{\Lambda_{\rm TC}^2}(S^* i\partial_\alpha S)\partial_\mu F^{\mu\alpha}. \tag{3.5}$$

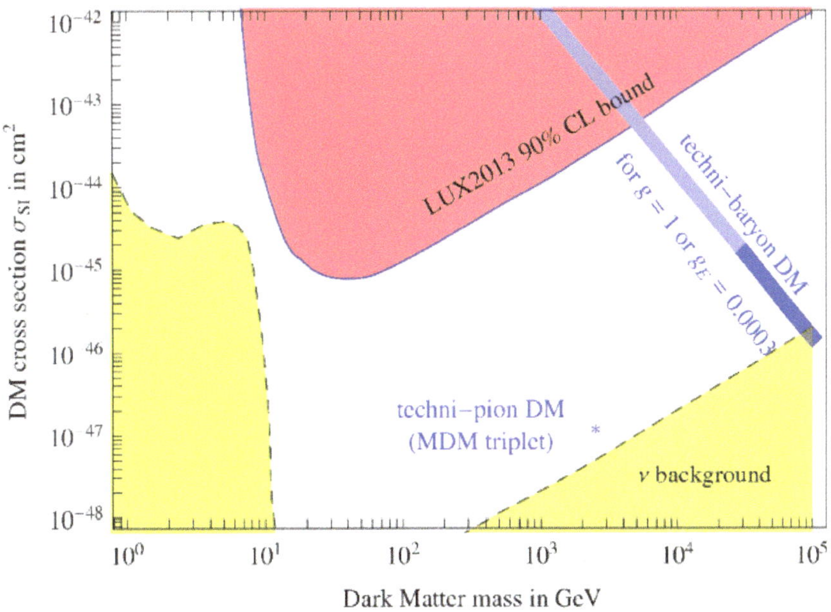

Figure 4: The signals in Dark Matter direct detection produced by a DM techni-baryon with magnetic or electric dipole moment (line) or from a Minimal-Dark-Matter-like techni-pion with thermal abundance (star), compared to the present experimental LUX bound [48] and to the background due to neutrinos.

The resulting cross section for direct detection is suppressed by four powers of the TC scale, and is negligible for $\Lambda_{TC} \sim$ few TeV. The situation is more promising if DM is a fermionic techni-baryon B, which generically has magnetic (and possibly electric) dipole moments, μ and d. They are described by the effective operator

$$\bar{B}\sigma_{\mu\nu}\frac{\mu + id\gamma_5}{2}B\ F_{\mu\nu}. \tag{3.6}$$

Electro-magnetic dipoles give sizeable direct detection signals with a characteristic testable enhancement at low recoil-energy E_R, given that the DM/matter scattering is mediated by the massless photon. Furthermore, in the relevant non-relativistic limit, the crosssection induced by the magnetic dipole μ is suppressed by two extra power of the relative DM/matter velocity v with respect to the cross section induced by the more speculative electric dipole d [49]

$$\frac{d\sigma}{dE_R} \approx \frac{e^2Z^2}{4\pi E_R}\left(\mu^2 + \frac{d^2}{v^2}\right). \tag{3.7}$$

For simplicity, we here assumed a nucleus with $Z \gg 1$, a recoil energy $ER \ll mNv2$ and approximated the nuclear charge form factor with unity.

The magnetic g-factor, defined by $\mu = ge/2m_B$, is expected to be of order one for a strongly-coupled particle (while it is loop suppressed for an elementary particle). We also define the electric g-factor as $d = gEe/2m_B$. In terms of such g-factors we find that the present direct detection bound is

$$g^2 + 1.2\ 10^7 g_E^2 < \left(\frac{m_B}{5.1\ \text{TeV}}\right)^3 \tag{3.8}$$

dominated by LUX data [48, 50]. This bound assumes that techni-baryons constitute all galactic DM, and must be rescaled otherwise. Figure 4 shows the resulting prediction in the usual plane (M_{DM}, σ_{SI}) used to describe spin-independent direct detection of Dark Matter. An electric dipole moment needs CP-violation. In our context, techni-quarks are strictly massless, such that the CP-violating techni-strong θ term is not physical. A small g_E could be generated if techni-quark masses are included.

A worked example

More quantitative predictions can be given in the QCD-like scenario with N = N_F = 3 [38]. In this case the spectrum can be obtained by rescaling known QCD results,

$$\frac{m_B}{m_\rho} \approx 1.3 \qquad \frac{m_\pi}{m_\rho} \approx 0.1\sqrt{J(J+1)} \tag{3.9}$$

where m ρ is the mass of the lightest techni-vector resonance and techni-pions π lie in the J representation of $SU(2)_L$. The second estimate is obtained from the electro-magnetic splitting of QCD pions, see the appendix. The lightest techni-baryons are an octet of flavour SU(3) and table 1 lists the two possible viable assignments for techni-quarks under $SU(2)_L \otimes U(1)_Y$:

$$\mathcal{Q} = \begin{cases} 1_{\mp 1/3} \oplus 2_{\pm 1/6} \\ 3_0 \end{cases} . \tag{3.10}$$

The hypercharges are determined by requiring that the lightest techni-baryon is neutral; in the first case their overall normalisation is determined by requiring that techni-quarks can have a Yukawa interaction with the Higgs. For this choice of quantum numbers the lightest techni-baryon is an electro-weak singlet with Y = Q = 0, avoiding direct detection constraints.

The lightest technibaryons decompose under $SU(2)_L$ as

$$8 = \begin{cases} \mathbf{2}(\mathrm{p},\mathrm{n}) \oplus \mathbf{3}(\Sigma^{\pm,0}) \oplus \mathbf{2}(\Xi^0,\Xi^-) \oplus \mathbf{1}(\Lambda_0) & \text{for } \mathcal{Q} = 1 \oplus 2 \\ \mathbf{3} \oplus \mathbf{5} & \text{for } \mathcal{Q} = 3 \end{cases} . \tag{3.11}$$

In the Q = 1 ⊕ 2 model we used the familiar names of the QCD octet. The lightest technibaryon is Λ_0, that is analogous to the QCD state $\Lambda_0^{\rm QCD} \sim s(ud - du)$. Its magnetic dipole moment can be estimated from QCD data: $\mu_{\Lambda_0^{\rm QCD}} = 0.61 e/2m_p$ [51]. Inserting g = −0.61 in eq. (3.8) we obtain the bound m_{Λ^0} > 3.7 TeV. The previous QCD-based estimate of the DM annihilation cross section becomes exact, such that the cosmological DM density is reproduced for m_{Λ^0} ≈ 200 TeV. In this model there are no stable techni-pions.

In the Q = 3Y model the lightest technibaryon is a triplet 3_{3Y} of $SU(2)_L$, so that neutral DM is obtained for Y = 0 and Y = ±1/3: the first possibility is allowed by direct detection constraints. Due to the absence of Yukawa and hypercharge interactions, the neutral member of the techni-pion triplet is the DM candidate, stable thanks to the accidental G-parity discussed in section 3.1. Its mass must be smaller than 2.5 TeV in order to avoid a thermal relic density bigger than the observed DM density. This implies that in this model the thermal relic density of the technibaryon dark matter is subdominant.

CONCLUSIONS

In conclusion, we presented a new class of models where the Standard Model is made dimension-less by dropping the mass term of the elementary Higgs and extended by adding techni-quarks with techni-color interactions arranged in such a way that they do not break the electro-weak gauge group nor generate a composite Higgs. Within the context of finite naturalness — the assumption that a QFT with no mass parameters nor power divergences might provide a revised concept of weak-scale naturalness and of the origin of mass scales — the simplest models of this type dynamically generate a mass term for the Higgs.

The elementary Higgs acquires a squared mass term m^2 entirely determined in terms of weak interactions of the techni-quarks and of the techni-color scale. Using various approximation techniques that allow to control the techni-color dynamics, in section 2.1 we found that the sign of m^2 is negative, such that $SU(2)_L \otimes U(1)_Y$ gets broken, and that the observed weak scale is obtained for a techni-color scale m ρ ≈ $4\pi M_h/\alpha_2$ ≈ 10–20 TeV. This is large enough that such models do not pose any phenomenological problem. Techni-pions are lighter, as determined by their electro-weak interactions, and could give observable signals at LHC; in particular techni-pions π_5 in the 5 of $SU(2)_L$ undergo anomalous decays into pairs of electro-weak vectors, $\pi_5 \to WW$. Such models can have the same number of free parameters as the Standard Model: all new physics is univocally predicted, up to theoretical uncertainties in the techni-strong dynamics, that could be reduced with respect to our estimates by performing dedicated lattice computations.[4]

Independently of the assumption of finite naturalness, the models studied in this paper contain two Dark Matter candidates: the lightest techni-baryon B with mass m_B ~ 50 TeV (section 3.2) and, in some models, the lightest techni-pion π_3, a triplet under $SU(2)_L$ with mass $m_{\pi 3} \sim 0.1 m_\rho \sim 1-2$ TeV (section 3.1). Their thermal relic abundance is also univocally predicted, with the result that the observed cosmological Dark Matter abundance is naturally reproduced in the techni-pion case, while the techni-baryon seems more likely to be a sub-dominant Dark Matter component, if a naive rescaling of the QCD $p\bar{p}$ cross-section holds, and ignoring possible techni-baryon asymmetries. The direct detection cross section of such DM candidates is predicted to be 2 − 3 orders of magnitude below present bounds. Magnetic moment interactions of techni-baryons would lead to recoil events with a distinctive energy spectrum (section 3.3).

[4]Techni-strong dynamics generates a negative vacuum energy of order $-\Lambda_{\rm TC}^4$. It can be canceled, compatibly with the scenario of dynamical mass generation in the SM sector, by adding another sector negligibly coupled to SM particles; this kind of sector is anyhow needed to account for the Planck mass. This cancellation is the usual huge fine-tuning associated with the cosmological constant problem, on which we have nothing to say.

Table 1 offers a panoramic of models that lead to DM candidates. In some models the quantum numbers allow for Yukawa interactions between techni-quarks and the elementary Higgs. Such Yukawas give extra negative contributions to the squared Higgs mass term (section 2.3), so that the techni-color scale needed to reproduce the weak scale gets lighter; in such models a singlet techni-pion is especially light. Models where techni-quarks only have QCD interactions or gravitational interactions do not seem to lead to a promising phenomenology, as discussed in section 2.2.

ACKNOWLEDGMENTS

We wish to thank Roberto Franceschini and Giovanni Villadoro for discussions and collaboration at the early stages of this work. We thank Roberto Contino, Riccardo Rattazzi, Slava Rychkov, Riccardo Torre for advice about strong dynamics and the authors of [50] (in particular Eugenio del Nobile) for help about using their code. The work of OA and MR is supported by the MIUR-FIRB grant RBFR12H1MW.

A EFFECTIVE POTENTIAL

The effective potential for the elementary Higgs and the techni-pions receives contributions at tree level in the Yukawa couplings and at loop level in the gauge and Yukawa couplings. It can be computed using the techniques reviewed in [41]: the relevant ingredients are the correlation functions of the composite operators of the theory. There are three main contributions: from SM gauge interactions at loop level (section A.1); from the possible Yukawa couplings at tree level (section A.2) and at loop level (section A.3). Summing these contributions, the full potential is studied in section A.4.

A.1 Gauge contribution at one loop level

At 1-loop the SM gauge interactions induce a techni-pion mass that can be computed in terms of correlators of the vector ($J^a_\mu = \sum_{\mathcal{Q}} \bar{\mathcal{Q}}\gamma_\mu T^a_{\mathcal{Q}}\mathcal{Q}$) and axial ($J^a_\mu = \sum_{\mathcal{Q}} \bar{\mathcal{Q}}\gamma_\mu T^{\hat{a}}_{\mathcal{Q}}\gamma_5\mathcal{Q}$) symmetry currents. On general grounds these have the form,

$$i\int d^4x\, e^{iq\cdot x}\langle 0|\mathrm{T}J^V_\mu(x)J^{V'}_\nu(0)|0\rangle \equiv \delta^{VV'}(q^2 g_{\mu\nu} - q_\mu q_\nu)\Pi_{VV}(q^2),$$
$$i\int d^4x\, e^{iq\cdot x}\langle 0|\mathrm{T}J^A_\mu(x)J^{A'}_\nu(0)|0\rangle \equiv \delta^{AA'}(q^2 g_{\mu\nu} - q_\mu q_\nu)\Pi_{AA}(q^2). \tag{A.1}$$

The one-loop techni-pion potential reads [41]:

$$V_{g1} \approx \frac{3}{2(4\pi)^2}\sum_i g_i^2 \mathrm{Tr}[UT^iU^\dagger T^i]\int_0^\infty Q^2dQ^2\left[\Pi_{AA}(-Q^2) - \Pi_{VV}(-Q^2)\right] \tag{A.2}$$

where $U = e^{i\pi^{\hat{a}}T^{\hat{a}}/f}$ is the Goldstone boson matrix, g_i are the SM couplings and T_i their generators. Gauge-charged techni-pions acquire positive squared masses, that, for the $SU(2)_L$ interactions, are estimated as

$$m_\pi^2 \approx \frac{3g_2^2}{(4\pi)^2}J(J+1)m_\rho^2 \tag{A.3}$$

where J is the weak isospin of the techni-pion representation.

A.2 Yukawa contribution at tree level

We now consider the potential generated by the Yukawa interactions. For concreteness we here focus on the case where techni-quarks $Q = 2 \oplus 1$ fill one doublet and one singlet of $SU(2)_L$ with hypercharges as in section 3.4. The 8 techni-pions decompose under $SU(2)_L \otimes U(1)_Y$ as

$$8 = 2_{\pm 1/2} + 3_0 + 1_0. \tag{A.4}$$

In general there are two Yukawa couplings:

$$yH\mathcal{Q}^1_L\mathcal{Q}^2_R + y'H^\dagger\mathcal{Q}^1_R\mathcal{Q}^2_L + \text{h.c.} = H\bar{\mathcal{Q}}_2\left(\frac{y+y'^*}{2} + \gamma_5\frac{-y+y'^*}{2}\right)\mathcal{Q}_1 + \text{h.c.} \tag{A.5}$$

where on the right hand side we used Dirac spinors $\mathcal{Q}_i = (\mathcal{Q}^i_L, \bar{\mathcal{Q}}^i_R)$. The phases of y and y' are not physical and can be chosen for convenience, for example real and positive. The terms above generate the tree level effective potential

$$V_{y0} = a_0\mathrm{Tr}[MU] + \text{h.c.} \tag{A.6}$$

where $a_0 \approx -m_\rho f^2$ and

$$M = \begin{array}{c} \\ \mathcal{Q}_L^1 \\ \mathcal{Q}_L^2 \end{array} \begin{array}{c} \begin{array}{cc} \mathcal{Q}_R^1 & \mathcal{Q}_R^2 \end{array} \\ \begin{pmatrix} 0 & yH \\ y'H^\dagger & 0 \end{pmatrix} \end{array}. \tag{A.7}$$

The explicit result for the potential of the doublet (π_2) and singlet (η) techni-pions is

$$V_{y0} = -8\sqrt{2}a_0 \,\mathrm{Im}\, e^{-\frac{i\eta}{4\sqrt{3}f}} \frac{\sin \Delta/f}{\Delta} [yH^\dagger \pi_2 + y'\pi_2^\dagger H], \qquad \Delta = \frac{1}{4}\sqrt{3\eta^2 + 8\pi_2^\dagger \pi_2}. \tag{A.8}$$

A.3 Yukawa contribution at one loop level

To compute the one-loop Yukawa correction to the effective potential we proceed similarly to the gauge interactions. We formally introduce sources $S_{\bar{i}j}$ for the techni-quark bilinears $\mathcal{Q}_L^i \mathcal{Q}_R^{\bar{j}}(x)$ (such that, in the real theory of interest, it contains yH in some of its components) and write the effective Lagrangian that describes the Higgs/techni-pion system after having integrated out the heavier techni-strong dynamics. For simplicity we consider vectorial couplings as in these case fewer invariants exist. In a constant techni-pion con- figuration and to quadratic order in the sources S, the effective action has the following structure determined by the symmetries,

$$\mathscr{L}_{\text{eff}}^{QQ} = a_0 \delta^4(q)(\mathrm{Tr}[SU] + \text{h.c.}) + \Pi_0^{QQ}(q^2)\mathrm{Tr}[SS^\dagger] + \Pi_1^{QQ}(q^2)|\mathrm{Tr}[SU]|^2. \tag{A.9}$$

The first term linear in S describes the $Q_L Q_R$ condensate. The form factors can be obtained integrating over the strong dynamics including techni-pion fluctuations. By construction they encode the two point functions of the techni-quark bilinears,

$$\langle 0|\bar{\mathcal{Q}}^i \mathcal{Q}^{\bar{j}}(q)\, \bar{\mathcal{Q}}^{\bar{k}} \mathcal{Q}^l(-q)|0\rangle = i\, G_{\text{Adj}}^{QQ}(q^2) \left(\delta^{i\bar{k}}\delta^{l\bar{j}} - \frac{1}{3}\delta^{i\bar{j}}\delta^{l\bar{k}}\right) + i\, G_S^{QQ}(q^2)\delta^{i\bar{j}}\delta^{l\bar{k}} \tag{A.10}$$

where G_S^{QQ} and G_{Adj}^{QQ} correspond to the singlet and adjoint channels (namely, the octet for $N_F = 3$). Matching eq.s (A.9) and (A.10) (for example choosing U = 1) one finds

$$\Pi_0^{QQ} = G_{\text{Adj}}^{QQ}, \qquad \Pi_1^{QQ} = G_S^{QQ} - \frac{1}{3}G_{\text{Adj}}^{QQ}. \tag{A.11}$$

At large N one has

$$G_{\text{Adj}}^{QQ}(q^2) = \frac{N}{16\pi^2}m_\rho^4 \sum_n \frac{c_{\text{Adj}_n}^2}{q^2 - m_{\text{Adj}_n}^2 + i\epsilon}, \qquad G_S^{QQ}(q^2) = \frac{N}{16\pi^2}m_\rho^4 \sum_n \frac{c_{S_n}^2}{q^2 - m_{S_n}^2 + i\epsilon}, \tag{A.12}$$

where the coefficients c are of order 1 and the sum is over the scalar resonances in the theory. The sum does not include techni-pions because we only consider vectorial Yukawa couplings that do not generate 1 techni-pion states.

To obtain the effective action for the scalars we just need to set to zero the non dynamical components of S and add kinetic terms for the components of S associated to the Higgs. This produces

$$\mathscr{L}_{\text{eff}}^{\text{H}} = a_0 \delta^4(q)(\mathrm{Tr}[MU] + \text{h.c.}) + (q^2 + y^2\Pi_0^{QQ}(q^2))H^\dagger H + \Pi_1^{QQ}(q^2)|\mathrm{Tr}[MU]|^2. \tag{A.13}$$

The first term describes the tree level contribution discussed above. The second term encodes the tree level effect of mixing with heavy scalar resonances that gives the HH* entry of the mass matrix in eq. (2.21),

$$m_H^2 = y^2\Pi_0^{QQ}(0) \sim -\frac{y^2 N}{(4\pi)^2}m_\rho^2. \tag{A.14}$$

Performing the path integral with respect to H we obtain the one-loop Yukawa contribution to the techni-pion potential,

$$V_{y1} = \frac{1}{2}\int \frac{d^4Q}{(2\pi)^4} \ln\left[Q^2 - y^2\Pi_0^{QQ}(-Q^2) - y^2\Pi_1^{QQ}(-Q^2)\sum_a \mathrm{Tr}[T^a U]\mathrm{Tr}[T^a U^\dagger]\right]$$

$$\approx v_0^4 - \frac{y^2}{2(4\pi)^2}\sum_a \mathrm{Tr}[T^a U]\mathrm{Tr}[T^a U^\dagger]\int_0^\infty dQ^2 \Pi_1^{QQ}(-Q^2) \tag{A.15}$$

where v_0 is the contribution to the vacuum energy and T^a are SU(3) matrices derived from (A.7).

One can prove that, similarly to the gauge contribution, the loop integral in (A.15) is finite: since Π_1^{QQ} is sensitive to the chiral symmetry breaking, the Operator Product Expansion demands that

$$\Pi_1^{QQ}(q^2) \overset{q^2 \gg \Lambda_{\rm TC}^2}{\simeq} \frac{\langle 0|(\bar{\mathcal{Q}}_L\Gamma_1\mathcal{Q}_R)(\bar{\mathcal{Q}}_R\Gamma_2\mathcal{Q}_L)|0\rangle}{q^4} \tag{A.16}$$

where $\Gamma_{1,2}$ are appropriate matrices in techni-color and flavour space, see [42].

Contrary to the gauge contribution we are not aware of any theorem that guarantees the sign of this contribution. As an estimate the contribution above gives

$$\delta m_\pi^2 \sim \frac{y^2 m_\rho^2}{(4\pi)^2}. \tag{A.17}$$

Summing up all the contributions we obtain a mass matrix with the structure of eq. (2.21).

A.4 Minimization of the potential

The vacuum is determined through the minimization of the potential

$$V_{\rm eff}(\pi,\eta,H) = V_{g1} + V_{y0} + V_{y1} + m^2|H|^2 + \lambda|H|^4 \tag{A.18}$$

where $m^2 < 0$ is induced by gauge loops (section 2.1). The gauge-charged techni-pions π acquire a large mass from gauge loops and can be integrated out, leaving an effective potential for the lighter scalars: the elementary Higgs doublet H and the gauge-neutral techni-pion η. In the parameter range of interest for us, g >> y, one has $V_{y1} \approx 0$ and $V_{g1} \approx \frac{1}{2}m_\pi^2\pi^2(1-\eta^2/16f^2)$, where, for simplicity, we expanded at second order in η/f sufficient to compute the mass of the singlet. We can freely redefine the phases of the Yukawa couplings y and y' so that yy' is real and negative. With this choice $\eta = 0$ indeed is a local minimum of the effective potential

$$V_{\rm eff}(\eta \ll f, H) \approx |H|^2\left[m^2 - 32\frac{m_\rho^2 f^2}{m_\pi^2}\left((|y|+|y'|)^2 - |yy'|\frac{\eta^2}{12f^2}\right)\right] + \lambda|H|^4. \tag{A.19}$$

Around the minimum η acquires a positive squared mass

$$M_\eta \sim |y|\frac{m_\rho}{m_\pi}v \tag{A.20}$$

without mixing with the Higgs, that receives a negative contribution to its m^2 parameter.

REFERENCES

1. C. Kilic, T. Okui and R. Sundrum, Vectorlike confinement at the LHC, JHEP 02 (2010) 018 [arXiv:0906.0577] .
2. M. Farina, D. Pappadopulo and A. Strumia, A modified naturalness principle and its experimental tests, JHEP 08 (2013) 022 [arXiv:1303.7244] .
3. A. de Gouvêa, D. Hernandez and T.M.P. Tait, Criteria for natural hierarchies, Phys. Rev. D 89 (2014) 115005 [arXiv:1402.2658] .
4. F. Englert, C. Truffin and R. Gastmans, Conformal invariance in quantum gravity, Nucl. Phys. B 117 (1976) 407 .
5. W. Bardeen, On naturalness in the standard model, FERMILAB-CONF-95-391-T (1995).
6. C.T. Hill, Conjecture on the physical implications of the scale anomaly, hep-th/0510177 .
7. A. Salvio and A. Strumia, Agravity, JHEP 06 (2014) 080 [arXiv:1403.4226] .
8. R. Hempfling, The next-to-minimal Coleman-Weinberg model, Phys. Lett. B 379 (1996) 153 [hep-ph/9604278] .
9. J.P. Fatelo, J.M. Gerard, T. Hambye and J. Weyers, Symmetry breaking induced by top loops, Phys. Rev. Lett. 74 (1995) 492 .
10. T. Hambye, Symmetry breaking induced by top quark loops from a model without scalar mass, Phys. Lett. B 371 (1996) 87 [hep-ph/9510266] .
11. W.-F. Chang, J.N. Ng and J.M.S. Wu, Shadow Higgs from a scale-invariant hidden U(1)s model, Phys. Rev. D 75 (2007) 115016 [hep-ph/0701254] .
12. R. Foot, A. Kobakhidze and R.R. Volkas, Electroweak Higgs as a pseudo-Goldstone boson of broken scale invariance, Phys. Lett. B 655 (2007) 156 [arXiv:0704.1165] .
13. R. Foot, A. Kobakhidze, K.L. McDonald and R.R. Volkas, A solution to the hierarchy problem from an almost decoupled hidden sector within a classically scale invariant theory, Phys. Rev. D 77 (2008) 035006 [arXiv:0709.2750] .
14. S. Iso, N. Okada and Y. Orikasa, The minimal B-L model naturally realized at TeV scale, Phys. Rev. D 80 (2009) 115007 [arXiv:0909.0128] .

15. S. Iso and Y. Orikasa, TeV Scale B-L model with a flat Higgs potential at the Planck scale — in view of the hierarchy problem , PTEP 2013 (2013) 023B08 [arXiv:1210.2848] .
16. C. Englert, J. Jaeckel, V.V. Khoze and M. Spannowsky, Emergence of the electroweak scale through the Higgs portal, JHEP 04 (2013) 060 [arXiv:1301.4224] .
17. E.J. Chun, S. Jung and H.M. Lee, Radiative generation of the Higgs potential, Phys. Lett. B 725 (2013) 158 [arXiv:1304.5815] .
18. T. Hambye and A. Strumia, Dynamical generation of the weak and dark matter scale, Phys. Rev. D 88 (2013) 055022 [arXiv:1306.2329] .
19. C.D. Carone and R. Ramos, Classical scale-invariance, the electroweak scale and vector dark matter, Phys. Rev. D 88 (2013) 055020 [arXiv:1307.8428] .
20. R. Foot, A. Kobakhidze, K.L. McDonald and R.R. Volkas, Poincaré protection for a natural electroweak scale, Phys. Rev. D 89 (2014) 115018 [arXiv:1310.0223]
21. A. Farzinnia, H.-J. He and J. Ren, Natural electroweak symmetry breaking from scale invariant Higgs mechanism, Phys. Lett. B 727 (2013) 141 [arXiv:1308.0295] .
22. C.T. Hill, Is the Higgs boson associated with Coleman-Weinberg dynamical symmetry breaking?, Phys. Rev. D 89 (2014) 073003 [arXiv:1401.4185] .
23. J. Guo and Z. Kang, Higgs naturalness and dark matter stability by scale invariance, arXiv:1401.5609 .
24. S. Benic and B. Radovcic, Electroweak breaking and dark matter from the common scale, Phys. Lett. B 732 (2014) 91 [arXiv:1401.8183] .
25. H. Davoudiasl and I.M. Lewis, Right-handed neutrinos as the origin of the electroweak scale, Phys. Rev. D 90 (2014) 033003 [arXiv:1404.6260] .
26. K. Allison, C.T. Hill and G.G. Ross, Ultra-weak sector, Higgs boson mass and the dilaton, Phys. Lett. B 738 (2014) 191 [arXiv:1404.6268] .
27. G.M. Pelaggi, Predictions of a model of weak scale from dynamical breaking of scale invariance, arXiv:1406.4104
28. W. Altmannshofer, W.A. Bardeen, M. Bauer, M. Carena and J.D. Lykken, Light dark matter, naturalness and the radiative origin of the electroweak scale, JHEP 01 (2015) 032 [arXiv:1408.3429] .
29. T. Hur and P. Ko, Scale invariant extension of the standard model with strongly interacting hidden sector, Phys. Rev. Lett. 106 (2011) 141802 [arXiv:1103.2571] .
30. M. Heikinheimo, A. Racioppi, M. Raidal, C. Spethmann and K. Tuominen, Physical naturalness and dynamical breaking of classical scale invariance, Mod. Phys. Lett. A 29 (2014) 1450077 [arXiv:1304.7006] .
31. T. Hambye and M.H.G. Tytgat, Confined hidden vector dark matter, Phys. Lett. B 683 (2010) 39 [arXiv:0907.1007] .
32. M. Holthausen, J. Kubo, K.S. Lim and M. Lindner, Electroweak and conformal symmetry breaking by a strongly coupled hidden sector, JHEP 12 (2013) 076 [arXiv:1310.4423] .
33. O. Antipin, M. Redi, A. Strumia and E. Vigiani, Thermal dark matter from strong interactions, work in progress.
34. K. Griest and M. Kamionkowski, Unitarity limits on the mass and radius of dark matter particles, Phys. Rev. Lett. 64 (1990) 615 .
35. B. von Harling and K. Petraki, Bound-state formation for thermal relic dark matter and unitarity, JCAP 12 (2014) 033 [arXiv:1407.7874] .
36. G.F. Giudice, G. Isidori, A. Salvio and A. Strumia, Softened gravity and the extension of the standard model up to infinite energy, arXiv:1412.2769 .
37. Y. Bai and R.J. Hill, Weakly interacting stable pions, Phys. Rev. D 82 (2010) 111701 [arXiv:1005.0008] .
38. R. Pasechnik, V. Beylin, V. Kuksa and G. Vereshkov, Vector-like technineutron dark matter: is a QCD-type technicolor ruled out by XENON100?, Eur. Phys. J. C 74 (2014) 2728 [arXiv:1308.6625] .
39. E. Witten, Some inequalities among hadron masses, Phys. Rev. Lett. 51 (1983) 2351 .
40. D. Marzocca, A. Parolini and M. Serone, Supersymmetry with a PNGB Higgs and partial compositeness, JHEP 03 (2014) 099 [arXiv:1312.5664] .
41. R. Contino, The Higgs as a composite Nambu-Goldstone boson, arXiv:1005.4269 .
42. M.A. Shifman, A.I. Vainshtein and V.I. Zakharov, QCD and resonance physics. Sum rules, Nucl. Phys. B 147 (1979) 385 .
43. M. Peskin and D.V. Schroeder, An introduction to quantum field theory, Westview Press, U.S.A. (1995).
44. M. Cirelli, N. Fornengo and A. Strumia, Minimal dark matter, Nucl. Phys. B 753 (2006) 178 [hep-ph/0512090] .
45. M. Cirelli, A. Strumia and M. Tamburini, Cosmology and astrophysics of minimal dark matter, Nucl. Phys. B 787 (2007) 152 [arXiv:0706.4071] .
46. J. Hisano, K. Ishiwata and N. Nagata, Gluon contribution to the dark matter direct detection, Phys. Rev. D 82 (2010) 115007 [arXiv:1007.2601] .
47. R.J. Hill and M.P. Solon, Universal behavior in the scattering of heavy, weakly interacting dark matter on nuclear targets, Phys. Lett. B 707 (2012) 539 [arXiv:1111.0016] .
48. LUX collaboration, C. Faham, First dark matter search results from the Large Underground Xenon (LUX) experiment, arXiv:1405.5906 .
49. V. Barger, W.-Y. Keung and D. Marfatia, Electromagnetic properties of dark matter: Dipole moments and charge form factor, Phys. Lett. B 696 (2011) 74 [arXiv:1007.4345] .
50. M. Cirelli, E. Del Nobile and P. Panci, Tools for model-independent bounds in direct dark matter searches, JCAP 10 (2013) 019 [arXiv:1307.5955] .
51. Particle Data Group collaboration, K.A. Olive et al., Review of particle physics, Chin. Phys. C 38 (2014) 090001 .

52. J. Kubo, K.S. Lim and M. Lindner, Electroweak symmetry breaking via QCD, Phys. Rev. Lett. 113 (2014) 091604 [arXiv:1403.4262] .

53. W.J. Marciano, Exotic new quarks and dynamical symmetry breaking, Phys. Rev. D 21 (1980) 2425 .

54. D. Lust, E. Papantonopoulos, K. Streng and G. Zoupanos, Phenomenology of high color fermions, Nucl. Phys. B 268 (1986) 49 .

CHAPTER 4

STRONG PHASE TRANSITION, DARK MATTER AND VACUUM STABILITY FROM SIMPLE HIDDEN SECTORS

Tommi Alanne, Kimmo Tuominen and Ville Vaskonen

INTRODUCTION

Extensions of the Standard Model (SM) of elementary particle interactions have been put under severe tests [1], [2], [3] and [4] after the discovery of the Higgs boson with mass m_h=126 GeV at the ATLAS and CMS experiments in the CERN Large Hadron Collider [5] and [6]. The requirement of a light Higgs scalar boson and no other obviously accessible states at the energy scales probed so far presents a challenge for traditional model paradigms like supersymmetry and technicolor which predict an extended spectrum beyond the SM. On the other hand, the cosmological observations on the dark matter abundance and matter–antimatter asymmetry clearly require, in the elementary particle physics context, the existence of new degrees of freedom not present in the SM.

One possible framework to address these aspects is to take the SM according to the current collider data and extend it with singlet fields communicating with the SM fields only through the scalar or vector portals. The singlet sector can consist of a single scalar [7] and [8], more complex scalar multiplets [9] and [10], fermions [11], [12] and [13] or vectors [14] and [15]. The resulting spectrum can contain particles stable over the timescales of the age of the universe and contribute to the observed abundance of the dark matter. Moreover, the extended scalar potential can modify the properties of the phase transitions in the early universe with respect to the results obtained in the SM. A strong first-order electroweak transition is a prerequisite for successful electroweak baryogenesis [16], and it is well known that the electroweak phase transition in the SM is not of first order but a smooth crossover [17] and [18]. If the electroweak sector of the SM were fully perturbative, a first-order phase transition would arise from a cubic term generated in the Higgs effective one-loop potential by the thermal effects of fields coupled to the Higgs. However, addition of a singlet scalar can sufficiently modify the picture already by tree level effects due to the presence of T-independent dimensional parameters appearing in the scalar potential and lead to a strong first-order transition[19] and [20]. Consequently, the ratio $v(T_c)/T_c$ which controls the sphaleron erasure of the baryon asymmetry can be large and lead to successful electroweak baryogenesis.

The phenomenologically interesting scenario would, thus, be the one where the strong electroweak phase transition is accompanied by an explanation of the dark matter relic density by a weakly coupled massive particle (WIMP). As a concrete model paradigm in this paper, we consider simple scalar portals between the hidden sector and the visible one. In addition to the Standard Model Higgs field, the scalar sector contains a real singlet scalar S.

As a simple limiting case, we consider a model where the singlet sector is solely constituted by S with a discrete Z_2 symmetry [7], [21], [22] and [23]. Then S can also act as a dark matter candidate provided that the global minimum of the potential at zero temperature does not spontaneously break this Z_2 symmetry. However, in this model it is not possible to simultaneously explain the strong electroweak phase transition and the observed dark matter relic abundance. Therefore, as a second example we consider a model where the scalar S is not assumed to have any discrete symmetries, but the singlet sector also contains a Dirac fermion [11], [12] and [24]. We find that in this model it is possible to realize simultaneously strong electroweak phase transition and the observed dark matter relic abundance.

Citation: Tommi Alanne, Kimmo Tuominen, Ville Vaskonen, Strong phase transition, dark matter and vacuum stability from simple hidden sectors, Nuclear Physics B, Volume 889, December 2014, Pages 692-711, ISSN 0550-3213, http://dx.doi.org/10.1016/j.nuclphysb.2014.11.001.

To establish our results, we perform a Monte Carlo analysis of the parameter space to search for viable models. We impose the constraints from LHC data and precision electroweak measurements, and from the dark matter direct searches [25] and [26]. To improve the earlier work on models of this type [22] and [12], we also require that the couplings remain perturbative up to scales of O(TeV). As a special case, we consider the possibility that the model could remain perturbative up to the Planck scale. This happens only on very specific couplings and allows to single out specific benchmark models where the dark matter constraints can be satisfied and a strong electroweak transition induced. Furthermore, since it is possible that multiple components contribute to the observed dark matter abundance [27], [28] and [29], we consider systematically the cases of subdominant dark matter and both freeze-out and freeze-in mechanisms to generate the dark matter abundance. In particular, we show that if the dark matter candidate is a singlet fermion, then its abundance can be generated either via the freeze-out or the freeze-in mechanism, and in both cases the model will also lead to a strong electroweak phase transition.

The paper is organized as follows. In Section 2, we introduce the model and the basic constraints. Then in Section 3, we discuss the numerical results for the model with only a Z_2-symmetric real singlet scalar and for the model with a real singlet scalar and a singlet Dirac fermion. In Section 4, we present our conclusions.

MODEL AND CONSTRAINTS

We consider an extended scalar sector described by the potential

$$V(H,S)=\mu_H^2 H^\dagger H+\lambda_H(H^\dagger H)^2+\frac{1}{2}\mu_S^2 S^2+\frac{\mu_3}{3}S^3+\frac{\lambda_S}{4}S^4 + \mu_{HS}(H^\dagger H)S+\frac{\lambda_{HS}}{2}(H^\dagger H)S^2, \tag{2.1}$$

which provides typical scalar portals between the hidden singlet sector and the Standard Model Higgs. In Eq. (2.1), the field H is the usual Standard Model Higgs doublet and S is a real singlet scalar. The Higgs doublet is written in terms of the components as

$$H=\begin{pmatrix}\phi^+\\ \frac{1}{\sqrt{2}}(v+\phi_r^0+i\phi_i^0)\end{pmatrix}, \tag{2.2}$$

where the superscript refers to the electroweak charge of the component and subscripts denote the real and imaginary parts. Note that the most general potential would have a linear term, $\mu_1^3 S$, but this can always be removed by shifting the field S. The stability of the potential requires that

$$\lambda_H>0,\qquad \lambda_S>0,\qquad \lambda_{HS}>-2\sqrt{\lambda_H\lambda_S}. \tag{2.3}$$

Depending on the values of the parameters, the vacuum structure can be rich [20]. We require that the extremum which leads to correct pattern of the electroweak symmetry breaking is the global minimum at zero temperature.

To study the consequences for the electroweak phase transition, we need to include the finite-temperature corrections. Since there are multiple vacua induced already at the tree level, it is sufficient to consider the finite-temperature corrections to the leading terms. In other words, we consider temperature dependent coefficients

$$\mu_1(T)^3=c_1T^2,\qquad \mu_S(T)^2=\mu_S^2+c_ST^2,\qquad \mu_H(T)^2=\mu_H^2+c_HT^2, \tag{2.4}$$

where

$$c_1=\frac{1}{12}(\mu_3+2\mu_{HS}),$$
$$c_S=\frac{1}{12}(2\lambda_{HS}+3\lambda_S),$$
$$c_H=\frac{1}{48}(9g^2+3g'^2+12y_t^2+24\lambda_H+2\lambda_{HS}), \tag{2.5}$$

in Eq. (2.1). We have assumed that the term linear in S , i.e. the one with coefficient μ_1 in the potential, has been shifted away at tree level at zero temperature and therefore arises only for T≠0. Starting from T=0, we monitor the evolutions of the electroweak symmetric and electroweak broken minima. If the symmetric minimum at some temperature T_c gets deeper than the asymmetric minimum, the phase transition is possible. For these cases we determine the critical temperature, T_c, and $v(T_c)/T_c$ to identify the parameter space domains where a strong electroweak transition can be realized.

For the dark matter candidate, there are several possibilities which can be built around this scalar sector. We will consider two. The simplest, and already much studied, case is to assume the singlet scalar to have a discrete symmetry, like Z_2, which renders it stable. This symmetry must then be imposed on the general potential of Eq. (2.1). The second alternative that we consider is to assume, in addition to the scalar, the existence of a singlet fermion, which then due to a conserved fermion number becomes a stable dark matter candidate. Then no symmetry restrictions need to be imposed on the scalar potential, leaving a richer possibility of patterns for the phase transition between the electroweak symmetric and broken vacua. The details and numerical result of these two model examples will be exposed more thoroughly in the next section. Here, we now discuss the general formulation of the computation of the dark matter abundance and direct detection limits.

To compute the relic abundance of dark matter, Ω_{DM}, we apply the standard freeze-out formalism.[1] The number density n of the thermal relic can be computed from the Lee–Weinberg equation [30],

$$\frac{\partial f(x)}{\partial x}=\frac{\langle v\sigma\rangle m^3x^2}{H}\left(f^2(x)-f_{eq}^2(x)\right), \tag{2.6}$$

written in terms of scaled variables $f(x)=n(x)/s_E$ and $x=s_E^{1/3}/m$. Here s_E is the entropy density at temperature T, m is the mass of the dark matter candidate and H is the Hubble parameter. For the averaged cross sections, we use the integral expression[31]

$$\langle v\sigma\rangle=\frac{1}{8m^4TK_2^2(m/T)}\int_{4m^2}^{\infty}ds\sqrt{s}(s-4m^2)K_1(\sqrt{s}/T)\sigma_{tot}(s), \tag{2.7}$$

where $K_i(y)$ are the modified Bessel functions of the second kind and s is the usual Mandelstam invariant. Given the cross sections, we can solve the Lee–Weinberg equation for f(0) which gives the present ratio of the number density of the dark matter candidate to its entropy density. The fractional density parameter, Ω_{DM}, can be computed from

$$\Omega_{\rm DM} \simeq 1.09 \cdot 10^6 m f(0). \tag{2.8}$$

Since there is no reason to expect that all of the dark matter abundance originates from a single source, we define the fraction

$$f_{\rm rel} = \Omega_{\rm DM} h^2 / (\Omega h^2)_c, \tag{2.9}$$

where $(\Omega h^2)_c$=0.12 from Planck [32].

Also, there are stringent constraints for this type of models arising both from the LHC data and direct searches for dark matter. If the singlet scalar is light enough, $2m_S \leq m_h$, then an obvious constraint on new light degrees of freedom from LHC data is the invisible width of the Higgs boson. According to present LHC data [33], [34], [35], [36], [37] and [38], the 2σ limit for the branching fraction to invisible channels is $Br_{inv} \leq 0.28$. On the other hand, after the symmetry breaking the scalar mass eigenstates are

$$h^0 = \phi_r^0 \cos\beta + S \sin\beta, \qquad H^0 = -\phi_r^0 \sin\beta + S \cos\beta, \tag{2.10}$$

and the mixing of the Higgs and the singlet scalar affects the couplings of the scalar to fermions and gauge bosons. These modifications can be constrained by the current collider data by fitting the mixing cosβ to the signal strength results. Moreover, compatibility with the electroweak precision measurements using the S and Tparameters [39] needs to be checked. Finally, considering the direct searches for dark matter, the scalar portal interactions in Eq. (2.1) contribute to the spin-independent scattering cross sections on nuclei for which the LUX experiment [26] provides currently the most stringent constraints.

In the cases we will consider, the WIMP couples to the nucleus via the scalars h^0 and $H^0$with strength depending on the mixing patter of the scalars and whether the WIMP is a scalar or a fermion. In both cases the Higgs–nucleon coupling is $f_N m_N/v$, where m_N=0.946 GeV, and we neglect the small differences between neutrons and protons. The effective Higgs–nucleon coupling,

$$f_N \equiv \frac{1}{m_N} \sum_q \langle N | m_q \bar{q} q | N \rangle, \tag{2.11}$$

describes the normalized total quark scalar current within the nucleon. The quark currents of the nucleon have been a subject of an intensive lattice research supplemented by efforts applying chiral perturbation theory methods and pion nucleon scattering. Consequently, f_N is fairly well determined currently. Following [23] we use f_N=0.345±0.016, where the uncertainty in f_N induces at most 20% error in the spin-independent direct detection limits.

The spin-independent cross section for a WIMP scattering on nuclei is computed by considering the t -channel exchange of h^0 and H^0 in the limit t→0. The matrix element for this process generally contains the factors

$$\left(\frac{\tilde{g}_h \cos\beta}{m_h^2} - \frac{\tilde{g}_H \sin\beta}{m_H^2} \right) f_N \frac{m_N}{v}, \tag{2.12}$$

where the coupling between the scalars h^0 and H^0 and the WIMP candidate is denoted by $\tilde{g}_h$ and $\tilde{g}_H$, respectively. In the explicit models we consider, we determine $\tilde{g}_h$ and $\tilde{g}_H$ and use the above formula when evaluating the spin-independent scattering cross section per nucleon, σ_{SI}^0. Since we consider the possibility that our WIMP candidate forms only a fraction of the total dark matter abundance, we need to take this into account when comparing with the direct searches. The direct search constraints on σ_{SI}^0 are given by the experimental collaborations under the assumption that f_{rel}=1. To apply the constraints under the assumption of subdominant WIMPs, we define an effective cross section

$$\sigma_{\rm SI}^{\rm eff} = f_{\rm rel} \sigma_{\rm SI}^0. \tag{2.13}$$

With these preliminaries we now turn to the numerical results of two explicit models based on the scalar sector described by the potential (2.1).

NUMERICAL RESULTS

We will now discuss two different concrete cases. First, assuming Z_2 symmetry for the singlet scalar simplifies the potential and allows the singlet scalar to act as dark matter. However, it is impossible to saturate even a modest fraction of the observed dark matter abundance and simultaneously induce a strong first-order electroweak phase transition. Second, we release the requirement of Z_2 symmetry on the singlet scalar but assume the existence of a singlet fermion. We find that in this case the scalar can induce a strong first-order electroweak transition, and the fermion can act as a dark matter candidate.

Scalar dark matter: Z_2-symmetric case

A simple benchmark model, which has been considered extensively in literature earlier [7], [21], [22] and [23], is provided by imposing a Z_2 symmetry on the potential in Eq. (2.1).

Consider the Lagrangian

$$\mathcal{L} = \mathcal{L}_{\rm kin} + \mathcal{L}_{\rm Yuk} - V(H,S)|_{Z_2}, \tag{3.1}$$

where L_{kin} contains the kinetic terms for the scalars, appropriately gauged under the SM charges, and for all the SM gauge and matter fields. The term L_{Yuk} contains the usual Yukawa couplings between the SM matter fields and the Higgs field H . The potential $V(H,S)|_{Z2}$ is obtained from Eq. (2.1) by setting $\mu HS=\mu_3=0$. Due to the Z_2symmetry, the singlet scalar in this model can act as a dark matter candidate. Depending on the model parameters, either the neutral component of H or the singlet can have a nonzero vacuum expectation value (vev). Since we assume the singlet to be a dark matter candidate, the T=0 vacuum has to be Z_2 symmetric to ensure the stability of the dark matter candidate. This means that at zero temperature the vev, w, of S must be zero, i.e. the global minimum of the potential must be at (v,w)=(246 GeV,0). This requires, in addition to the bounds from vacuum stability in Eq. (2.3), that

$$\mu_S^2 > -v^2\sqrt{\lambda_H\lambda_S}. \tag{3.2}$$

In the notation of Eq. (2.10), the mass eigenstates are directly $h^0=\phi_5$ and $H^0=S$, i.e. $\cos\beta=1$. We trade the mass parameter μ_S^2 with the physical mass of the singlet, $m_S^2=\mu_S^2+v^2\lambda_{HS}/2$. Then the above constraint and the requirement of the stability can be combined into a bound

$$-2\sqrt{\lambda_H\lambda_S} < \lambda_{HS} < \frac{2m_S^2}{v^2} + 2\sqrt{\lambda_H\lambda_S}. \tag{3.3}$$

The known values of the electroweak scale, v=246 GeV, and the Higgs mass, m_h=126 GeV, fix the model parameters λ_H=0.131 and $\mu_H^2=-v^2\lambda_H$. The remaining parameters, m_S^2, λ_S and λ_{HS} are free but subject to the constraints listed above. We scan the parameter space by performing a simple Monte Carlo analysis generating a random distribution of points with

$$0 < \lambda_S \le \pi, \qquad 5\text{ GeV} \le m_S \le 650\text{ GeV}. \tag{3.4}$$

As a further constraint, we require perturbativity of the couplings reasonably far out in energy. The renormalization group (RG) equations are solved at one loop, and points which remain perturbative up to scales μ~10 TeV are accepted. The RG equations for the gauge couplings and the Yukawa coupling of the top quark are as in the Standard Model. The β functions of the couplings are defined as

$$\beta_g = \frac{dg}{d\ln\mu}, \tag{3.5}$$

and in the Z_2-symmetric case they read

$$16\pi^2\beta_{\lambda_H} = 24\lambda_H^2 + \frac{1}{2}\lambda_{HS}^2 - 3(3g^2+g'^2-4y_t^2)\lambda_H + \frac{3}{8}(3g^4+2g^2g'^2+g'^4) - 6y_t^4, \tag{3.6}$$

$$16\pi^2\beta_{\lambda_{HS}} = 4\lambda_{HS}^2 + (12\lambda_H+6\lambda_S)\lambda_{HS} - 3\left(\frac{3}{2}g^2+\frac{1}{2}g'^2-2y_t^2\right)\lambda_{HS}, \tag{3.7}$$

$$16\pi^2\beta_{\lambda_S} = 18\lambda_S^2 + 2\lambda_{HS}^2. \tag{3.8}$$

To determine how much of the dark matter relic abundance the singlet scalar S can provide for, we carry out the standard freeze-out calculation. There are three annihilation channels, $SS\to h^0h^0$, VV and $\bar{f}f$, where V denotes the electroweak gauge bosons, V=W,Z. The annihilation cross sections to these three distinct final states are

$$\sigma_{hh} = \frac{v_h}{32\pi s v_S}\left|\lambda_{HS} + \frac{3m_h^2\lambda_{HS}}{s-m_h^2+im_h\Gamma_h} - \frac{4v^2\lambda_{HS}^2}{s-2m_h^2}\right|^2,$$
$$\sigma_{VV} = \frac{v_V}{4\pi s v_S}\frac{M_V^4\lambda_{HS}^2}{|s-m_h^2+im_h\Gamma_h|^2}\left(3+\frac{s(s-4M_V^2)}{4M_V^4}\right)\delta_V,$$
$$\sigma_{ff} = \frac{v_f X_f}{16\pi s v_S}\frac{m_f^2(s-4m_f^2)\lambda_{HS}^2}{|s-m_h^2+im_h\Gamma_h|^2}. \tag{3.9}$$

Here $v_X=\sqrt{1-4m_X^2/s}$, $\delta_{W,Z}$=1,1/2 and for QCD colour singlet fermions X_f=1while for quarks

$$X_f = 3\left(1+\left(\frac{3}{2}\ln\frac{m_q^2}{s}+\frac{9}{4}\right)\frac{4\alpha_s}{3\pi}\right), \tag{3.10}$$

where α_s=0.12 is the strong coupling constant.[2]

In the range $m_h/2<m_S<m_h$, we factorize the annihilation to fermion and gauge boson final states to the $SS\to h^0$ fusion part times the virtual Higgs decay using the full width of the Higgs [40], which also takes the 4-body final states into account.

To see if this model can provide for the strong electroweak transition, $v(T_c)/T_c>1$, we consider the finite-temperature corrections to the quadratic terms; the linear term in S does not arise in this Z_2-symmetric case. At very high temperatures, the potential has a unique minimum at (S,H)=(0,0). As the temperature decreases, the potential generates two minima: one at nonzero S and the other at nonzero value of the neutral component of H. At some intermediate temperature these two become degenerate, and the possibility for a strong electroweak phase transition arises.

To take into account the constraints from LHC, we consider the decay width of the Higgs to two singlets, $h^0\to SS$,

$$\Gamma_{h^0\to SS} = \frac{\lambda_{HS}^2v^2}{32\pi m_h}\sqrt{1-\frac{4m_S^2}{m_h^2}}. \tag{3.11}$$

The Higgs total decay width to the visible Standard Model channels is Γ_h=4.07 MeV for m_h=126 GeV[40], and this implies a bound for the Higgs portal coupling.

The basic result of the model is shown in the left panel of Fig. 1. The color coding shows the relative dark matter relic abundance. Above the solid black curve, $\lambda_{HS}=2m_S^2/v^2$, $\mu_S^2$is negative and modifications to the electroweak transition are possible. The figure illustrates that the parameter space of the model where (a reasonable fraction of) the relic density can be explained does not overlap with the parameter space where a strong first-order phase transition may arise.

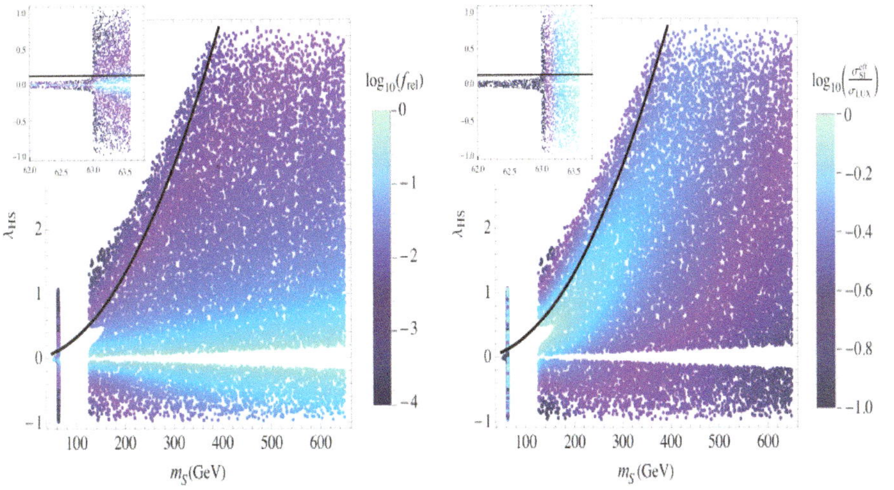

Figure 1: Left panel: the color coding shows the value of f_{rel} in the (m_S,λHS) plane. Above the solid black line $\mu_S^2 < 0$, and modification to the electroweak transition are possible. The constraint from the perturbativity of couplings, LHC invisible Higgs width and direct dark matter searches have been imposed as explained in the text. Right panel: results from dark matter direct searches. The color coding shows how the parameter space is constrained further if the sensitivity of the direct search experiments rises. For the black points $\sigma^{eff}_{SI}/\sigma_{LUX} \leq 0.1$. (For interpretation of the references to color in this figure, the reader is referred to the web version of this article.)

The points in the left panel of Fig. 1 have also been constrained to be compatible with the direct search results from the LUX experiment [26], which provides currently the best constraints for this type of model where only spin-independent cross section between the WIMP and ordinary matter arises. In the model we consider here, the effective couplings $\tilde{g}_h$ and $\tilde{g}_H$ in Eq. (2.12) are

$$\tilde{g}_h = \frac{\lambda_{HS}}{2} v, \qquad \tilde{g}_H = 0. \tag{3.12}$$

Using these, we evaluate the cross section for elastic WIMP nucleon scattering and obtain

$$\sigma^0_{SI} = \frac{1}{4\pi} \frac{\lambda_{HS}^2 \mu_N^2 f_N^2 m_N^2}{m_h^4 m_S^2}, \tag{3.13}$$

where μ_N^2 is the reduced mass of the WIMP–nucleon pair. To illustrate the distance of our results from the LUX bound, we show in the right panel of Fig. 1 the same points as in the left panel, with the color coding now showing the spin-independent cross section in relative to the LUX constraint.

Of course the dark matter candidate need not be produced by the freeze-out mechanism as we have assumed so far. Since it is impossible to explain both the strong electroweak phase transition and the dark matter abundance with the scalar sector considered here, let us leave the phase transition for a while and focus only on dark matter. Given the fact that no trace of WIMPs has been observed in the direct searches, we consider briefly the possibility of freezing in the relic abundance [41]. Assuming that the scalar is light, it can be produced from the thermal bath of Higgs bosons. The number density of S is described by the Boltzmann equation formally similar to the one governing the density in the freeze-out case. Analogously to the well known approximate result in the freeze-out case, the dark matter abundance in this case is given by

$$\Omega_S h^2 \simeq \frac{1.09 \cdot 10^{27}}{g_S \sqrt{g_\rho}} \frac{m_S \Gamma_{h^0 \to SS}}{m_h^2}, \tag{3.14}$$

where $g_{s,\rho}$ denotes the effective degrees of freedom for the entropy and energy density, respectively. The essential feature of this mechanism is that the coupling required for the production of sufficient relic abundance is superweak; one obtains a parametric estimate

$$\lambda_{HS} \simeq 10^{-12} \left(\frac{\Omega_S h^2}{0.12}\right)^{1/2} \left(\frac{\mathrm{GeV}}{m_S}\right)^{1/2}. \tag{3.15}$$

The result on the relic density in the case of freeze-in is shown in Fig. 2 as a function of the scalar mass, m_S, and the portal coupling, λHS. The contours show the values of f_{rel}, defined in Eq. (2.9), as indicated explicitly in the figure.

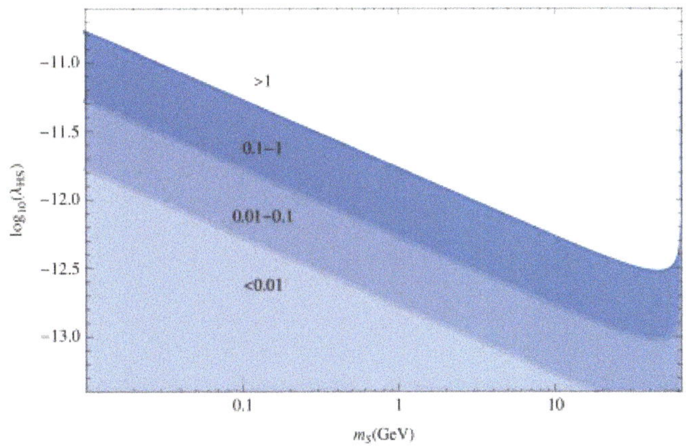

Figure 2: The contours, from top to bottom, show the values of λHS and m_S required to produce relic density f_{rel}=1,0.1,0.01, respectively, via the freeze-in mechanism. The slope of the curves at constant $\Omega_S h^2$ is 1/2, corresponding to the scaling $\lambda_{HS} \sim m_S^{-1/2}$; see Eq. (3.15).

We do not consider the phenomenology of this freeze-in scenario in more detail here, but rather turn to a model where a strong electroweak phase transition can be obtained simultaneously with the observed dark matter relic density. Moreover, the relic density can be produced either via thermal freeze-out or out-of-equilibrium freeze-in scenarios.

Fermion dark matter

In this section we consider a model where the SM matter content is extended with the singlet sector containing a real scalar and a fermion [12], [24] and [42]. The scalar sector is given by the full potential in Eq. (2.1), as we are not assuming Z_2 symmetry for the singlet scalar. Rather, in this model the dark matter candidate is the singlet fermion, which enters through the Lagrangian

$$\mathcal{L}_{\rm DM} = \bar{\psi}(i\partial\!\!\!/ - m)\psi + g_S S\bar{\psi}\psi. \tag{3.16}$$

We assume that the singlet fermion carries an exact global fermion number symmetry or that its mixing with SM neutrinos is otherwise forbidden. The main motivation for this model over the simple scalar case considered in previous section is that since the properties of the electroweak phase transition are affected by the scalar sector, and the dark matter relic density is determined by the properties of the singlet fermion, one can simultaneously explain both. As we will discuss in the following, the dark matter relic density can arise either via freeze-out or freeze-in mechanisms.

At T=0, the potential (2.1) has several local extrema, which either break or conserve electroweak symmetry. There are altogether three symmetric extrema (v=0), two of which are minima. We denote one of these by $S=w_0$. Then the two other extrema are at S=0 and $S=-w_0-\mu_3/\lambda_S$. Since we do not assume Z_2 symmetry for the singlet scalar, the electroweak broken minimum is at (v=246 GeV,w) where the vev of the singlet scalar, w, is not necessarily zero.

We trade the parameters μ_H^2, μ_S^2 and μHS with v,w and w_0 using the extremization conditions

$$\begin{aligned}\mu_H^2 &= \frac{-2\lambda_H v^4 + \lambda_{HS} v^2 w^2 + 4w^2(w-w_0)(\mu_3 + \lambda_S(w+w_0))}{2v^2},\\ \mu_S^2 &= -w_0(\mu_3 + \lambda_S w_0),\\ \mu_{HS} &= -\frac{w(\lambda_{HS}v^2 + 2(w-w_0)(\mu_3 + \lambda_S(w+w_0)))}{v^2}.\end{aligned} \tag{3.17}$$

In the electroweak symmetric minimum, the requirement that the eigenvalues of the Hessian matrix have to be positive gives

$$\begin{aligned}&w_0(\mu_3 + 2\lambda_S w_0) \geq 0,\\ &\lambda_{HS}v^2 + 4w\big(\mu_3 + \lambda_S(w+w_0)\big) \geq \frac{2\lambda_H v^4}{(w-w_0)^2}.\end{aligned} \tag{3.18}$$

Moreover, we trade the parameters w , λ_H and m with physical masses m_h=126 GeV, m_H and m_ψ so that finally the free parameters are m_ψ,m_H,λ_S,λHS,w_0,$\mu_3$and g_S. We perform Monte Carlo scan of the parameter space with

$$\begin{aligned}&0 < \lambda_S < \pi, \qquad -2\sqrt{\lambda_H\lambda_S} < \lambda_{HS} < \pi, \qquad -\pi < g_S < \pi,\\ &-4000\ \text{GeV} < \mu_3 < 4000\ \text{GeV}, \qquad -2000\ \text{GeV} < w_0 < 2000\ \text{GeV},\\ &0\ \text{GeV} < m_\psi < 800\ \text{GeV}, \qquad 200\ \text{GeV} < m_H < 1400\ \text{GeV}.\end{aligned} \tag{3.19}$$

In addition to the constraints listed above, we check that the electroweak broken minimum is the deepest one at T=0 and require perturbativity up to scales μ~1.5 TeV. The RG equations for the dimensionless couplings of the model are as follows. The β function of the Yukawa coupling, g_S, is

$$16\pi^2\beta_{g_S} = 5g_S^3, \tag{3.20}$$

while the β functions of the quartic couplings are

$$\begin{aligned}16\pi^2\beta_{\lambda_H} &= 24\lambda_H^2 + \frac{1}{2}\lambda_{HS}^2 - 3(3g^2 + g'^2 - 4y_t^2)\lambda_H\\ &\quad + \frac{3}{8}(3g^4 + 2g^2g'^2 + g'^4) - 6y_t^4,\end{aligned} \tag{3.21}$$

$$\begin{aligned}16\pi^2\beta_{\lambda_{HS}} &= 4\lambda_{HS}^2 + (12\lambda_H + 6\lambda_S)\lambda_{HS}\\ &\quad - 3\left(\frac{3}{2}g^2 + \frac{1}{2}g'^2 - 2y_t^2 - \frac{4}{3}g_S^2\right)\lambda_{HS},\end{aligned} \tag{3.22}$$

$$16\pi^2\beta_{\lambda_S} = 18\lambda_S^2 + 2\lambda_{HS}^2 + 8\lambda_S g_S^2 - 8g_S^4. \tag{3.23}$$

Note that these reproduce the results for the Z_2 symmetric scalar case considered in the previous section in the limit $g_S \to 0$.

Since we are not assuming Z_2 symmetry for the singlet S, fields S and ϕ_r^0 are not mass eigenstates. Rather the mass eigenstates are of the form (2.10), and the mixing pattern is constrained by the LHC data. As already discussed in Section 2, since the lighter mass eigenstate is identified with the m_h=126 GeV boson observed at the LHC, the presence of the singlet component constrains the allowed values of the mixing angle. Due to the simple mixing pattern, all couplings of the lighter mass eigenstate are suppressed by cosβ relative to the couplings of the SM Higgs boson. We perform a global fit to the current data taking signal strengths from the ATLAS, CMS and Tevatron experiments [33], [34], [35], [36], [37] and [38]. We then allow only parameters which are within 2σ limit, cosβ>0.85, of the best fit value, cosβ=0.95. We also impose constraints from the precision electroweak measurements on the oblique corrections, i.e. S and T parameters, using formulae given in [43]. For the experimental input we use S=0.04±0.09 and T=0.07±0.08 with correlation of 0.88 from [44], and accept only points which are within the 2σ ellipsis around the central value quoted above. The constraint from the invisible width is similar as in the case of Z_2 symmetric scalar. The essential difference is that now there are two channels contributing to the invisible width, $h^0 \to H^0H^0$ and $h^0 \to \psi\psi$,

$$\Gamma_{h^0 \to H^0 H^0} = \frac{\lambda_{hHH}^2}{32\pi m_h}\sqrt{1-\frac{4m_H^2}{m_h^2}}, \qquad \Gamma_{h^0 \to \psi\psi} = \frac{g_h^2 m_h}{8\pi}\left(1-\frac{4m_\psi^2}{m_h^2}\right)^{\frac{3}{2}}, \tag{3.24}$$

where $g_h = g_S \sin\beta$ and the coupling λ_{hHH} is given by Eq. (A.3).

To calculate the relic abundance of the singlet fermion, we again apply first the usual freeze-out formalism. The annihilation channels are similar to those of the Z_2-symmetric scalar case, as the annihilation to SM fields can again proceed only via the scalar portal interactions. The cross sections are given by the fairly complicated and unilluminating expressions collected in Appendix A.

The thermal corrections are given by Eqs. (2.4) and (2.5), with the additional contributions $g_S^2/12$ and $-g_S m/6$ to c_S and c_1, respectively, arising from the coupling between the singlet fermion and scalar.[3] Using the temperature dependent effective potential we again monitor the evolution of the electroweak symmetric and broken minima as T is increased from T=0. The critical temperature, T_c, is obtained from the condition that the symmetric minimum becomes deeper than the broken one. To check the strength of the phase transition, we evaluate $v(T_c)/T_c$.

The results on the dark matter abundance are shown in Fig. 3 as a function of the mass of the WIMP candidate, m_ψ, and its coupling with the singlet scalar, g_S. We see that a sizable fraction of the dark matter relic abundance can be realized with practically any mass larger than m_h, but the viable values of g_S are confined within narrow interval, $g_S \simeq O(0.1)$. On the other hand, the model now easily produces strong electroweak phase transition. This is shown in the inset of the left panel of Fig. 3, where the shaded lower (green) parts of the histogram correspond to points which yield f_{rel}>0.5.

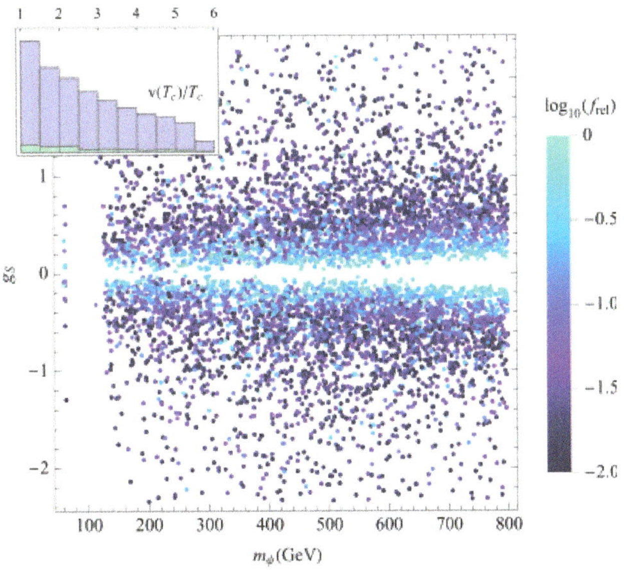

Figure 3: Dark matter density as a function of the dark matter mass m_ψ and the Yukawa coupling g_S. The inset shows the distribution of values of $v(T_c)/T_c$ corresponding to the points in the plot. The shaded lower portion of the histogram bars correspond to points which yield f_{rel}>0.5. The constraints from the perturbativity of couplings and LHC invisible Higgs width have been taken into account. Also, we show only the points which give $0.01<f_{rel}\leq 1$, $v(T_c)/T_c>1$ and T_c>40 GeV. All shown points are also compatible with the LUX constraints. (For interpretation of the references to color in this figure, the reader is referred to the web version of this article.)

All the points shown in Fig. 3 are compatible with the direct search constraint from LUX experiment. To compare with the direct searches, we again evaluate the spin-independent WIMP–nucleon cross section. Now, the interaction between the WIMP and nucleons contains the factors

$$\tilde{g}_h = g_S \sin\beta, \qquad \tilde{g}_H = g_S \cos\beta, \tag{3.25}$$

which leads to

$$\sigma_{SI}^0 = \frac{\mu_N^2 f_N^2 m_N^2}{\pi v^2} g_S^2 \sin^2\beta \cos^2\beta \left(\frac{1}{m_h^2} - \frac{1}{m_H^2}\right)^2. \tag{3.26}$$

A detailed account of the direct search limits from LUX experiment are shown in Fig. 4. The left panel shows the points from Fig. 3 in the (m_ψ, m_H) plane, with the color coding now corresponding to the magnitude of the relative cross section $\sigma_{SI}^{eff}/\sigma_{LUX}$. We see that the points furthest below the LUX bound are mostly concentrated around the region $m_H \sim 2m_\psi$. The right panel of Fig. 4 shows the distribution of the results with respect to the values of f_{rel} and $\sigma_{SI}^{eff}/\sigma_{LUX}$, with the color coding now corresponding to the values of g_S. Here we see the result already observed in Fig. 3 that the large values of the relic density are confined to values of $g_S \sim O(0.1)$ and that, in addition, the compatibility with the LUX bound is controlled by the mixing pattern in the scalar sector.

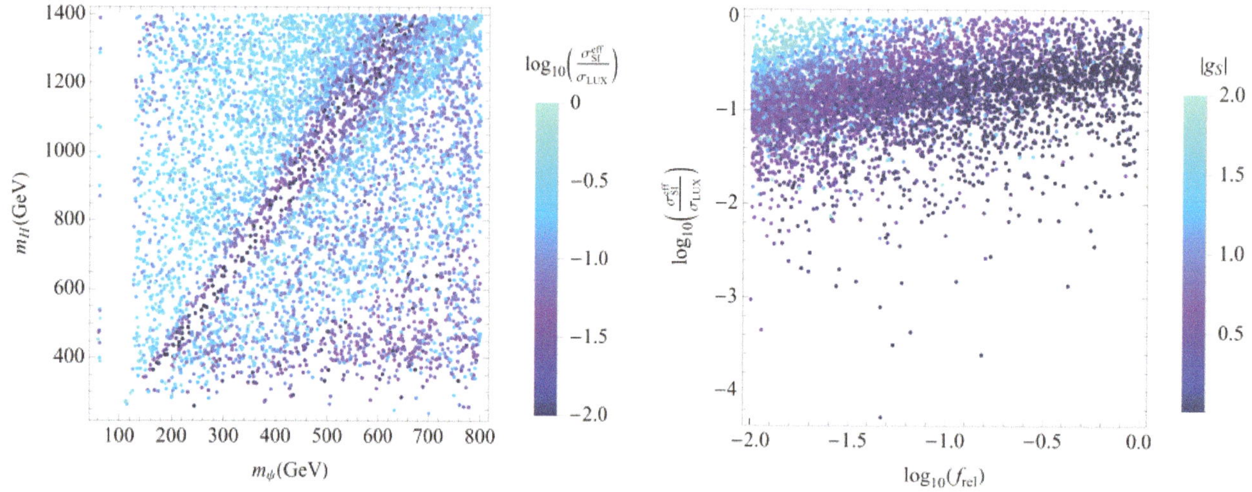

Figure 4: Left panel: the same data points as in Fig. 3 in the (m_ψ, m_H) plane. The color code now shows the magnitude of relative cross section $\sigma_{SI}^{eff}/\sigma_{LUX}$. Right panel: the data points as a function of f_{rel} and $\sigma_{SI}^{eff}/\sigma_{LUX}$. The color coding corresponds to the magnitude of the coupling g_S. In both panels the constraints from the perturbativity of couplings and LHC invisible Higgs width have been taken into account. Also, we show only the points which give $0.01 < f_{rel} \leq 1$, $v(T_c)/T_c > 1$ and $T_c > 40$ GeV and which are not excluded by the LUX data. (For interpretation of the references to color in this figure, the reader is referred to the web version of this article.)

To conclude this phenomenological analysis, let us again consider the dark matter relic density in the case of freeze in. Now the dark matter number density is produced from h^0 and H^0, which both contain the component of the singlet S which couples to the dark matter candidate ψ. In Fig. 5 we show the values of mass m_ψ and coupling g_S for which $\Omega_\psi h^2 = 0.12$ can be obtained. Of course, with such weak coupling the model in this part of the parameter space remains unconstrained by the dark matter direct search experiments. At all points shown in Fig. 5, also a strong electroweak phase transition can be realized. This is possible since the relic density is determined by the coupling g_S while the properties of the phase transition are controlled by the couplings in the scalar sector. Similarly to the scalar case considered earlier, also here the characteristic feature of the mechanism is superweak coupling, $\Omega_\psi h^2 \sim 10^{-12}\sqrt{\mathrm{GeV}/m_\psi}$, and the region where the mechanism can be applied is bounded by $m_\psi = m_H/2$.

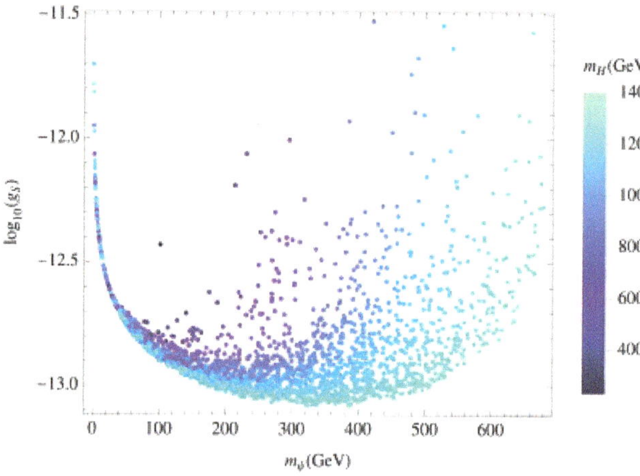

Figure 5: The relic density $f_{rel}=1$ produced via freeze-in mechanism as a function of dark matter mass m_ψ and Yukawa coupling g_S. For all points $f_{rel}=1$, $v(T_c)/T_c > 1$ and $T_c > 40$ GeV.

Stability

One of the intriguing properties of the Standard Model with a light Higgs, m_h=126 GeV, is that it is very near of being stable up to the Planck scale. In the SM the Higgs self-coupling becomes negative at one-loop level around energies $\mu \simeq 10^8$ GeV, but only so slightly that the theory remains in the region of metastability.[4] It is well known that extra singlets affect this situation, see e.g. [47]. For example, in our model we find that setting $\lambda_S(m_t)=g_S(m_t)=0$, the Higgs self-coupling remains positive for $\lambda HS(m_t) \gtrsim 0.4$ all the way up to the Planck scale. Currently there is some motivation for such stabilization from the measurement of the polarization of the cosmic microwave background by the BICEP-2 experiment [48]. The observation seems to be consistent with the tensor–scalar ratio $r = 0.16^{+0.06}_{-0.05}$ which, if due to gravitational waves produced during inflation, sets the scale of the energy density to be very large, of the order of $(10^{16}\ \mathrm{GeV})^4$, during inflation. This makes the computations of metastability with respect to quantum tunneling futile, since such large energy densities allow the Higgs field to classically roll into the instability; see e.g. [49], [50] and [51]. Possible solutions arising from modifying the inflationary dynamics were recently investigated in [52]. Of course another obvious solution arises if the beyond SM degrees of freedom force the scalar couplings to guarantee positivity of the potential at large field values up to the Planck scale. Let us, therefore, investigate the possibilities further in the model we have studied here.

The results are shown in Fig. 6. The shaded regions in the lower right corner correspond to parameter domains where the Higgs self-coupling becomes negative and signal the vacuum instability. The shaded regions in the upper right corner show the parameter domains where some of the couplings become non-perturbative. Finally, the horizontal lines show the values where the coupling λ_S is driven negative. For simplicity, the results of Fig. 6 depict the Z_2-symmetric case for which the Higgs self-coupling has the SM value at the electroweak scale, $\lambda_H(m_t)=0.128$. The left panel of Fig. 6 shows the dependence on $\lambda_S(m_t)$ at value of $g_S(m_t)=0.4$. The figure shows, in particular, how sensitive the perturbativity of the model is on the value of λ_S: if perturbativity is to be required all the way to the Planck scale, we must have $\lambda_S \lesssim 0.2$. The right panel of Fig. 6illustrates similarly the dependence on $g_S(m_t)$: increasing $g_S(m_t)$ above ca. 0.5 forces an instability on λ_S unless λHS becomes large, which in turn forces the theory into a non-perturbative domain.

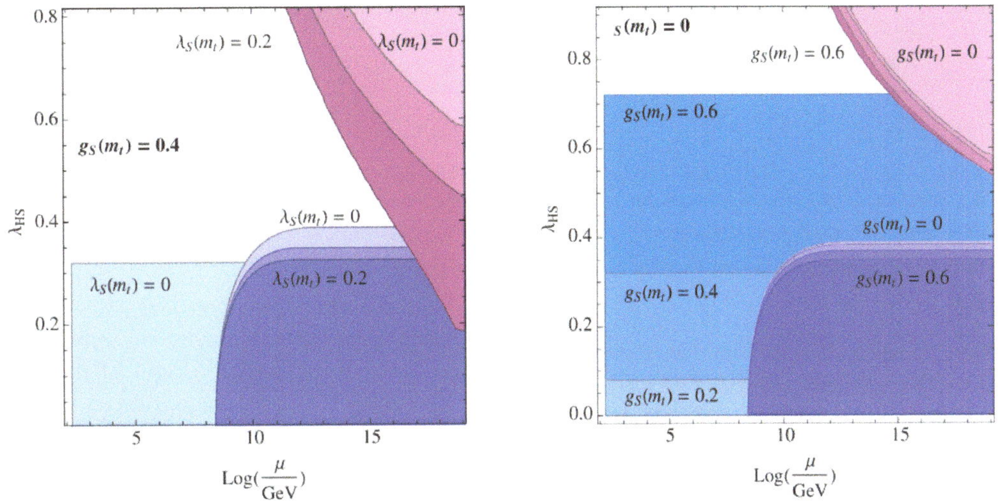

Figure 6: The figures show constraints from vacuum stability and perturbativity of the couplings. The contours in the lower right corner show the regions where the Higgs self-coupling becomes negative, while the contours in the upper right corner show the regions where one or more of the couplings become large. Finally, the horizontal contours correspond to λ_S becoming negative. In the left panel $g_S(m_t)=0.4$, and the contours show the dependence on $\lambda_S(m_t)$, while the right panel shows the dependence on $g_S(m_t)$ at $\lambda_S(m_t)=0$.

These results lead to a narrow range of couplings which correspond to a model which remains perturbative and stable up to the Planck scale. These ranges are $\lambda_S(m_t) \lesssim 0.2$, $g_S(m_t) \lesssim 0.6$ and $0.35 \lesssim \lambda HS \lesssim 0.55$.

A further effect on the stability is obtained by relaxing the Z_2 symmetry of the potential. Although allowing trilinear terms in the scalar potential does not affect the β functions of the quartic couplings, it does change the vacuum structure thereby mixing the singlet and the doublet scalars at electroweak scale. This in turn changes the value of the coupling λ_H at the electroweak scale when compared with the SM value. We do a random scan over the parameter space to illustrate that within our model it is possible to obtain a considerable amount of the dark matter and simultaneously to have a strong electroweak phase transition with a stable potential all the way up to the Planck scale. As a result, we find a benchmark point with

$$\lambda_H(m_t) = 0.197, \qquad \lambda_S(m_t) = 0.053,$$
$$\lambda_{HS}(m_t) = 0.376, \qquad g_S(m_t) = 0.247, \tag{3.27}$$

for which $f_{rel}=0.94$ and $v(T_c)/T_c=1.3$.

Finally, let us discuss possible implications for the LHC. The above benchmark point serves as a concrete example. At this point the singlet fermion mass is m_ψ=679 GeVand the heavy scalar mass is m_H=371 GeV. Both of these are above $m_h/2$ so the hidden sector will remain invisible with zero branching fraction. However, even if the new degrees of freedom are not expected to manifest at LHC, the model predicts deviations with respect to SM. As already mentioned above, the mixing of the singlet and doublet scalars affects the coupling λ_H, and this is directly connected with the scalar potential. In the SM the Higgs self-coupling has value $\lambda_{SM}(m_t)=0.128$, while for example for the parameter space point in Eq. (3.27) we have $\lambda_H(m_t)=0.197=1.5\lambda_{SM}(m_t)$.

To search for deviations from the SM all the Higgs couplings should be measured and compared with the predicted SM values. The cubic coupling of three physical Higgs bosons, λ_{hhh}, could be measured at LHC in the production of two Higgs bosons, see e.g. [53]. For the benchmark point, Eq. (3.27), we have $\lambda_{hhh}/\lambda_{hhh}^{SM}=2.00$. Generally, the dependence of $\lambda_{hhh}/\lambda_{hhh}^{SM}$ on m_S and singlet-doublet mixing cosβ within the allowed 2σrange in the model is shown in the left panel of Fig. 7. We find that sizable deviations from the SM value are indeed expected. The growth of the deviation can be traced to the dimensionful couplings, in particular to μ_3, whose values can be large in comparison to the electroweak scale, v. On the other hand, the large intrinsic scales will also affect the Sand T parameters through the mixing of singlet and doublet scalars. This is illustrated by the right panel of Fig. 7, which shows the trilinear Higgs coupling with respect to the deviation, Δ, from the best fit value of the oblique parameters S and T , the value Δ=1corresponding to the 2σ limit. All our data points are naturally within this constraint by construction.

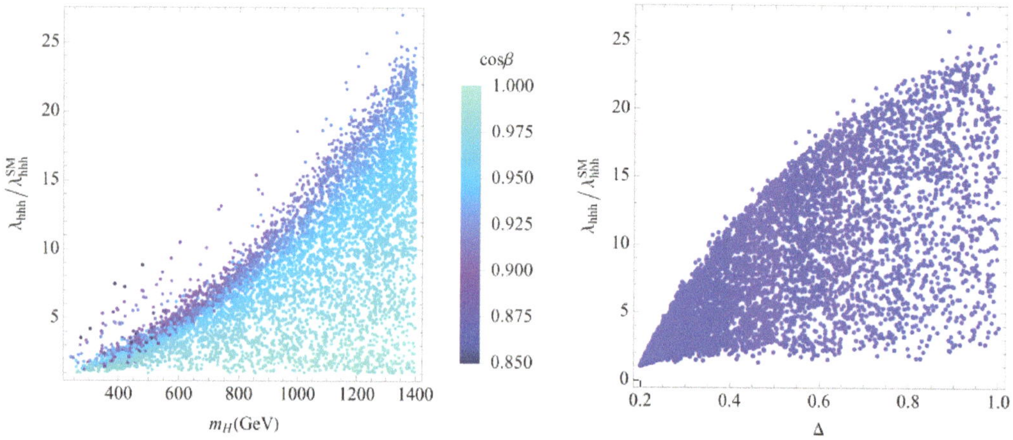

Figure 7: Left panel: the value of the trilinear Higgs coupling, λ_{hhh}, in the model as a function of m_H and cos β. The points correspond to the ones shown in Fig. 3. Right panel: the value of the trilinear Higgs coupling as a function of the deviation Δ from the optimal values of the fit to the S and T parameters. The value Δ = 1 corresponds to the 2σ limit.

CONCLUSIONS

In this paper we have considered simple singlet extensions of the SM motivated by the possibility to explain cosmological paradigms of dark matter and electroweak baryogenesis. For the latter we have focused on the possibility to realize a strong electroweak phase transition which is a necessary general condition for any particular realization of creating the matter–antimatter asymmetry.

We considered two concrete realizations. First we studied a model where the dark matter is a singlet scalar protected by a discrete Z_2 symmetry in the zero temperature vacuum. Second, we considered a model where the dark matter candidate is a singlet fermion coupled with the SM via a singlet scalar. In this case the scalar potential can be arranged to lead to strong electroweak phase transition while the singlet fermion saturates the observed dark matter abundance. In this case, the dark matter abundance can be produced either via freeze-out or freeze-in mechanisms.

Between these two models, the main difference which is relevant for the phenomenological aspects of dark matter and strong first-order phase transition is that in the Z_2-symmetric-scalar case both aspects arise from the same field, whereas in the fermion case the dark matter abundance is due to the fermion while the modifications to the electroweak transition are entirely due to the portal scalar. One can also imagine alternative realizations based on similar dichotomy. For example, one could consider two different scalar fields. Imposing a discrete symmetry on only one of them would make the field a plausible origin of dark matter similar to the Z_2-symmetric case we considered here, while the other field would provide the necessary modifications to strengthen the electroweak phase transition.

We also studied the implications of the RG equations on the stability and perturbativity of the model. We found that due to the singlet scalar, the Higgs self-coupling can remain positive at all energies below the Planck scale. With suitable values of the portal coupling, λHS, the self-coupling of the singlet scalar, λ_S, and the scalar Yukawa coupling, g_S, the entire model was found to remain perturbative and stable all the way up to the Planck scale.

A possible experimental access to the singlet sector is provided via the couplings of the Higgs boson, in particular the trilinear coupling of three physical Higgs bosons. If measured at LHC, and deviations from the corresponding value in the SM are found, this can provide, if no new states are directly observed at the LHC experiments, important clues about a possible hidden sector akin to the models studied in this paper.

ACKNOWLEDGEMENTS

This work has been financially supported by Finnish Cultural Foundation and by the Academy of Finland project 267842.

APPENDIX A. CROSS SECTIONS

Here we give the formulae for the computation of the annihilation cross section for the model considered in Section 3.2. To make the equations more concise, it is useful to define the couplings

$$\lambda_{hhh} = -6\lambda_H v\cos^3\beta - 3\lambda_{HS}v\sin^2\beta\cos\beta - 2\sin^3\beta(\mu_3 + 3\lambda_S w) \\ - 3\sin\beta\cos^2\beta(\mu_H + \lambda_{HS}w), \tag{A.1}$$

$$\lambda_{hhH} = -2v\sin\beta\cos^2\beta(\lambda_{HS} - 3\lambda_H) + \lambda_{HS}v\sin^3\beta \\ + \sin\beta\sin(2\beta)(-\mu_3 + \mu_H - 3\lambda_S w + \lambda_{HS}w) - \cos^3\beta(\mu_H + \lambda_{HS}w), \tag{A.2}$$

$$\lambda_{hHH} = v\sin\beta\sin(2\beta)(\lambda_{HS} - 3\lambda_H) - \lambda_{HS}v\cos^3\beta \\ - 2\sin\beta\cos^2\beta(\mu_3 - \mu_H + 3\lambda_S w - \lambda_{HS}w) - \sin^3\beta(\mu_H + \lambda_{HS}w), \tag{A.3}$$

$$\lambda_{HHH} = 6v\sin^2\beta\cos\beta(\lambda_{HS} - 3\lambda_H) - 3\lambda_{HS}v\cos^3\beta \\ - 6\sin\beta\cos^2\beta(\mu_3 - \mu_H + 3\lambda_S w - \lambda_{HS}w) - 3\sin^3\beta(\mu_H + \lambda_{HS}w), \tag{A.4}$$

and

$$g_h = g_S\sin\beta, \qquad g_H = g_S\cos\beta, \qquad Y_h = \frac{m_f}{v}\cos\beta, \qquad Y_H = -\frac{m_f}{v}\sin\beta,$$
$$g_{hZ} = \frac{v(g^2+g'^2)}{2}\cos\beta, \qquad g_{hW} = \frac{vg^2}{2}\cos\beta,$$
$$g_{HZ} = -\frac{v(g^2+g'^2)}{2}\sin\beta, \qquad g_{HW} = -\frac{vg^2}{2}\sin\beta. \tag{A.5}$$

The squared amplitude, averaged over the initial states and summed over the final states, for the dark matter, ψ , annihilating to two scalars, $h_i,h_j=h^0,H^0$, is given by

$$|T(\psi\bar\psi\to h_ih_j)|^2 = \left(1-\frac{\delta_{ij}}{2}\right)\left(2(s-4m_\psi^2)\left|\sum_{k=h,H}\frac{g_k\lambda_{ijk}}{s-m_k^2+im_k\Gamma_k}\right|^2\right. \\ + 2g_i^2g_j^2(tu - m_\psi^2(t+u) + m_\psi^4 - m_i^2m_j^2)\left(\frac{1}{t-m_\psi^2} - \frac{1}{u-m_\psi^2}\right)^2 \\ \left. + 4m_\psi(t-u)\left(\frac{1}{t-m_\psi^2} - \frac{1}{u-m_\psi^2}\right)\sum_{k=h,H}\frac{g_ig_jg_k\lambda_{ijk}(s-m_k^2)}{|s-m_k^2+im_k\Gamma_k|^2}\right). \tag{A.6}$$

Similarly the squared amplitudes, averaged over the initial states and summed over the final states, to fermion and gauge boson final states are

$$|T(\psi\bar\psi\to ff)|^2 = s^2v_\psi^2v_f^2X_f\left|\sum_{k=h,H}\frac{g_kY_k}{s-m_k^2+im_k\Gamma_k}\right|^2, \tag{A.7}$$

and

$$|T(\psi\bar\psi\to VV)|^2 = 2sv_\psi^2\left(3 - \frac{s}{M_V^2} + \frac{s^2}{4M_V^4}\right)\left|\sum_{k=h,H}\frac{g_kg_{kV}}{s-m_k^2+im_k\Gamma_k}\right|^2\delta_V. \tag{A.8}$$

In the expression for the fermion channel the factor $X_f=1$ for leptons in the final state, while for quarks

$$X_f = 3\left(1 + \left(\frac{3}{2}\ln\frac{m_q^2}{s} + \frac{9}{4}\right)\frac{4\alpha_s}{3\pi}\right), \tag{A.9}$$

where $\alpha_s=0.12$ is the strong coupling constant. Similarly as in the scalar case [23], also for fermion dark matter the region where this correction is important is ruled out by the invisible width of the Higgs.

Then, the cross section for the process $\psi\bar\psi\to ij$ reads

$$\sigma(\psi\psi \to ij) = \frac{1}{16\pi s^2 v_\psi^2} \int_{t_-^{(ij)}}^{t_+^{(ij)}} dt \left|T(\psi\psi \to ij)\right|^2, \tag{A.10}$$

where

$$t_\pm^{(ij)} = m_\psi^2 + \frac{1}{2}(m_i^2 + m_j^2 - s) + \frac{1}{2}\sqrt{\left(1 - \frac{4m_\psi^2}{s}\right)\left(s^2 + (m_i^2 - m_j^2)^2 - 2s(m_i^2 + m_j^2)\right)}. \tag{A.11}$$

REFERENCES

1. D. Carmi, A. Falkowski, E. Kuflik, T. Volansky, Interpreting LHC Higgs results from natural new physics perspective, J. High Energy Phys. 1207 (2012) 136, arXiv:1202.3144.
2. J. Espinosa, C. Grojean, M. Muhlleitner, M. Trott, Fingerprinting Higgs suspects at the LHC, J. High Energy Phys. 1205 (2012) 097, arXiv:1202.3697.
3. P.P. Giardino, K. Kannike, M. Raidal, A. Strumia, Reconstructing Higgs boson properties from the LHC and Tevatron data, J. High Energy Phys. 1206 (2012) 117, arXiv:1203.4254.
4. T. Alanne, S. Di Chiara, K. Tuominen, LHC data and aspects of new physics, J. High Energy Phys. 1401 (2014) 041, arXiv:1303.3615.
5. ATLAS Collaboration, G. Aad, et al., Observation of a new particle in the search for the Standard Model Higgs boson with the ATLAS detector at the LHC, Phys. Lett. B 716 (2012) 1–29, arXiv:1207.7214.
6. CMS Collaboration, S. Chatrchyan, et al., Observation of a new boson at a mass of 125 GeV with the CMS experiment at the LHC, Phys. Lett. B 716 (2012) 30–61, arXiv:1207.7235.
7. J. McDonald, Gauge singlet scalars as cold dark matter, Phys. Rev. D 50 (1994) 3637–3649, arXiv:hep-ph/0702143.
8. C. Burgess, M. Pospelov, T. ter Veldhuis, The minimal model of nonbaryonic dark matter: a singlet scalar, Nucl. Phys. B 619 (2001) 709–728, arXiv:hep-ph/0011335.
9. L. Lopez Honorez, E. Nezri, J.F. Oliver, M.H. Tytgat, The inert doublet model: an archetype for dark matter, J. Cosmol. Astropart. Phys. 0702 (2007) 028, arXiv:hep-ph/0612275.
10. J. Ruiz-Alvarez, C. de S. Pires, F.S. Queiroz, D. Restrepo, P. Rodrigues da Silva, On the connection of gamma-rays, dark matter and Higgs searches at LHC, Phys. Rev. D 86 (2012) 075011, arXiv:1206.5779.
11. L. Lopez-Honorez, T. Schwetz, J. Zupan, Higgs portal, fermionic dark matter, and a Standard Model like Higgs at 125 GeV, Phys. Lett. B 716 (2012) 179–185, arXiv:1203.2064.
12. M. Fairbairn, R. Hogan, Singlet fermionic dark matter and the electroweak phase transition, J. High Energy Phys. 1309 (2013) 022, arXiv:1305.3452.
13. A. Alves, S. Profumo, F.S. Queiroz, The dark Z portal: direct, indirect and collider searches, J. High Energy Phys. 1404 (2014) 063, arXiv:1312.5281.
14. T. Hambye, Hidden vector dark matter, J. High Energy Phys. 0901 (2009) 028, arXiv:0811.0172.
15. H. Davoudiasl, I.M. Lewis, Dark matter from hidden forces, Phys. Rev. D 89 (2014) 055026, arXiv:1309.6640.
16. V. Kuzmin, V. Rubakov, M. Shaposhnikov, On the anomalous electroweak Baryon number nonconservation in the early universe, Phys. Lett. B 155 (1985) 36.
17. K. Kajantie, M. Laine, K. Rummukainen, M.E. Shaposhnikov, Is there a hot electroweak phase transition at m(H) larger or equal to m(W)? Phys. Rev. Lett. 77 (1996) 2887–2890, arXiv:hep-ph/9605288.
18. K. Rummukainen, M. Tsypin, K. Kajantie, M. Laine, M.E. Shaposhnikov, The universality class of the electroweak theory, Nucl. Phys. B 532 (1998) 283–314, arXiv:hep-lat/9805013.
19. S. Profumo, M.J. Ramsey-Musolf, G. Shaughnessy, Singlet Higgs phenomenology and the electroweak phase transition, J. High Energy Phys. 0708 (2007) 010, arXiv:0705.2425.
20. J.R. Espinosa, T. Konstandin, F. Riva, Strong electroweak phase transitions in the standard model with a singlet, Nucl. Phys. B 854 (2012) 592–630, arXiv:1107.5441.
21. J. McDonald, Thermally generated gauge singlet scalars as selfinteracting dark matter, Phys. Rev. Lett. 88 (2002) 091304, arXiv:hep-ph/0106249.
22. J.M. Cline, K. Kainulainen, Electroweak baryogenesis and dark matter from a singlet Higgs, J. Cosmol. Astropart. Phys. 1301 (2013) 012, arXiv:1210.4196.
23. J.M. Cline, K. Kainulainen, P. Scott, C. Weniger, Update on scalar singlet dark matter, Phys. Rev. D 88 (2013) 055025, arXiv:1306.4710.
24. T. Li, Y.-F. Zhou, Strongly first order phase transition in the singlet fermionic dark matter model after LUX, arXiv:1402.3087.
25. XENON100 Collaboration, E. Aprile, et al., Dark matter results from 100 live days of XENON100 data, Phys. Rev. Lett. 107 (2011) 131302, arXiv:1104.2549.
26. LUX Collaboration, D. Akerib, et al., First results from the LUX dark matter experiment at the Sanford underground research facility, Phys. Rev. Lett. 112 (2014) 091303, arXiv:1310.8214.

27. G. Duda, G. Gelmini, P. Gondolo, Detection of a subdominant density component of cold dark matter, Phys. Lett. B 529 (2002) 187–192, arXiv:hep-ph/0102200.

28. S. Profumo, K. Sigurdson, L. Ubaldi, Can we discover multi-component WIMP dark matter? J. Cosmol. Astropart. Phys. 0912 (2009) 016, arXiv:0907.4374.

29. M. Aoki, M. Duerr, J. Kubo, H. Takano, Multi-component dark matter systems and their observation prospects, Phys. Rev. D 86 (2012) 076015, arXiv:1207.3318.

30. B.W. Lee, S. Weinberg, Cosmological lower bound on heavy neutrino masses, Phys. Rev. Lett. 39 (1977) 165–168.

31. P. Gondolo, G. Gelmini, Cosmic abundances of stable particles: improved analysis, Nucl. Phys. B 360 (1991) 145–179.

32. Planck Collaboration, P. Ade, et al., Planck 2013 results. XVI. Cosmological parameters, arXiv:1303.5076.

33. ATLAS Collaboration, Updated coupling measurements of the Higgs boson with the ATLAS detector using up to 25 fb−1 of proton–proton collision data, ATLAS-CONF-2014-009, 2014.

34. CMS Collaboration, S. Chatrchyan, et al., Search for the standard model Higgs boson produced in association with a W or a Z boson and decaying to bottom quarks, Phys. Rev. D 89 (2014) 012003, arXiv:1310.3687.

35. CMS Collaboration, S. Chatrchyan, et al., Evidence for the 125 GeV Higgs boson decaying to a pair of τ leptons, J. High Energy Phys. 1405 (2014) 104, arXiv:1401.5041.

36. CMS Collaboration, Updated measurements of the Higgs boson at 125 GeV in the two photon decay channel, CMS-PAS-HIG-13–CMS-PAS-HIG-001, 2013.

37. CMS Collaboration, S. Chatrchyan, et al., Measurement of Higgs boson production and properties in the WW decay channel with leptonic final states, J. High Energy Phys. 1401 (2014) 096, arXiv:1312.1129.

38. CMS Collaboration, S. Chatrchyan, et al., Measurement of the properties of a Higgs boson in the four-lepton final state, Phys. Rev. D 89 (2014) 092007, arXiv:1312.5353.

39. M.E. Peskin, T. Takeuchi, A new constraint on a strongly interacting Higgs sector, Phys. Rev. Lett. 65 (1990) 964–967.

40. LHC Higgs Cross Section Working Group Collaboration, S. Dittmaier, et al., Handbook of LHC Higgs cross sections: 1. Inclusive observables, arXiv:1101.0593.

41. L.J. Hall, K. Jedamzik, J. March-Russell, S.M. West, Freeze-in production of FIMP dark matter, J. High Energy Phys. 1003 (2010) 080, arXiv:0911.1120.

42. M.A. Fedderke, J.-Y. Chen, E.W. Kolb, L.-T. Wang, The fermionic dark matter Higgs portal: an effective field theory approach, arXiv:1404.2283.

43. W. Grimus, L. Lavoura, O. Ogreid, P. Osland, The oblique parameters in multi-Higgs-doublet models, Nucl. Phys. B 801 (2008) 81–96, arXiv:0802.4353.

44. Particle Data Group Collaboration, J. Beringer, et al., Review of particle physics (RPP), Phys. Rev. D 86 (2012) 010001.

45. G. Degrassi, S. Di Vita, J. Elias-Miro, J.R. Espinosa, G.F. Giudice, et al., Higgs mass and vacuum stability in the Standard Model at NNLO, J. High Energy Phys. 1208 (2012) 098, arXiv:1205.6497.

46. O. Antipin, M. Gillioz, J. Krog, E. Mølgaard, F. Sannino, Standard model vacuum stability and Weyl consistency conditions, J. High Energy Phys. 1308 (2013) 034, arXiv:1306.3234.

47. N. Haba, H. Ishida, K. Kaneta, R. Takahashi, Vanishing Higgs potential at the Planck scale in singlets extension of the standard model, arXiv:1406.0158.

48. BICEP2 Collaboration, P. Ade, et al., Detection of B-mode polarization at degree angular scales by BICEP2, Phys. Rev. Lett. 112 (2014) 241101, arXiv:1403.3985.

49. J. Espinosa, G. Giudice, A. Riotto, Cosmological implications of the Higgs mass measurement, J. Cosmol. Astropart. Phys. 0805 (2008) 002, arXiv:0710.2484.

50. A. Kobakhidze, A. Spencer-Smith, Electroweak vacuum (in)stability in an inflationary universe, Phys. Lett. B 722 (2013) 130–134, arXiv:1301.2846.

51. A. Spencer-Smith, Higgs vacuum stability in a mass-dependent renormalisation scheme, arXiv:1405.1975.

52. M. Fairbairn, R. Hogan, Electroweak vacuum stability in light of BICEP2, Phys. Rev. Lett. 112 (2014) 201801, arXiv:1403.6786.

53. V. Barger, L.L. Everett, C. Jackson, G. Shaughnessy, Higgs-pair production and measurement of the triscalar coupling at LHC(8, 14), Phys. Lett. B 728 (2014) 433–436, arXiv:1311.2931.

CHAPTER 5

GRAVITATIONAL FORM FACTORS AND NUCLEON SPIN STRUCTURE

O. V. Teryaev

Keywords: Gravity, form factors, equivalence principle

INTRODUCTION

The gravitational form factors (GFFs) are the matrix elements of the energy-momentum tensor (EMT). These objects describing the interaction with fermions probably first appeared in the seminal paper of Kobzarev and Okun in which the equivalence principle (EP) for spin motion was first identified [1]. Note that as spin is essentially a quantum concept, this paper was probably (one of) the first discussions of the interaction of classical gravity with quantum objects. According to the EP, the anomalous gravitomagnetic moment (AGM), which is the gravitational analog of the anomalous magnetic moment, goes to zero.

This result was soon derived starting from conservation of momentum and angular momentum [2], which manifests the appearance of the EP as a low-energy theorem [3]. Here the axiomatic aspect of the EP is in fact transferred to the postulation that the EMT is a current coupled to gravity.

The particular case of strongly interacting particles was considered by Pagels [4]. Later, when energymomentum matrix elements were considered as the moments of generalized parton distributions (GPDs), the conservation laws mentioned above led to Ji's sum rules [5, 6], which control the partition of momentum and total angular momentum between quarks and gluons.

The consideration of EMT hadronic matrix elements as couplings to gravity [7, 8] provided another interpretation of these sum rules and opened the possibility of probing the gravitational couplings of quarks and gluons separately. This, with special emphasis on the applications of GFFs to gravitational and cosmological problems, is the main subject of the current paper.

GFFS OF NUCLEON AND PARTON ANGULAR MOMENTA

Let us define the matrix elements of the EMT as

$$\langle p'|T^{\mu\nu}_{q,g}|p\rangle = \bar{u}(p')[A_{q,g}(\Delta^2)\gamma^{(\mu}p^{\nu)} + B_{q,g}(\Delta^2)P^{(\mu}i\sigma^{\nu)\alpha}\Delta_\alpha/2M]u(p), \tag{1}$$

where $P^\mu = (p^\mu + (p')^\mu)/2$, $\Delta^\mu = (p')^\mu - p^\mu$, and u(p) is the nucleon spinor. We omit, for the moment, the terms of higher order in Δ, as well as those containing $g^{\mu\nu}$, which will be discussed later. The parton momenta and total angular momenta are just

$$P_{q,g} = A_{q,g}(0),$$
$$J_{q,g} = \frac{1}{2}\left[A_{q,g}(0) + B_{q,g}(0)\right]. \tag{2}$$

Taking into account conservation of momentum and angular momentum,

$$A_q(0) + A_g(0) = 1, \tag{3}$$

$$A_q(0) + B_q(0) + A_g(0) + B_g(0) = 1, \tag{4}$$

one can see that the difference between partition of the momentum and that of the orbital angular momentum arises entirely from "anomalous" form factors [$B_q(0) = -B_g(0)$].

Citation: O. V. Teryaev, Gravitational form factors and nucleon spin structure, DOI 10.1007/s11467-016-0573-6.

This property has a seemingly unexpected counterpart in another field of physics [7]. Namely, using the fact that the matrix element (1) describes the interaction of a nucleon with the classical external gravitational field, one arrives at the interpretation of B as an AGM that is the straightforward analog of its electromagnetic (EM) counterpart. The natural extension of the well-known Einstein EP results in zero AGM.

NUCLEONS IN AN EXTERNAL GRAVITATIONAL FIELD

It is instructive to examine the GFFs (which are straightforward analogs of EM ones) by considering [7] the action of EM and gravity fields uniformly and comparing them. The presentation follows the textbook derivation for the EM case (see [9, 10], section 116).

Let us start with the more common case of the interaction with an electromagnetic field, which is described by the matrix element of the EM current,

$$M = -\langle P'|J^{\mu}_{q}|P\rangle A_{\mu}(\Delta). \tag{5}$$

This matrix element at zero momentum transfer is fixed by the fact that the interaction is due to the local U(1) symmetry, whose global counterpart produces the conserved charge [and of course depends on the normalization of eigenvectors $\langle P|P'\rangle = (2\pi)^3 2E\delta(\boldsymbol{P} - \boldsymbol{P}')$].

$$\langle P|J^{\mu}_{q}|P\rangle = 2e_q P^{\mu}. \tag{6}$$

Thus, in the rest frame, the interaction is defined completely by the scalar potential:

$$M_0 = -\langle P|J^{\mu}_{q}|P\rangle A_{\mu} = -2e_q M\phi(\Delta). \tag{7}$$

At the same time, the interaction with the weak classical gravitational field is

$$M = -\frac{1}{2}\sum_{q,G}\langle p'|T^{\mu}_{q,G}|p\rangle h_{\mu\nu}(\Delta), \tag{8}$$

where h is the deviation of the metric tensor from its Minkowski value. The relative factor 1/2, which will play a crucial role, arises from the fact that the variation of the action with respect to the metric produces an EMT with the coefficient 1/2, whereas the variation with respect to a classical source A^{μ} produces a current without this coefficient. It is this coefficient that guarantees the correct value in the Newtonian limit, which is fixed by the global translational invariance

$$\sum_{q,G}\langle P|T^{\mu\nu}_{i}|P\rangle = 2P^{\mu}P^{\nu}, \tag{9}$$

which, together with the approximation for h (where the factor of 2 has a geometrical origin) [11, 12],

$$h_{00}(x) = 2\phi(x), \tag{10}$$

results in the rest frame expression:

$$M_0 = -\frac{1}{2}\sum_{q,G}\langle P|T^{\mu\nu}_{i}|P\rangle h_{\mu\nu}(\Delta) = -2M\cdot M\phi(\Delta), \tag{11}$$

where we used the same notation for the gravitational and scalar EM potentials and identified the normalization factor as 2M in order to make the similarity between (7) and (11) obvious. One can see that the interaction with a gravitational field is described by the charge, which is equal to the particle mass; this is just the EP. It appears here as a low-energy theorem rather than a postulate. The similarity with the EM case allows one to clarify the origin of this theorem, suggesting that the interaction with gravity is due to the local counterpart of global symmetry, although it may be proved starting just from the Lorentz invariance of the soft graviton approximation [3]. Here the axiomatic aspect of the EP is in fact transferred to the postulation of the EMT as a current coupled to gravity.

The situation for the terms linear in Δ is different for electromagnetism and gravity. Whereas these terms are defined by the specific dynamics in the EM case, producing the anomalous magnetic moment, the similar contribution in the gravitational case is entirely fixed by angular momentum conservation (4).1) This means that, in terms of the gravitational interaction, the AGM of any particle is identically equal to zero.

One can generalize this pioneering result of Kobzarev and Okun and show [7] that it is not restricted to the nucleon or spin-1/2 Dirac particle. The presence of Dirac spinors in the parametrization (1) is actually not crucial. To show this, it is convenient to use the equation of motion in order to attribute all the Δ dependence to the anomalous form factor $P^{\mu}\bar{u}\sigma^{\nu\alpha}u\Delta_{\alpha}$. After the linear Δ dependence is extracted, the spinors can be taken at the same momentum, which is a convenient choice for the average one, P, and calculation of the matrix element is reduced to the trace of the density matrix:

$$\begin{aligned}\bar{u}(P)\sigma^{\nu\alpha}\Delta_{\alpha}u(p) &= Tr\rho(P)i\sigma^{\nu\alpha}\Delta_{\alpha}\\ &= Tr\frac{1}{2}(\hat{P}+M)(1+\hat{S}\gamma_5)i\sigma^{\nu\alpha}\Delta_{\alpha}\\ &= 2i\epsilon^{\rho\sigma\nu\alpha}P_{\rho}S^{\sigma}\Delta^{\alpha}.\end{aligned} \tag{12}$$

By considering the matrix element of the projection of the Pauli–Lubanski operator, the constraint (4) may now be easily generalized to a particle of any spin; thus, the total conserved EMT of all the constituents is

$$\langle p'|\sum T^{\mu\nu}|p\rangle = 2P^{\mu}P^{\nu} + iP^{(\mu}\epsilon^{\nu)\sigma\rho\alpha}P_{\rho}S^{\sigma}\Delta^{\alpha}/M. \tag{13}$$

As in the spin-1/2 case, S is the average spin of the states $|P\rangle, |P'\rangle$ (they should enter in the symmetric way because of the positive charge parity of the EMT). As the form factors in the spin-1/2 case differ from those for the matrix element of the vector current J^{μ} by the common factor P^{ν}, one may define the gyrogravitomagnetic ratio in the same way as the common gyromagnetic ratio, and it should have the Dirac value g = 2 for a particle of any spin J:

$$\mu_G = J, \tag{14}$$

which coincides with the standard Dirac magnetic moment up to the interchange e ↔ M, making the Bohr magneton equal to 1/2.

However, the situation changes if one defines the gyrogravitomagnetic moment as a response to an external gravitomagnetic field. The ε tensor in coordinate space produces the curl, and the gravitomagnetic field acting on the particle spin is equal to

$$\boldsymbol{H}_J = \frac{1}{2} rot \boldsymbol{g}; \ \boldsymbol{g}_i \equiv g_{0i}, \tag{15}$$

where the factor 1/2 is just the normalization factor in Eq. (10) mentioned above. The relevant off-diagonal components of the metric tensor may be generated by rotation of a massive gravity source [11, 12].

This field also induces another effect: the straightforward analog of the Lorentz force [11, 12] produced by the first (spin-independent) term in (14). In that case, the gravitomagnetic field for low-velocity particles (this restriction is actually inessential, as we can always perform a Lorentz boost, reducing the particle velocity sufficiently) is

$$\boldsymbol{H}_L = rot \boldsymbol{g} = 2\boldsymbol{H}_G. \tag{16}$$

Consider now the motion of the particle in the gravitomagnetic field. The effect of the Lorentz force is reduced according to the Larmor theorem (which is also valid for low velocities) to rotation with the Larmor frequency

$$\omega_L = \frac{H_L}{2}. \tag{17}$$

This is also the frequency of the macroscopic gyroscope dragging. At the same time, the microscopic particle dragging frequency is

$$\omega_J = \frac{\mu_G}{J} H_J = \frac{H_L}{2} = \omega_L. \tag{18}$$

The common frequency for microscopic and macroscopic gyroscopes is just the Larmor frequency; thus, the effect of the gravitomagnetic field is equivalent to frame rotation. This should be considered as a (post-Newtonian) manifestation of the EP.

Our approach clarifies the origin of this equality as a cancellation of the geometrical factor 1/2 in Eq. (12) and the "quantum" value 2 of the gyrogravitomagnetic ratio. Note that for a free particle, the latter coincides with the usual gyromagnetic ratio, and such a cancellation provides an interesting connection between geometry, the EP, and the special renormalization properties (cancellation of the strongest divergencies) for particles with g = 2. 2) Another interesting connection is provided by the fact that it is just the deviation from g = 2 that determines the Gerasimov–Drell–Hearn sum rule for a particle with arbitrary spin [14].

The equality of precession of classical and quantum rotators was confirmed by analysis of the Dirac equations in external gravitational and inertial fields starting from the rotating frame [15] and in weak gravity [16] and later by generalization to the case of rotating bodies in strong and even arbitrary fields [17–19]. During this analysis, it was observed [16] that the definition of the EP as equal motion of a test body in a gravity field and the noninertial frame, which is frequently given in textbooks, is actually valid only for nonrelativistic motion. This can easily be shown using the GFF method.

The particle momentum evolution is fully determined by the forward matrix element fixed by momentum conservation (9), and the matrix element (11) for a particle at rest takes the form

$$\mathcal{M} = M^2 h_{00}(q). \tag{19}$$

The coincidence of the (00) components of the metric in the gravitational field and accelerated frame proves the EP.

At the same time, for a moving particle, the space components of the metric $h_{zz} = h_{xx} = h_{yy}$ = h00 (see, e.g., [11, 12])3) also contribute. As a result, the matrix elements in the gravitational field (M_g) and in the accelerated frame (M_a) differ by the obvious kinematic factor:

$$M_g = (\epsilon^2 + (\boldsymbol{p})^2) h_{00}(q), \quad M_a = \epsilon^2 h_{00}(q). \tag{20}$$

The ratio of these matrix elements is

$$R = \frac{\epsilon^2 + (\boldsymbol{p})^2}{\epsilon^2}. \tag{21}$$

This expression for ultrarelativistic particles provides just the well-known factor of 2 that is the difference between the Einsteinian and Newtonian expressions for light deflection. This means that the problem being discussed, despite some misleading statements in the literature, is familiar to experts. This also demonstrates that the coupling to the EMT is the real content of the EP.

HELICITY IN A GRAVITATIONAL FIELD AND POSSIBLE COSMOLOGICAL IMPLICATIONS

The crucial factor of 1/2 makes the evolution of the particle helicity rather different in magnetic and gravitomagnetic fields. The spin of a (Dirac) particle in a magnetic field is dragged with the cyclotron frequency, which is twice the Larmor frequency. It coincides with the frequency of the velocity precession; thus, helicity is conserved. At the same time, the gravitomagnetic field causes the velocity to be dragged twice as fast as the spin, changing the helicity. This factor of 2, however, is precisely that required to realize the possibility of reducing the entire effect of the gravitomagnetic field to frame rotation. Whereas the spin vector is the same in the rotating frame and is dragged only by rotation of the coordinate axis, the velocity vector is transformed and makes an additional contribution, providing the factor of 2 to Coriolis acceleration. The geometrical factor of 1/2 connects these phenomena with the EP.

Note that all of these considerations are essentially based on the smallness of the particle velocity achieved by the Lorentz boost as mentioned above and therefore do not lead to loss of generality. Let us consider a massive particle scattered by a rotating astrophysical object. The effect of the gravitomagnetic field is reduced to rotation of the local comoving frame, which becomes inertial at large distances before and after scattering. Consequently, the helicity is not changed by the gravitomagnetic field, which is confirmed by explicit calculation of the Born helicity-flip matrix element in the case of a massive neutrino [7, 17].

It may seem that the EP should entirely exclude the possibility of the helicity flip in scattering by a gravity source. This is, however, not the case if the usual Newtonian-type "gravitoelectric" force is considered [7]. Its action is also reduced to local acceleration of the comoving frame, in which the helicity of the particle is not altered. However, the comoving frame after scattering differs from the initial one by the corresponding velocity $\delta v = \int a dt$. The corresponding boost to the original frame generally changes the helicity of the massive particle (the similar effect for the gravitomagnetic field is just rotation by the angle $\delta \Omega = \int \omega dt$ and does not affect the helicity). The same boost may be considered as the source of the well-known deflection of particle momentum $\delta\phi \approx |\delta v|/|v|$ leading to the Newtonian expression, as this consideration applies to a nonrelativistic particle. The average helicity of the completely polarized beam after such scattering may be estimated in the semiclassical approximation as $\langle P\rangle \approx cos\phi \approx 1 - \phi^2/2$. According to the correspondence principle, this quantity may be expressed as

$$\langle P\rangle = \frac{d\sigma_{++} - d\sigma_{+-}}{d\sigma_{++} + d\sigma_{+-}} \approx 1 - 2\frac{d\sigma_{+-}}{d\sigma_{++}}, \quad (22)$$

where $d\sigma_{+-} \ll d\sigma_{++}$ are the helicity-flip and non-flip cross sections, respectively. Comparing the classical and quantum expressions for $\langle P\rangle$, one obtains

$$\frac{d\sigma_{+-}}{d\sigma_{++}} \approx \frac{\phi^2}{4}. \quad (23)$$

To check this simple approach, one may calculate this ratio for a Dirac particle scattered by a gravitational source. In the Born approximation, the result is easy to find:

$$\frac{d\sigma_{+-}}{d\sigma_{++}} = \frac{\tan^2 \frac{\phi}{2}}{(2\gamma - \gamma^{-1})^2}. \quad (24)$$

This expression coincides with the estimate (23) when the deflection angle is small and the particle is slow ($\gamma = E/m \to 1$), whereas for fast particles,

$$\frac{d\sigma_{+-}}{d\sigma_{++}} \approx \frac{\phi^2}{16\gamma^2}. \quad (25)$$

This effect should, in particular, lead to the helicity flip of any massive neutrino. It is very small when scattering by a single object is considered but may be enhanced when a neutrino propagates in the Universe. This effect can be especially strong for anisotropic Bianchi universes [21]. The anisotropy of the Universe leads to precession of spin (and, according to the EP, as discussed above, to the helicity flip) with a characteristic timescale on the order of a very early Universe age. This should turn Majorana neutrinos into antineutrinos and Dirac ones into their sterile counterparts. The resulting fraction of sterile Dirac neutrinos should be, generally speaking, of order 1. At the same time, modest fine-tuning on the order of a percent may lead to an excess of sterile neutrinos over active ones by two orders of magnitude, which might be sufficient to explain the deficit of neutrinos that prevents them from constituting dark matter.

Manifestation of the post-Newtonian EP is especially interesting when the gravitoelectric component is absent. In contrast to the EM case, one cannot realize this situation by canceling the contributions of positive and negative charges. At the same time, one may consider instead the interior of a rotating shell (the Lense–Thirring effect). Especially interesting is the case of the shell constituting a model of the Universe (see Ref. [22], Section 21.12; Ref. [23], Section 9.7), the mass and radius of which are of the same order (which probably should categorize such models as open universes), when the dragging frequency may be equal to the shell rotation frequency, which is just Mach's principle. Note that the low-energy theorem and emerging EP, which guarantee a unique precession frequency for all quantum and classical rotators, is a necessary counterpart of Mach's principle. Otherwise, rotation of the Universe as a whole may be detected from the disagreement between classical and quantum rotators.

EXTENSION OF EP

The relation of GFFs to the moments of GPDs opens, in principle, the possibility of extracting the separate couplings of quarks and gluons to gravity using experimental data and nonperturbative quantum chromodynamics (NPQCD) calculations [7]. One may pass then to the extended EP (ExEP) [24, 25], which states that the individual contributions of quarks and gluons to the AGM of the nucleon are zero in NPQCD.

This property is violated in perturbation theory (PT) [26–29]. This is not surprising qualitatively, as quarks and gluons correspond to the matrix elements of nonconserved operators that cannot be separated owing to operator mixing, whereas beyond the PT, one may consider the dominant nonperturbative (up to PT corrections) separation.

The reason for PT violation of the ExEP value was attributed [26, 27] to the use of the unsubtracted dispersion relations for the relevant form factors. In the case of performing the subtraction, it results in zero AGM but leads to the absence of a smooth transition of the electron mass $m_e \to 0$. At the same time, such a limit is unphysical in the case of (almost) massless quarks because of spontaneous chiral symmetry breaking, and the ExEP may be restored [24, 25].

This picture is already supported by the well-known fact of approximate cancellation of the proton and neutron anomalous magnetic moments. Their singlet combination differs from the AGM because they correspond to the first and second moments, respectively, of the GPDs E. This picture is also supported by the lattice data, including the most recent data [30]. However, interesting support is coming from the consideration of spin-1 hadrons.

EP for deuterons and vector mesons

As we discussed above, the concepts of the AGM and (Ex) EP may be easily generalized [7] to particles of any spin. As in the case of spin-1/2 particles, the smallness of the AGM should correspond to the smallness of the isoscalar magnetic moment; for ρ mesons, it should lead [31] to $g \approx 2$. Recently, the gyromagnetic ratio of charged ρ mesons was calculated in lattice QCD [32], and a value fairly close to the ExEP-supported factor of 2 was obtained.

There are more spin degrees of freedom for vector mesons (and, say, deuterons), which leads to additional GFFs [33]. The tensor polarization is P-even, unlike the vector ones. It does not require the Levi-Civita tensor; thus, EP is also demonstrated [34] for forward matrix elements.

These considerations lead to the zero sum rule for the sum of the second moments of the quark and gluon tensor spin structure functions derived many years ago [35, 36], and the suggested demonstration [34] of the EP is related to them as that of Kobzarev and Okun is related to Ji's sum rules:

$$\sum_q \langle P,S|T_i^{\mu\nu}|P,S\rangle_{\mu^2} = 2P^\mu P^\nu(1-\delta(\mu^2)) + 2M^2 S^{\mu\nu}\delta_1(\mu^2), \quad (26)$$

$$\langle P,S|T_g^{\mu\nu}|P,S\rangle_{\mu^2} = 2P^\mu P^\nu \delta(\mu^2) - 2M^2 S^{\mu\nu}\delta_1(\mu^2). \quad (27)$$

The ExEP, in turn, indicates that it is valid separately for quarks and for gluons; thus, $\delta(\mu^2) \approx 0$. The available data on the deuteron tensor spin structure function [37] provide just the required singlet combination for quarks and indicate that the second moment is more compatible with zero than the first one. This means that collective gluons in the deuteron are suppressed in comparison to the collective sea, and the ExEP explains this fact.

Cosmological implications of ExEP

One may of course think that the smallness of the individual quark and gluon AGMs for nucleons and the closeness of the quark gravigyromagnetic ratio to 2 for vector mesons is occasional. It is possible, nevertheless, to speculate [39] about their possible relevance to very general problems of gravity and cosmology

Consider the very strong gravity field near a black hole. Assume also the semiclassical picture in which the average angular momenta of quarks and gluons precess independently. If their angular velocities differ, the field may deconfine hadrons. If, however, they precess in accordance with the ExEP, the confinement may be gravityproof, which is attractive because of the well-known black hole complementarity [39].

Another possibility may be the closeness of the ExEP values to the asymptotic ones. Assuming that QCD evolution is growing backward in the expanding Universe, one should recall that this process is "antikinetic" [40]. As a result, too large deviations from the asymptotic values at large scales should result in violation of positivity at lower scales. Assuming the existence of a multiverse, one may expect that the probability of obtaining a particular deviation from the asymptotic values should be its decreasing function. As a result, Vilenkin's mediocrity principle (see, e.g., Ref. [41]) should favor smaller deviations from the asymptotic values.

QUADRUPOLE AND TIME-LIKE GFFS

Let us now discuss the quadrupole GFF, which is related in terms of the GPDs to the elegant Polyakov–Weiss Dterm [42] appearing in analyticity-based analysis as a subtraction constant [43, 44]:

$$\langle P+q/2|T^{\mu\nu}|P-q/2\rangle = C(q^2)(g^{\mu\nu}q^2 - q^\mu q^\nu) + \dots , \quad (28)$$

where the gravitoelectric and gravitomagnetic form factors (1) are now dropped. C has definite (positive) sign in all known cases, including hadrons (see [42], which also discusses general stability arguments), photons [45], and Q-balls [46, 47].

It is interesting to compare [48] this decomposition to the vacuum matrix element when one has the well-known cosmological constant:

$$\langle 0|T^{\mu\nu}|0\rangle = \Lambda g^{\mu\nu}. \quad (29)$$

One may relate the matrix element (29) to the vacuum one in two-dimensional transverse space orthogonal to P and q to obtain the effective two-dimensional "cosmological constant":

$$\Lambda = C(q^2)q^2. \quad (30)$$

The positive C leads to a negative cosmological constant in the scattering process (when $q^2 < 0$) and to a positive one in the annihilation process described by time-like GFFs when $q^2 > 0$. There seems to be some relation between annihilation and inflation!

It is of course very interesting whether the real cosmological constant in our Universe may be understood as emerging from annihilation at extra dimensions. This is qualitatively similar to brane cosmology, and one should stress that in this scenario, the Big Bang is due to one-graviton annihilation. The specification of extradimensional states that provide the cosmological constant in our space, which has the mass dimension 4, remains to be investigated.

DISCUSSION AND CONCLUSIONS

The GFFs provide a complementary interpretation, stemming from the EP, of some basic elements of the nucleon's spin structure, such as Ji's sum rules.

Scattering amplitude analysis involving these form factors constitutes a complementary approach to gravitational phenomena involving elementary particles. In particular, the helicity dynamics of massive Dirac neutrinos in an anisotropic Universe may lead to formation of dark matter from their sterile components. From the other side, the GFFs provide a unique option for extracting information about the separate couplings of quarks and gluons to gravity from experiments and (mostly) from NPQCD. This knowledge is not incompatible with the approximate validity of the ExEP separately for quarks and gluons. It may be valid only in NPQCD and may possibly lead to gravity-proof confinement.

It is quite interesting to consider GFFs in time-like regions, where they indicate some similarity between inflation and annihilation. Further tests of the ExEP may include lattice calculations and experiments, in particular more accurate measurement of the deuteron tensor spin structure function.

ACKNOWLEDGEMENTS

I would like to thank Haiyan Gao and Bo-Qiang Ma for invitation to write this paper. This work was supported in part by RFBR Grant 14-01-00647.

REFERENCES

1. I. Y. Kobzarev and L. B. Okun, Gravitational interaction of fermions, Zh. Eksp. Teor. Fiz. 43, 1904 (1962) [Sov. Phys. JETP 16, 1343 (1963)]

2. I. Yu. Kobzarev and V. I. Zakharov, Spin precession in a gravitational field, Ann. Phys. 37(1), 1 (1966)
3. S. Weinberg, Photons and gravitons in s matrix theory: Derivation of charge conservation and equality of gravitational and inertial mass, Phys. Rev. 135(4B), B1049 (1964)
4. H. Pagels, Energy-momentum structure form factors of particles, Phys. Rev. 144(4), 1250 (1966)
5. X. Ji, J. Tang, and P. Hoodbhoy, Spin structure of the nucleon in the asymptotic limit, Phys. Rev. Lett. 76(5), 740 (1996)
6. X. Ji, Gauge invariant decomposition of the nucleon spin, Phys. Rev. Lett. 78(4), 610 (1997)
7. O. V. Teryaev, Spin structure of nucleon and equivalence principle, arXiv: hep-ph/9904376 (1999)
8. S. J. Brodsky, M. Diehl, and D. S. Hwang, Light cone wave function representation of deeply virtual Compton scattering, Nucl. Phys. B 596(1–2), 99 (2001), arXiv: hepph/0009254
9. L. D. Landau and E. M. Lifshitz, Theoretical Physics, Vol. 4
10. L. D. Landau, L. P. Pitaevskii, E. M. Lifshitz, and V. B. Berestetskii, Relativistic Quantum Theory, Pergamon, 1971
11. L. D. Landau and E. M. Lifshitz, Theoretical Physics, Vol. 2
12. L. D. Landau and E. M. Lifshitz, The Classical Theory of Fields, Elsevier, 2013
13. I. B. Khriplovich, A spinning particle in a gravitational field, Zh. Eksp. Teor. Fiz. 96, 385 (1989)
14. S. B. Gerasimov, Dispersion sum rule for magnetic moments and spin dependence of photoabsorption cross-section, Yad. Fiz. 5, 1263 (1967)
15. F. W. Hehl and W. T. Ni, Inertial effects of a Dirac particle, Phys. Rev. D 42(6), 2045 (1990)
16. A. J. Silenko and O. V. Teryaev, Semiclassical limit for Dirac particles interaction with a gravitational field, Phys. Rev. D 71(6), 064016 (2005), arXiv: gr-qc/0407015
17. Y. N. Obukhov, A. J. Silenko, and O. V. Teryaev, Spin dynamics in gravitational fields of rotating bodies and the equivalence principle, Phys. Rev. D 80(6), 064044 (2009), arXiv: 0907.4367 [gr-qc]
18. Y. N. Obukhov, A. J. Silenko, and O. V. Teryaev, Dirac fermions in strong gravitational fields, Phys. Rev. D 84(2), 024025 (2011), arXiv: 1106.0173 [hep-th]
19. Y. N. Obukhov, A. J. Silenko, and O. V. Teryaev, Spin in an arbitrary gravitational field, Phys. Rev. D 88(8), 084014 (2013), arXiv: 1308.4552 [gr-qc]
20. D. Choudhury, N. D. Hari Dass, and M. V. N. Murthy, Gravitational helicity flip for massive neutrinos and SN 1987 A, Class. Quantum Gravity 6(9), L167 (1989)
21. A. Y. Kamenshchik and O. V. Teryaev, Chaotic spin precession in anisotropic universes and fermionic dark matter, arXiv: 1510.08253 [gr-qc] (2015)
22. C. Misner, K. Thorne, and J. Wheeler, Gravitation, Freeman, 1973
23. S. Weinberg, Gravitation and Cosmology, John Wiley and Sons, 1972
24. O. V. Teryaev, Spin structure of nucleon in QCD: Inclusive and exclusive processes, Prepared for International Conference on New Trends in High-Energy Physics: Experiment, Phenomenology, Theory, Yalta, Crimea, 22–29 Sep., 2001
25. O. V. Teryaev, Sources of time reversal odd spin asymmetries in QCD, Czech. J. Phys. 53, 47 (2003), arXiv: hepph/0306301
26. K. A. Milton, Quantum corrections to stress tensors and conformal invariance, Phys. Rev. D 4(12), 3579 (1971)
27. K. A. Milton, Quantum-electrodynamic corrections to the gravitational interaction of the electron, Phys. Rev. D 15(2), 538 (1977)
28. F. A. Berends and R. Gastmans, Quantum electrodynamical corrections to graviton matter vertices, Ann. Phys. 98(1), 225 (1976)
29. S. J. Brodsky, D. S. Hwang, B. Q. Ma, and I. Schmidt, Light-cone representation of the spin and orbital angular momentum of relativistic composite systems, Nucl. Phys. B 593(1–2), 311 (2001)
30. M. Deka, T. Doi, Y. B. Yang, B. Chakraborty, S. J. Dong, T. Draper, M. Glatzmaier, M. Gong, H. W. Lin, K. F. Liu, D. Mankame, N. Mathur, and T. Streuer, Lattice study of quark and glue momenta and angular momenta in the nucleon, Phys. Rev. D 91(1), 014505 (2015), arXiv: 1312.4816 [hep-lat]
31. O. V. Teryaev, Equivalence principle and partition of angular momenta in the nucleon, AIP Conf. Proc. 915, 260 (2007), arXiv: hep-ph/0612205
32. E. V. Luschevskaya, O. A. Kochetkov, O. V. Teryaev, and O. E. Solovjeva, $\pi\pm$ and $\rho0,\pm$ mesons in a strong magnetic field on the lattice, JETP Lett. 101(10), 674 (2015)
33. Z. Abidin and C. E. Carlson, Gravitational form factors of vector mesons in an AdS/QCD model, Phys. Rev. D 77(9), 095007 (2008), arXiv: 0801.3839 [hep-ph]
34. O. V. Teryaev, Nucleon spin and orbital structure: 20 years later, Mod. Phys. Lett. A 24, 2831 (2009)
35. A. V. Efremov and O. V. Teryaev, On high P(T) vector mesons spin alignment, Sov. J. Nucl. Phys. 36, 557 (1982) [Yad. Fiz. 36, 950 (1982)]
36. A. V. Efremov and O. V. Teryaev, On the oscillations of the tensor spin structure function, arXiv: hep-ph/9910555 (1999)
37. A. Airapetian, et al. (HERMES Collaboration), First measurement of the tensor structure function b1 of the deuteron, Phys. Rev. Lett. 95(24), 242001 (2005), arXiv: hepex/0506018
38. O. V. Teryaev, Hadron spin and external fields, Int. J. Mod. Phys. Conf. Ser. 39 1560083 (2015)
39. L. Susskind, L. Thorlacius, and J. Uglum, The Stretched horizon and black hole complementarity, Phys. Rev. D 48(8), 3743 (1993), arXiv: hep-th/9306069
40. X. Artru, M. Elchikh, J. M. Richard, J. Soffer, and O. V. Teryaev, Spin observables and spin structure functions:

inequalities and dynamics, Phys. Rep. 470(1–2), 1 (2009), arXiv: 0802.0164 [hep-ph]

41. A. Vilenkin, The principle of mediocrity, arXiv: 1108.4990 [hep-th] (2011)
42. K. Goeke, M. V. Polyakov, and M. Vanderhaeghen, Hard exclusive reactions and the structure of hadrons, Prog. Part. Nucl. Phys. 47(2), 401 (2001), arXiv: hep-ph/0106012
43. O. V. Teryaev, Analytic properties of hard exclusive amplitudes, arXiv: hep-ph/0510031 (2005)
44. I. V. Anikin and O. V. Teryaev, Dispersion relations and subtractions in hard exclusive processes, Phys. Rev. D 76, 056007 (2007), arXiv: 0704.2185 [hep-ph]
45. I. R. Gabdrakhmanov and O. V. Teryaev, Analyticity and sum rules for photon GPDs, Phys. Lett. B 716(3-5), 417 (2012), arXiv: 1204.6471 [hep-ph]
46. M. Mai and P. Schweitzer, Energy momentum tensor, stability, and the D-term of Q-balls, Phys. Rev. D 86(7), 076001 (2012), arXiv: 1206.2632 [hep-ph]
47. M. Mai and P. Schweitzer, Radial excitations of Q-balls, and their D-term, Phys. Rev. D 86(9), 096002 (2012), arXiv: 1206.2930 [hep-ph]
48. O. Teryaev, Analyticity and higher twists, Nucl. Phys. B 245, 195 (2013), arXiv: 1309.1985 [hep-ph]

CHAPTER 6

EXCEPTIONAL QUANTUM GEOMETRY AND PARTICLE PHYSICS

Michel Dubois-Violette

INTRODUCTION

It is well known that the Standard Model of particles is very successful but contains several inputs which ought to have explanations at a fundamental level. Some of these inputs like the existence of the Higgs particles, which seem to have recently acquired experimental confirmations, appear naturally for instance in the (almost commutative) noncommutative geometric formulations [8], [9], [11], [13], [15] and [16] or in the superconnection formulations [37]. Some other structural inputs come almost directly from the experimental observation. For instance the quark–lepton symmetry or the existence of 3 generations belong to these latter inputs and it is one of our aims here to connect these 2 inputs and to suggest some theoretical explanations for these facts.

By the quark–lepton symmetry, we mean the fact that to each quark corresponds one lepton and conversely. For instance to the quark u corresponds the electronic neutrino ν_e, to the quark d corresponds the electron e, etc.

On the theoretical side, it is worth noticing here that the cancellation of anomalies is a very strong argument for the quark–lepton symmetry.

The classical quark field is a spinor field with values in a complex 3-dimensional space acted by the color group SU(3). Thus, forgetting the spin etc., at each point of spacetime, the internal space E for a quark is a complex 3-dimensional Hilbert space (since $SU(3)\subset U(3)$) which is endowed with a complex volume that is with an antisymmetric $\mathbb{C}$-trilinear form (since elements of SU(3) are of determinant 1). By using the Hilbertian scalar product of E, one can transform the volume into an antilinear antisymmetric product $\times$ on E. Thus E is equipped with a product

$\times : E\times E\rightarrow E$

and the Hilbertian scalar product

$\langle , \rangle : E\times E\rightarrow C$

and SU(3) is the group of the $\mathbb{C}$-linear transformations which preserve these structures. It is then natural to combine these products into an SU(3)-invariant product on the Hilbertian direct sum $A=\mathbb{C}\oplus E$ such that $1\in\mathbb{C}\subset A$ is a unit denoted by 1 for the product of A and such that the norm of the product of 2 elements of A is the product of the norms of these 2 elements. The product on A is then not C-bilinear but is only real bilinear and the corresponding real algebra is isomorphic to the algebra of octonions [4]. The group SU(3) identifies then with the group of $\mathbb{C}$-linear isomorphisms of A $(\simeq C^4)$ which preserve the product. Under this action the invariant subspace E of $A=\mathbb{C}\mathbb{1}\oplus E$ corresponds to the fundamental representation of SU(3) while its orthogonal corresponds to the trivial representation. Thinking of E as the internal quark space, it is then natural to identify the component $\mathbb{C}\mathbb{1}$ of A as the trivial internal space of the corresponding lepton. In this way one connects the quark–lepton symmetry to the S of SU(3), i.e. to the unimodularity condition.

It is worth noticing here that this interpretation of the quark–lepton symmetry leads directly to the original interpretation given in [22] of the color group SU(3) as a subgroup of the automorphism group G_2 of the octonion algebra. Indeed it is classical (see e.g. in [4], [35] and [41]) that the automorphism group of the real algebra $\mathbb{O}$ of octonions is the first exceptional group G_2 and that SU(3) identifies with the subgroup of G_2 of automorphisms which preserve a given imaginary unit of $\mathbb{O}$. This is of course directly connected to the above construction

Citation: Michel Dubois-Violette, Exceptional quantum geometry and particle physics, Nuclear Physics B, Available online 25 April 2016, ISSN 0550-3213, http://dx.doi.org/10.1016/j.nuclphysb.2016.04.018.

which is explained in details in Section 2; the corresponding imaginary unit of $\mathbb{O}$ being then the imaginary unit i of $\mathbb{C}$.

Since SU(3) is a gauge group, that is the structure group of a bundle over spacetime, this construction means that at each point of spacetime one has a complex 4-dimensional vector space $A=\mathbb{C}\oplus E$ which is a real algebra (for the underlying structure of real 8-dimensional vector space) isomorphic to the octonion algebra (SU(3) being the automorphism group of this structure). In other words associated to the SU(3)-bundle we have a complex vector bundle A of rank 4 which is also a bundle in octonion algebras. A natural question is what is the role of this local algebra structure if it is relevant?

Let us observe that there are 3 generations and that the principle of triality combined with octonions leads directly to the exceptional Jordan algebra $J_3^8=\mathcal{H}_3(\mathbb{O})$ of octonionic hermitian 3 by 3 matrices, see e.g. in [1]. This algebra was introduced in [28] where it was pointed out that it is a quantum version of an algebra of "real functions" on a "finite quantum space" (see also [23]) as any formally real (Euclidean) finite-dimensional Jordan algebra.

The above discussion suggests to put over each spacetime point a finite quantum space corresponding to the exceptional Jordan algebra J_3^8. As before this corresponds to a bundle in algebras J_3^8 which is associated to the above (color) SU(3)-bundle. In other words this suggests to replace the real algebra $C^\infty(M)$ of smooth functions on spacetime M by the real Jordan algebra $J_3^8(M)$ of smooth sections of this bundle in algebras J_3^8. Since in this paper we assume that M is the usual Minkowski space $\mathbb{R}^4$, this algebra $J_3^8(M)$ is isomorphic to the algebra $C^\infty(M,J_3^8)=J_3^8\otimes C^\infty(M)$ of smooth J_3^8-valued functions on M . That is, we take $J_3^8(M)=J_3^8\otimes C^\infty(M)$ which is interpreted as the "algebra of real functions" on the quantum space which is the product of the "finite quantum space" corresponding to J_3^8 with the spacetime M . Since the "quantum part" is "finite", we refer to such a quantum space as an "almost classical quantum space". Classical matter fields are then elements of $J_3^8(M)$-modules and gluon fields are part of connections on these modules. However none of these two items, modules and connections, are straightforward for Jordan algebras.

There is a notion of (bi)module [27] and [38] for Jordan algebras (see also in [14]) which will be used here. It seems that no differential calculus has been developed so far for Jordan algebras, a step which is certainly necessary to speak of connections. We define such a differential calculus over Jordan algebras and describe several examples and their properties. A particular attention is devoted to the case of the exceptional Jordan algebra J_3^8. An appropriate associated notion of connection on Jordan modules is introduced.

It is our aim to develop in the future these notions in the present context in order to get a natural description of the field content of the Standard Model of particles and of some possible generalizations.

The present approach is close in spirit to the approach of noncommutative geometry. However here it is not a finite noncommutative space corresponding to a finite-dimensional associative algebra but a finite quantum space corresponding to a finite-dimensional Euclidean Jordan algebra which is added to spacetime. Furthermore since the exceptional Albert algebra is involved, one cannot use directly the technics of noncommutative geometry (i.e. of associative algebra) and some care must be taken to handle such theory. Nevertheless as in the noncommutative geometric approaches, it is expected that the apparition of the Higgs fields is a consequence of the quantum nature of the almost classical spacetime (i.e. components of connections in the quantum directions).

Finally let us recall that occurrence in physics of octonions and exceptional structures has a long history (see e.g. [12], [21], [22], [23], [24], [28] and [36]). They appear very naturally in extensions of the Standard Model. More recently, in [7], [19] and [20], an interesting tentative approach to incorporate nonassociativity in the framework of the spectral action principle has been carried over.

Our notations are standard and we use Einstein convention of summation over repeated up-down indices in the formulas.

UNIMODULARITY OF SU(3) AND THE QUARK–LEPTON SYMMETRY

SU(3)-spaces

In the following, an SU(3)-space will be a complex 3-dimensional Hilbert space E equipped with an antisymmetric complex 3-linear form $v:\Lambda^3E\to C$ of norm $\|v\|=1$, that is such that one has

$|v(e_1,e_2,e_3)|=1$

for any orthonormal basis (e_k) of E.

An SU(3)-basis of E will be an orthonormal basis (e_1,e_2,e_3) of E such that one has $v(e_1,e_2,e_3)=1$. Given such an SU(3)-basis (e_1,e_2,e_3), one defines a bijection of the group SU(3) onto the set of all SU(3)-basis $g\mapsto(e_1^{(g)},e_2^{(g)},e_3^{(g)})$ by setting

$$e_k^{(g)}=e_\ell g_k^\ell \tag{2.1}$$

for $k\in\{1,2,3\}$ where g_k^ℓ are the matrix elements of $g\in SU(3)$ (in the fundamental representation).

Given an SU(3)-basis (e_1,e_2,e_3) of E one defines the components $Z^k\in\mathbb{C}$ of a vector $Z\in E$ by

$Z=e_kZ^k$

and E can then be identified with $\mathbb{C}^3$ with Hilbertian scalar product given by

$$\langle Z_1,Z_2\rangle=\sum_{k=1}^{3}\bar{Z}_1^kZ_2^k \tag{2.2}$$

and v given by

$$v(Z_1,Z_2,Z_3)=\varepsilon_{k\ell m}Z_1^kZ_2^\ell Z_3^m \tag{2.3}$$

for $Z_1,Z_2,Z_3\in E$. An element $Z\in E$ is identified with the column $[Z^k]\in\mathbb{C}^3$ of its components and a change of SU(3)-basis corresponds to the left action of an SU(3)-matrix on $[Z^k]$ (which leaves invariant the forms (2.2) and (2.3)).

SU(3)-algebra version of octonion algebra

One can combine the 3-linear form v on E with the Hilbertian scalar product to define an antibilinear antisymmetric product $(Z_1,Z_2)\mapsto Z_1\times Z_2$ on E by

$$v(Z_1, Z_2, Z_3) = \langle Z_1 \times Z_2, Z_3 \rangle \quad (2.4)$$

for any $Z_1, Z_2, Z_3 \in E$. In the SU(3)-basis (e_1, e_2, e_3) and the corresponding identification of E with $\mathbb{C}^3$ this product reads

$$(Z_1 \times Z_2)^k = \varepsilon_{k\ell m} \bar{Z}_1^\ell \bar{Z}_2^m \quad (2.5)$$

for $k \in \{1,2,3\}$. Notice that this product on E is nonassociative and is invariant by the action of SU(3). Thus one has a product

$$\times : E \times E \to E, \quad (Z, Z') \mapsto Z \times Z'$$

and the scalar product

$$\langle , \rangle : E \times E \to \mathbb{C}, (Z, Z') \mapsto \langle Z, Z' \rangle$$

which are invariant by the action of SU(3) on E . This representation of SU(3) is the fundamental representation and SU(3) is exactly the subgroup of the complex linear group of E which preserves these two products; $\mathbb{C}$ being equipped with the trivial representation of SU(3)

$$(g, z) \mapsto gz = z$$

for $g \in SU(3)$, $z \in \mathbb{C}$. It is therefore natural to combine these products into an SU(3)-invariant product

$$((z, Z), (z', Z')) \mapsto (z, Z)(z', Z')$$

on $\mathbb{C} \oplus E$ in such a way that

$$(0, Z)(0, Z') = (\alpha \langle Z, Z' \rangle, \beta Z \times Z') \quad (2.6)$$

for some $\alpha, \beta \in \mathbb{C}$ and that $\mathbb{1} = (1,0)$ is a unit, i.e. that one has in particular

$(1,0)(0,Z)=(0,Z)(1,0)=(0,Z)$

for any $Z \in E$ with of course $(1,0)(1,0)=(1,0)$ and more generally

$(z,0)(z',0)=(zz',0)$

for $z, z' \in \mathbb{C}$.

If $\mathbb{C} \oplus E$ is endowed with its natural structure of 4-dimensional Hilbert space (Hilbertian direct sum), one has

$\|(0,Z)(0,Z')\|^2 = (|\alpha|^2 - |\beta|^2)| \langle Z,Z' \rangle |^2 + |\beta|^2 \|Z\|^2 \|Z'\|^2$

so that one has

$\|(0,Z)(0,Z')\|^2 = \|(0,Z)\|^2 \|(0,Z')\|^2$

by choosing

$$|\alpha|^2 = |\beta|^2 = 1 \quad (2.7)$$

as normalization. One has also of course

$\|(z,0)(z',0)\|^2 = \|(z,0)\|^2 \|(z',0)\|^2$

so it is natural to require

$\|(z,Z)(z',Z')\|^2 = \|(z,Z)\|^2 \|(z',Z')\|^2$

with $\|(z,Z)\|^2 = |z|^2 + \|Z\|^2$ and the real bilinearity of the product. A solution is given by

$$(z, Z)(z', Z') = (zz' - \langle Z, Z' \rangle, \bar{z}Z' + z'Z + iZ \times Z') \quad (2.8)$$

This product is probably essentially unique up to irrelevant normalizations. Setting

$$\overline{(z, Z)} = (\bar{z}, -Z) \quad (2.9)$$

one obtains in particular

$$\overline{(z, Z)}(z, Z) = (z, Z)\overline{(z, Z)} = (|z|^2 + \| Z \|^2)(1, 0)$$

$$\overline{(z, Z)}(z, Z) = (z, Z)\overline{(z, Z)} = \| (z, Z) \|^2 \mathbb{1} \quad (2.10)$$

which implies that as a real 8-dimensional algebra for the product (2.8), $A = \mathbb{C}\mathbb{1} \oplus E$ is a normed division algebra which is (necessarily) isomorphic to the algebra $\mathbb{O}$ of octonions. The group SU(3) is the group of automorphisms of this algebra which preserves the structure of 4-dimensional complex vector space of $\mathbb{C} \oplus E$ or, equivalently, the group of all complex automorphisms of $\mathbb{C} \oplus E$ which preserve the product given by (2.8).

The quark–lepton symmetry

The representation of SU(3) on $A = \mathbb{C}\mathbb{1} \oplus E$ is the direct sum of the fundamental representation on E with the trivial representation on its orthogonal complement $\mathbb{C}\mathbb{1} \simeq \mathbb{C}$. If one interprets E as the internal space for a quark, it is therefore natural to identify $\mathbb{C}\mathbb{1}$ with the trivial internal space of the corresponding lepton in the quark–lepton symmetry. This identification connects the quark–lepton symmetry to the unimodularity of the color group SU(3).

This raises the following question: Is there a reason for the occurrence of the division algebra $A \simeq \mathbb{O}$ in the present context?

Gauge theoretical aspect

Up to now, the discussion is over each spacetime point. One knows however that the color group SU(3) is the structure group of an SU(3)-gauge theory of strong interactions. Indeed the gluon field which induces the strong interactions is an SU(3)-connection.

This means that one has a complex vector bundle over spacetime with fiber $\mathbb{C} \oplus \mathbb{C}^3$ associated with a principal SU(3)-bundle which is the direct sum of a trivial complex vector bundle of rank 1 with a complex vector bundle of rank 3 corresponding to the fundamental representation of SU(3). This bundle is also an associated bundle in algebras with product given by (2.8) in each fiber. The gluon field being then a connection on this bundle.

Let A(M) be the algebra of smooth sections of this bundle and let us denote by Der(A)the Lie algebra of derivations of an algebra A. The covariant derivative ∇_X along the vector field $X \in \Gamma(TM) = Der(C^\infty(M))$ associated to the gluon field SU(3)-connection is a derivation of A(M) and the restriction ρ of this derivation to $C^\infty(M) = C^\infty(M)1 \subset A(M)$ satisfies

$\rho(\nabla_X) = X$

which means that $X \mapsto \nabla_X = s_\nabla(X)$ is a splitting

$s_\nabla : Der(C^\infty(M)) \to Der(A(M))$

of the exact sequence of $C^\infty(M)$-modules (compare with [18])

$$0 \to \mathrm{Int}(A(M)) \overset{\subset}{\to} \mathrm{Der}(A(M)) \overset{\rho}{\to} \mathrm{Der}(C^\infty(M)) \to 0 \quad (2.11)$$

where $Int(A(M))=C^\infty(M,g_2)$ with the exceptional Lie algebra g_2 identified with the Lie algebra $Der(\mathbb{O})$ of derivations of the octonion algebra which are all inner (see e.g. in [38]) and thus $Der(\mathbb{O})=Int(\mathbb{O})=g_2$ and one has $Int(A(M))=C^\infty(M,g_2)$. This splitting of (2.11) gives a splitting of the exact sequence of $C^\infty(M)$-modules

$$0 \to C^\infty(M, su(3)) \overset{\subset}{\to} Der_0(A(M)) \overset{\rho}{\to} Der(C^\infty(M)) \to 0 \tag{2.12}$$

where $Der_0(A(M))$ is the $C^\infty(M)$-submodule and Lie subalgebra of the derivation of A(M) compatible with the representation $\mathbb{C} \oplus \mathbb{C}^3$ of $\mathbb{O}$. Notice that (2.11) and (2.12) are also exact sequences of Lie algebras but that the obstruction for s_∇ to be a Lie algebra homomorphism is the curvature of ∇, that is the field strength of the gluon field. Corresponding to s_∇ one defines a projection π_∇ of Der(A(M)) onto $Int(A(M))=C^\infty(M,g_2)=Ker(\rho)$ by setting (as in [18])

$$\pi_\nabla(\delta) = \delta - \nabla_{\rho(\delta)} \tag{2.13}$$

which restricts as a projection of $Der_0(A(M))$ onto $C^\infty(M,su(3))$.

In the following it is assumed that the spacetime is the ordinary Minkowski space $\mathbb{R}^4$. This simplifying assumption implies that all the bundles over spacetime can be considered as products so their sections identify with functions with values in fixed spaces (like $\mathbb{C}\oplus\mathbb{C}^3$, etc.). However the whole discussion extends easily to the case where spacetime is a non-trivial pseudo-Riemannian manifold and where the bundles are non-trivial.

THE 3-GENERATIONS, TRIALITY AND THE EXCEPTIONAL FINITE QUANTUM SPACE

The 3-generations

There are 6 flavors of quark–lepton

$$(u, \nu_e),\ (d, e),\ (c, \nu_\mu),\ (s, \mu),\ (t, \nu_\tau),\ (b, \tau)$$

which can be grouped in 3 generations of doublets of quark–lepton which are the columns of the following table

generations			
quarks Q=2/3	u	c	t
leptons Q=0	νe	νμ	ντ
quarks Q=−1/3	d	s	b
leptons Q=−1	e	μ	τ

where Q denotes the electric charge.

This is the present experimental situation which reveals a sort of "triality". This triality combined with the above interpretation of the quark–lepton symmetry is the starting point for the following analysis which suggests to add over each spacetime point an exceptional "finite quantum space" corresponding to the exceptional Jordan algebra J_3^8 of the hermitian 3×3 octonionic matrices and to take the internal spaces of the basic fermions as elements of appropriate modules over this algebra.

In fact the combination of octonions with triality leads naturally to the exceptional Jordan algebra J_3^8 (see e.g. in [1], [27] and [41]) and it turns out that this Jordan algebra can be considered as the "algebra of real functions" on a "finite quantum space".

Let us first make precise what is meant here by a finite quantum space.

Finite quantum spaces

The notion of finite-dimensional Euclidean (or formally real) Jordan algebra was introduced and analyzed in [28] under the name r-number algebra in order to formalize the properties of the observables of finite quantum systems. A minimal set of requirement for the observables of a finite quantum system is that it is a finite-dimensional real vector space J such that one can consistently define powers of any element and that if a sum of squares of elements vanishes it implies that each of these elements vanishes. Given the square $x\mapsto x^2$, one defines a symmetric bilinear product on J by setting $x\cdot y=1/2((x+y)^2-x^2-y^2)$ for $x,y\in J$.

So let J be a finite-dimensional real vector space endowed with a symmetric bilinear product

$$(x, y) \mapsto x.y = y.x \tag{3.1}$$

for $x,y\in J$ and assume that the following condition is satisfied

$$\sum_{i\in I}(x_i)^2 = 0 \Rightarrow x_i = 0,\quad \forall i \in I \tag{3.2}$$

for any finite family $(x_i)_{i\in I}$ in J, where $(x)^2=x.x$ for $x\in J$. The above condition for the product is referred to as formal reality condition. One has the following result [28].

Theorem 1. Let J be as above and define $x^n\in J$ for $x\in J$ and $n\in \mathbb{N}^+$ by induction on n as $x^1=x$ and $x^{n+1}=x.x^n$.

Then the following conditions (i) and (ii) are equivalent:

(i) $x^r.x^s = x^{r+s}$, *for any* $x \in J$ and $r,s\in\mathbb{N}^+$

(ii) $(x^2.y).x = x^2.(y.x)$, *for any* $x, y \in J$

A finite-dimensional real commutative algebra satisfying (3.2) and the equivalent conditions (i) and (ii) of the theorem is referred to as a finite-dimensional Euclidean (or formally real) Jordan algebra, [6], [26] and [34].

Condition (i) of Theorem 1 is referred to as power associativity while Condition (ii) is referred to as the Jordan identity. It is well known and easy to show that the Jordan identity implies the power associativity but it is a very nontrivial result that in the above context the power associativity combined with the formal reality condition imply the Jordan identity [28]. One can find in [3] other natural conditions on commutative power-associative algebras leading to the Jordan identity.

It is worth noticing that Condition (3.2) and Condition (i) of Theorem 1 (i.e. power associativity) are exactly the conditions needed to have spectral resolutions of the elements of J. Indeed a finite-dimensional Euclidean Jordan algebra J has a unit $\mathbb{1}\in J$ such that $\mathbb{1}.x=x$ for any $x\in J$ and any $x\in J$ has a finite spectral resolution

$$x = \sum_{r\in I_x} \lambda_r e_r \tag{3.3}$$

where I_x is a finite set, $\lambda_r \in \mathbb{R}$ and $e_r^2 = e_r \in J$ for any $r \in I_x$, $e_r.e_s = 0$ for $r, s \in I_x$ with $r \neq s$

and

$$\mathbb{1} = \sum_{r \in I_x} e_r \tag{3.4}$$

i.e. (e_r) is a finite family of orthogonal idempotents in J the sum of which is the unit 1∈J. Furthermore the number of elements of I_x, card(I_x), is bounded by a finite number which only depends on J that is one has $\sup_{x \in J}(\text{card}(I_x)) < \infty$.

This means that one can set

$$x^0 = \mathbb{1} \tag{3.5}$$

and that

$$P \mapsto P(x) = \sum_r P(\lambda_r) e_r \tag{3.6}$$

defines a homomorphism of unital algebras from the algebra $\mathbb{R}$ [X] of real polynomials into J for any x∈J.

These finite-dimensional Euclidean Jordan algebras were classified in [28] where the following theorem is proved.

Theorem 2. Any finite-dimensional Euclidean Jordan algebra is a direct sum of a finite number of simple ideals. A finite-dimensional simple Euclidean Jordan algebra is isomorphic to one of

$\mathbb{R}$ and $JSpin_{n+2} = \mathbb{R}\mathbb{1} \oplus \mathbb{R}^{n+2}$,
$J^1_{n+3} = \mathcal{H}_{n+3}(\mathbb{R})$, $J^2_{n+3} = \mathcal{H}_{n+3}(\mathbb{C})$,
$J^4_{n+3} = \mathcal{H}_{n+3}(\mathbb{H})$, with $n \in \mathbb{N}$
and $J^8_3 = \mathcal{H}_3(\mathbb{O})$.

In the above statement

$JSpin_{n+1} = \mathbb{R}\mathbb{1} \oplus \mathbb{R}^{n+1}$ is endowed with the product

$$(r\mathbb{1} \oplus v).(r'\mathbb{1} \oplus v') = (rr' + \langle v, v'\rangle)\mathbb{1} \oplus (rv' + r'v)$$

where ⟨,⟩ is the scalar product of the Euclidean space $\mathbb{R}^{n+1}$ and, for any involutive algebra A, H_n(A) denotes the space of hermitian n×n-matrices with entries in Aendowed with the product

$$A.B = \frac{1}{2}(AB + BA) \tag{3.7}$$

where (A,B)↦AB is the product in M_n(A).

The absence of $JSpin_1$ in the above list comes from the fact that $JSpin_1$ is not simple but is isomorphic to the commutative associative algebra $\mathbb{R} \oplus \mathbb{R}$ of the real functions on a two-points set. Furthermore, notice that one has the following coincidence

$\mathcal{H}_1(\mathbb{R}) = \mathcal{H}_1(\mathbb{C}) = \mathcal{H}_1(\mathbb{H}) = \mathcal{H}_1(\mathbb{O}) = \mathbb{R}$,
$\mathcal{H}_2(\mathbb{R}) = JSpin_2$, $\mathcal{H}_2(\mathbb{C}) = JSpin_3$, $\mathcal{H}_2(\mathbb{H}) = JSpin_5$,
$\mathcal{H}_2(\mathbb{O}) = JSpin_9$

and that the $H_n(\mathbb{O})$ are not Jordan algebras for n≥4.

In each finite-dimensional simple Euclidean Jordan algebra J , its unit $\mathbb{1}$ admits a decomposition

$$\mathbb{1} = \sum_{i=1}^{c(J)} e_i$$

with $e_i.e_j=0$ for i≠j (orthogonality) where the e_i are primitive idempotents. The number c(J) does only depend on J and is referred to as the capacity (or the degree) of J . One has c($\mathbb{R}$)=1, c($JSpin_{n+2}$)=2, $c(J^1_3) = c(J^2_3) = c(J^4_3) = c(J^8_3) = 3$ and $c(J^1_n) = c(J^2_n) = c(J^4_n) = n$ for n≥4.

Note the obvious fact that a finite-dimensional Euclidean Jordan algebra J which is associative is the algebra of real functions on a finite space K which can be identified to the set of characters of J, that is the set of homomorphisms

$$\chi : J \to \mathbb{R}$$

of unital algebras. More generally, it will be convenient in the following to consider that a finite-dimensional Euclidean Jordan algebra J is the algebra of "real function" on some associated "finite quantum space" (a dual object). This is justified by the above spectral properties as well as by the fact that J is the set of observable of a finite quantum system (see below). Indeed the pure states on J are the primitive idempotents in J and the transition probabilities between two states are the appropriate traces of the corresponding product of idempotents.

Recall finally that a Jordan algebra over a commutative field $\mathbb{K}$ is a $\mathbb{K}$-vector space endowed with a symmetric bilinear product (x,y)↦x.y satisfying Property (ii) of Theorem 1. For infinite-dimensional real Jordan algebras, there are various generalizations of the formal reality condition (3.2). In this frame there are (bounded) generalizations of Theorem 1, see [25] for the JB-algebra version and [39] for a weak version.

The exceptional finite quantum space

Apart J^8_3, all the other finite-dimensional simple Euclidean Jordan algebras can be realized as real vector spaces of self-adjoint operators in finite-dimensional Hilbert spaces, stable by the anticommutator of the associative composition of linear operators. They can therefore be considered as algebras of observables of some finite quantum systems. These special algebras enter in the frame of noncommutative geometry based on associative *-algebras and each nonexceptional simple Lie algebra occurs as Lie algebra of derivations of one of these special simple Euclidean Jordan algebras.

However, as proved in [2] the Jordan algebra J^8_3 is exceptional which means that it cannot be inbedded in an associative algebra as a subspace stable by the anticommutator of the associative product as Jordan product.

Let us recall that if A is an associative algebra with product (x,y)↦xy, then the anticommutator $(x, y) \mapsto x.y = \frac{1}{2}(xy + yx)$ is a Jordan product on A and one denotes by A^{Jord} the corresponding Jordan algebra. A Jordan algebra which is isomorphic to a Jordan subalgebra of some A^{Jord} as above is said to be special, otherwise it is said to be exceptional.

In spite of the fact that J^8_3 is exceptional it can be consistently considered as the algebra of observable of a finite quantum system, see e.g. [23] and [28]. Thus it corresponds to an exceptional quantum space which is in fact unique [42] and [43]. Notice that since $\dim_{\mathbb{R}}(J^8_3) = 27 < \infty$, we consider that the "corresponding quantum space" is "finite".

The group of all automorphisms of J_3^8 is the exceptional Lie group F_4 with Lie algebra Lie(F_4)=f_4 which is the Lie algebra of all derivations of J_3^8. Therefore at first sight F_4plays the role of diffeomorphism group of the exceptional quantum space while the exceptional simple Lie algebra f_4 plays the role of the Lie algebra of vector fields on this exceptional quantum space.

Remark.

Notice that these actions are inner. It follows that F_4 and its Lie algebra f_4 admit an alternative interpretation in terms of gauge group and Lie algebra. This applies as well for their subgroups and Lie subalgebras. Of particular interest in view of the analysis of Section 2 is the subgroup of F_4 which preserves the representation of the octonion algebra $\mathbb{O}$ as $\mathbb{C}\oplus\mathbb{C}^3$ (see in Section 4).

Exceptional almost classical quantum space-time

As in §2.4 these algebras J_3^8 over each spacetime point are the fiber of a bundle in algebras J_3^8 which is associated to the principal (color) SU(3)-bundle on which the gluon field is a connection. This connection is compatible with the identification of the octonionic matrix elements of J_3^8 as elements of $\mathbb{C}\oplus\mathbb{C}^3$. Let $J_3^8(M)$ be the real Jordan algebra of the sections of this bundle in J_3^8 algebras. The matrix elements of $J_3^8(M)$ are then elements of A(M) (see in §2.4). We denote by the same symbol ∇_X the covariant derivative along the vector field X∈Der(C^∞(M)) corresponding to the SU(3)-connection and by ρ the restriction of $\mathrm{Der}(J_3^8(M))$ to $C^\infty(M)=C^\infty(M)\otimes\mathbb{1}\subset J_3^8(M)$. Again $X\mapsto\nabla_X=s_\nabla(X)$ is a splitting of the exact sequence of C^∞(M)-modules

$$0\to\mathrm{Int}(J_3^8(M))\overset{\subset}{\to}\mathrm{Der}(J_3^8(M))\overset{\rho}{\to}\mathrm{Der}(C^\infty(M))\to 0 \tag{3.8}$$

where now $\mathrm{Int}(J_3^8(M))=C^\infty(M,\mathfrak{f}_4)$ with the exceptional Lie algebra f_4 identified with the Lie algebra $\mathrm{Der}(J_3^8)$ of derivations of the exceptional Jordan algebra J_3^8. The derivations of J_3^8 being inner, one has $\mathrm{Der}(J_3^8)=\mathrm{Int}(J_3^8)$ so $\mathrm{Int}(J_3^8(M))=C^\infty(M,\mathfrak{f}_4)$. Setting for $\delta\in\mathrm{Der}(J_3^8(M))$

$\pi_\nabla(\delta)=\delta-\nabla_{\rho(\delta)}$

as in §2.4, one defines a projection π_∇ of $\mathrm{Der}(J_3^8(M))$ onto $\mathrm{Int}(J_3^8(M))=C^\infty(M,\mathfrak{f}_4)=\mathrm{Ker}(\rho)$.

Assuming as before that $M\cong R^4$, one makes the identifications

$$J_3^8(M)=C^\infty(M,J_3^8)=C^\infty(M)\otimes J_3^8 \tag{3.9}$$

and, by an abuse of language, one defines the exceptional almost classical quantum spacetime by saying that its algebra of real "smooth functions" is $J_3^8(M)$. With the identification (3.9), it is the product of the classical spacetime M with the exceptional finite quantum space corresponding to J_3^8. More generally, the notion of almost classical quantum spacetime, or simply of almost classical spacetime, is obtained by replacing in (3.9)J_3^8 by an arbitrary finite-dimensional Euclidean Jordan algebra.

The automorphism group of J_3^8 is the exceptional group F_4 while the automorphism group of C^∞(M) is the group Diff(M) of diffeomorphisms of M . The automorphism group $\mathrm{Aut}(C^\infty(M,J_3^8))$ of $C^\infty(M,J_3^8)=C^\infty(M)\otimes J_3^8$ is the semidirect product $C^\infty(M,F_4)\rtimes$Diff(M) of the group $C^\infty(M,F_4)$ with Diff(M). Thus in a sense this automorphism group plays the role of the diffeomorphism group of the exceptional almost classical quantum spacetime. Correspondingly the Lie algebra $\mathrm{Der}(C^\infty(M,J_3^8))$of derivations of the Jordan algebra $C^\infty(M,J_3^8)$ is the semidirect sum of the Lie algebra $C^\infty(M,f_4)$ and the Lie algebra Der(C^∞(M)) of vector fields on M and plays in the same above sense the role of the Lie algebra of vector fields on the exceptional almost classical quantum spacetime.

However Section 2.4 and the remark of §3.3 suggest to give to F_4 and its subgroups an alternative interpretation as structure groups for a gauge theory on J_3^8-modules.

We wish to develop a field theory on this quantum spacetime. For this one has to introduce the appropriate notion of (bi-)module over $C(M,J_3^8)$, the relevant matter's fields being then elements of such bimodules. It is clear that one needs first a description of modules over J_3^8. Furthermore, in view of the previous interpretation of the quark–lepton symmetry, one has to describe the subgroup of F_4 acting on such modules which is compatible with the representation of $\mathbb{O}$ as $\mathbb{C}\oplus\mathbb{C}^3$.

MODULES AND FUNDAMENTAL FERMIONS

Modules over Jordan algebras

There are general definitions of bimodules for classes of algebras [27] and [38] or for categories of algebras [14]. For Jordan algebras this reduces to the following (see also in [31]). Let J be a Jordan algebra and let M be a vector space equipped with left and right actions of J

$$J\otimes M\to M,\quad x\otimes\Phi\mapsto x\Phi\in M$$
$$M\otimes J\to M,\quad \Phi\otimes x\mapsto \Phi x\in M$$

for x∈J, Φ∈M. Define a bilinear product on J⊕M by setting (x+Φ)(x'+Φ')=xx'+(xΦ'+Φx').

Then M is said to be a Jordan bimodule over J if J⊕M endowed with the above product is a Jordan algebra. This is equivalent to the following conditions (i), (ii) and (iii):

$$\begin{cases}\text{(i)}\ \ x\Phi=\Phi x\\ \text{(ii)}\ \ x(x^2\Phi)=x^2(x\Phi)\\ \text{(iii)}\ \ (x^2y)\Phi-x^2(y\Phi)=2((xy)(x\Phi)-x(y(x\Phi)))\end{cases}$$

for x,y∈J and Φ∈M. When J is unital with unit denoted by $\mathbb{1}$, , the Jordan bimodule M will be said to be unital whenever it satisfies the further condition (iv)

$$\text{(iv)}\qquad \mathbb{1}\Phi=\Phi$$

for any Φ∈M, which implies that J⊕M is again a unital Jordan algebra with unit $\mathbb{1}$.

It follows from Condition (i) that one can restrict attention to the left action (for instance), this is why Jordan bimodules are usually referred to as Jordan modules.

Let us define the linear mapping $x\mapsto L_x$ of J into the algebra L(M) of linear endomorphisms of the J-module M by setting

$$L_x\Phi = x\Phi \tag{4.1}$$

for x∈J and Φ∈M. Then (ii) reads

$$[L_x, L_{x^2}] = 0 \tag{4.2}$$

for x∈J, while (iii) reads

$L_{x}^{2}{}_{y} - L_{x}{}^{2}L_{y} - 2L_{xy}L_{x} + 2L_{x}L_{y}L_{x} = 0$

for x,y∈J which is equivalent to

$$L_{x^3} - 3L_{x^2}L_x + 2L_x^3 = 0 \tag{4.3}$$

and

$$[[L_x, L_y], L_z] + L_{[x,z,y]} = 0 \tag{4.4}$$

for x,y,z∈J where [x,z,y]=(xz)y−x(zy) is the associator.

Thus Conditions (ii) and (iii) can be replaced by Conditions (4.2), (4.3) and (4.4) above while the unital condition (iv) reads

$$L_{\mathbb{1}} = \mathbb{1} \tag{4.5}$$

where on the right-hand side $\mathbb{1}$ denotes the unit of the algebra L(M).

Notice that if p∈J is an idempotent, i.e. satisfies $p^2=p$, then (4.3) reads

$$L_p - 3L_p^2 + 2L_p^3 = 0$$

that is

$$L_p(L_p - \frac{1}{2})(L_p - 1) = 0 \tag{4.6}$$

from which follows the associated Pierce decomposition of the module M.

J_3^8-modules and SU(3)×SU(3)-action

Any Jordan algebra J is canonically a Jordan module over itself which is unital whenever J has a unit. The list of the unital irreducible Jordan modules over the finite-dimensional simple Euclidean Jordan algebras can be found in [27]. In the case of the exceptional Jordan algebra J_3^8, it turns out that any unital Jordan module is a product or a direct sum of modules isomorphic to J_3^8, [27]. In particular any finite module M over J_3^8 is a finite product (or direct sum) of modules isomorphic to J_3^8 so that one has $M = J_3^8 \otimes E$ for some finite-dimensional real vector space E.

In view of our interpretation of the quark–lepton symmetry it is important to describe the subgroup of F_4 which preserves the representations of the octonions occurring in the elements of J_3^8 as elements of $\mathbb{C}\oplus\mathbb{C}^3$. For this, following [41], one associates to the element

$$\begin{pmatrix} \zeta_1 & x_3 & \bar{x}_2 \\ \bar{x}_3 & \zeta_2 & x_1 \\ x_2 & \bar{x}_1 & \zeta_3 \end{pmatrix} \tag{4.7}$$

of J_3^8 the following element

$$\begin{pmatrix} \zeta_1 & z_3 & \bar{z}_2 \\ \bar{z}_3 & \zeta_2 & z_1 \\ z_2 & \bar{z}_1 & \zeta_3 \end{pmatrix} + (Z_1, Z_2, Z_3) \tag{4.8}$$

of $J_3^2 \oplus M_3(\mathbb{C})$ where

$$x_i = z_i + Z_i \in \mathbb{C}\oplus\mathbb{C}^3 \tag{4.9}$$

are the representation in $\mathbb{C}\oplus\mathbb{C}^3$ of the 3 elements x_i of O. The desired subgroup of F_4 is the group SU(3)×SU(3)/ $\mathbb{Z}_3$ with action induced by the action of SU(3)×SU(3) on $J_3^2 \oplus M_3(\mathbb{C})$ given by [41]

$$H \mapsto VHV^*, \quad M \mapsto UMV^* \tag{4.10}$$

for (U,V)∈SU(3)×SU(3) and $H \in J_3^2 = \mathcal{H}_3(\mathbb{C}), M \in M_3(\mathbb{C})$. Notice that the action of the first factor U is the one which corresponds to the previous action of SU(3) on $\mathbb{C}\oplus\mathbb{C}^3(\simeq O)$ described in Section 2. Thus with the previous interpretation, the first factor SU(3) is the "color group". The second factor SU(3) mixes the generations and furthermore mixes the complexes z_i corresponding to generations of leptons with the reals ζ_i of the diagonal which correspond in this picture to new particles of spin 1/2. Since the mass scales of the 3 generations are very different this latter factor SU(3) must be strongly broken, however there could remain some unbroken finite subgroup of this SU(3) which should then play a role in this approach for the see-saw mechanism and the structure of the fermion mass matrices.

Modules for the two families by generation

As recalled in §3.1, there are two families by generation. In view of §4.2 it seems, at first sight, reasonable to take as J_3^8-module the product of two copies J^u and J^d of J_3^8 as module with the particle assignment

$$J^u = \begin{pmatrix} \alpha_1 & \nu_\tau + t & \bar{\nu}_\mu - c \\ \bar{\nu}_\tau - t & \alpha_2 & \nu_e + u \\ \nu_\mu + c & \bar{\nu}_e - u & \alpha_3 \end{pmatrix} \tag{4.11}$$

$$J^d = \begin{pmatrix} \beta_1 & \tau + b & \bar{\mu} - s \\ \bar{\tau} - b & \beta_2 & e + d \\ \mu + s & \bar{e} - d & \beta_3 \end{pmatrix} \tag{4.12}$$

or the representation (4.8)

$$J^u = \begin{pmatrix} \alpha_1 & \nu_\tau & \bar{\nu}_\mu \\ \bar{\nu}_\tau & \alpha_2 & \nu_e \\ \nu_\mu & \bar{\nu}_e & \alpha_3 \end{pmatrix} + (u, c, t) \tag{4.13}$$

$$J^d = \begin{pmatrix} \beta_1 & \tau & \bar{\mu} \\ \bar{\tau} & \beta_2 & e \\ \mu & \bar{e} & \beta_3 \end{pmatrix} + (d, s, b) \tag{4.14}$$

where we have identified the fundamental fermions with there internal spaces ($\mathbb{C}^3$ for the quarks and $\mathbb{C}$ for the leptons). The diagonals correspond then to new spin1/2 fermions with $\mathbb{R}$ as internal spaces. It is therefore natural to consider these particles α_k, β_k (k∈{1,2,3}) as described by Majorana spinors.

The interest of this latter identifications is that the addition of such particles does not spoil the anomalies cancellation of the Standard Model. In this context the corresponding module over $C^\infty(M, J_3^8)$ would be the tensor product of the $C^\infty(M)$-module of Majorana spinor fields over spacetime by the above J_3^8-module.

Problem of the U(1)×SU(2)-symmetry

Up to now the analysis deals only with the color symmetry SU(3), its tentative connection with the quark–lepton symmetry and the 3 generations. This leads naturally to the above assignment of internal spaces for fundamental fermions (of §4.3) by (4.13)and (4.14) together with an action of $SU(3)\times SU(3)/\mathbb{Z}_3$ described in §4.2 by (4.10)where the first factor SU(3) is the color group.

However then, the description of the U(1)×SU(2)-symmetry may become problematic. Either one can solve the problem by γ_5-combinatorics or one should work with a product of four factors $J_L^u, J_R^u, J_L^d, J_R^d$. But in this latter option the cancellation of anomalies could become a non-obvious task.

In view of the above framework, there is another natural way to explore which consists in adding over each spacetime point to the exceptional finite quantum space another finite quantum space in which the group U(1)×SU(2) is involved. That is to add to J_3^8another finite-dimensional Euclidean Jordan algebra which admits U(1)×SU(2) as automorphism group preserving some additional structure (like $SU(3)\times SU(3)/\mathbb{Z}_3$ for J_3^8). There is such an algebra, namely the simple Euclidean Jordan algebra

$$J_2^4 = \mathcal{H}_2(\mathbb{H}) = J\mathrm{Spin}(5)$$

of the hermitian 2×2-matrices with entries in the field of quaternions $\mathbb{H}$. Let us explain this.

A quaternion q can be viewed as two complexes z_1 and z_2

$$q = (z_1, z_2) = z_1 + z_2 j \tag{4.15}$$

and the subgroup of $\mathrm{Aut}(\mathbb{H})=SU(2)/\mathbb{Z}_2$ which is complex linear, i.e. which preserves the imaginary i , is the group U(1) acting as $(z_1,z_2)\mapsto(z_1,e^{i\theta}z_2)$, that is

$$z_1 + z_2 j \mapsto z_1 + e^{i\theta} z_2 j \tag{4.16}$$

for $e^{i\theta}\in U(1)$. Thus U(1) is the analog for $\mathbb{H}$ to SU(3) for $\mathbb{O}$.

Let now consider an element of J_2^4 written in terms of $q=z_1+z_2j$

$$\begin{pmatrix} \xi_1 & q \\ \bar{q} & \xi_2 \end{pmatrix} = \begin{pmatrix} \xi_1 & z_1 + z_2 j \\ \bar{z}_1 - z_2 j & \xi_2 \end{pmatrix} = \begin{pmatrix} \xi_1 & z_1 \\ \bar{z}_1 & \xi_2 \end{pmatrix} + z_2 \begin{pmatrix} 0 & j \\ -j & 0 \end{pmatrix}$$

where $\xi_1,\xi_2\in\mathbb{R}$. Let $U\in SU(2)$ be an arbitrary unitary 2×2 complex matrix of determinant 1. By using $jz = \bar{z}j$ for $z\in C$ one gets

$$U \begin{pmatrix} 0 & j \\ -j & 0 \end{pmatrix} U^* = \begin{pmatrix} 0 & j \\ -j & 0 \end{pmatrix} \tag{4.17}$$

from which it follows that the action of $(e^{i\theta},U)\in U(1)\times SU(2)$

$$\begin{pmatrix} \xi_1 & z_1 \\ \bar{z}_1 & \xi_2 \end{pmatrix} + z_2 \begin{pmatrix} 0 & j \\ -j & 0 \end{pmatrix} \mapsto U \begin{pmatrix} \xi & z_1 \\ \bar{z}_1 & \xi_2 \end{pmatrix} U^* + e^{i\theta} z_2 \begin{pmatrix} 0 & j \\ -j & 0 \end{pmatrix}$$

is an automorphism of J_2^4 which preserves the representation of the quaternions occurring in the elements of J_2^4 as elements of $\mathbb{C}\oplus\mathbb{C}$. Furthermore any such an automorphism is of this form. Thus U(1) is the analog for J_2^4 of the color SU(3) for J_3^8 while SU(2) is the analog for J_2^4 of the other SU(3) for J_3^8 (with the action described by (4.4) in §4.2). Notice that the SU(2)-action is an action of $SU(2)/\mathbb{Z}_2=SO(3)$.

It is too early to know whether the addition of the factor J_2^4 is relevant but if it occurs it is tempting to add the factor $J_1^2 = \mathbb{R}$ since then we would have

$$J = \oplus_{k=1}^{3} J_k^{2^k} \tag{4.18}$$

as finite-dimensional Euclidean Jordan algebra which looks like the semi-simple part of some more elaborated object. Notice that J is the real Jordan algebra of hermitian elements of

$$\mathbb{C} \oplus M_2(\mathbb{H}) \oplus M_3(\mathbb{O}) \tag{4.19}$$

which is not associative in view of the occurrence of the last term $M_3(\mathbb{O})$.

Charge conjugation

In the identification of the octonion algebra $\mathbb{O}$ as $\mathbb{C}\oplus\mathbb{C}^3$, the octonionic conjugate of (z,Z)=z+Z is $\overline{(z,Z)} = (\bar{z},-Z) = \bar{z} - Z$.

On the other hand if one interprets $Z\in\mathbb{C}^3$ as an internal quark state and $z\in\mathbb{C}$ as an internal state of the corresponding lepton, the charge conjugation C induces the complex conjugation in $\mathbb{C}^3$ and in C that is

$$C(z + Z) = \bar{z} + \bar{Z} \tag{4.20}$$

for $z\in\mathbb{C}$, $Z\in\mathbb{C}^3$.

It is easy to verify by using Formula (2.8) that C is an automorphism of $\mathbb{O}$ ($C\in G_2$) with $C^2=\mathbb{1}$ by construction. This automorphism of $\mathbb{O}$ induces canonically an involutive automorphism of the Jordan algebra J_3^8 as well as of the J_3^8-modules. These automorphisms will be all denoted by C.

DIFFERENTIAL CALCULUS OVER JORDAN ALGEBRAS

First order differential calculi

Given a category **C** of algebras over some field $\mathbb{K}$ and an algebra A in **C**, a pair $(\Omega^1,d)$of an A-bimodule Ω^1 for **C**[14] and a derivation d of A into Ω^1, that is a $\mathbb{K}$-linear mapping satisfying (the Leibniz rule)

$$d(xy) = d(x)y + xd(y)$$

for any x,y∈A, will be referred to as a first order differential calculus over A for **C** or simply a first order calculus over A when no confusion arises. There is an obvious notion of homomorphism of first order differential calculi over A. This terminology as well as the one for higher order differential calculi was introduced in [40] for the case of the category of associative algebras on a field $\mathbb{K}$.

In the following we shall be concerned with the category of real unital Jordan algebras so, if J is such a unital Jordan algebra, a J-bimodule is a unital J-module which will be simply referred to as a J-module when no confusion arises.

Let J be a unital Jordan algebra and let us define its center Z(J) by

$$Z(J)=\{z\in J\,|\,[x,y,z]=[x,z,y]=0,\forall x,y\in J\} \quad (5.1)$$

where [x,y,z]=(xy)z−x(yz) denotes the associator of x,y,z∈J. The center Z(J)is a unital associative subalgebra of J . The Lie algebra Der(J) of all derivations of J into itself is also a Z(J)-module and we denote by $\Omega^1_{\mathrm{Der}}(J)$ the J -module of all the Z(J)-homomorphisms of Der(J) into J. One defines a derivation

$$d_{\mathrm{Der}}:J\to\Omega^1_{\mathrm{Der}}(J)$$

by setting

$d_{\mathrm{Der}}(x)(X)=X(x)$

for any x∈J and X∈Der(J).

This first-order differential calculus $(\Omega^1_{\mathrm{Der}}(J),d_{\mathrm{Der}})$ will be referred to as the derivation-based first order differential calculus over J.

First order differential calculi over J_3^8

The Lie algebra $\mathrm{Der}(J_3^8)$ of all derivations of the exceptional Jordan algebra J_3^8 is the exceptional compact simple Lie algebra f_4. Since J_3^8 is simple $Z(J_3^8)=\mathbb{R}$ so the unital J_3^8-module $\Omega^1_{\mathrm{Der}}(J_3^8)$ is the vector space of all linear mappings of $\mathrm{Der}(J_3^8)$ into J_3^8 that is

$$\Omega^1_{\mathrm{Der}}(J_3^8)=J_3^8\otimes\mathrm{Der}(J_3^8)^*$$

where $\mathrm{Der}(J_3^8)^*=\mathfrak{f}_4^*$ denotes the dual vector space of $\mathrm{Der}(J_3^8)$.

One defines as before the derivation $d_{\mathrm{Der}}:J_3^8\to\Omega_{\mathrm{Der}}(J_3^8)$ by setting $d_{\mathrm{Der}}(x)(X)=X(x)$for any $x\in J_3^8$ and $X\in\mathrm{Der}(J_3^8)$. The first order differential calculus $(\Omega_{\mathrm{Der}}(J_3^8),d_{\mathrm{Der}})$ over J_3^8 is characterized by the following universal property.

Proposition 3. For any first order calculus (Ω^1,d)over J_3^8, there is a unique (unital) Jordan module homomorphism $i_d:\Omega^1_{\mathrm{Der}}(J_3^8)\to\Omega^1$ such that one has $d=i_d\circ d_{\mathrm{Der}}$.

Proof.

From the general structure of the J_3^8-modules, Ω^1 is of the form $\Omega^1=J_3^8\otimes E$ for some vector space E . Let (e^α) be a basis of E then $d(x)=X_\alpha(x)\otimes e^\alpha$ with $X_\alpha\in\mathrm{Der}(J_3^8)$ for any α . Let (∂_k) be a basis of $\mathrm{Der}(J_3^8)$ with dual basis θ^k then one has $d(x)=\partial_k(x)\otimes C^k_\alpha e^\alpha$ with the $C^k_\alpha\in\mathbb{R}$. Define i_d by $i_d(x\otimes\theta^k)=x\otimes C^k_\alpha e^\alpha$, then i_d is a J_3^8-module homomorphism satisfying $d=i_d\circ d_{\mathrm{Der}}$ which is clearly unique.

Remark.

Proposition 3 is very specific to the exceptional Jordan algebra J_3^8 and is a direct consequence of the fact that J_3^8 is, as J_3^8-module, the only irreducible module. For instance if J is a finite-dimensional simple Euclidean Jordan algebras of Theorem 2 which is distinct of $\mathbb{R}$ and of J_3^8, the first order differential calculus $(\Omega^1_{\mathrm{Der}}(J),d_{\mathrm{Der}})$ is not universal since then there several inequivalent irreducible J-modules and not only J itself. In all these cases one has $\Omega^1_{\mathrm{Der}}(J)=J\otimes\mathrm{Der}(J)^*$ and Der(J) is a real compact form of a classical simple Lie algebra. Furthermore all the classical simple Lie algebras are realized in this way as Lie algebras of derivations of the finite-dimensional simple special Euclidean Jordan algebras.

Differential graded Jordan algebras

Let us define a differential graded Jordan algebra to be a $\mathbb{N}$-graded algebra

$$\Omega=\oplus_{n\in\mathbb{N}}\Omega^n$$

which is a Jordan superalgebra (see e.g. in [29] and [33]), for the induced $\mathbb{Z}/2\mathbb{Z}$ degree and which is equipped with a differential, that is with an antiderivation d of degree 1 and of square 0.

Thus Ω is a graded-commutative algebra, that is one has

$$ab=(-1)^{|a||b|}ba\in\Omega^{|a|+|b|} \quad (5.2)$$

for $a\in\Omega^{|a|}$, $b\in\Omega^{|b|}$, and one has the graded Jordan identity

$$(-1)^{|a||c|}[L_{ab},L_c]_{\mathrm{gr}}+(-1)^{|b||a|}[L_{bc},L_a]_{\mathrm{gr}}+(-1)^{|c||b|}[L_{ca},L_b]_{\mathrm{gr}}=0 \quad (5.3)$$

for $a\in\Omega^{|a|}$,$b\in\Omega^{|b|}$,$c\in\Omega^{|c|}$ where L_a is the left-multiplication operator by a∈Ωdefined by $L_a(x)=ax$ for any x∈Ω and where $[\bullet,\bullet]_{\mathrm{gr}}$ denotes the graded commutator

$$[A,B]_{\mathrm{gr}}=AB-(-1)^{|a||b|}BA$$

for A of degree |a| and B of degree |b|. The differential d satisfies

$$d^2=0$$

$$d\Omega^n\subset\Omega^{n+1}$$

for any n∈$\mathbb{N}$, and the graded Leibniz rule

$$d(ab)=d(a)b+(-1)^{|a|}ad(b)$$

for $a\in\Omega^{|a|}$, b∈Ω.

In the following the real unital differential graded Jordan algebras will be our models for the "algebras of differential forms" on "quantum spaces".

Let J be a Jordan algebra. A differential graded Jordan algebra $\Omega=\oplus_{n\geq0}\Omega^n$ such that $\Omega^0=J$ will be called a differential calculus over J.

Higher order derivation-based differential calculi

Let J be a unital Jordan algebra with center Z(J) and let $\Omega^n_{\mathrm{Der}}(J)$ be the J-module of all n -Z(J)-linear antisymmetric mappings of Der(J) into J , that is $\omega\in\Omega^n_{\mathrm{Der}}(J)$ is a Z(J)-linear mapping

$$\omega:\wedge^n_{Z(J)}\mathrm{Der}(J)\to J$$

of the n -th exterior power over Z(J) of the Z(J)-module Der(J) into J as a Z(J)-module. Then $\Omega_{\mathrm{Der}}(J)=\oplus_{n\geq 0}\Omega^n_{\mathrm{Der}}(J)$ is canonically a differential graded Jordan algebra with differential given by the Chevalley–Eilenberg formula

$$d\omega(X_0,\cdots,X_n)=\sum_{0\leq k\leq n}(-1)^k X_k\,\omega(X_0,\overset{k}{\check{\cdots}},X_n)$$
$$+\sum_{0\leq r<s\leq n}(-1)^{r+s}\,\omega([X_r,X_s],X_0,\overset{r}{\check{\cdots}}\overset{s}{\check{\cdots}},X_n) \quad (5.4)$$

for $\omega\in\Omega^n_{\mathrm{Der}}(J)$ and $X_p\in\mathrm{Der}(J)$. Thus $\Omega_{\mathrm{Der}}(J)$ is a differential calculus over Jwhich will be referred as the derivation-based differential calculus over J.

If J is a finite-dimensional simple Euclidean Jordan algebra (i.e. one of the list of Theorem 2), one has

$$\Omega_{\mathrm{Der}}(J)=J\otimes\wedge\mathrm{Der}(J)^* \quad (5.5)$$

where $\wedge\mathrm{Der}(J)^*$ is the exterior algebra of the dual of the finite-dimensional real Lie algebra Der(J). In the case where J is the exceptional Jordan algebra J^8_3, this differential calculus is characterized by the following universal property.

Proposition 4.

Any homomorphism φ of unital Jordan algebra of J^8_3into the Jordan subalgebra $\Omega^0$of a unital differential graded Jordan algebra $\Omega=\oplus\Omega^n$has a unique extension $\tilde{\varphi}:\Omega_{\mathrm{Der}}(J^8_3)\to\Omega$as a homomorphism of differential graded Jordan algebras.

Proof (sketch).

By using the fact that the J^8_3-modules are of the form $J^8_3\otimes\mathcal{E}$ for some vector spaces Eand the definition of the $\mathbb{N}$-graded Jordan superalgebras, Formulae (5.2) and (5.9), it follows that any unital differential graded Jordan algebra $\Omega=\oplus\Omega^n$ which contains J^8_3 as unital Jordan subalgebra of Ω^0 is as algebra of the form

$$\Omega=J^8_3\otimes\Omega_{(0)}$$

where $\Omega_{(0)}=\oplus\Omega^n_{(0)}$ is an associative graded-commutative algebra (this generalizes a classical result in the non-graded case). The proof follows then from Proposition 3, from $d^2=0$ with the graded Leibniz rule for d and from the simplicity of J^8_3.

A more detailed proof will appear in a forthcoming paper.

Remark.

Notice that as Proposition 3, Proposition 4 is very specific to the exceptional Jordan algebra J^8_3.

As a consequence all differential calculus over J^8_3 which are generated as differential graded Jordan algebra by J^8_3 are quotients of $\Omega_{\mathrm{Der}}(J^8_3)$, in particular among these, one has all the differential graded Jordan algebras of the form

$$\Omega=J^8_3\otimes\wedge\mathfrak{g}^* \quad (5.6)$$

where g is a Lie subalgebra of $\mathrm{Der}(J^8_3)=\mathfrak{f}_4$, the differential being the Chevalley–Eilenberg differential (g acting by derivations on J^8_3).

Differential calculi as A_∞-algebras

Let A be a complex unital *-algebra and let $\Omega_u(\mathcal{A})=\oplus_{n\geq 0}\Omega^n_u(\mathcal{A})$ be the universal differential calculus over A, [30]. It is shown in [40] (see also in [14]) that there is a unique involution on Ω_u(A) which extends the involution of A for which it is a differential graded *-algebra. This means that one has

$$\begin{cases}(\alpha\beta)^*=(-1)^{ab}\beta^*\alpha^*\\ d(\gamma^*)=(d\gamma)^*\end{cases} \quad (5.7)$$

for any $\alpha\in\Omega^a_u$ (A), $\beta\in\Omega^b_u$ (A) and $\gamma\in\Omega_u$ (A). The cohomology of Ω_u (A) is trivial, i.e. Hn(Ω_u (A)) = 0 for n ≥ 1 and H0(Ω_u (A)) = C, and in fact by using a linear form ω on Asuch that ω(1) =1, one constructs a linear mapping K of degree -1 on Ω_u (A)such that

$$dK+Kd=I \quad (5.8)$$

on $\oplus_{n\geq 1}\Omega^n_u(\mathcal{A})$ (see in [14]). By replacing K by $\frac{1}{2}(K+K^*)$ one can assume that $K(\alpha^*)=K(\alpha)^*$ for any $\alpha\in\Omega_u$(A).

Let J=H(A) be the real Jordan algebra of the hermitian elements of A, i.e. $J=\{h\in A|h^*=h\}$ endowed with the product

$$h\circ h'=\frac{1}{2}(hh'+h'h)$$

for h,h'∈J. Then the real graded subspace Ω(J) of Ω_u(A) of hermitian elements is a real differential graded Jordan algebra, that is a differential calculus over J for the graded Jordan product defined by

$$\alpha\circ\beta=\frac{1}{2}(\alpha\beta+(-1)^{ab}\beta\alpha) \quad (5.9)$$

for $\alpha\in\Omega^a(J),\beta\in\Omega^b(J)$.

The contracting homotopy K restricts to Ω(J). By using K, one constructs by induction on n a sequence of product $m_n(\alpha_1,\cdots,\alpha_n)$ of respective degrees 2−n on Ω(J)starting with $m_1(\alpha)=d\alpha$ and $m_2(\alpha,\beta)=\alpha\circ\beta$ such that, endowed with these products, Ω(J) is an A_∞-algebra [32]. This implies in particular that the graded Jordan product α∘β of Ω(J) is associative up to homotopy. In fact, by taking

$$m_3(\alpha,\beta,\gamma)=K((\alpha\circ\beta)\circ\gamma-\alpha\circ(\beta\circ\gamma)) \quad (5.10)$$

for $\alpha\in\Omega^a(J),\beta\in\Omega^b(J),\gamma\in\Omega^c(J)$ with a+b+c≥1, one gets

$$(\alpha\circ\beta)\circ\gamma-\alpha\circ(\beta\circ\gamma)=d(m_3)(\alpha,\beta,\gamma)$$

i.e. the associativity up to homotopy of the graded Jordan product.

The occurrence of A_∞-structures in the present context is not accidental. For instance it will be shown in another paper that, for the simple finite dimensional Euclidean Jordan algebras of Theorem 2, the derivation-based differential calculus admits such A_∞-structures with m_1 given by the differential and m_2 given by the graded Jordan product.

CONNECTIONS ON JORDAN MODULES

Derivation-based connections. First definition

Let J be a unital Jordan algebra with center (= centroid) Z(J) and let Der(J) be the Lie algebra and Z(J)-module of derivations of J. A (derivation-based) connection ∇ on a unital J-module M is a linear mapping $X \mapsto \nabla_X$ of Der(J) into the linear endomorphisms of M such that

$$\begin{cases} \nabla_X(xm) = X(x)m + x\nabla_X(m) \\ \nabla_{zX}(m) = z\nabla_X(m) \end{cases} \tag{6.1}$$

for any m∈M, x∈J and z∈Z(J).

It follows from this definition that the difference $\nabla - \nabla'$ between 2 connections on M is a Z(J)-linear mapping of Der(J) into the Z(J)-module of all the J-module endomorphisms of M. It also follows that

$$R_{X,Y} = [\nabla_X, \nabla_Y] - \nabla_{[X,Y]} \tag{6.2}$$

satisfies

$$\begin{cases} R_{X,Y}(xm) = xR_{X,Y}(m) \\ R_{zX,Y}(m) = zR_{X,Y}(m) \end{cases} \tag{6.3}$$

in other words that R is a Z(J)-linear mapping of $\wedge^2_{Z(J)}\mathrm{Der}(J)$ into the Z(J)-module of all the J-module endomorphism of M which will be referred to as the curvature of ∇.

It is clear that if g is a Lie subalgebra and a Z(J)-submodule of Der(J), the restriction of (6.1) to g, i.e. for X∈g, makes sense and the corresponding notion will be referred to as a derivation-based g-connection on M.

Derivation-based connections. Second definition

Let J , Z(J), Der(J) and M be as in §6.1 and let $\Omega^n_{\mathrm{Der}}(M)$ be the J-module of all n -Z(J)-linear antisymmetric mapping of Der(J) into M , which means that $\Phi \in \Omega^n_{\mathrm{Der}}(M)$ is a Z(J)-linear mapping

$$\Phi : \wedge^n_{Z(J)}\mathrm{Der}(J) \to M$$

with the notations of §5.4. The graded J -module $\Omega_{\mathrm{Der}}(M) = \oplus_{n\geq 0}\Omega^n_{\mathrm{Der}}(M)$ in naturally a graded Jordan module over the graded Jordan algebra $\Omega_{\mathrm{Der}}(J)$: The product of $\omega \in \Omega^m_{\mathrm{Der}}(J)$ with $\Phi \in \Omega^n_{\mathrm{Der}}(M)$ is the element $\omega\Phi \in \Omega^{m+n}_{\mathrm{Der}}(M)$ obtained by product of evaluations on derivations followed by antisymmetrization in the derivations. A (derivation-based) connection on M is a linear endomorphism ∇ of $\Omega_{\mathrm{Der}}(M)$ such that

$$\begin{cases} \nabla(\Omega^n_{\mathrm{Der}}(M)) \subset \Omega^{n+1}_{\mathrm{Der}}(M) \\ \nabla(\omega\Phi) = d(\omega)\Phi + (-1)^m\omega\nabla(\Phi) \end{cases} \tag{6.4}$$

for any m,n∈$\mathbb{N}$, $\omega \in \Omega^m_{\mathrm{Der}}(J)$ and $\Phi \in \Omega_{\mathrm{Der}}(M)$. Let ∇ be such a connection and define $\nabla_X m$ by

$$\nabla(m)(X) \tag{6.5}$$

for $m \in M = \Omega^0_{\mathrm{Der}}(M)$ and X∈Der(J). Then $X \mapsto \nabla_X$ is a connection on M in the sense of § 6.1. Conversely if $X \mapsto \nabla_X$ is a connection in the sense of §6.1 one defines ∇ on $\Omega^n_{\mathrm{Der}}(M)$ by

$$\begin{aligned} \nabla(\Phi)(X_0, \cdots, X_n) &= \textstyle\sum_{p=0}^n (-1)^p \nabla_{X_p}(\Phi(X_0, \cdots \overset{p}{\vee} \cdots, X_n)) \\ &+ \textstyle\sum_{0\leq r<s\leq n} (-1)^{r+s}\, \Phi([X_r, X_s], X_0, \cdots \overset{r}{\vee} \cdots \overset{s}{\vee} \cdots, X_n) \end{aligned} \tag{6.6}$$

for $\Phi \in \Omega^n_{\mathrm{Der}}(M)$, X_p∈Der(J). This ∇ satisfies the axioms (6.4). We shall refer to ∇ as the covariant differential while ∇_X for X∈Der(J) will be referred to as the covariant derivative along X.

From the axioms (6.4), it follows that one has

$$\nabla^2(\omega\Phi) = \omega\nabla^2(\Phi) \tag{6.7}$$

for any $\omega \in \Omega_{\mathrm{Der}}(J)$ and $\Phi \in \Omega_{\mathrm{Der}}(M)$ furthermore

$$\nabla^2(m)(X, Y) = R_{X,Y}(m)$$

for m∈M. Thus ∇^2 is an homomorphism of $\Omega_{\mathrm{Der}}(J)$-module which is also called the curvature of ∇ since it is expressible in terms of the $R_{X,Y}$ and conversely.

General connections

Let $\Omega = \oplus_{n\geq 0}\Omega^n$ be a differential graded Jordan algebra and let $\Gamma = \oplus_{n\geq 0}\Gamma^n$ be a graded Jordan module over the graded Jordan algebra Ω (the axioms for this notion are easy to guess). A connection on Γ will be defined to be a linear endomorphism ∇ of Γ satisfying

$$\begin{cases} \nabla(\Gamma^n) \subset \Gamma^{n+1} \\ \nabla(\omega\Phi) = d(\omega)\Phi + (-1)^m\omega\nabla(\Phi) \end{cases} \tag{6.8}$$

for m,n∈$\mathbb{N}$, $\omega \in \Omega^m$ and $\Phi \in \Gamma$.

The axioms (6.8) imply that one has

$$\nabla^2(\omega\Phi) = \omega\nabla^2(\Phi) \tag{6.9}$$

for any $\omega \in \Omega$ and $\Phi \in \Gamma$, so ∇^2 is an homogeneous Ω-module homomorphism of degree 2 which will be refereed to as the curvature of the connection ∇.

Notice that the formalism of 6.2 is a particular case of this formalism and that within it, the Bianchi identity reduces to the trivial identity

$\nabla\nabla^2 = \nabla^2\nabla$

that is to the associativity of the composition of the endomorphism ∇.

TENTATIVE CONCLUSION

From a physical point of view, it is clear that what is described in these notes is quite incomplete: One should write some dynamics. Before that, one must develop several points.

Firstly one must understand the formulation of the U(1)×SU(2)-symmetry in this frame. Since the new suspected particles have $\mathbb{R}$ as internal spaces, it is suggested that they are described by Majorana fermions in a 4-Lorentzian approach where the charge conjugation is represented by complex conjugation of components. Notice that, as shown in [5], the KO-dimension 6 is completely natural in the Lorentzian framework. Introducing such new particles does not spoil the anomalies cancellation of the standard model. However it is not clear that this is consistent with the usual formulation of the U(1)×SU(2)-symmetry.

A second point to understand is the formulation of the first-order condition. Indeed the definition of the first-order operators between bimodules over associative algebras is clear [10] as well as the noncommutative generalization of their symbols [17]. This notion of first-order operator is fundamental in the noncommutative geometry approach to the standard model of [8] and [9]. In contrast, the definition of first-order operator between Jordan (bi)modules is not straightforward in spite of the fact that one can write explicitly what it is for modules over the exceptional algebra J_3^8.

It seems that one should make some progress on these two points before tentative formulations of the dynamics. Then there are two natural ways to take and I think that both should be pushed. The first one is to try to adapt the spectral action principle to the present frame, the second one is to directly use the differential calculus and the connections over Jordan modules. The first approach, if it is possible, would probably be the most economical one. However two difficulties are that the spectral action principle is formulated in the frame of the Euclidean signature (instead of the Lorentzian one) and that there is the occurrence of the exceptional Jordan algebra here which needs some care. The second approach which is closer to the usual formulation of gauge theory will fully use the development of Sections 4, 5 and 6.

This work is currently in progress. In any case it is an occasion for starting to develop the differential calculus over Jordan algebras.

ACKNOWLEDGEMENTS

From the beginning of my interest in the subject, Raymond Stora gave me useful advices and related references.

I thank Jean Iliopoulos for his kind interest and the discussions we have had for several years.

I also thank Shane Farnsworth, Gianni Landi, Todor Popov and Ivan Todorov for discussions.

REFERENCES

1. J.F. Adams, Spin (8), triality, F4 and all that, in: Hawking, Rocek (Eds.), Superspace and Supergravity, Cambridge University Press, 1980.
2. A.A. Albert, On a certain algebra of quantum mechanics, Ann. Math. 35 (1934) 65–73.
3. A.A. Albert, A theory of power-associative commutative algebras, Trans. Am. Math. Soc. 69 (1950) 503–527.
4. J. Baez, The octonions, Bull. Am. Math. Soc. 39 (2002) 145–205.
5. J.W. Barrett, Lorentzian version of the noncommutative geometry of the standard model of particle physics, J. Math. Phys. 48 (2007) 012303.
6. W. Bertram, The Geometry of Jordan and Lie Structures, Springer, 2000.
7. L. Boyle, S. Farnsworth, Non-commutative geometry, non-associative geometry and the standard model of particle physics, New J. Phys. 16 (2014) 123027.
8. A. Chamseddine, A. Connes, The spectral action principle, Commun. Math. Phys. 186 (1997) 731–750.
9. A.H. Chamseddine, A. Connes, M. Marcolli, Gravity and the standard model with neutrino mixing, Adv. Theor. Math. Phys. 11 (2007) 991–1089.
10. A. Connes, Non-Commutative Geometry, Academic Press, 1994.
11. A. Connes, J. Lott, Particle models and noncommutative geometry, Nucl. Phys. B, Proc. Suppl. 18 (1990) 29–47.
12. G.M. Dixon, Division Algebras: Octonions, Quaternions, Complex Numbers and the Algebraic Design of Physics, Kluwer Academic Publishers, 1994.
13. M. Dubois-Violette, Non-commutative differential geometry, quantum mechanics and gauge theory, in: C. Bartocci, U. Bruzzo, R. Cianci (Eds.), Differential Geometric Methods in Theoretical Physics, Springer Verlag, 1991, pp. 13–24.
14. M. Dubois-Violette, Lectures on graded differential algebras and noncommutative geometry, in: Y. Maeda, et al. (Eds.), Noncommutative Differential Geometry and Its Applications to Physics, Shonan, Japan, 1999, pp. 245–306, Kluwer Academic Publishers, 2001.
15. M. Dubois-Violette, R. Kerner, J. Madore, Gauge bosons in a non-commutative geometry, Phys. Lett. B 217 (1989) 485–488.
16. M. Dubois-Violette, R. Kerner, J. Madore, Non-commutative differential geometry and new models of gauge theory, J. Math. Phys. 31 (1990) 323–329.
17. M. Dubois-Violette, T. Masson, On the first-order operators in bimodules, Lett. Math. Phys. 37 (1996) 467–474.
18. M. Dubois-Violette, T. Masson, SU(n)-gauge theories in noncommutative differential geometry, J. Geom. Phys. 25 (1998) 104–118.
19. S. Farnsworth, L. Boyle, Rethinking Connes' approach to the standard model of particle physics via noncommutative geometry, New J. Phys. 17 (2014) 023021.
20. S. Farnsworth, L. Boyle, Non-associative geometry and the spectral action principle, J. High Energy Phys. 1507 (2015) 023.
21. M. Günaydin, F. Gürsey, Quark structure and octonions, J. Math. Phys. 14 (1973) 1651–1667.
22. M. Günaydin, F. Gürsey, Quark statistics and octonions, Phys. Rev. D 9 (1974) 3387–3391.
23. M. Günaydin, C. Piron, H. Ruegg, Moufang plane and octionic quantum mechanics, Commun. Math. Phys. 61 (1978) 69–85.
24. L.P. Horwitz, L.C. Biedenharn, Exceptional gauge groups and quantum theory, J. Math. Phys. 20 (1979) 269–298.
25. B. Iochum, G. Loupias, Banach power-associative algebras and J-B algebras, Ann. Inst. Henri Poincaré 43 (1985) 211–225.
26. R. Iordanescu, Jordan Structures in Analysis, Geometry and Physics, Ed. Acad. Romane, 2009.

27. N. Jacobson, Structure and Representations of Jordan Algebras, American Mathematical Society, 1968.
28. P. Jordan, J. von Neumann, E. Wigner, On an algebraic generalization of the quantum mechanical formalism, Ann. Math. 36 (1934) 29–64.
29. V.G. Kac, Classification of simple Z-graded Lie superalgebras and simple Jordan superalgebras, Commun. Algebra 5 (1977) 1375–1400.
30. M. Karoubi, Homologie cyclique des groupes et algèbres, C. R. Acad. Sci. Paris, Sér. I 297 (1983) 381–384.
31. I. Kashuba, S. Ovsienko, I. Shestakov, Representation type of Jordan algebras, Adv. Math. 226 (2011) 385–418.
32. B. Keller, Introduction to $A\infty$-algebras and modules, Homol. Homotopy Appl. 3 (2001) 1–35.
33. C. Martinez, E. Zelmanov, Representations of Jordan superalgebras, Trans. Am. Math. Soc. 362 (2010) 815–846.
34. K. McCrimmon, A Taste of Jordan Algebras, Springer, 2004.
35. P. Ramond, Introduction to exceptional Lie groups and algebras, 1976, CALTECH-68-577 (unpublished).
36. P. Ramond, Exeptional groups in physics, in: J.P. Gazeau, R. Kerner, J.P. Antoine, S. Melens, J.Y. Thibon (Eds.), 24th Int. Coll. Group Theoretical Methods in Physics, in: Inst. Phys. Conf. Ser., vol. 173, Institute of Physics, 2003, pp. 129–138.
37. G. Roepstorff, Superconnections: an interpretation of the standard model, in: Mathematical Physics and Quantum Field Theory, Electron. J. Differ. Equ., Conf. 04 (2000) 165–174.
38. R.D. Schafer, An Introduction to Nonassociative Algebras, Dover Publication, 1995.
39. J. von Neumann, On an algebraic generalization of the quantum mechanical formalism (Part I), Mat. Sb. 1 (1936) 415–484.
40. S.L. Woronowicz, Differential calculus on compact matrix pseudogroups (quantum groups), Commun. Math. Phys. 122 (1989) 129–170.
41. I. Yokota, Exceptional Lie groups, arXiv:0902.0431, 2009.
42. E. Zelmanov, Primary Jordan algebras, Algebra Log. 18 (1979) 103–111.
43. E. Zelmanov, On prime Jordan algebras II, Sib. Math. J. 24 (1983) 73–85.

CHAPTER 7

PHOTON AND PHOTINO AS NAMBU–GOLDSTONE ZERO MODES IN AN EMERGENT SUSY QED

J. L. Chkareuli

INTRODUCTION AND OVERVIEW

It has long been believed that spontaneous Lorentz invariance violation (SLIV) may lead to the emergence of massless Nambu–Goldstone (NG) zero modes [1, 2], which are identified with photons and other gauge fields appearing in the Standard Model. This old idea [3–5], supported by a close analogy with the dynamical origin of massless particle excitations for spontaneously broken internal symmetries, has gained new impetus in recent years. On the other hand, besides its generic implication for the possible origin of physical gauge fields [6–12] in a conventional quantum field theory (QFT) framework, there are many different contexts in the literature where Lorentz violation may stem itself from string theory [13, 14], quantum gravity [15] or any unspecified dynamics at an ultraviolet scale perhaps related to the Planck scale [16–22]. Though we are mainly focused on the spontaneous Lorentz violation in QFT, particularly in QED and the Standard Model, we give below some brief comments on other approaches and we make clearer the aims and results of the present work.

Vector NG bosons in gauge theories. Inactive SLIV

When speaking about SLIV, one important thing to notice is that, in contrast to the spontaneous violation of internal symmetries, it seems not to necessarily imply a physical breakdown of Lorentz invariance. Rather, when appearing in a gauge theory framework, this may ultimately result in a non-covariant gauge choice in an otherwise gauge invariant and Lorentz invariant theory. In substance, the SLIV ansatz, due to which the vector field develops a vacuum expectation value (VEV)

$$\langle A_\mu(x)\rangle = n_\mu M \tag{1}$$

(where n_μ is a properly oriented unit Lorentz vector, $n^2= n_\mu n^\mu = \pm 1$, while MM is the proposed SLIV scale) may itself be treated as a pure gauge transformation with a gauge function linear in the coordinates, $\omega(x) = n_\mu x^\mu M$. From this viewpoint gauge invariance in QED leads to the conversion of SLIV into gauge degrees of freedom of the massless photon emerged.

A good example for such a kind of SLIV, which we call the "inactive" SLIV hereafter, is provided by the nonlinearly realized Lorentz symmetry for the underlying vector field $A_\mu(x)$ through the length-fixing constraint

$$A_\mu A^\mu = n^2 M^2. \tag{2}$$

This constraint in the gauge invariant QED framework was first studied by Nambu a long ago [23], and in more detail in recent years [24–28]. The constraint (2) is in fact very similar to the constraint appearing in the nonlinear σ-model for pions [29], $\sigma^2 + \pi^2 = f_\pi^2$, where f_π, where fπfπ is the pion decay constant. Rather than imposing it by postulate, the constraint (2) may be implemented into the standard QED Lagrangian extended by the invariant Lagrange multiplier term

$$\mathcal{L} = L_{\text{QED}} - \frac{\lambda}{2}\left(A_\mu A^\mu - n^2 M^2\right) \tag{3}$$

provided that the initial values for all fields (and their momenta) involved are chosen so as to restrict the phase space to values with a vanishing multiplier function λ(x), λ=0. Otherwise, as

Citation: J. L. Chkareuli, Photon and photino as Nambu–Goldstone zero modes in an emergent SUSY QED, DOI: 10.1140/epjc/s10052-014-2906-3.

was shown in [30] (see also [27]), it might be problematic to have a ghost-free QED model with a positive Hamiltonian.[1]

One way or the other, the constraint (2) means in essence that the vector field A_μ develops the VEV (1) and Lorentz symmetry SO(1, 3) breaks down to SO(3) or SO(1, 2),, depending on whether the unit vector n_μ is time-like ($n^2<0$) or space-like ($n^2<0$). The point, however, is that, in sharp contrast to the nonlinear σ-model for pions, the nonlinear QED theory, due to gauge invariance in the starting Lagrangian L_{QED}, leaves physical Lorentz invariance intact. Indeed, the nonlinear QED contains a plethora of Lorentz and CPT violating couplings when it is expressed in terms of the pure vector NG boson modes (a_μ) associated with a physical photon,

$$A_\mu = a_\mu + n_\mu (M^2 - n^2 a^2)^{\frac{1}{2}}, \quad n_\mu a_\mu = 0 \quad (a^2 \equiv a_\mu a^\mu), \tag{4}$$

including the effective Higgs mode given by the second term in (4) properly expanded in a power series of a^2. However, the contributions of all these couplings to physical processes completely cancel out among themselves, as was shown in the tree [23] and one-loop approximations [24]. Actually, the nonlinear constraint (2) implemented as a supplementary condition can be interpreted in essence as a possible gauge choice for the starting vector field A_μ. Meanwhile the SS-matrix remains unaltered under such a gauge convention unless gauge invariance in the theory turns out to be really broken (see the next subsection) rather than merely being restricted by gauge condition (2). A later similar result concerning the inactive SLIV in gauge theories was also confirmed for spontaneously broken massive QED [25], non-Abelian theories [26], and tensor field gravity [28].

Remarkably enough, the nonlinear QED model (3) may be considered in some sense as originating from a conventional QED Lagrangian extended by vector field potential energy terms,

$$\mathcal{L}' = L_{\mathrm{QED}} - \frac{\lambda}{4}\left(A_\mu A^\mu - n^2 M^2\right)^2 \tag{5}$$

(where λ is a coupling constant) rather than by the Lagrange multiplier term. This is the simplest example of a theory sometimes referred to as the "bumblebee" model (see [11, 12] and references therein) where physical Lorentz symmetry could in principle be spontaneously broken due to the presence of an active Higgs mode in the model. On the other hand, the Lagrangian (5) taken in the limit $\lambda\to\infty$ can formally be regarded as the nonlinear QED. Actually, both models are physically equivalent in the infrared energy domain, where the Higgs mode is considered infinitely massive. However, as was argued in [30], a bumblebee-like model appears generally unstable; its Hamiltonian is not bounded from below unless the phase space sector is not limited by the nonlinear vector field constraint $A_\mu A^\mu = n^2 M^2$ (2). With this condition imposed, the massive Higgs mode never appears, the Hamiltonian is positive, and the model is physically equivalent to the constraint-based nonlinear QED (3) with the inactive SLIV, which does not lead to physical Lorentz violation.[2]

To summarize, we have considered above the standard QED with vector field constraint (2) being implemented into the Lagrangian through the Lagrange multiplier term (3). In crucial contrast to internal symmetry breaking (say, the breaking of a chiral SU(2)× SU(2) symmetry in the nonlinear σ-model for pions), SLIV caused by a similar σ-model type vector field constraint, (2), does not lead to physical Lorentz violation. Indeed, though SLIV induces the vector Goldstone-like states (4), all observable SLIV effects appear to be completely canceled out among themselves due to the generic gauge invariance of QED. We call it the inactive SLIV in the sense that one may have Goldstone-like states in a theory but may have not a non-zero symmetry breaking effect. This is a somewhat new and unusual situation that just happens with SLIV in gauge invariant theories (and never in an internal symmetry breaking case). More precisely there are, in essence, two different aspects regarding the inactive SLIV (different, though related to each other). The first is the generation of Goldstone modes which inevitably happens once the nonlinear σ-model type constraint (2) is put on the vector field. The second is that gauge invariance, even being restricted by this constraint (interpreted as a gauge condition), provides a cancelation mechanism for physical Lorentz violation. As a consequence, emergent gauge theories induced by the inactive SLIV mechanism are in fact indistinguishable from conventional gauge theories. Their emergent nature can only be seen when a gauge condition is taken to be the vector field length-fixing constraint (2). Any other gauge, e.g. the Coulomb gauge, is not in line with the emergent picture, since it explicitly breaks Lorentz invariance. As to the observational evidence in favor of emergent theories, the only way for SLIV to be activated may appear if gauge invariance in these theories turns out to be broken in an explicit rather than spontaneous way. As a result, the SLIV cancelation mechanism does not work any longer and one inevitably ends up with physical Lorentz violation.

Activating SLIV by gauge symmetry breaking

Looking for some appropriate examples of physical Lorentz violation in a QFT framework one necessarily comes across the problem of proper suppression of gauge non-invariant high-dimension couplings where such a violation can in principle occur. Remarkably enough, for QED type theories with the supplementary vector field constraint (2) gauge symmetry breaking naturally appears only for five- and higher-dimensional couplings. Indeed, all dimension-four couplings are generically gauge invariant, if the vector field kinetic term has a standard $F_{\mu\nu}F^{\mu\nu}$ and, apart from relativistic invariance, the restrictions related to the conservation of parity, charge-conjugation symmetry, and fermion number conservation are generally imposed on the theory [31, 32]. With these restrictions taken, one can easily confirm that all possible dimension-five couplings are also combined by themselves in some would-be gauge invariant form provided that the vector field is constrained by the SLIV condition (2). Indeed, for charged matter fermions interacting with vector field such couplings generally amount to

$$L_{\dim 5} = \frac{1}{\mathcal{M}} \check{D}^*_\mu \overline{\psi} \cdot \check{D}^\mu \psi + \frac{G}{\mathcal{M}} A_\mu A^\mu \overline{\psi}\psi, \quad A_\mu A^\mu = n^2 M^2. \tag{6}$$

Such couplings could presumably become significant at an ultraviolet scale MM, probably close to the Planck scale MPMP. They, besides covariant derivative terms, also include an independent "sea-gull" fermion–vector field term with the coupling constant GG being in general of the order 11. The main point regarding the Lagrangian (6) is that, while it is gauge invariant in itself, the coupling constant $\check{e}$ in the covariant derivative $\check{D}^\mu = \partial^\mu + i\check{e}A^\mu$ differs in general from the coupling e in the covariant derivative $D^\mu=\partial^\mu+ieA^\mu$ in the standard Dirac Lagrangian (3)

$$L_{\rm QED} = -\frac{1}{4}F_{\mu\nu}F^{\mu\nu} + \overline{\psi}(i\gamma_\mu D^\mu - m)\psi \ . \tag{7}$$

Therefore, gauge invariance is no longer preserved in the total Lagrangian $L_{\rm QED}+L_{\rm dim\,5}$. It is worth noting that, though the high-dimension Lagrangian part $L_{\rm dim\,5}$ (6) usually only gives some small corrections to a conventional QED Lagrangian (7), the situation may drastically change when the vector field A_μ develops a VEV and SLIV occurs.

Actually, putting the SLIV parameterization (4) into the basic QED Lagrangian (7) one obtains the truly emergent model for QED being essentially nonlinear in the vector Goldstone modes a_μ associated with photons. This model contains, among other terms, the inappropriately large (while false; see below) Lorentz-violating fermion bilinear $-eM\overline{\psi}(n_\mu\gamma^\mu)\psi$. This term appears when the effective Higgs mode expansion in Goldstone modes a_μ [as is given in the parametrization (4)] is applied to the fermion current interaction term $-e\overline{\psi}\gamma_\mu A^\mu\psi$ in the QED Lagrangian (7). However, due to local invariance this bilinear term can be gauged away by making an appropriate redefinition of the fermion field $\psi \to e^{-ie\omega(x)}\psi$ with a gauge function $\omega(x)$ linear in coordinates, $\omega(x) = (n_\mu x^\mu)M$. Meanwhile, the dimension-five Lagrangian $L_{\rm dim\,5}$ (6) is substantially changed under this redefinition, which significantly modifies the fermion bilinear terms

$$L_{\overline{\psi}\psi} = \overline{\psi}(i\gamma_\mu\partial^\mu - m)\psi - i\Delta e\frac{M}{\mathcal{M}}n_\mu\overline{\psi}\overleftrightarrow{\partial^\mu}\psi$$
$$+[G + (\Delta e)^2 n^2]\frac{M^2}{\mathcal{M}}\overline{\psi}\psi \ , \tag{8}$$

where we retained the notation ψ for the redefined fermion field and denoted, as usually, $\overline{\psi}\overleftrightarrow{\partial^\mu}\psi = \overline{\psi}(\partial^\mu\psi)-(\partial^\mu\overline{\psi})\psi$. . Note that the extra fermion derivative terms given in (8) are produced just due to the gauge invariance breaking that is determined by the electromagnetic charge difference $\Delta e = \check{e} - e$ in the total Lagrangian $L_{\rm QED}+L_{\rm dim\,5}$. As a result, there appears the entirely new, SLIV inspired, dispersion relation for a charged fermion (taken with 4-momentum $p_\mu p_\mu$) of the type

$$p_\mu^2 \cong [m_f + 2\delta(p_\mu n^\mu/n^2)]^2, \quad m_f = \left(m - G\frac{M^2}{\mathcal{M}}\right) - \delta^2 n^2\mathcal{M}, \tag{9}$$

given to an accuracy of $O(m_f^2/\mathcal{M}^2)$ with a properly modified total fermion mass m_f. Here δ stands for the small characteristic, positive or negative, parameter $\delta = (\Delta e)M/\mathcal{M}$ of physical Lorentz violation that reflects the joint effect as is given, on the one hand, by the SLIV scale M and, on the other hand, by the charge difference Δe being a measure of an internal gauge non-invariance. Notably, the spacetime in itself still possesses Lorentz invariance, however, fermions with SLIV contributing into their total mass m_f (9) propagate and interact in it in the Lorentz non-covariant way. At the same time, the photon dispersion relation is still retained in the order 1/M considered.

So, we have shown in the above that SLIV caused by the vector field VEV (1), while being superficial in gauge invariant theory, becomes physically significant for some high value of the SLIV scale M close to the scale MM, which is proposed to be located near the Planck scale M_P. This may happen even at relatively low energies provided the gauge non-invariance caused by high-dimension couplings of the matter and vector fields is not vanishingly small. This leads, as was demonstrated in [31, 32], through special dispersion relations appearing for matter, charged fermions, to a new class of phenomena, which could be of distinctive observational interest in particle physics and astrophysics. They include a significant change in the GZK cutoff for UHE cosmic-ray nucleons, stability of high-energy pions and W bosons, modification of nucleon beta decays, and some others just in the presently accessible energy area in cosmic-ray physics.

However, though one could speculate about some generically broken or partial gauge symmetry in a QFT framework [31, 32], this seems to be too high a price for an actual Lorentz violation which may stem from SLIV. What is more: is there really any strong theoretical reason left for Lorentz invariance to be physically broken, if emergent gauge fields are anyway generated through the "safe" inactive SLIV models which recover conventional Lorentz invariance?

Direct Lorentz non-invariant extensions of SM and gravity

Nevertheless, it must not be ruled out that physical Lorentz invariance might be explicitly, rather than spontaneously, broken at high energies. This has attracted considerable attention in recent years as an interesting phenomenological possibility appearing in direct Lorentz non-invariant extensions of SM [16–22]. They are generically regarded as originating in a more fundamental theory at some large scale, probably related to the Planck scale M_P. These extensions are in a certain measure motivated [13, 14] by string theory, according to which an explicit (from a QFT point of view) Lorentz violation might be in essence a spontaneous Lorentz violation related to hypothetical tensor-valued fields acquiring non-zero VEVs in some non-perturbative vacuum. These VEVs appear effectively as a set of external background constants so that interactions with these coefficients have preferred spacetime directions in an effective QFT framework. The full SM extension (SME) [18–20] is then defined as the effective gauge invariant field theory obtained when all such Lorentz-violating vector and tensor field backgrounds are contracted term by term with SM (and gravitational) fields. However, without a completely viable string theory, it is not possible to assign definite numerical values to these coefficients. Moreover, not to have disastrous consequences (especially when these coefficients are contracted with non-conserved currents) one also has to additionally propose that observable violating effects in the low-energy theory with a laboratory scale mm should be suppressed by some power of the ratio m/M_P, dependent on the dimension of the Lorentz breaking couplings. Therefore, one has in this sense a pure phenomenological approach treating the above arbitrary coefficients as quantities to be bounded in experiments as if they would simply appear due to explicit

Lorentz violation. Actually, in sharp contrast to the above formulated SLIV in a pure QFT framework, there is nothing in the SME itself that requires that these Lorentz-violation coefficients emerge due to a process of a spontaneous Lorentz violation. Indeed, neither the corresponding massless vector (tensor) NG bosons are required to be generated, nor do these bosons have to be associated with photons or any other gauge fields of SM.

Apart from Lorentz violation in the Standard Model, one can generally think that the vacuum in quantum gravity may also determine a preferred rest frame at the microscopic level. If such a frame exists, it must be very much hidden in low-energy physics since, as was mentioned above, numerous observations severely limit the possibility of Lorentz-violating effects for the SM fields [16–22]. However, the constraints on Lorentz violation in the gravitational sector are generally far weaker. This allows one to introduce a pure gravitational Lorentz violation having no significant impact on the SM physics. An elegant way, close in spirit to our SLIV model (3, 4), seems to appear in the so-called Einstein-aether theory [15]. This is in essence a general covariant theory in which local Lorentz invariance is broken by some vector "aether" field u_μ defining the preferred frame. This field is similar to our constrained vector field A_μ, apart from that this field is taken to be unit, $u_\mu u^\mu=1$. It spontaneously breaks Lorentz symmetry down to a rotation subgroup, just like as our constrained vector field A_μ does it for a time-like Lorentz violation. So, they both give a nonlinear realization of Lorentz symmetry thus leading to its spontaneous violation and inducing the corresponding Goldstone-like modes. The crucial difference is that, while modes related to the vector field A_μ are collected into the physical photon, modes associated with the unit vector field u_μ (one helicity-0 and two helicity-1 modes) exist by them own appearing in some effective SM and gravitational couplings. Some of them might disappear being absorbed by the corresponding spin-connection fields related to local Lorentz symmetry in the Einstein-aether theory. In any case, while the aether field u_μ can significantly change the dispersion relations of fields involved, thus leading to many gravitational and cosmological consequences of preferred frame effects, it certainly cannot be a physical gauge field candidate (say, the photon in QED).

Lorentz violation and supersymmetry. The present paper

There have been a few active attempts [33, 34] over the last decade to construct Lorentz-violating operators for matter and gauge fields in the supersymmetric standard model through their interactions with external vector and tensor field backgrounds. These backgrounds, according to the SME approach [18–20] discussed above, are generated by some Lorentz-violating dynamics at an ultraviolet scale of order the Planck scale. As some advantages over the ordinary SME, it was shown that in the supersymmetric standard model the lowest possible dimension for such operators is five, just as we had above in the high-dimensional SLIV case (6). Therefore, they are suppressed by at least one power of an ultraviolet energy scale, providing a possible explanation for the smallness of Lorentz violation and its stability against radiative corrections. All possible dimension-five and -six Lorentz-violating operators in the SUSY QED were classified [34], their properties at the quantum level analyzed, and their observational consequences in this theory described. These operators, as was confirmed, do not induce destabilizing DD-terms, gauge anomaly, and the Chern–Simons term for the photons. Dimension-five Lorentz-violating operators were shown to be constrained by low-energy precision measurements at 10^{-10}–10^{-5} level in units of the inverse Planck scale, while the Planck-scale suppressed dimension-six operators are allowed by the observational data.

Also, the supersymmetric extension has been constructed of the Einstein-aether theory [35] discussed above. It has been found that the dynamics of the super-aether is somewhat richer than of its non-SUSY counterpart. In particular, the model possesses a family of inequivalent vacua exhibiting different symmetry breaking patterns while remaining stable and ghost free. Interestingly enough, as long as the aether VEV preserves spatial supersymmetry (SUSY algebra without boosts), the Lorentz breaking does not propagate into the SM sector at the renormalizable level. The eventual breaking of SUSY, which must be incorporated in any realistic model, is unrelated to the dynamics of the aether. It is assumed to come from a different source, characterized by a lower energy scale. However, in spite of its own merits, a significant final step which would lead to natural accommodation of this super-aether model into the supergravity framework has not yet been done.

In contrast, we strictly focus here on a spontaneous Lorentz violation in an actual gauge QFT framework related to the Standard Model rather than in an effective low-energy theory with some hypothetical remnants in terms of external tensor-valued backgrounds originating somewhere around the Planck scale. In essence, we try to extend to their supersymmetric analogs the emergent gauge theories with SLIV and the associated emergence of gauge bosons as massless vector Nambu–Goldstone modes studied earlier [6–12] (see also [24–28]). Generally speaking, it may turn out that SLIV is not the only reason why massless photons could dynamically appear, if spacetime symmetry is further enlarged. In this connection, special interest may be in supersymmetry, as was recently argued in [36]. Actually, the situation is changed remarkably in the SUSY inspired emergent models which, in contrast to non-SUSY theories, could naturally have some clear observational evidence. Indeed, as we discussed above (Sect. 1.2), ordinary emergent theories admit some experimental verification only if gauge invariance is properly broken being caused by some high-dimension couplings. Their SUSY counterparts, and primarily emergent SUSY QED, generically appear with supersymmetry being spontaneously broken in a visible sector to ensure stability of the theory. Therefore, the verification is now related to the inevitable emergence of a goldstino-like photino state in the SUSY particle spectrum at low energies, while physical Lorentz invariance is still left intact.[3] In this sense, a generic source for the massless photon to appear may be spontaneously broken supersymmetry rather than physically manifest spontaneous Lorentz violation.

To see how such a scenario may work, we consider the supersymmetric QED model extended by an arbitrary polynomial potential of a massive vector superfield that induces the spontaneous SUSY violation in the visible sector. As a consequence, a massless photino emerges as the fermion NG mode in the broken SUSY phase, and a photon as a photino companion, also massless in the tree approximation (Sect. 2). However, the requirement of vacuum stability in such a class of models renders Lorentz invariance spontaneously broken as well. As a consequence, the massless photon has now appeared as the vector NG mode, and also a special gauge invariance is simultaneously generated in an emergent SUSY QED. This invariance is only restricted by the supplemented vector field constraint being invariant under supergauge transformations (Sect. 3). Due to this invariance all observable SLIV effects appear to be completely canceled out among themselves, and physical Lorentz invariance is restored. Meanwhile, the photino being mixed with another goldstino appearing from a spontaneous SUSY violation in the hidden sector largely appears in the form of a light pseudo-goldstino, whose physics seems to be of special observational interest (Sect. 4). Finally, we conclude (Sect. 5).

EXTENDED SUPERSYMMETRIC QED

We start by considering a conventional SUSY QED extended by an arbitrary polynomial potential of a general vector superfield $V(x,\theta,\overline{\theta})$, which in the standard parametrization [37–39] has the form

$$V(x,\theta,\overline{\theta}) = C(x) + i\theta\chi - i\overline{\theta}\overline{\chi} + \frac{i}{2}\theta\theta S - \frac{i}{2}\overline{\theta\theta}S^* - \theta\sigma^{\mu}\overline{\theta}A_{\mu} + i\theta\theta\overline{\theta}\overline{\lambda'} - i\overline{\theta\theta}\theta\lambda' + \frac{1}{2}\theta\theta\overline{\theta\theta}D', \tag{10}$$

where its vector field component A_{μ} is usually associated with a photon. Note that, apart from the conventional photino field λ and the auxiliary D field, the superfield (10) contains in general the additional degrees of freedom in terms of the dynamical C and χ fields and non-dynamical complex scalar field S (we have used the brief notations, $\lambda' = \lambda + \frac{i}{2}\sigma^{\mu}\partial_{\mu}\overline{\chi}$ and $D' = D + \frac{1}{2}\Box C$ with $\sigma^{\mu} = (1, \overrightarrow{\sigma})$ and $\overline{\sigma}^{\mu} = (1, -\overrightarrow{\sigma})$). The corresponding SUSY invariant Lagrangian may be written as

$$\mathcal{L} = L_{\text{SQED}} + \sum_{n=1} b_n V^n|_D \tag{11}$$

where the terms in this sum (bnbn are some constants) for the vector superfield (10) are given through the polynomial DD-term $V^n|D$ expansion into the component fields. It can readily be checked that the first term in this expansion appears to be the known Fayet–Iliopoulos D-term, while the other terms only contain bilinear, trilinear, and quadrilinear combination of the superfield components A_{μ}, S, λ, and χ, respectively.[4] Actually, there appear higher-degree terms only for the scalar field component C(x). Expressing them all in terms of the C field polynomial

$$P(C) = \sum_{n=1} \frac{n}{2} b_n C^{n-1}(x) \tag{12}$$

and its first three derivatives with respect to the CC field

$$P' \equiv \frac{\partial P}{\partial C}, \quad P'' \equiv \frac{\partial^2 P}{\partial C^2}, \quad P''' \equiv \frac{\partial^3 P}{\partial C^3}, \tag{13}$$

one has for the whole Lagrangian L

$$\mathcal{L} = -\frac{1}{4}F^{\mu\nu}F_{\mu\nu} + i\lambda\sigma^{\mu}\partial_{\mu}\overline{\lambda} + \frac{1}{2}D^2 + P\left(D + \frac{1}{2}\Box C\right) + P'\left(\frac{1}{2}SS^* - \chi\lambda' - \overline{\chi}\overline{\lambda'} - \frac{1}{2}A_{\mu}A^{\mu}\right) + \frac{1}{2}P'' \times \left(\frac{i}{2}\overline{\chi\chi}S - \frac{i}{2}\chi\chi S^* - \chi\sigma^{\mu}\overline{\chi}A_{\mu}\right) + \frac{1}{8}P'''(\chi\chi\overline{\chi\chi}), \tag{14}$$

where, for more clarity, we still omitted matter superfields in the model reserving them for Sect. 4. One can see that the superfield component fields C and χ become dynamical due to the potential terms in (14) rather than from the properly constructed supersymmetric field strengths, as appear for the vector field A_{μ} and its gaugino companion λ. A very remarkable point is that the vector field A_{μ} may only appear with bilinear mass terms in the polynomially extended Lagrangian (14). Hence it follows that the "bumblebee" type model mentioned above (5) with nontrivial vector field potential containing both a bilinear mass term and a quadrilinear stabilizing term can in no way be realized in a SUSY context. Meanwhile, the nonlinear QED model, as will become clear below, successfully matches supersymmetry.

Varying the Lagrangian L with respect to the D field we obtain

$$D = -P(C), \tag{15}$$

which finally gives the standard potential energy for the field system considered

$$U(C) = \frac{1}{2}P^2 \tag{16}$$

provided that the other superfield components do not develop VEVs. The potential (16) may lead to the spontaneous SUSY breaking in the visible sector if the polynomial P (12) has no real roots, while its first derivative has,

$$P \neq 0, \quad P' = 0. \tag{17}$$

This requires P(C) to be an even degree polynomial with properly chosen coefficients b_n in (12), which will force its derivative P′P′ to have at least one root, $C=C_0$, in which the potential (16) is minimized and supersymmetry is spontaneously broken. As an immediate consequence, one can readily see from the Lagrangian L (14) that a massless photino λ being a Goldstone fermion in the broken SUSY phase make all the other component fields in the superfield $V(x,\theta,\overline{\theta})$ including the photon also massless. However, the question then arises whether this masslessness of the photon will be stable against radiative corrections since gauge invariance is explicitly broken in the Lagrangian (14). We show below that it could be the case if the vector superfield $V(x,\theta,\overline{\theta})$ would appear to be properly constrained.

CONSTRAINED VECTOR SUPERFIELD

Instability of superfield polynomial potential

Let us first analyze the possible vacuum configurations for the superfield components in the polynomially extended QED case taken above. In general, besides the "standard" potential energy expression (16) determined solely by the scalar field component C(x) of the vector superfield (10), one also has to consider other field component contributions into the potential energy. A possible extension of the potential energy (16) seems to appear only due to the pure bosonic field contributions, namely due to the couplings of the vector and auxiliary scalar fields, A_μ and S, in (14)

$$\mathcal{U} = \frac{1}{2}P^2 + \frac{1}{2}P'(A_\mu A^\mu - SS^*), \tag{18}$$

rather than due to the potential terms containing the superfield fermionic components.[5] It can immediately be seen that these new couplings in (18) can make the potential unstable since the vector and scalar fields mentioned may in general develop any arbitrary VEVs. This happens, as emphasized above, due the fact that their bilinear term contributions are not properly compensated by appropriate four-linear field terms, which are generically absent in a SUSY theory context.

For more details we consider the extremum conditions for the entire potential (18) with respect to all fields involved: C, A_μ, and S. They are given by the appropriate first partial derivative equations

$$\mathcal{U}'_C = PP' + \tfrac{1}{2}P''(A_\mu A^\mu - SS^*) = 0,$$
$$\mathcal{U}'_{A_\mu} = P'A^\mu = 0, \; \mathcal{U}'_S = -P'S^* = 0, \tag{19}$$

where all the VEVs are denoted by the corresponding field symbols (supplied below with the lower index 0). One can see that there can occur a local minimum for the potential (18) with the unbroken SUSY solution.[6]

$$C = C_0, \; P(C_0) = 0, \; P'(C_0) \neq 0; \; A_{\mu 0} = 0, \; S_0 = 0 \tag{20}$$

with the vanishing potential energy

$$\mathcal{U}^s_{\min} = 0, \tag{21}$$

provided that the polynomial P (12) has some real root $C=C_0$. Otherwise, a local minimum with the broken SUSY solution can occur for some other C field value (though denoted by the same letter C_0)

$$C = C_0, \quad P(C_0) \neq 0, \quad P'(C_0) = 0;$$
$$A_{\mu 0} \neq 0, \quad S_0 \neq 0, \quad A_{\mu 0}A_0^\mu - S_0 S_0^* - 0. \tag{22}$$

In this case one has the non-zero potential energy

$$\mathcal{U}^{as}_{\min} = \frac{1}{2}[P(C_0)]^2, \tag{23}$$

as directly follows from the extremum equations (19) and potential energy expression (18).

However, as shown by the standard second partial derivative test, the fact is that the local minima mentioned above are minima with respect to the C field VEV (C_0) only. Actually, for all three fields VEVs, the potential (18) has indeed saddle points with "coordinates" indicated in (20) and (22), respectively. For testing convenience this potential can be rewritten in the form

$$\mathcal{U} = \tfrac{1}{2}P^2 + \tfrac{1}{2}P'g^{\Theta\Theta'}B_\Theta B_{\Theta'},$$
$$g^{\Theta\Theta'} = diag\,(1, -1, -1, -1, -1, -1) \tag{24}$$

with only two variable fields C and B_Θ, where the new field BΘBΘ unifies the A_μ and S field components, $B_\Theta=(A_\mu,S_a)$ (Θ=μ,a;Θ=μ,a;μ=0,1,2,3).[7] The complex SS field is now taken in a real basis, $S_1 = (S + S^*)/\sqrt{2}$ and $S_2 = (S - S^*)/i\sqrt{2}$, so that the "vector" B_Θ field has one time and five space components. As a result, one finally comes to the following Hessian 7×7 matrix [being in fact the second-order partial derivatives matrix taken in the extremum point (C_0, $A_{\mu 0}$, S_0) (20)]:

$$H(\mathcal{U}^s) = \begin{bmatrix} [P'(C_0)]^2 & 0 \\ 0 & P'(C_0)g^{\Theta\Theta'} \end{bmatrix},$$
$$|H(\mathcal{U}^s)| = -[P'(C_0)]^8\,. \tag{25}$$

This matrix clearly has the negative determinant $|H(U^s)|$, as is indicated above, which confirms that the potential definitely has a saddle point for the solution (20). This means the VEVs of the A_μ and S fields can take in fact any arbitrary value making the potential (18, 24) to be unbounded from below in the unbroken SUSY case that is certainly inaccessible.

One might think that in the broken SUSY case the situation would be better since due to the conditions (22) the B_Θ term completely disappears from the potential U (18, 24) in the ground state. Unfortunately, the direct second partial derivative test in this case is inconclusive, since the determinant of the corresponding Hessian 7×7 matrix appears to vanish,

$$H(\mathcal{U}^{as}) = \begin{bmatrix} P(C_0)P''(C_0) & P''(C_0)g^{\Theta\Theta'}B_{\Theta'} \\ & P''(C_0)g^{\Theta\Theta'}B_{\Theta'}0 \end{bmatrix},$$
$$|H(\mathcal{U}^{as})| = 0\,. \tag{26}$$

Nevertheless, since in general the B_Θ term can take both positive and negative values in small neighborhoods around the vacuum point (C_0, $A_{\mu 0}$, S_0) where the conditions (22) are satisfied, this point also turns out to be a saddle point. Thus, the potential U (18, 24) appears generically unstable both in the SUSY invariant and the SUSY broken phase.

Stabilization of vacuum by constraining vector superfield

The only possible way to stabilize the ground states (20) and (22) seems to seek the proper constraints on the superfield component fields (C, A_μ, S) themselves rather than on their expectation values. Indeed, if such (potential bounding) constraints are physically realizable, the vacua (20) and (22) will be automatically stabilized.

In a SUSY context a constraint can only be put on the entire vector superfield $V(x,\theta,\overline{\theta})$ (10), rather than individually on its field components. Actually, we can constrain our vector superfield $V(x,\theta,\overline{\theta})$ by analogy with the constrained vector field in the nonlinear QED model [see (3)]. This will be done again through some invariant Lagrange multiplier coupling simply adding its D term to the above Lagrangian (11, 14)

$$\mathcal{L}_{\rm tot} = \mathcal{L} + \frac{1}{2}\Lambda(V - C_0)^2|_D, \tag{27}$$

where $V(x,\theta,\overline{\theta})$ is some auxiliary vector superfield, while C_0 is the constant background value of the C field for which the potential UU (16) vanishes as is required for the supersymmetric minimum or has some non-zero value corresponding to the SUSY breaking minimum (17) in the visible sector. We will consider both cases simultaneously using the same notation C0C0 for either of the potential minimizing the values of the C field.

Note first of all, that the Lagrange multiplier term in (27) has in fact the simplest possible form that leads to some nontrivial constrained superfield $V(x,\theta,\overline{\theta})$. The alternative minimal forms, such as the bilinear form $\Lambda(V - C_0)$ or trilinear one $\Lambda(V^2 - C_0^2)$, appear to be too restrictive. One can easily confirm that they eliminate most component fields in the superfield $V(x,\theta,\overline{\theta})$ including the physical photon and photino fields, which is definitely inadmissible. As to the appropriate non-minimal high linear multiplier forms, they basically lead to the same consequences as follow from the minimal multiplier term taken in the total Lagrangian (27). Writing down its invariant D term through the component fields, one finds

$$\begin{aligned}
&\Lambda(V - C_0)^2|_D \\
&= C_\Lambda\left[\widetilde{C}D' + \left(\frac{1}{2}SS^* - \chi\lambda' - \overline{\chi}\overline{\lambda}' - \frac{1}{2}A_\mu A^\mu\right)\right] \\
&+\chi_\Lambda\left[2\widetilde{C}\lambda' + i(\chi S^* + i\sigma^\mu\overline{\chi}A_\mu)\right] \\
&+\overline{\chi}_\Lambda[2\widetilde{C}\overline{\lambda}' - i(\overline{\chi}S - i\chi\sigma^\mu A_\mu)] \\
&+\frac{1}{2}S_\Lambda\left(\widetilde{C}S^* + \frac{i}{2}\overline{\chi}\overline{\chi}\right) + \frac{1}{2}S_\Lambda^*\left(\widetilde{C}S - \frac{i}{2}\chi\chi\right) \\
&+2A_\Lambda^\mu(\widetilde{C}A_\mu - \chi\sigma_\mu\overline{\chi}) + 2\lambda'_\Lambda(\widetilde{C}\chi) + 2\overline{\lambda}'_\Lambda(\widetilde{C}\overline{\chi}) \\
&+\frac{1}{2}D'_\Lambda\widetilde{C}^2
\end{aligned} \tag{28}$$

where

$$C_\Lambda,\ \chi_\Lambda,\ S_\Lambda,\ A_\Lambda^\mu,\ \lambda'_\Lambda = \lambda_\Lambda + \frac{i}{2}\sigma^\mu\partial_\mu\overline{\chi}_\Lambda, \quad D'_\Lambda = D_\Lambda + \frac{1}{2}\Box C_\Lambda \tag{29}$$

are the component fields of the Lagrange multiplier superfield $\Lambda(x,\theta,\overline{\theta})$ in the standard parametrization (10) and $\widetilde{C}$ stands for the difference $C(x)-C_0$. Varying the Lagrangian (27) with respect to these fields and properly combining their equations of motion

$$\frac{\partial\mathcal{L}_{\rm tot}}{\partial\left(C_\Lambda, \chi_\Lambda, S_\Lambda, A_\Lambda^\mu, \lambda_\Lambda, D_\Lambda\right)} = 0 \tag{30}$$

we find the constraints which appear to be put on the V superfield components

$$C = C_0, \quad \chi = 0, \quad A_\mu A^\mu = SS^*. \tag{31}$$

Again, as before in non-SUSY case (3), we only take a solution with initial values for all fields (and their momenta) chosen so as to restrict the phase space to vanishing values of the multiplier component fields (29), which will provide a ghost-free theory with a positive Hamiltonian.[8]

Remarkably, the constraints (31) do not touch on the physical degrees of freedom of the superfield $V(x,\theta,\overline{\theta})$ related to the photon and photino fields. The point is, however, that apart from the constraints (31), one has the equations of motion for all fields involved in the basic superfield $V(x,\theta,\overline{\theta})$. With vanishing multiplier component fields (29), as was proposed above, these equations appear in fact as extra constraints on components of the superfield $V(x,\theta,\overline{\theta})$. Indeed, the equations of motion for the fields C, S and χ obtained by the corresponding variations of the total Lagrangian L (14) turned out to be, respectively,

$$P(C_0)P'(C_0) = 0, \quad S(x)P'(C_0) = 0, \quad \lambda(x)P'(C_0) = 0 \tag{32}$$

where the basic constraints (31) emerging at the potential extremum point $C=C_0$ have also been used. One can immediately see now that these equations turn out to become trivial identities in the broken SUSY case, in which the factor $P'(C_0)$ in each of them appears to be identically vanished, $P'(C_0)=0$ (22). In the unbroken SUSY case, in which the potential (16) vanishes instead, i.e. $P(C_0)=0$ (20), the situation is drastically changed. Indeed, though the first equation in (32) still automatically turns into an identity at the extremum point $C(x)=C_0$, the other two equations require that the auxiliary field S and photino field λ have to be identically vanishing as well. This causes in turn the photon field to vanish, according to the basic constraints (31). Besides, the D field component in the vector superfield is also vanished in the unbroken SUSY case according to (15), $D=-P(C_0)=0$. Thus, one is ultimately left with a trivial superfield $V(x,\theta,\overline{\theta})$, which only contains the constant C field component C_0, and that is unacceptable. So, we have to conclude that the unbroken SUSY fails to provide stability of the potential (18) even by constraining the superfield $V(x,\theta,\overline{\theta})$. In contrast, in the spontaneously broken SUSY case extra constraints do not appear at all, and one has a physically meaningful theory; this is what we basically consider in what follows.

Actually, substituting the constraints (31) into the total Lagrangian L_{tot} (27, 14) we eventually come to the emergent SUSY QED appearing in the broken SUSY phase

$$\begin{aligned}
\mathcal{L}_{\rm tot}^{\rm em} &= -\frac{1}{4}F^{\mu\nu}F_{\mu\nu} + i\lambda\sigma^\mu\partial_\mu\overline{\lambda} + \frac{1}{2}D^2 + P(C_0)D, \\
A_\mu A^\mu &= SS^*
\end{aligned} \tag{33}$$

supplemented by the vector field constraint as its vacuum stability condition. Remarkably, for the constrained vector superfield involved, we have

$$\begin{aligned}
\widehat{V}(x,\theta,\overline{\theta}) &= C_0 + \frac{i}{2}\theta\theta S - \frac{i}{2}\overline{\theta\theta}S^* \\
&\quad -\theta\sigma^\mu\overline{\theta}A_\mu + i\theta\theta\overline{\theta\lambda} - i\overline{\theta\theta}\theta\lambda + \frac{1}{2}\theta\theta\overline{\theta\theta}D,
\end{aligned} \tag{34}$$

we have the almost standard SUSY QED Lagrangian with the same states—photon, photino and an auxiliary scalar D field—

in its gauge supermultiplet, while another auxiliary complex scalar field S gets only involved in the vector field constraint. The linear (Fayet–Iliopoulos) DD-term with the effective coupling constant $P(C_0)$ in (33) shows that the supersymmetry in the theory is spontaneously broken, due to which the D field acquires the VEV, $D=-P(C_0)$. Taking the non-dynamical S field in the constraint (31) to be some constant background field (for a more formal discussion, see below) we come to the SLIV constraint (2) which we discussed above regarding an ordinary non-supersymmetric QED theory (Sect. 1). As is seen from this constraint in (33), one may only have the time-like SLIV in a SUSY framework but never the space-like one. There also may be a light-like SLIV, if the S field vanishes.[9] So, any possible choice for the S field corresponds to the particular gauge choice for the vector field A_μ in an otherwise gauge invariant theory.

Constrained superfield: a formal view

We conclude this section by showing that the extended Lagrangian L_{tot} (27, 14), underlying the emergent QED model described above, as well as the vacuum stability constraints on the superfield component fields (31) appearing due to the Lagrange multiplier term in (27) are consistent with supersymmetry. The first part of this assertion is somewhat immediate, since the Lagrangian L_{tot}, aside from the standard supersymmetric QED part L_{SQED} (11), only contains D-terms of various vector superfield products. They are, by definition, invariant under conventional SUSY transformations [37–39], which for the component fields (10) of a general superfield $V(x,\theta,\overline{\theta})$ (10) are written as

$$\begin{aligned}
&\delta_\xi C = i\xi\chi - i\overline{\xi}\overline{\chi}, \quad \delta_\xi \chi = \xi S + \sigma^\mu\overline{\xi}(\partial_\mu C + iA_\mu),\\
&\frac{1}{2}\delta_\xi S = \overline{\xi}\overline{\lambda} + \overline{\sigma}_\mu\partial^\mu\chi,\\
&\delta_\xi A_\mu = \xi\partial_\mu\chi + \overline{\xi}\partial_\mu\overline{\chi} + i\xi\sigma_\mu\overline{\lambda} - i\lambda\sigma_\mu\overline{\xi},\\
&\delta_\xi\lambda = \frac{1}{2}\xi\sigma^\mu\overline{\sigma}^\nu F_{\mu\nu} + \xi D,\\
&\delta_\xi D = -\xi\sigma^\mu\partial_\mu\overline{\lambda} + \overline{\xi}\sigma^\mu\partial_\mu\lambda.
\end{aligned} \tag{35}$$

However, there may still be left a question as to whether supersymmetry remains in force when the constraints (31) on the field space are "switched on", thus leading to the final Lagrangian $\mathcal{L}_{tot}^{em}$ (33) in the broken SUSY phase with both dynamical fields CC and χ eliminated. This Lagrangian appears similar to the standard supersymmetric QED taken in the Wess–Zumino gauge, except that the supersymmetry is spontaneously broken in our case. In both cases the photon stress tensor $F_{\mu\nu}$, the photino λ, and the non-dynamical scalar DD field form an irreducible representation of the supersymmetry algebra [the last two lines in (35)]. Nevertheless, any reduction of the component fields in the vector superfield is not consistent in general with the linear superspace version of supersymmetry transformations, whether it is the Wess–Zumino gauge case or our constrained superfield $\widehat{V}$ (34). Indeed, a general SUSY transformation does not preserve the Wess–Zumino gauge: a vector superfield in this gauge acquires some extra terms when being SUSY transformed. The same occurs with our constrained superfield $\widehat{V}$ as well. The point, however, is that in both cases a total supergauge transformation,

$$V \to V + i(\Omega - \Omega^*), \tag{36}$$

where Ω is a chiral superfield gauge transformation parameter, can always restore the superfield initial form. Actually, the only difference between these two cases is that whereas the Wess–Zumino supergauge leaves the ordinary gauge freedom untouched, in our case this gauge is unambiguously fixed in terms of the above vector field constraint (31). However, this constraint remains under the supergauge transformation (36) applied to our superfield $\widehat{V}$ (34). Indeed, the essential part of this transformation which directly acts on the constraint (31) has the form

$$\widehat{V} \to \widehat{V} + i\theta\theta F - i\overline{\theta}\overline{\theta}F^* - 2\theta\sigma^\mu\overline{\theta}\partial_\mu\varphi, \tag{37}$$

where the real and complex scalar field components, φ and F, in the chiral superfield parameter Ω are properly activated. As a result, the vector and scalar fields, A_μ and S, in the supermultiplet $\widehat{V}$ (34) transform as

$$A_\mu \to A'_\mu = A_\mu - \partial_\mu(2\varphi), \quad S \to S' = S + 2F\,. \tag{38}$$

It can be immediately seen that our basic Lagrangian $\mathcal{L}_{tot}^{em}$ (33), being gauge invariant and containing no scalar field SS, is automatically invariant under either of these two transformations individually. In contrast, the supplementary vector field constraint (31), though it also turned out to be invariant under the supergauge transformations (38), but only if they are made jointly. Indeed, for any choice of the scalar φ in (38) there can always be found such a scalar F (and vice versa) that the constraint remains invariant,

$$A_\mu A^\mu = SS^* \to A'_\mu A'^\mu = S'S'^*. \tag{39}$$

In other words, the vector field constraint is invariant under supergauge transformations (38) but not invariant under an ordinary gauge transformation. As a result, in contrast to the Wess–Zumino case, the supergauge fixing in our case will also lead to the ordinary gauge fixing. We will use this supergauge freedom to reduce the S field to some constant background value and find the final equation for the gauge function $\varphi(x)$. So, for the parameter field F chosen in such a way to have

$$S' = S + 2F = Me^{i\alpha(x)}, \tag{40}$$

where M is some constant mass parameter (and $\alpha(x)$ is an arbitrary phase), we end by (39) in

$$(A_\mu - 2\partial_\mu\varphi)(A^\mu - 2\partial^\mu\varphi) = M^2, \tag{41}$$

which is precisely our old SLIV constraint (2) being varied by the gauge transformation (38). Recall that this constraint, as was thoroughly discussed in the Introduction (Sect. 1.1), only fixes the gauge (which such a gauge function $\varphi(x)$ has to satisfy), rather than that it physically breaks gauge invariance. Notably, in contrast to the non-SUSY case where this constraint was merely postulated, it now follows from the vacuum stability and supergauge invariance in the emergent SUSY QED. Besides, this constraint, as mentioned above, may only be time-like (and light-like if the mass parameter M is taken to be zero). When such inactive time-like SLIV is properly developed one ends with the essentially nonlinear emergent SUSY QED, in which the physical photon arises as a three-dimensional Lorentzian NG mode (just as is the case in non-SUSY for the time-like SLIV; see Sect. 1.1).

To finalize, it was shown that the vacuum stability constraints (31) on the allowed configurations of the physical fields in a general polynomially extended Lagrangian (27) appear entirely consistent with supersymmetry. In the broken SUSY phase one eventually comes to the standard SUSY QED type Lagrangian (33) being supplemented by the vector field constraint which is invariant under supergauge transformations. One might think that, unlike the gauge invariant linear (Fayet–Iliopoulos) superfield term, the quadratic and higher order superfield terms in the starting Lagrangian (27) would seem to break gauge invariance. However, this fear proves groundless. Actually, as was shown above, this breaking amounts to the gauge fixing determined by the nonlinear vector field constraint (39). It is worth noting that this constraint formally follows from the SUSY invariant Lagrange multiplier term in (27) for which the phase space is required to be restricted to vanishing values of all the multiplier component fields (29). The total vanishing of the multiplier superfield provides the SUSY invariance of such restrictions. Any non-zero multiplier component field left in the Lagrangian would immediately break supersymmetry and, even worse, would eventually lead to ghost modes in the theory and a Hamiltonian unbounded from below.

BROKEN SUSY PHASE: PHOTINO AS PSEUDO-GOLDSTINO

Let us now turn to matter superfields, which have not yet been included in the model. In their presence spontaneous SUSY breaking in the visible sector, which fundamentally underlies our approach, might be phenomenologically ruled out by the well-known supertrace sum rule [37–39] for actual masses of quarks and leptons and their superpartners.[10] However, this sum rule is acceptably relaxed when taking into account large radiative corrections to the masses of the supersymmetric particles, which, according to the proposal, stem from the hidden sector. This is just what one may expect in conventional supersymmetric theories with the standard two-sector paradigm, according to which SUSY breaking entirely occurs in a hidden sector; then this spontaneous breaking is mediated to the visible sector by some indirect interactions whose nature depends on a particular mediation scenario [37–39]. The emergent QED approach advocated here requires some modification of this idea in such a way that, while the hidden sector is largely responsible for spontaneous SUSY breaking, supersymmetry can also be spontaneously broken in the visible sector, which ultimately leads to a double spontaneous SUSY breaking pattern.

We may suppose, just for uniformity, only DD-term SUSY breaking both in the visible and hidden sectors.[11] Properly stated, our supersymmetric QED model may be further extended by some extra local U′(1) symmetry which is proposed to be broken at the very high-energy scale M′ (for some appropriate anomaly mediated scenarios, see [40] and references therein). It is natural to think that due to the decoupling theorem all effects of the U′(1) are suppressed at energies E<<M′ by powers of 1/M′ and only the D′-term of the corresponding vector superfield $V(x,\theta,\bar{\theta})$ remains in essence when going down to low energies. Actually, this term with a proper choice of messenger fields and their couplings naturally provides the M_{SUSY} order contributions to the masses of the scalar superpartners.

As a result, the simplified picture discussed above (in Sects. 2 and 3) is properly changed: a strictly massless fermion eigenstate, the true goldstino ζ_g, should now be some mix of the visible sector photino λ and the hidden sector goldstino λ′

$$\zeta_g = \frac{\langle D\rangle \lambda + \langle D'\rangle \lambda'}{\sqrt{\langle D\rangle^2 + \langle D'\rangle^2}}, \tag{42}$$

where ⟨D⟩ and ⟨D′⟩ are the corresponding D-component VEVs in the visible and hidden sectors, respectively. Another orthogonal combination of them may be referred to as the pseudo-goldstino ζ_{pg},

$$\zeta_{pg} = \frac{\langle D'\rangle \lambda - \langle D\rangle \lambda'}{\sqrt{\langle D\rangle^2 + \langle D'\rangle^2}}. \tag{43}$$

In the supergravity context, the true goldstino ζ_g is eaten through the super-Higgs mechanism to form the longitudinal component of the gravitino, while the pseudo-goldstino ζ_{pg} gets some mass proportional to the gravitino mass from supergravity effects. Due to the large soft masses required to be mediated, one may generally expect that SUSY is much stronger broken in the hidden sector than in the visible one, ⟨D′⟩>>⟨D⟩, which means in turn that the pseudo-goldstino ζ_{pg} is largely the photino λ,

$$\zeta_{pg} \simeq \lambda. \tag{44}$$

These pseudo-goldstonic photinos seem to be of special observational interest in the model, which, apart from some indication of the QED emergence nature, may shed light on SUSY breaking physics. The possibility that the supersymmetric Standard Model visible sector might also spontaneously break SUSY, thus giving rise to some pseudo-goldstino state, was also considered, though in a different context, in [41, 42].

Interestingly enough, our polynomially extended SQED Lagrangian (11) is not only SUSY invariant but also generically possesses a continuous R-symmetry U(1)R [37–39]. Indeed, vector superfields always have zero R-charge, since they are real. Accordingly, it follows that the physical field components in the constrained vector superfield $\widehat{V}$ (34) transform as

$$A_\mu \to A_\mu, \quad \lambda \to e^{i\alpha}\lambda, \quad D \to D \tag{45}$$

and so have R charges 0, 1, and 0, respectively. Along with that, we assume a suitable R-symmetric matter superfield setup as well making a proper R-charge assignment for basic fermions and scalars (and messenger fields) involved. This will lead to the light pseudo-goldstino in the gauge-mediated scenario. Indeed, if the visible sector possesses an RR-symmetry which is preserved in the course of mediation the pseudo-goldstino mass is protected up to the supergravity effects which violate an R-symmetry. As a result, the pseudo-goldstino mass appears proportional to the gravitino mass, and, eventually, the same region of parameter space simultaneously solves both gravitino and pseudo-goldstino overproduction problems in the early universe [42].

Apart from cosmological problems, many other sides of new physics related to pseudo-goldstinos appearing through the multiple SUSY breaking were also studied recently (see [41–46] and references therein). The point, however, is that there have been exclusively used non-vanishing F-terms as the only mechanism of the visible SUSY breaking in the models considered. In this connection, our pseudo-goldstonic photinos solely caused by non-vanishing DD-terms in the visible SUSY sector may lead to somewhat different observational consequences. One of the most serious differences may be related to the Higgs boson decays when the present SUSY QED is further extended to the supersymmetric Standard Model. For the cosmologically safe masses of pseudo-goldstino and gravitino ($\lesssim$1keV, as typically follows from the R-symmetric gauge mediation) these decays are appreciably modified. Actually, the dominant channel becomes the conversion of the Higgs boson (say, the lighter CP-even Higgs boson h^0) into a conjugated pair of corresponding pseudo-sgoldstinos ϕ_{pg} and $\overline{\phi}_{pg}$ (being superpartners of pseudo-goldstinos ζ_{pg} and $\overline{\zeta}_{pg}$, respectively),

$$h^0 \rightarrow \phi_{pg} + \overline{\phi}_{pg} \, , \tag{46}$$

once it is kinematically allowed. This means that the Higgs boson will dominantly decay invisibly for F-term SUSY breaking in a visible sector [42]. By contrast, for the D-term SUSY breaking case considered here the roles of pseudo-goldstino and pseudo-sgoldstino are just played by photino and photon, respectively, which could enhance the standard two-photon decay channel of the Higgs boson even somewhat. In the light of the recent discovery of the Higgs-like state [47, 48] just through its visible decay modes, the F-term SUSY breaking in the visible sector seems to be disfavored by data, while the SUSY breaking D-term is not in trouble with them.

CONCLUDING REMARKS

It is well known that spontaneous Lorentz violation in general vector field theories may lead to an appearance of massless Nambu–Goldstone modes, which are identified with photons and other gauge fields in the Standard Model. Nonetheless, it may turn out that SLIV is not the only reason for emergent massless photons to appear, if the spacetime symmetry is further enlarged. In this connection, special interest may be in supersymmetry and its possible theoretical and observational relation to SLIV.

To see how such a scenario may work we have considered supersymmetric QED model extended by an arbitrary polynomial potential of a general vector superfield $V(x,\theta,\overline{\theta})$ whose pure vector field component $A_\mu(x)$ is associated with a photon in the Lorentz invariant phase. Gauge non-invariant couplings other than potential terms are not included into the theory. For the theory in which gauge invariance is not required from the outset this is in fact the simplest generalization of a conventional SUSY QED. This superfield potential (18) is turned out to be generically unstable unless SUSY is spontaneously broken. However, it appears not to be enough. To provide an overall stability of the potential one additionally needs the special direct constraint being put on the vector superfield itself that is made by an appropriate SUSY invariant Lagrange multiplier term (27). Remarkably enough, when this term is written in field components it leads precisely to the nonlinear σ-model type constraint of type (2) which one has in the non-SUSY case. So, we come again to the picture, which we called the inactive SLIV, with a Goldstone-like photon and special (SLIV restricted) gauge invariance providing the cancelation mechanism for physical Lorentz violation. But now this picture follows from the vacuum stability and supergauge invariance in the extended SUSY QED rather than being postulated as is in the non-SUSY case. This allows one to think that a generic trigger for massless photons to dynamically emerge happens to be spontaneously broken supersymmetry rather than physically manifested Lorentz non-invariance.

In more exact terms, in the broken SUSY phase one eventually comes to the almost standard SUSY QED Lagrangian (33) possessing some special gauge invariance emerged. This invariance is only restricted by the gauge condition put on the vector field, $A_\mu A^\mu = |S|^2$, which appears to be invariant under supergauge transformations. One can use this supergauge freedom to reduce the non-dynamical scalar field S to some constant background value so as to eventually come to the nonlinear vector field constraint (2). As a result, the inactive time-like SLIV is properly developed, thus leading to essentially nonlinear emergent SUSY QED in which the physical photon arises as a three-dimensional Lorentzian NG mode. So, figuratively speaking, the photon passes through three evolution stages being initially the massive vector field component of a general vector superfield (14), then the three-level massless companion of an emergent photino in the broken SUSY stage (17) and finally a generically massless state as an emergent Lorentzian mode in the inactive SLIV stage (31).

As to an observational status of emergent SUSY theories, one can see that, as in an ordinary QED, physical Lorentz invariance is still preserved in the SUSY QED model at the renormalizable level and can only be violated if some extra gauge non-invariant couplings [being supersymmetric analogs of the high-dimension couplings (6)] are included into the theory. However, one may have some specific observational evidence in favor of the inactive SLIV even in the minimal (gauge invariant) supersymmetric QED and Standard Model. Indeed, since as mentioned above the vacuum stability is only possible in spontaneously broken SUSY case, this evidence is related to an existence of an emergent goldstino–photino type state in the SUSY visible sector. Being mixed with another goldstino appearing from a spontaneous SUSY violation in the hidden sector this state largely turns into the light pseudo-goldstino. Its study seem to be of special observational interest for this class of models that, apart from some indication of an emergence nature of QED and the Standard Model, may appreciably extend the scope of SUSY breaking physics being actively studied in recent years. We may return to this important issue elsewhere.

ACKNOWLEDGMENTS

I thank Colin Froggatt, Alan Kostelecky, Rabi Mohapatra and Holger Nielsen for stimulating discussions and correspondence. Discussions with the participants of the International Workshop "What Comes Beyond the Standard Models?"(14–21 July 2013, Bled, Slovenia) are also appreciated. This work was supported in part by the Georgian National Science Foundation under Grant No. 31/89.

REFERENCES

1. Y. Nambu, G. Jona-Lasinio, Phys. Rev. 122, 345 (1961)
2. J. Goldstone, Nuovo Cimento 19, 154 (1961)
3. J.D. Bjorken, Ann. Phys. (N. Y.) 24, 174 (1963)
4. P.R. Phillips, Phys. Rev. 146, 966 (1966)
5. T. Eguchi, Phys. Rev. D 14, 2755 (1976)
6. J.L. Chkareuli, C.D. Froggatt, H.B. Nielsen, Phys. Rev. Lett. 87, 091601 (2001). hep-ph/0106036
7. J.L. Chkareuli, C.D. Froggatt, H.B. Nielsen, Nucl. Phys. B 609, 46 (2001). hep-ph/0103222
8. J.D. Bjorken. hep-th/0111196
9. P. Kraus, E.T. Tomboulis, Phys. Rev. D 66, 045015 (2002)
10. Jenkins, Phys. Rev. D 69, 105007 (2004)
11. V.A. Kostelecky, Phys. Rev. D 69, 105009 (2004)
12. R. Bluhm, V.A. Kostelecky, Phys. Rev. D 71, 065008 (2005)
13. V.A. Kostelecky, S. Samuel, Phys. Rev. D 39, 683 (1989)
14. V.A. Kostelecky, R. Potting, Nucl. Phys. B 359, 545 (1991)
15. T. Jacobson, D. Mattingly, Phys. Rev. D 64, 024028 (2001)
16. S. Chadha, H.B. Nielsen, Nucl. Phys. B 217, 125 (1983)
17. S.M. Carroll, G.B. Field, R. Jackiw, Phys. Rev. D 41, 1231 (1990)
18. D. Colladay, V.A. Kostelecky, Phys. Rev. D 55, 6760 (1997)
19. D. Colladay, V.A. Kostelecky, Phys. Rev. D 58, 116002 (1998)
20. V.A. Kostelecky, R. Lehnert, Phys. Rev. D 63, 065008 (2001)
21. S. Coleman, S.L. Glashow, Phys. Lett. B 405, 249 (1997)
22. S. Coleman, S.L. Glashow, Phys. Rev. D 59, 116008 (1999)
23. Y. Nambu, Progr. Theor. Phys. Suppl. Extra, 190 (1968)
24. A.T. Azatov, J.L. Chkareuli, Phys. Rev. D 73, 065026 (2006). arXiv:hep-th/0511178
25. J.L. Chkareuli, Z.R. Kepuladze, Phys. Lett. B 644, 212 (2007). hep-th/0610277
26. J.L. Chkareuli, J.G. Jejelava, Phys. Lett. B 659, 754 (2008). arXiv:0704.0553 [hep-th]
27. O.J. Franca, R. Montemayor, L.F. Urrutia, Phys. Rev. D 85, 085008 (2012)
28. J.L. Chkareuli, J.G. Jejelava, G. Tatishvili, Phys. Lett. B 696, 124 (2011). arXiv:1008.3707 [hep-th]
29. S. Weinberg, The Quantum Theory of Fields, vol. 2 (Cambridge University Press, Cambridge, 2000)
30. R. Bluhm, N.L. Gagne, R. Potting, A. Vrublevskis, Phys. Rev. D 77, 125007 (2008)
31. J.L. Chkareuli, Z. Kepuladze, G. Tatishvili, Eur. Phys. J. C 55, 309 (2008). arXiv:0802.1950 [hep-th]
32. J.L. Chkareuli, Z. Kepuladze, Eur. Phys. J. C 72, 1954 (2012). arXiv:1108.0399 [hep-th]
33. S.G. Nibbelink, M. Pospelov, Phys. Rev. Lett. 94, 081601 (2005)
34. P.A. Bolokhov, S.G. Nibbelink, M. Pospelov, Phys. Rev. D 72, 015013 (2005)
35. O. Pujolas, S. Sibiryakov, JHEP 1201, 062 (2012)
36. J.L. Chkareuli, Phys. Lett. B 721, 146 (2013) arXiv:1212.6939 [hep-th]
37. H.P. Nilles, Phys. Rep. 110, 1 (1984)
38. J. Wess, J. Bagger, Supersymmetry and Supergravity, 2nd edn. (Princeton University Press, Princeton, 1992)
39. S.P. Martin, A Supersymmetry Primer. hep-ph/9709356
40. R. Hodgson, I. Jack, D.R.T. Jones, G.G. Ross, Nucl. Phys. B 728, 192 (2005)
41. ADSCrossRefMATHMathSciNet
42. K. Izawa, Y. Nakai, T. Shimomura, JHEP 1103, 007 (2011)
43. D. Bertolini, K. Rehermann, J. Thaler, JHEP 1204, 130 (2012)
44. C. Cheung, Y. Nomura, J. Thaler, JHEP 1003, 073 (2010)
45. N. Craig, J. March-Russell, M. McCullough, JHEP 1010, 095 (2010)
46. H.-C. Cheng, W.-C. Huang, I. Low, A. Menon, JHEP 1103, 019 (2011)
47. R. Argurio, Z. Komargodski, A. Mariotti, Phys. Rev. Lett. 107, 061601 (2011)
48. ATLAS Collaboration, G. Aad et al., Phys. Lett. B 716, 1 (2012)
49. CMS Collaboration, S. Chatrchyan et al., Phys. Lett. B 716, 30 (2012)

CHAPTER 8

OCCAM'S RAZOR IN QUARK MASS MATRICES

Morimitsu Tanimoto and Tsutomu T. Yanagida

INTRODUCTION

The standard model is now well established by the recent discovery of the Higgs boson. In spite of the success of the standard model, the underlying physics determining the quark and lepton mass matrices is still unknown. Because of this, a number of models based on flavor symmetries have been proposed, but no convincing model has yet been suggested.

A long time ago, Weinberg [1] considered a mass matrix for the down-quark sector on the basis of the up-quark mass matrix being diagonal. He assumed a vanishing (1,1) element in the 2×2 matrix and imposed a symmetric form of the matrix in order to reduce the number of free parameters. Thus, the number of free parameters is reduced to only two and hence he successfully predicted the Cabibbo angle to be $\sqrt{m_d/m_s}$,, which is called the Gatto–Sartori–Tonin relation [2].1

The success of the Weinberg approach encouraged many authors to consider various flavor symmetries in the quark matrices by extending it to the three-family case. In this paper, however, we point out that the Cabibbo angle is predicted successfully in the framework of the "Occam's razor" approach proposed in the lepton sector [6].2 In other words, we show that the Weinberg matrix can be obtained without any symmetry.

From the standpoint of the Occam's razor approach, we consider the minimum number of parameters needed for successful CKM mixing and CP violation without assuming the symmetric or Hermitian mass matrix of down quarks. We impose three zeros in the down-quark mass matrix. We always take the up-quark mass matrix to be diagonal since we do not need any off-diagonal element to explain the observation. Therefore, the down-quark mass matrix is given with six complex parameters. Among them, five phases can be removed by the phase redefinition of the three right-handed and three left-handed down-quark fields. After the field-phase rotation, there remain six real parameters and one CP-violating phase, which is the minimal number needed to reproduce the seven observed data, i.e., the three down-quark masses and the four CKM parameters. It is emphasized that the three–zero texture keeps one CP-violating phase in the down-quark mass matrix. In the present Occam's razor approach with three families, we show that a successful prediction of the Cabibbo angle is obtained. It is surprising that Weinberg's anzatz, which needs the (1,2)(1,2) and (2,1)(2,1) elements to be symmetric, is derived in our Occam's razor approach.

In order to reproduce the bottom-quark mass and the CKM mixing angles, V_{us} and V_{cb}, we take the elements (3, 3),(2, 3),(1, 2) of the 3 × 3 down-quark mass matrix to be nonvanishing. Then, we have ${}_6C_3$=20 textures with three zeros. For convenience, we classify them into two categories: (A) where a (2,2) element is nonvanishing and (B) where a (2,2) element is vanishing. In category (A), the six textures with three zeros are viable for the down-quark mass matrix. In category (B), there are seven viable textures with three zeros. Finally, we have found that these thirteen textures are all consistent with the present experimental data on quark masses and the CKM parameters. However, we should note that some textures are equivalent to each other due to the freedom of the unitary transformation of the right-handed down quarks.

In Sect. 2, we show viable down-quark mass matrices with three zeros in categories (A) and (B), and explain how to obtain the CKM mixing angles and the CP-violating phase. In Sect. 3, we show numerical results for those mass matrices. Section 4 is devoted to the discussion and summary. In Appendices A and B, we show unfavored down-quark mass matrices and the above redundancy of our textures, respectively.

Citation: Morimitsu Tanimoto and Tsutomu T. Yanagida, Occam's razor in quark mass matrices, DOI: 10.1093/ptep/ptw024.

QUARK MASS MATRIX IN THE OCCAM'S RAZOR APPROACH

Three texture zeros for down-quarks

Let us discuss the down-quark mass matrix. We always take the basis where the up-quark mass matrix is diagonal. The number of free parameters in the down-quark mass matrix is reduced by putting zero at several elements in the matrix. We consider the three texture zeros, which provide the minimum number of parameters needed for successful CKM mixing and CP violation. We never assume any flavor symmetry. We call this the "Occam's razor" approach.

Before investigating the quark mass matrix, we present our setup in more detail. The Lagrangian for the quark Yukawa sector is given by

$$\mathcal{L}_Y = y^u_{\alpha\beta}\bar{Q}_{L\alpha}u_{R\beta}\tilde{h} + y^d_{\alpha\beta}\bar{Q}_{L\alpha}d_{R\beta}h, \tag{1}$$

where $Q_{L\alpha}$, $u_{R\beta}$, $d_{R\beta}$, and h denote the left-handed quark doublets, the right-handed up-quark singlet, the right-handed down-quark singlet, and the Higgs doublet, respectively. The quark mass matrices are given as $m_{\alpha\beta} = y_{\alpha\beta}v_H$ with $v_H = 174.104$ GeV. In order to reproduce the observed quark masses and the CKM matrix with the minimal number of parameters, we take the diagonal basis in the up-quark sector:

$$M_u = \begin{pmatrix} m_u & 0 & 0 \\ 0 & m_c & 0 \\ 0 & 0 & m_t \end{pmatrix}_{LR}. \tag{2}$$

For the down-quark mass matrix, we impose the three texture zeros. Then, the texture of the down-quark mass matrix M_d is given with six complex parameters. The five phases can be removed by the phase rotation of the three right-handed and three left-handed down-quark fields. Therefore, there remain six real parameters and one CP-violating phase, which is the minimal number needed to reproduce the observed data of masses and the CKM parameters.

Now, we can discuss textures for the down-quark mass matrix. Let us start with taking (3, 3), (2, 3), (1, 2) elements of M_d to be nonvanishing values to reproduce the observed bottom-quark mass and the CKM mixing angles, V_{us} and V_{cb}.[3] Then, we have ${}_6C_3 = 20$ textures with three zeros for the down-quark mass matrix. For convenience, we classify them into two categories, (A) and (B), as explained in the introduction. In (A), we have ${}_5C_2 = 10$ textures with a nonvanishing (2, 2) element and in (B) we also have ten textures with a vanishing (2, 2) element.

We first discuss the textures in category (A). The ten textures are written as follows:

$$\begin{pmatrix} 0 & A & 0 \\ A' & B & C \\ 0 & C' & D \end{pmatrix}, \begin{pmatrix} A' & A & 0 \\ 0 & B & C \\ 0 & C' & D \end{pmatrix}, \begin{pmatrix} 0 & A & 0 \\ 0 & B & C \\ A' & C' & D \end{pmatrix}, \begin{pmatrix} 0 & A & C' \\ A' & B & C \\ 0 & 0 & D \end{pmatrix}, \begin{pmatrix} A' & A & C' \\ 0 & B & C \\ 0 & 0 & D \end{pmatrix},$$
$$\begin{pmatrix} 0 & A & C' \\ 0 & B & C \\ A' & 0 & D \end{pmatrix}, \begin{pmatrix} 0 & A & 0 \\ A' & B & C \\ C' & 0 & D \end{pmatrix}, \begin{pmatrix} A' & A & 0 \\ C' & B & C \\ 0 & 0 & D \end{pmatrix}, \begin{pmatrix} 0 & A & C' \\ 0 & B & C \\ 0 & A' & D \end{pmatrix}, \begin{pmatrix} A' & A & 0 \\ 0 & B & C \\ C' & 0 & D \end{pmatrix}, \tag{3}$$

where A, A' , B, C, C' , and D are complex parameters. We will show that the first six textures are consistent with the present experimental data. Those six down-quark mass matrices are parametrized after removing five phases by the phase rotation of quark fields as follows:

$$M_d^{(1)} = \begin{pmatrix} 0 & a & 0 \\ a' & b\,e^{-i\phi} & c \\ 0 & c' & d \end{pmatrix}_{LR}, \quad M_d^{(2)} = \begin{pmatrix} a' & a & 0 \\ 0 & b\,e^{-i\phi} & c \\ 0 & c' & d \end{pmatrix}_{LR}, \quad M_d^{(3)} = \begin{pmatrix} 0 & a & 0 \\ 0 & b\,e^{-i\phi} & c \\ a' & c' & d \end{pmatrix}_{LR},$$
$$M_d^{(4)} = \begin{pmatrix} 0 & a & c' \\ a' & b\,e^{-i\phi} & c \\ 0 & 0 & d \end{pmatrix}_{LR}, \quad M_d^{(5)} = \begin{pmatrix} a' & a & c' \\ 0 & b\,e^{-i\phi} & c \\ 0 & 0 & d \end{pmatrix}_{LR}, \quad M_d^{(6)} = \begin{pmatrix} 0 & a & c' \\ 0 & b\,e^{-i\phi} & c \\ a' & 0 & d \end{pmatrix}_{LR}, \tag{4}$$

where a, a' , b, c, c' , and d are real parameters, and φ is the CP-violating phase. It should be stressed that our matrices are not symmetric at all. The CP-violating phase φ is put in the (2, 2) entry.4 In the next section, we examine these seven parameters numerically to reproduce the three down-quark masses, three CKM mixing angles, and one CP-violating phase.

We discuss briefly why the last four textures (the seventh–tenth ones) in Eq. (3) are excluded by the experimental data. The seventh and eighth textures in Eq. (3) give a vanishing CKM mixing angle. The ninth one gives us one massless quark because the first column is a zero vector in the flavor space. The last one cannot reproduce the magnitude of the CP violation. The details are shown in Appendix A.

Now, we discuss the textures in category (B), in which the (2,2) element is zero. We can write the ten textures as follows:

$$\begin{pmatrix} A' & A & B \\ 0 & 0 & C \\ 0 & C' & D \end{pmatrix}, \begin{pmatrix} 0 & A & B \\ A' & 0 & C \\ 0 & C' & D \end{pmatrix}, \begin{pmatrix} 0 & A & B \\ 0 & 0 & C \\ A' & C' & D \end{pmatrix}, \begin{pmatrix} A & A' & C' \\ B & 0 & C \\ 0 & 0 & D \end{pmatrix}, \begin{pmatrix} A & A' & B \\ 0 & 0 & C \\ C' & 0 & D \end{pmatrix},$$
$$\begin{pmatrix} 0 & A & B \\ A' & 0 & C \\ C' & 0 & D \end{pmatrix}, \begin{pmatrix} A & A' & 0 \\ B & 0 & C \\ C' & 0 & D \end{pmatrix}, \begin{pmatrix} A' & A & 0 \\ 0 & 0 & C \\ B & C' & D \end{pmatrix}, \begin{pmatrix} 0 & A & 0 \\ A' & 0 & C \\ B & C' & D \end{pmatrix}, \begin{pmatrix} A' & A & 0 \\ B & 0 & C \\ 0 & C' & D \end{pmatrix}. \tag{5}$$

The first seven textures are also consistent with the present experimental data. After removing five phases by the phase rotation of quark fields, these seven down-quark mass matrices are parametrized as:

$$M_d^{(11)} = \begin{pmatrix} a' & a\,e^{-i\phi} & b \\ 0 & 0 & c \\ 0 & c' & d \end{pmatrix}_{LR}, \quad M_d^{(12)} = \begin{pmatrix} 0 & a\,e^{-i\phi} & b \\ a' & 0 & c \\ 0 & c' & d \end{pmatrix}_{LR}, \quad M_d^{(13)} = \begin{pmatrix} 0 & a\,e^{-i\phi} & b \\ 0 & 0 & c \\ a' & c' & d \end{pmatrix}_{LR},$$

$$M_d^{(14)} = \begin{pmatrix} a\,e^{i\phi} & a' & c' \\ b & 0 & c \\ 0 & 0 & d \end{pmatrix}_{LR}, \quad M_d^{(15)} = \begin{pmatrix} a\,e^{-i\phi} & a' & b \\ 0 & 0 & c \\ c' & 0 & d \end{pmatrix}_{LR},$$

$$M_d^{(16)} = \begin{pmatrix} 0 & a & b \\ a' & 0 & c\,e^{-i\phi} \\ c' & 0 & d \end{pmatrix}_{LR}, \quad M_d^{(17)} = \begin{pmatrix} a & a' & 0 \\ b & 0 & c\,e^{i\phi} \\ c' & 0 & d \end{pmatrix}_{LR}. \tag{6}$$

The last three textures (the eighth–tenth ones) in Eq. (5) are also excluded by the experimental data. The eighth and ninth textures in Eq. (5) give a vanishing CKM mixing angle. The tenth one cannot reproduce the magnitudes of the CP violation. The details are discussed in Appendix A.

Finally, we comment on the freedoms of the unitary transformation of the right-handed quarks. Since the CKM matrix is the flavor mixing among the left-handed quarks, some textures in Eqs. (4) and (6) are equivalent to each other due to the freedom of the unitary transformation of the right-handed quarks. We show the redundancy among them in Appendix B.

CKM parameters for $M_d^{(1)}$

Let us show how to predict the CKM mixing angles and the CP violation, taking the case of $M_d^{(1)}$ in Eq. (4) as an example. Since the up-type quark mass matrix is diagonal, the CKM matrix is obtained by diagonalizing the down-quark mass matrix $M_d^{(1)}$. In order to determine the left-handed quark mixing angles, we study $M_d^{(1)} M_d^{(1)\dagger}$:

$$M_d^{(1)} M_d^{(1)\dagger} = \begin{pmatrix} a^2 & ab\,e^{i\phi} & ac' \\ ab\,e^{-i\phi} & a'^2+b^2+c^2 & bc'\,e^{-i\phi}+cd \\ ac' & bc'\,e^{i\phi}+cd & c'^2+d^2 \end{pmatrix}_{LL}. \tag{7}$$

By solving the eigenvalue equation of $M_d^{(1)} M_d^{(1)\dagger}$, we obtain

$$\begin{aligned} & m_d^2+m_s^2+m_b^2 = a^2+a'^2+b^2+c^2+c'^2+d^2,\ m_d^2 m_s^2+m_s^2 \\ & m_b^2+m_b^2 m_d^2 = a^2a'^2+a^2\left(c^2+d^2\right)+a'^2\left(c'^2+d^2\right)+c^2c'^2+b^2d^2-2bcc'd\cos\phi, \\ & m_d^2 m_s^2 m_b^2 = a^2a'^2d^2. \end{aligned} \tag{8}$$

On the other hand, the eigenvectors lead to the CKM elements V_{ij} and the CKM phase δ_{CP}, which is given in the Particle Data Group (PDG) parametrization [8]. These are given in the leading order as follows:

$$|V_{us}| \simeq \frac{ab}{m_s^2}\left|\sin\frac{\phi}{2}\right|, \quad |V_{cb}| \simeq \sqrt{2}\frac{c}{m_b}\left|\cos\frac{\phi}{2}\right|, \quad |V_{ub}| \simeq \frac{ac'}{m_b^2}, \quad \delta_{CP} \simeq \frac{1}{2}(\pi-\phi), \tag{9}$$

where we adopt the approximate relations b ~ c and c' ~ d, which will be justified in our numerical results. Although there is only one source of the CP violation, we have various measurements of the CP-violating phase. We can also define the other CP-violating quantity, β (or φ_1), which is one angle of the unitarity triangle:

$$\beta\ (\phi_1) = \arg\left(-\frac{V_{cd}V_{cb}^*}{V_{td}V_{tb}^*}\right). \tag{10}$$

Actually, sin2β has been measured precisely in the B-factory [8]. It is given concisely in the leading order by

$$\sin 2\beta \simeq \sin\phi. \tag{11}$$

There is another CP-violating observable, the Jarlskog invariant J_{CP} [9], which is derived from the following relation:

$$\begin{aligned} & iC \equiv \left[M_u M_u^\dagger, M_d M_d^\dagger\right], \\ & \det C = -2J_{CP}\left(m_t^2-m_c^2\right)\left(m_c^2-m_u^2\right)\left(m_u^2-m_t^2\right)\left(m_b^2-m_s^2\right)\left(m_s^2-m_d^2\right)\left(m_d^2-m_b^2\right). \end{aligned} \tag{12}$$

The predicted one is exactly expressed in terms of the parameters of the mass matrix elements as:

$$J_{CP} = \frac{-1}{\left(m_b^2 - m_s^2\right)\left(m_s^2 - m_d^2\right)\left(m_d^2 - m_b^2\right)} j_{CP}, \tag{13}$$

where

$$j_{CP} = a^2 bcc' d \sin\phi. \tag{14}$$

For $M_d^{(k)}$ $(k = 1–6)$ and $M_d^{(k)}$ $(k = 11–17)$, we summarize three CKM elements V_{ij}, the CP-violating phase δ_{CP}, and j_{CP} in Table 1.

Table 1: The predicted CKM elements, the CP phase, and the CP-violating measure j_{CP}, where $|V_{ij}|$ and δ_{CP} are given in the leading order, and j_{CP} is the exact one

	$\|V_{us}\|$	$\|V_{cb}\|$	$\|V_{ub}\|$	δ_{CP}	j_{CP}
$M_d^{(1)}, M_d^{(2)}, M_d^{(3)}, M_d^{(16)}, M_d^{(17)}$	$\frac{ab}{m_s^2}\left\|\sin\frac{\phi}{2}\right\|$	$\frac{\sqrt{2}c}{m_b}\left\|\cos\frac{\phi}{2}\right\|$	$\frac{ac'}{m_b^2}$	$\frac{1}{2}(\pi-\phi)$	$a^2bcc'd\sin\phi$
$M_d^{(4)}, M_d^{(5)}, M_d^{(6)}, M_d^{(14)}$	$\frac{ab}{m_s^2}$	$\frac{c}{m_b}$	$\frac{c'}{m_b}$	ϕ	$abcc'd^2\sin\phi$
$M_d^{(11)}, M_d^{(12)}, M_d^{(13)}, M_d^{(15)}$	$\frac{ac}{m_s^2}\frac{c'}{m_b}$	$\frac{c}{m_b}$	$\frac{b}{m_b}$	$\pi-\phi$	$abc^2c'd\sin\phi$

NUMERICAL STUDIES OF TEXTURES

Derivation of Weinberg's mass matrix for $M_d^{(1)}$

Since the texture of $M_d^{(1)}$ is the most familiar one among the thirteen textures in Eqs. (4) and (6), we show how to obtain the numerical result by taking $M_d^{(1)}$ as a typical representative. In order to reduce the number of free parameters, we use the three observed down-quark masses as inputs. We adopt the data of the down-quark Yukawa couplings at the M_Z scale [10] as

$$y_d = \left(1.58^{+0.23}_{-0.10}\right) \times 10^{-5}, \quad y_s = \left(3.12^{+0.17}_{-0.16}\right) \times 10^{-4}, \quad y_b = (1.639 \pm 0.015) \times 10^{-2}, \tag{15}$$

which give quark masses as $m_q = y_q v_H$ with $v_H = 174.104$ GeV.

For $M_d^{(1)}$, it is convenient to eliminate the parameters a′, d, and ϕ by using Eq. (8), where the three quark masses in Eq. (15) are put at the 90%90% C.L., respectively. Then, there remain four parameters a, b, c, and c′. We calculate the CKM parameters by scanning these parameters randomly in the region of {a,b,c,c′,d}={0,mb} with a,b,c,c′≤d.

In Fig. 1, we show the frequency distribution of the predicted value of $|V_{us}|$. The large angle is allowed as well as the small angle with almost same weight. In Fig. 2, we show $|V_{us}|$ versus a/a′a/a′. Since we do not assume any relation among the parameters a, b, c, and c′, larger values than 1/√2 are allowed. The observed Cabibbo angle favors a value around a/a′=1, which was assumed in Weinberg›s approach. Since we do not assume a/a′=1, the allowed $|V_{us}|$ lies in the range 0–1 at this stage. We will show that the desired relation a/a′≃1 is finally derived without using the observed Cabibbo angle in our approach.

In Figs. 3 and 4, we show the frequency distributions of the predicted value of $|V_{cb}|$ and $|V_{ub}|$, respectively. $|V_{cb}|$ is allowed up to the maximal mixing 1/2√1/2. On the other hand, $|V_{ub}|$ is predicted to be smaller than 0.020.02because of the quark mass hierarchy, as seen in Eqs. (8) and (9). However, it is still much larger than the observed value.

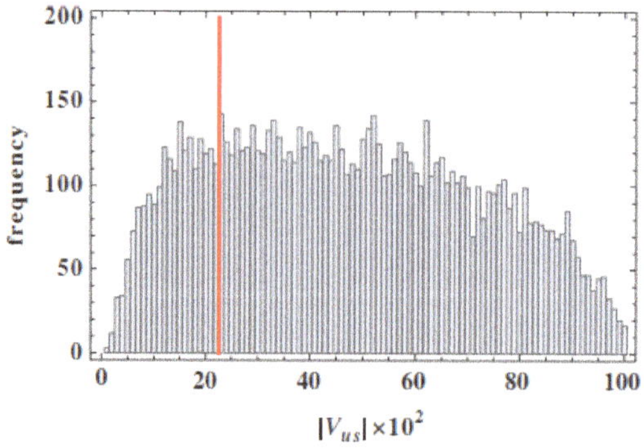

Figure 1: The frequency distribution of the predicted $|V_{us}|$ in arbitrary units without the CKM constraints for $M_d^{(1)}$. The red line denotes the experimental value.

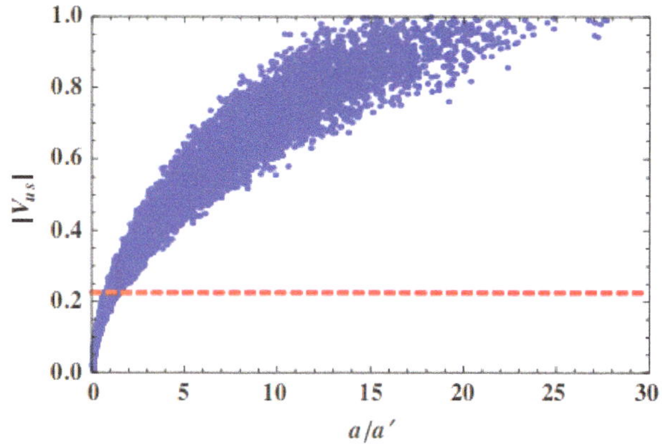

Figure 2: The prediction of $|V_{us}|$ versus a/a′ without the CKM constraints for $M_d^{(1)}$. The red dashed line denotes the experimental value.

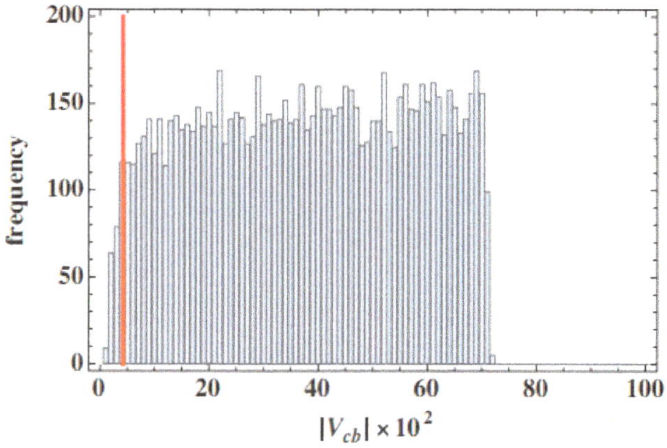

Figure 3: The frequency distribution of the predicted $|V_{cb}|$ in arbitrary units without the CKM constraints for $M_d^{(1)}$. The red line denotes the experimental value.

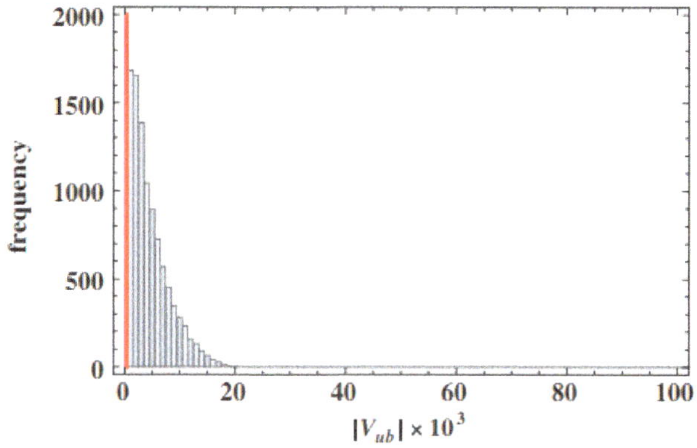

Figure 4: The frequency distribution of the predicted $|V_{ub}|$ in arbitrary units without the CKM constraints for $M_d^{(1)}$. The red line denotes the experimental value.

Here, we show the present experimental data on the CKM mixing angles and the CP-violating phase at the M_Z scale [10],

$$\theta_{12} = 0.227\,35 \pm 0.000\,72, \quad \theta_{23} = (4.208 \pm 0.064) \times 10^{-2},$$
$$\theta_{13} = (3.64 \pm 0.13) \times 10^{-3}, \quad \delta_{CP} = 1.208 \pm 0.054 \text{ [rad]}, \tag{16}$$

where θ_{ij} is defined in the PDG parametrization [8]. We can also use the other CP-violating measure, sin2β, as an input parameter. The experimental data on sin2β are given by [8]:

$$\sin 2\beta = 0.682 \pm 0.019. \tag{17}$$

We will discuss the predicted J_{CP} in comparison with the experimental data,

$$J_{CP} = \left(3.06^{+0.21}_{-0.20}\right) \times 10^{-5}, \tag{18}$$

which is obtained by the global fit of the CKM parameters [8].

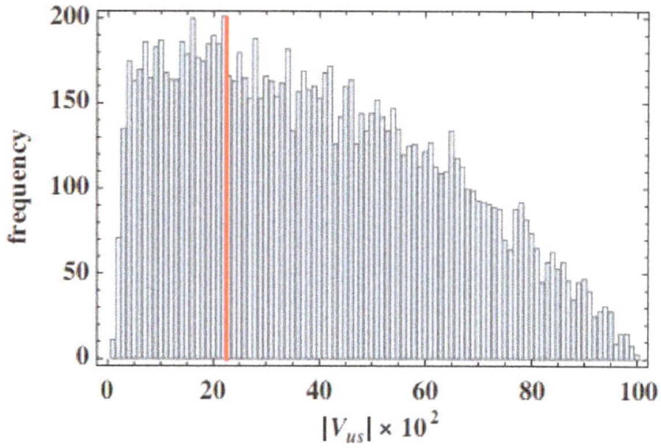

Figure 5: The frequency distribution of $|V_{us}|$ in arbitrary units with the constraint of sin2β for $M_d^{(1)}$. The red line denotes the experimental value.

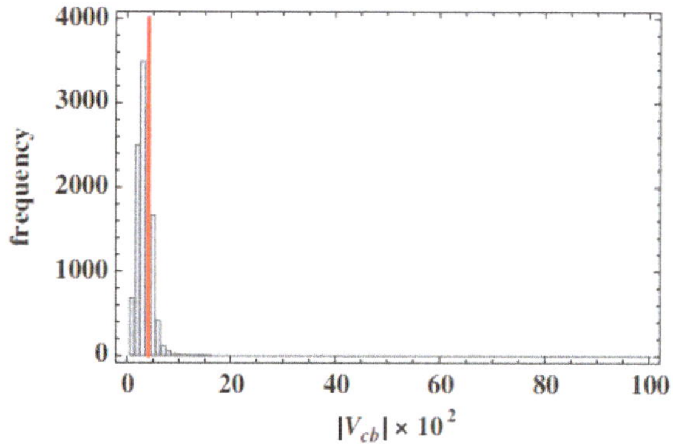

Figure 6: The frequency distribution of $|V_{cb}|$ in arbitrary units with the constraint of sin2β for $M_d^{(1)}$. The red line denotes the experimental value.

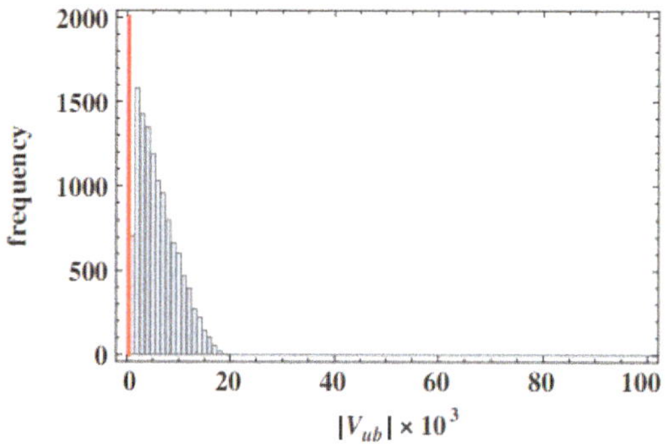

Figure 7: The frequency distribution of $|V_{ub}|$ in arbitrary units with the constraint of sin2β for $M_d^{(1)}$. The red line denotes the experimental value.

Let us first use the CP-violating phase in addition to the three quark masses. We adopt the observed data on sin2β in Eq. (17), which are more sensitive for constraining the parameters compared with the data on δ_{CP}. In Figs. 5 and 6, we show the frequency distributions of the predicted values of $|V_{us}|$ and $|V_{cb}|$, respectively. The predicted $|V_{us}|$ is still broad at 0–1. However, the frequency distribution of $|V_{cb}|$ is impressive because the peak is very sharp around the observed value. Thank to the constraint of the experimental data of sin2β, the predicted $|V_{cb}|$ is restricted to be in a narrow region at one push. In Fig. 7, we show the frequency distribution of the predicted value of $|V_{ub}|$. This distribution has not changed a great deal compared with the case in Fig. 4.

In the next step, we add the constraint of the experimental data θ_{13} in Eq. (16). In Fig. 8, we show the frequency distribution of the predicted value of $|V_{us}|$. The peak of this distribution is around the experimental value, although large mixing is still allowed. The frequency distribution of the predicted value of $|V_{cb}|$ is presented in Fig. 9. This distribution has not changed a great deal compared with the case in Fig. 6.

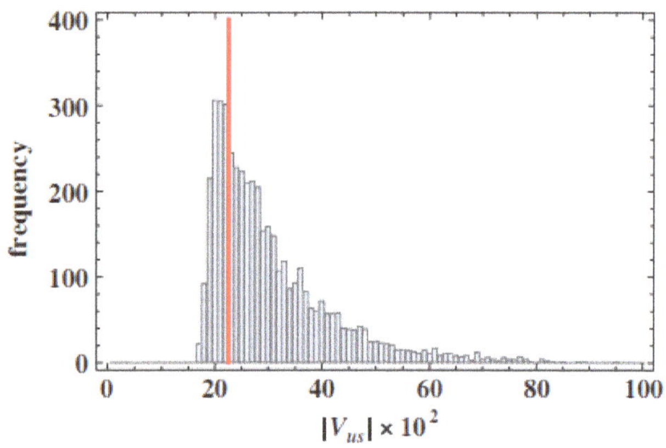

Figure 8: The frequency of $|V_{us}|$ in arbitrary units with the constraints of sin2β and $|V_{ub}|$ for $M_d^{(1)}$. The red line denotes the experimental value.

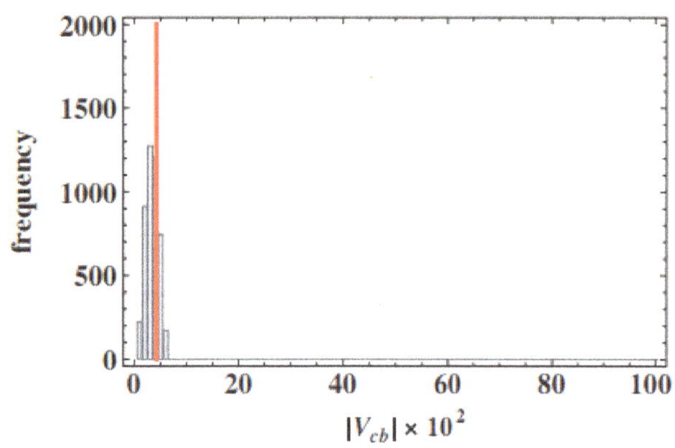

Figure 9: The frequency of $|V_{cb}|$ in arbitrary units with the constraints of sin2β and $|V_{ub}|$ for $M_d^{(1)}$. The red line denotes the experimental value.

In the last step, we impose the constraint of the experimental data θ23 in Eq. (16). Then, we obtain the predicted Cabibbo angle with good accuracy. In Fig. 10, the frequency distribution of the predicted value of $|V_{us}|$ is shown. The predicted $|V_{us}|$ is shown versus the parameter a/a′ in Fig. 11. The prediction is completely consistent with the experimental data, and the ratio a/a′=0.7–1.9 is given. Thus, Weinberg›s anzatz a/a′=1 is obtained by using the experimental data sin2β, θ_{13}, and θ_{23} without assuming any flavor symmetry. Now, we can determine the parameter a (or a/a′) precisely by using the experimental data θ_{12} in Eq. (16).

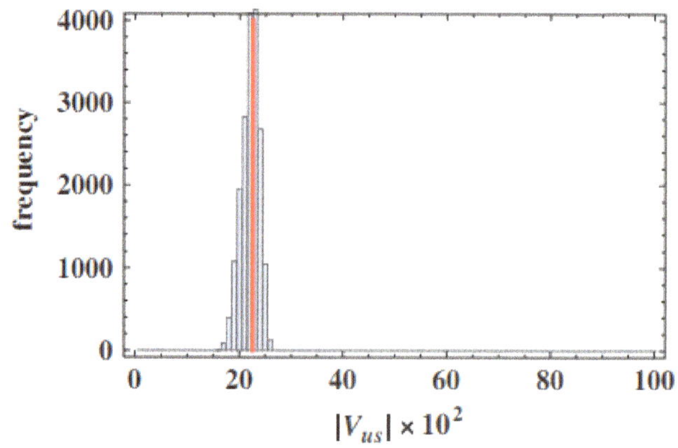

Figure 10: The frequency of $|V_{us}|$ in arbitrary units with the constraints of sin2β, $|V_{ub}|$, and $|V_{cb}|$ for $M_d^{(1)}$. The red line denotes the experimental value.

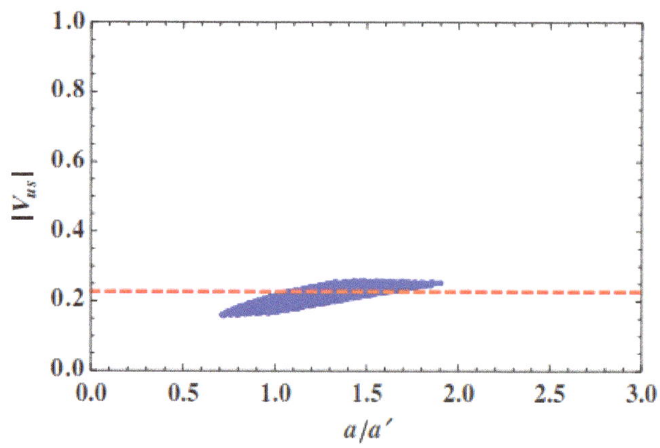

Figure 11: The prediction of $|V_{us}|$ versus a/a′ with the constraints of sin2β, $|V_{ub}|$, and $|V_{cb}|$ for $M_d^{(1)}$. The red dashed line denotes the experimental value.

The seven parameters of $M_d^{(1)}$ are determined completely by the experimental data. However, we can add further investigations since there are still large error bars in the experimental data $|V_{ub}|$. We show the predicted ratio $|V_{ub}/V_{cb}|$ versus sin2β in Fig. 12. The upper bound of the predicted ratio is 0.090, while the experimental data allow up to 0.0940. Thus, precise measurements of the ratio $|V_{ub}/V_{cb}|$ and sin2β are a crucial test of our textures.

There is another CP-violating parameter, J_{CP}, in addition to sin2β and δ_{CP}. We also show the predicted J_{CP} versus $|V_{ub}|$ in Fig. 13, where the red dashed lines denote the experimental bounds in Eqs. (16) and (17). An experimentally allowed region is excluded in our texture. The precise measurements of $|V_{ub}|$ and J_{CP} also provide us with a test of our textures.

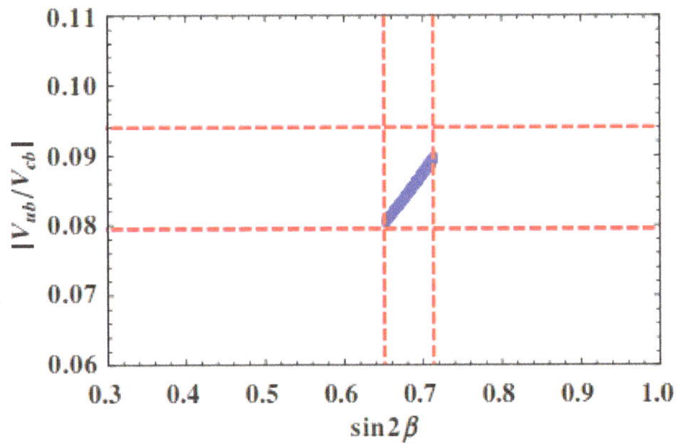

Figure 12: The predicted ratio $|V_{ub}|/|V_{cb}|$ versus sin2β in $M_d^{(1)}$. The red dashed lines denote the upper and lower bounds of the experimental data with 90% C.L.

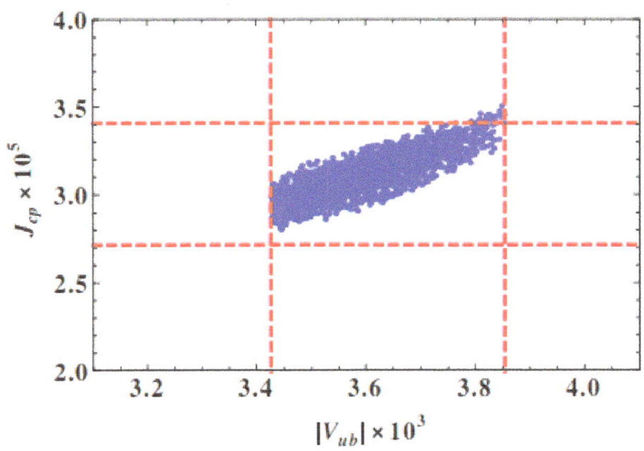

Figure 13: The predicted JCPJCP versus $|V_{ub}|$ in $M_d^{(1)}$. The red dashed lines denote the upper and lower bounds of the experimental data with 90% C.L.

Other textures $M_d^{(k)}$ (k=2–6, 11–17)

The predictions of the mixing angles and the CP-violating phase in our thirteen textures are classified into three groups, as seen in Table 1. The predictions of the first group have been presented for $M_d^{(1)}$ in the previous subsection. Other mass matrices $M_d^{(k)}$ (k=2–6, 11–17) can also be studied in the same way. We have checked numerically that the Cabibbo angle is predicted with good accuracy by using the experimental data sin2β, θ_{13}, and θ_{23} without assuming any flavor symmetry. We summarize the allowed region of the parameters in Tables 2 and 3, where is the exceptional case with c'>d. We have found that only six of the thirteen textures are independent, as shown $M_d^{(16)}$ in Appendix B.

Table 2: The allowed regions of parameters for each $M_d^{(k)}$ (k=1–6)

	a [MeV]	a' [MeV]	b [MeV]	c [MeV]	c' [GeV]	d [GeV]	ϕ [°]
$M_d^{(1)}$	15–17.5	10–15	92–104	78–95	1.65–2.0	2.0–2.3	37–48
$M_d^{(2)}$	15–17	2–4	94–106	78–95	1.65–2.0	2.0–2.3	40–49
$M_d^{(3)}$	15–17.5	250–380	92–104	78–95	1.65–2.0	2.0–2.3	37–48
$M_d^{(4)}$	11–14	9–17	45–58	115–128	0.009–0.011	2.8–2.9	63–75
$M_d^{(5)}$	11–14	2–4	45–58	115–128	0.009–0.011	2.8–2.9	63–75
$M_d^{(6)}$	11–14	220–420	45–58	115–128	0.009–0.011	2.8–2.9	63–75

Table 3: The allowed regions of parameters for each $M_d^{(k)}$ (k=11–17)

	a [MeV]	a' [MeV]	b [MeV]	c [MeV]	c' [GeV]	d [GeV]	ϕ [°]
$M_d^{(11)}$	10–12	2.5–3.5	11–13	125–135	1.0–1.2	2.5–2.7	104–118
$M_d^{(12)}$	10–12	11–18	11–13	125–135	1.0–1.2	2.5–2.7	106–120
$M_d^{(13)}$	10–12	260–390	11–13	125–135	1.0–1.2	2.5–2.7	104–118
$M_d^{(14)}$	11–14	2–4	45–58	115–128	0.009–0.011	2.8–2.9	63–75
$M_d^{(15)}$	10–12	2.5–3.5	11–13	125–135	1.0–1.2	2.5–2.7	104–118
$M_d^{(16)}$	2–4	78–95	15–17	94–106	2.0–2.3	1.65–2.0	40–49
$M_d^{(17)}$	15–17	2–4	94–106	78–95	1.65–2.0	2.0–2.3	40–49

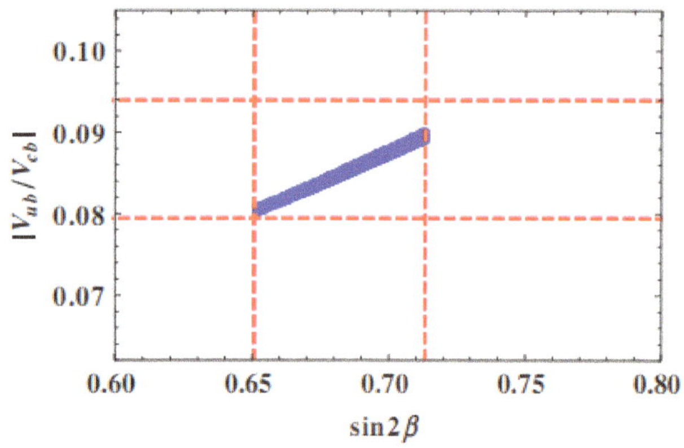

Figure 14: The predicted ratio |Vub|/|Vcb| versus sin2β in $M_d^{(5)}$. The red dashed lines denote the upper and lower bounds of the experimental data with 90% C.L. in Eq. (16).

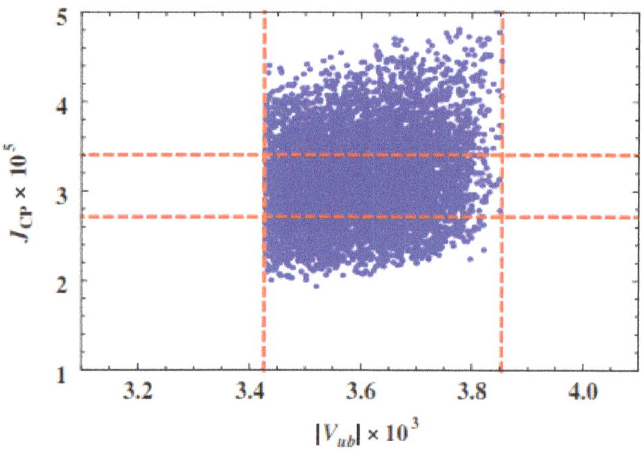

Figure 15: The predicted J_{CP} versus $|V_{ub}|$ in $M_d^{(5)}$. The red dashed lines denote the upper and lower bounds of the experimental data with 90% C.L. in Eq. (16).

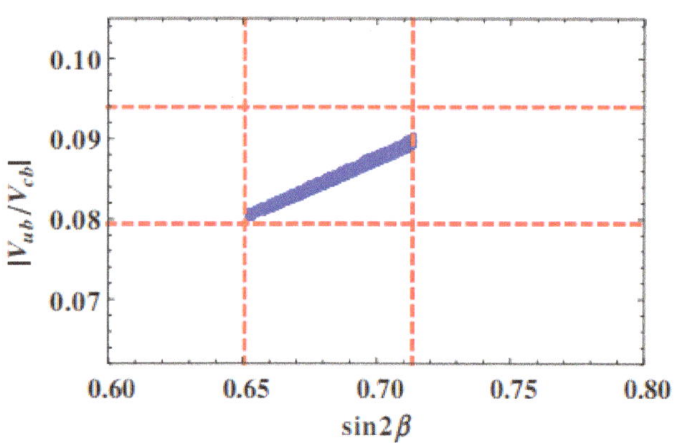

Figure 16: The predicted ratio $|V_{ub}|/|V_{cb}|$ versus sin2β in $M_d^{(11)}$. The red dashed lines denote the upper and lower bounds of the experimental data with 90% C.L. in Eq. (16).

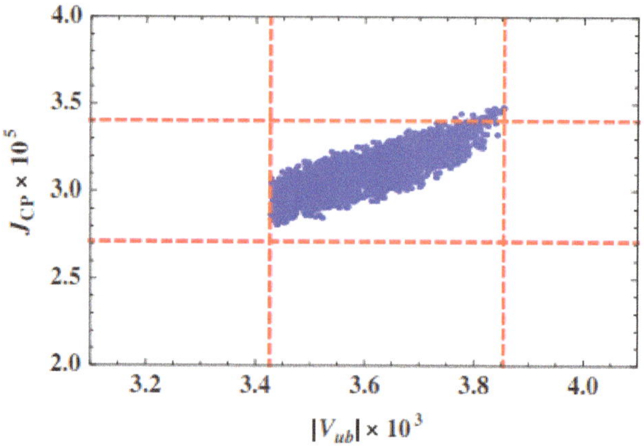

Figure 17: The predicted JCPJCP versus |Vub||Vub| in $M_d^{(11)}$. The red dashed lines denote the upper and lower bounds of the experimental data with 90%90% C.L. in Eq. (16).

We show here predictions for $M_d^{(5)}$ and $M_d^{(11)}$ since the numerical results are almost the same in the same group in Table 1. In Fig. 14, we show the predicted ratio $|V_{ub}/V_{cb}|$ versus sin2β for $M_d^{(5)}$ as a representative of the second group in Table 1. The upper bound of the predicted ratio is also 0.090, which is the same as the one in Fig. 12 for $M_d^{(1)}$. The prediction is smaller than the experimental upper bound, 0.0940. We show J_{CP} versus $|V_{ub}|$ in Fig. 15 for $M_d^{(5)}$, where the red dashed lines denote experimental bounds in Eqs. (16) and (17). The predicted J_{CP} is different from the one in Fig. 13 for $M_d^{(1)}$.

In Fig. 16, we show the predicted ratio $|V_{ub}/V_{cb}|$ versus sin2β for $M_d^{(11)}$ as a representative of the third group in Table 1. The prediction is the same as the one for $M_d^{(1)}$ and $M_d^{(5)}$, as seen in Figs. 12 and 14. We show J_{CP} versus $|V_{ub}|$ for $M_d^{(11)}$ in Fig. 17. The predicted J_{CP} is almost the same as the one in Fig. 13, but different from the one in Fig. 15. Therefore, precise measurements of $|V_{ub}/V_{cb}|$, J_{CP}, and sin2β are important to distinguish the textures.

DISCUSSIONS AND SUMMARY

In our work, we have considered the minimum number of parameters of the quark mass matrices needed for successful CKM mixing angles and CP violation. We impose three zeros in the down-quark mass matrix by taking the diagonal up-quark mass matrix. The three zeros are maximal zeros to keep the CP-violating phase in the quark mass matrix. Thus, there remain six real parameters and one CP-violating phase, which is the minimal number needed to reproduce the observed data of the down-quark masses and the CKM parameters. In order to reproduce the bottom-quark mass and the CKM mixing, V_{us} and V_{cb}, we take the (3, 3), (2, 3), (1, 2) elements of the 3 × 3 down-quark mass matrix to be nonvanishing. Therefore, we have ${}_6C_3$=20 textures with three zeros. We have found that thirteen of the twenty textures are viable for the down-quark mass matrix. They are classified into three groups in Table 1. It is remarked that these textures have freedoms of the unitary transformation of the right-handed quarks and some textures are transformed to other ones. By using such transformations, we have found that six of the thirteen textures are independent (see Appendix B).

As a representative of the above six textures, we have discussed the texture $M_d^{(1)}$ in detail to see how well the Cabibbo angle is predicted. By imposing the experimental data on sin2β, θ_{13}, and θ_{23}, the Cabibbo angle is predicted to be close to the experimental data. We have found that this surprising result remains unchanged in all other viable textures. Thus, the Occam's razor approach is very powerful for obtaining the successful Cabibbo angle.

After fixing all parameters by using the experimental data of the three down-quark masses, the three CKM mixing angles, and one CP-violating phase, we have investigated the correlation between $|V_{ub}/V_{cb}|$ and sin2β. For all textures, the maximal value of the ratio $|V_{ub}/V_{cb}|$ is 0.09, which is smaller than the upper bound of the experimental data, 0.094. We have also discussed J_{CP} versus $|V_{ub}|$. The predicted J_{CP} is almost the same among the first and third groups of Table 1, but different from the one in the second group. The precise data $|V_{ub}/V_{cb}|$, J_{CP}, and sin2β provide us with an important test of our textures.

Our textures have been analyzed at the electroweak scale in this paper. The stability of texture zeros of the quark mass matrix has been examined against the renormalization-group evolution from the GUT scale to the electroweak scale by Xing and Zhao [11]. They found that texture zeros of the quark mass matrix are essentially stable against the evolution. Thus, we expect that the conclusions derived in this paper would not change a great deal even if we considered the textures in Eqs. (4) and (6) at the GUT scale.

Previous SectionNext Section

ACKNOWLEDGMENTS

This work is supported by a JSPS Grand-in-Aid for Scientific Research (15K05045; M.T.) and Scientific Research B (No. 26287039; T.T.Y.). This work is also supported by the World Premier International Research Center Initiative (WPI Initiative), MEXT, Japan.

APPENDIX A. UNFAVORED DOWN-QUARK MASS MATRICES

As discussed in Sect. 3, the four textures (the seventh–tenth ones) in Eq. (3) are excluded by the experimental data. The following seventh, eighth, and ninth textures are easily shown to be inconsistent with the experimental data:

$$M_d^{(7)} = \begin{pmatrix} 0 & a & 0 \\ a' & b\,e^{-i\phi} & c \\ c' & 0 & d \end{pmatrix}, \quad M_d^{(8)} = \begin{pmatrix} a' & a & 0 \\ c' & b\,e^{-i\phi} & c \\ 0 & 0 & d \end{pmatrix}, \quad M_d^{(9)} = \begin{pmatrix} 0 & a & c' \\ 0 & b\,e^{-i\phi} & c \\ 0 & a' & d \end{pmatrix}. \tag{A1}$$

Both matrices $M_d^{(7)} M_d^{(7)\dagger}$ and $M_d^{(8)} M_d^{(8)\dagger}$ give the vanishing (13), (31) elements. These zeros are consistent only when $\theta_{13} = 0$, $\theta_{23} = 90°$, or $\delta_{CP} = 0$. Actually, the Jarlskog invariant J_{CP} of these textures vanishes. The matrix $M_d^{(9)}$ has one zero eigenvalue; i.e, the lightest d quark is massless.

The last texture in Eq. (3) is a nontrivial one. We parametrize the texture as follows:

$$M_d^{(10)} = \begin{pmatrix} a' & a & 0 \\ 0 & be^{-i\phi} & c \\ c' & 0 & d \end{pmatrix}, \tag{A2}$$

which gives the CKM elements in the leading order as follows:

$$|V_{us}| \simeq 2\frac{ab}{m_s^2}\left|\sin\frac{\phi}{2}\right|, \quad |V_{cb}| \simeq \frac{c}{m_b}, \quad |V_{ub}| \simeq \frac{a'c'}{m_b^2}, \quad \delta_{CP} \simeq \frac{1}{2}(\pi + \phi), \tag{A3}$$

where we adopt the approximate relations

$a' \sim a$, $c \sim 2b$, and $c' \sim d$. J_{CP} is given as

$$J_{CP} = \frac{1}{\left(m_b^2 - m_s^2\right)\left(m_s^2 - m_d^2\right)\left(m_b^2 - m_d^2\right)} aa'bcc'd \sin\phi. \tag{A4}$$

Therefore, sinϕ should be positive. Thus, δ_{CP} is larger than 90° or negative, which is completely excluded by the experimental data in Eq. (16). Actually, we have obtained $\delta_{CP} \geq 90°$ under the experimental constraints of $|V_{us}|$, $|V_{cb}|$, and $|V_{ub}|$ numerically.

The following textures (the seventh–tenth ones) in Eq. (5) are also excluded by the experimental data:

$$M_d^{(18)} = \begin{pmatrix} a' & a\,e^{i\phi} & 0 \\ 0 & 0 & c \\ b & c' & d \end{pmatrix}, \quad M_d^{(19)} = \begin{pmatrix} 0 & a & 0 \\ a' & 0 & c \\ b & c'\,e^{-i\phi} & d \end{pmatrix}, \quad M_d^{(20)} = \begin{pmatrix} a' & a & 0 \\ b & 0 & c\,e^{i\phi} \\ 0 & c' & d \end{pmatrix} \tag{A5}$$

Both matrices $M_d^{(18)} M_d^{(18)\dagger}$ and $M_d^{(19)} M_d^{(19)\dagger}$ give the vanishing (12), (21) elements. These zeros are consistent only when $\theta_{13} = 0$, $\theta_{23} = 0$, or $\delta_{CP} = 0$. The Jarlskog invariant J_{CP} of these textures vanishes as well as $M_d^{(7)}$ and $M_d^{(8)}$. Since $M_d^{(20)}$ becomes $M_d^{(10)}$ by the rotation of the right-handed quarks, $M_d^{(20)}$ is excluded for the same reason as $M_d^{(10)}$, $\delta_{CP} \geq 90°$.

APPENDIX B. REDUNDANCY OF OUR TEXTURES

Since the CKM matrix is the flavor mixing among the left-handed quarks, the textures in Eqs. (4) and (6) have freedoms of the unitary transformation of the right-handed quarks. Under this transformation, $M_d M_d^\dagger$ is invariant. We can easily see that some textures are transformed to other ones, as follows:

$$M_d^{(1)} \equiv M_d^{(3)}, \quad M_d^{(2)} \equiv M_d^{(16)} \equiv M_d^{(17)}, \quad M_d^{(4)} \equiv M_d^{(5)} \equiv M_d^{(14)},$$
$$M_d^{(11)} \equiv M_d^{(13)} \equiv M_d^{(15)}, \quad M_d^{(10)} \equiv M_d^{(20)}, \tag{B1}$$

where the notation "≡" means the equivalent up to the unitary transformation of the right-handed quarks. In addition, we can easily see that $M_d^{(7)}, M_d^{(8)}, M_d^{(18)}$, and $M_d^{(19)}$ can be transformed to textures with four zeros by the right-handed quark rotation. Therefore, these ones are excluded. In conclusion, we have six viable independent textures for the down-quark mass matrix.

REFERENCES

1. Weinberg S., Trans. New York Acad. Sci. 38, 185 (1977). (doi:10.1111/j.2164-0947.1977.tb02958.x)
2. Gatto R., Sartori G., and Tonin M., Phys. Lett. B 28, 128 (1968). (doi:10.1016/0370-2693(68)90150-0)
3. Fritzsch H., Phys. Lett. B 73, 317 (1978). (doi:10.1016/0370-2693(78)90524-5)
4. Fritzsch H., Nucl. Phys. B 155, 189 (1979). (doi:10.1016/0550-3213(79)90362-6)
5. Ramond P., Roberts R. G., and Ross G. G., Nucl. Phys. B 406, 19 (1993)]. (doi:10.1016/0550-3213(93)90159-M)
6. Harigaya K., Ibe M., and Yanagida T. T., Phys. Rev. D 86, 013002 (2012). (doi:10.1103/PhysRevD.86.013002)
7. Gonzalez M. C., Malton M., and Schwetz T..
8. Olive K. A., et al. [Particle Data Group], Chin. Phys. C 38, 090001 (2014). (doi:10.1088/1674-1137/38/9/090001)
9. Jarlskog C., Phys. Rev. Lett. 55, 1039 (1985). (doi:10.1103/PhysRevLett.55.1039)
10. Antusch S. and Maurer V., J. High Energy Phys. 1311, 115 (2013). (doi:10.1007/JHEP11(2013)115)
11. Xing Z. z and Zhao Z. h, Nucl. Phys. B 897, 302 (2015). (doi:10.1016/j.nuclphysb.2015.05.027)

CHAPTER 9

BOUNDARY TERMS IN QUANTUM FIELD THEORY AND THE SPIN STRUCTURE OF QCD

Peter Lowdon

INTRODUCTION

When classical and quantum field theories are discussed it is often assumed that spatial boundary terms do not contribute [1], [2], [3], [4] and [5]. The standard reasoning given for this is that the dynamical fields in the theory vanish at spatial infinity. Although there may be instances in classical field theory where this boundary condition is applicable, generally this condition is too restrictive. Plane waves [6], or cases where the space of field configurations has a non-trivial topology [7], are two such examples where field solutions may not vanish asymptotically. In the quantum case, axiomatic formulations of field theory assert that fields φ are operator-valued distributions [8]. Distributions are continuous linear functionals which map a space of test functions T onto the complex numbers: $\varphi: T \rightarrow \mathbb{C}$. In quantum field theory (QFT), T is chosen to be some set of space–time functions; usually either the space of continuous functions with compact support $D(\mathbb{R}^{1,3})$, or the space of Schwartz functions[1] $S(\mathbb{R}^{1,3})$. In either of these cases one can represent the image of the map φ on a space–time function f as:

$$\varphi(f) = (\varphi, f) := \int d^4x\, \varphi(x) f(x) \tag{1.1}$$

which gives meaning to the x -dependent field expression $\varphi(x)$. Since φ is an operator -valued distribution in QFT, only the smeared expression $\varphi(f)$ is guaranteed to correspond to a well-defined operator. The derivative of a distribution φ' is defined by:

$$(\varphi', f) := -(\varphi, f') \tag{1.2}$$

and is itself also a distribution [8]. By applying the integral representation in Eq. (1.1), one can interpret this definition as an integration by parts where the boundary terms have been 'dropped':

$$\int d^4x\, \varphi'(x) f(x) = -\int d^4x\, \varphi(x) f'(x) \tag{1.3}$$

Although this shorthand notation is useful, and will be used for the calculations in this paper, it can also be slightly misleading. Sometimes it is incorrectly stated that integration by parts of quantum fields can be performed, and the boundary terms neglected. However, distributions are generally not point-wise defined, so boundary expressions like: $\int_{\partial\mathbb{R}^3} \varphi(x) f(x)$ are often ill-defined. Therefore, when manipulations like this are performed one is really just applying the definition of the derivative of a distribution, there are no boundary contributions. This makes the question of whether spatial boundary term operators vanish a more subtle issue in QFT than in the classical case.

The physical rationale behind using operator-valued distributions as opposed to operator-valued functions in QFT is because operators inherently imply a measurement, and this is not well-defined at a single (space–time) point since this would require an infinite amount of energy [9]. Instead, one can perform a measurement over a space–time region U, and model the corresponding operator A(f) as a distribution A smeared with some test function f which has support in U. If one were to smear A with another test function g, which has different support to f , then in general the operators A(f) and A(g) would be different. But the interpretation is that these operators measure the same quantity, just within the different space–time regions: supp(f) and supp(g).

Citation: Peter Lowdon, Boundary terms in quantum field theory and the spin structure of QCD, http://dx.doi.org/10.1016/j.nuclphysb.2014.11.004.

As well as differentiation it is also possible to extend the notion of multiplication by a function to distributions. Given a distribution φ, a test function f, and some function g, this is defined as:

$$(g\varphi, f) := (\varphi, gf) \tag{1.4}$$

In order that gφ defines a distribution in the case where f∈D, it suffices that g be an infinitely differentiable function. For tempered distributions, in which f∈S, it is also necessary that g and all of its derivatives are bounded by polynomials [8].

Besides the assumption that fields are operator-valued distributions, axiomatic approaches to QFT usually postulate several additional conditions that the theory must satisfy. Although different axiomatic schemes have been proposed, these schemes generally contain a common core set of axioms.[2] For the purpose of the calculations in this paper, the core axioms which play a direct role are:

1. Local (anti-)commutativity

If the support of the test functions f, g, of the fields Ψ, Φ, are space-like separated then:

$$[\Psi(f), \Phi(g)]_{\pm} = \Psi(f)\Phi(g) \pm \Phi(g)\Psi(f) = 0$$

holds when applied to any state vector, for any fields Ψ, Φ.

2. Non-degeneracy

The inner product $\langle\cdot|\cdot\rangle$ on the space of states Vis non-degenerate:

$$\langle\Psi|\omega\rangle = 0, \quad \forall|\Psi\rangle \in \mathcal{V} \quad \Longrightarrow \quad |\omega\rangle = 0$$

The physical motivation for the local (anti-)commutativity axiom is that it imposes a causality restriction on the theory. Since the action of field operators Ψ(f) on states can be interpreted as the performance of a particular measurement in the space–time region supp(f)[9], if another measurement Φ(g) is performed in a space–time region supp(g) which is space-like separated to supp(f), the axiom states that these two measurements must either commute or anti-commute with one another. In physical terms this means that measurements which are performed a space-like distance apart cannot be causally related to one another. A space of fields F which satisfies this property is called a local field algebra [11]. By contrast, the non-degeneracy axiom can be imposed without any real loss of generality since any vector |ω⟩ whose inner product with any other state vanishes will not introduce physical effects into the theory that are describable using the inner product [10]. Such states |ω⟩ are therefore physically trivial with regards to the quantum theory, and can hence be set to zero.

Additional complications arise when defining quantised gauge field theories. This is because the restriction of a theory to be invariant under a gauge group symmetry G, which corresponds to local invariance under some global symmetry group G, leads to a strengthened form of the Noether current conservation condition called the local Gauss law [11]:

$$J^a_\mu = \partial^\nu G^a_{\mu\nu}, \quad G^a_{\mu\nu} = -G^a_{\nu\mu} \tag{1.5}$$

where J^a_μ is the Noether current associated with invariance under the global group G. Because of this condition it turns out that there are essentially two quantisation strategies [11]:

1. One demands that Eq. (1.5) holds as an operator equation, which implies that the algebra of fields F is no longer local. In particular, if a field transforms non-trivially with respect to the group G (i.e. has a non-zero G-charge), the field must be non-local.
2. One adopts a local gauge quantisation in which the local Gauss law is modified. This modification ensures that the field algebra remains local (even for charged fields), but necessitates the introduction of an indefinite inner product on the space of states V, and a condition: $\langle\Psi|J^a_\mu - \partial^\nu G^a_{\mu\nu}|\Psi\rangle = 0$ for identifying the physical states $|\Psi\rangle \in V_{phys} \subset V$ (the weak Gauss law).

The advantage with the latter approach is that it allows one keep a local field algebra, and so all of the results from local field theory remain applicable. For the purposes of discussion in this paper we will only consider local quantisations, and in particular we will focus on the local BRST quantisation of Yang–Mills theory. A key feature of BRST quantisation is that a gauge-fixing term LGF is added to the Lagrangian density L. The modified Lagrangian L+LGF is no longer gauge-invariant, but remains invariant under a residual BRST symmetry with a conserved charge Q_B. By defining the physical space of states $V_{phys} \subset V$ to be the states which satisfy the subsidiary condition : $Q_B V_{phys} = 0$, this ensures that the weak Gauss law is satisfied and that the field algebra F is local. The introduction of an indefinite inner product on V also leads to unphysical negative norm states, which are generated by the Faddeev–Popov ghost degrees of freedom in LGF. In terms of these extended state spaces, the physical Hilbert space is a quotient space of the form[3] $\mathcal{H} := \overline{\mathcal{V}_{phys}/\mathcal{V}_0}$, where V_0 contains the zero norm states in V[10].

The remainder of this paper is structured as follows; in Section 2 we apply these general QFT properties in order to establish a necessary and sufficient condition for spatial boundary terms to vanish, in Section 3 we give a theoretical overview of the angular momentum decomposition problem in QCD and why spatial boundary terms are of particular relevance, and in Section 4 we apply the results of Section 2 to this problem. Finally, in Section 5 we summarise our results and discuss their interpretation.

SPATIAL BOUNDARY TERMS IN QFT

The aim of this section will be to demonstrate that the general properties of a quantum field theory, some of which were outlined in Section 1, are enough to establish a necessary and sufficient condition for spatial boundary terms of the form: $\int d^3x\, \partial_i B^i$ to vanish in the Hilbert space of physical states H[10], where $B^i(x)$ is some arbitrary local[4] field. However, before fully discussing this condition it is important to first outline the differences between the classical and quantum field theory approaches to conserved (and non-conserved) charges. Classically, charges Q are defined to be the spatial integral of the temporal component of some (not necessarily conserved) current density $j^\mu(x)$:

$$Q = \int d^3x\, j^0(x) \tag{2.1}$$

However, in QFT $j^\mu(x)$ is typically some product of fields, and is therefore an operator-valued distribution.[5] This means that because no smearing with a test function has been performed, the quantised version of the classically motivated definition in Eq. (2.1) will generally not correspond to a proper operator in QFT [11]. But given some space–time test function f, a well-defined quantum representation of Q can be written:

$$Q = \int d^4x\, f(x) j^0(x) = j^0(f) \tag{2.2}$$

Following the discussion in Section 1, Q is interpreted as acting on the space–time region supp(f). In order to extend the action of Q to the whole of space, by analogy with the classical case, one can choose the following test function [6]: $f := \alpha(x_0) f_R(\mathbf{x})$, with $\alpha \in D(\mathbb{R})$ ($\mathrm{supp}(\alpha) \subset [-\delta, \delta]$, $\delta > 0$) and $f_R \in D(\mathbb{R}^3)$ where:

$$\int dx_0\, \alpha(x_0) = 1, \qquad f_R(\mathbf{x}) = \begin{cases} 1, & |\mathbf{x}| < R \\ 0, & |\mathbf{x}| > R(1+\varepsilon) \end{cases} \tag{2.3}$$

with $\varepsilon > 0$. Because of the way f_R is defined this means that $\partial_i f_R$ vanishes for $|\mathbf{x}| < R$. Using these test functions one can then construct the following well-defined (localised) charge operator Q_R:

$$Q_R = \int d^4x\, \alpha(x_0) f_R(\mathbf{x}) j^0(x) \tag{2.4}$$

The reason why α and f_R are chosen to have this form is so that in the special case where the quantised version of the charge is genuinely conserved, this definition is in agreement with the classically-motivated form of Q (Eq. (2.1)) in the limit $R \to \infty$.

Setting: $j^0 = \partial_i B^i$ it is clear that spatial boundary terms: $\int d^3x\, \partial_i B^i$ are simply a special class of charges. Therefore, when: $\int d^3x\, \partial_i B^i$ is written in the proceeding discussion (for brevity), this actually corresponds to the smeared expression: $\lim_{R\to\infty} \int d^4x\, \alpha(x_0) f_R(\mathbf{x}) \partial_i B^i(x)$. With this notation in mind, one has the following theorem:

Theorem 1.

$\int d^3x\, \partial_i B^i$ vanishes in $\mathcal{H}$ $\Longleftrightarrow$ $\int d^3x\, \partial_i B^i |0\rangle = 0$.

Proof ($\Leftarrow$).

Let $\varphi \in F(O)$ be some (smeared) local operator[7] and let $\alpha \in D(\mathbb{R})$ and $f_R \in D(\mathbb{R}^3)$ be the test functions of compact support defined in Eq. (2.3). Then:

$$\begin{aligned} \int d^3x\, (\partial_i B^i)\varphi|0\rangle &= \int d^3x\, [\partial_i B^i, \varphi]_\pm |0\rangle \\ &= \lim_{R\to\infty} \int d^4x\, \alpha(x_0) f_R(\mathbf{x}) [\partial_i B^i(x), \varphi]_\pm |0\rangle \\ &= -\lim_{R\to\infty} \int d^4x\, \alpha(x_0) (\partial_i f_R(\mathbf{x})) [B^i(x), \varphi]_\pm |0\rangle \end{aligned}$$

where in the first line one uses the assumption that: $\int d^3x\, \partial_i B^i |0\rangle = 0$ (with $|0\rangle$ the vacuum state), and in the last line the definition for the derivative of a distribution is used. Now because of the way f_R is defined it follows that: $\mathrm{supp}(\partial_i f_R) = \{x \in \mathbb{R}^3; R \le |\mathbf{x}| \le R(1+\varepsilon)\}$, and so the support of $\alpha \partial_i f_R$ will be restricted to the space–time points: $(x_0, \mathbf{x})$, where $|\mathbf{x}| \ge R$ and $|x_0| \le \delta$. So in the limit[8] $R \to \infty$ the supports of $\alpha \partial_i f_R$, and the test function for which φ is implicitly smeared, will become space-like separated. But by the local (anti-)commutativity axiom this implies that the (anti-) commutator in the last line must vanish exactly, and therefore: $\int d^3x\, (\partial_i B^i)\varphi|0\rangle = 0$. The vanishing of the (anti-)commutator is independent of the explicit form for α because α is continuous, has compact support, and is therefore bounded, which means that $\alpha \partial_i f_R$ will vanish wherever $\partial_i f_R$ does. Moreover, it follows directly from local (anti-)commutativity that the vanishing of the (anti-)commutator is also independent of the explicit form of f_R [13]. Because of the vanishing of: $\int d^3x\, (\partial_i B^i)\varphi|0\rangle$, one has:

$$\langle \Psi | \int d^3x\, \partial_i B^i (\varphi|0\rangle) = 0, \quad \forall |\Psi\rangle \in \mathcal{H}$$

and applying the Reeh–Schlieder Theorem [9] implies:

$$\langle \Psi | \int d^3x\, \partial_i B^i |\Phi\rangle = 0, \quad \forall |\Psi\rangle, |\Phi\rangle \in \mathcal{H}$$

which is precisely the statement that the spatial boundary term: $\int d^3x \partial_i B^i$ vanishes in H.

Proof ($\Longrightarrow$).

Conversely, if one assumes that the spatial boundary term vanishes in H then:

$$\langle \Psi | \int d^3x\, \partial_i B^i |0\rangle = 0, \quad \forall |\Psi\rangle \in \mathcal{H}$$

since $|0\rangle \in H$. But this inner product between states is taken to be non-degenerate, so the above condition implies that: $\int d^3x\, \partial_i B^i |0\rangle = 0$.

One subtlety in establishing Theorem 1 comes from a property which is well-established in Quantum Electrodynamics (QED) [14], as well as other gauge theories [15] – charged states are non-local. This means that it is not possible to create a charged state by applying a local operator to the vacuum. However, by virtue of the Reeh–Schlieder Theorem, a charged state can always be approximated by local states as closely as one likes in the sense of convergence in some allowed topology on H. Often this topology is chosen to be the weak topology [10] and so convergence means weak convergence . Therefore, given that $|\Phi\rangle \in H$ is a charged state, there exists a sequence of local operators $\{\varphi_n\}$ such that $\lim_{n\to\infty} \langle \Psi|\varphi_n|0\rangle = \langle \Psi|\Phi\rangle$, $\forall |\Psi\rangle \in H$. In QED it has in fact been explicitly shown that physical charged states can be constructed from weak limits of local states [16]. With this convergence property in mind, the proof of Theorem 1 can then still be shown to hold in the case where $|\Phi\rangle$ is a charged state, since one can apply the same steps as before with φ replaced by φ_n, conclude that: $\langle \Psi | \int d^3x\, \partial_i B^i (\varphi_n|0\rangle) = 0, \forall |\Psi\rangle \in \mathcal{H}$, and then take the limit $n \to \infty$. Moreover, because the (anti-)commutator in the proof is shown to vanish regardless of the explicit form of both α and f_R, this demonstrates that Theorem 1 holds independently of the specific test functions in the smearing, and can hence be applied to all spatial boundary term operators of the form: $\int d^3x\, \partial_i B^i$, where $B^i(x)$ is any local field.

The proof of Theorem 1 is based on similar discussions in [10], [11], [12] and [13], which address the issue of defining a consistent local charge operator and its action on states in H. The surprising conclusion of this theorem is that the vanishing of a spatial boundary term only requires that the corresponding operator annihilates the vacuum state – it is independent of how this operator acts on the full space of states. This result has interesting physical consequences for any QFT, but in particular its relevance to the angular momentum decomposition problem in Quantum Chromodynamics (QCD) will be discussed in Sections 3 and 4.

Generally, Theorem 1 demonstrates that a spatial boundary term operator annihilating the vacuum state is both a necessary and a sufficient condition for that boundary term itself to vanish in the physical Hilbert space H. However, in order to practically determine whether this operator annihilates the vacuum or not, it is easier to instead relate these conditions to equivalent conditions involving matrix elements. This connection is given by the following simple relations:

$$\text{If } \langle\Psi| \int d^3x\, \partial_i B^i |0\rangle = 0, \quad \forall|\Psi\rangle \in \mathcal{H} \quad \Longrightarrow \quad \int d^3x\, \partial_i B^i |0\rangle = 0 \tag{2.5}$$

$$\text{If } \exists|\Psi\rangle \in \mathcal{H} \text{ s.t. } \langle\Psi| \int d^3x\, \partial_i B^i |0\rangle \neq 0 \quad \Longrightarrow \quad \int d^3x\, \partial_i B^i |0\rangle \neq 0 \tag{2.6}$$

Eq. (2.5) follows immediately from the assumption that the inner product in H is non-degenerate, and Eq. (2.6) is the logical negation of the statement that the boundary operator acting on the vacuum state is the null vector in H. These relations imply that if one can find any state $|\Psi\rangle \in$H such that: $\langle\Psi| \int d^3x\, \partial_i B^i |0\rangle \neq 0$, then this definitively proves: $\int d^3x\, \partial_i B^i |0\rangle \neq 0$, and hence by Theorem 1 that: $\int d^3x\, \partial_i B^i \neq 0$. Otherwise, it must be the case that: $\int d^3x\, \partial_i B^i |0\rangle = 0$, and thus: $\int d^3x\, \partial_i B^i = 0$. To determine the explicit form of these matrix elements one can use the transformation property of fields under translations [9]:

$$B^i(x) = e^{iP_\mu x^\mu} B^i(0) e^{-iP_\mu x^\mu} \tag{2.7}$$

and insert this between the states $|\Psi\rangle$ and $|0\rangle$ to obtain:

$$\begin{aligned} \langle\Psi| \int d^3x\, \partial_i \left[e^{iP_\mu x^\mu} B^i(0) e^{-iP_\mu x^\mu}\right] |0\rangle &= \langle\Psi| \int d^3x \left[(iP_i) e^{iP_\mu x^\mu} B^i(0) e^{-iP_\mu x^\mu} \right. \\ &\quad \left. + e^{iP_\mu x^\mu} B^i(0) e^{-iP_\mu x^\mu} (-iP_i)\right] |0\rangle \\ &= \langle\Psi| \int d^3x \left[(iP_i) e^{iP_\mu x^\mu} B^i(0)\right] |0\rangle \end{aligned}$$

where the second term in the first line vanishes because $P_\mu|0\rangle = 0$. Now if one takes $|\Psi\rangle$ to be some momentum eigenstate $|p\rangle$, the above relation simplifies to:

$$\langle p| \int d^3x\, \partial_i B^i |0\rangle = \begin{cases} 0, & p = 0 \\ \lim_{R\to\infty} \int d^4x\, ip_i \alpha(x_0) f_R(\mathbf{x}) e^{ip_\mu x^\mu} \langle p|B^i(0)|0\rangle, & p \neq 0 \end{cases} \tag{2.8}$$

It is interesting to note here that the vacuum expectation value of the spatial boundary operator: $\int d^3x\, \partial_i B^i$ completely vanishes, whereas the off-diagonal matrix element depends on the local term: $\langle p|B^i(0)|0\rangle$.

THE PROTON ANGULAR MOMENTUM DECOMPOSITION

Theoretical investigations into the spin structure of nucleons have been ongoing ever since the inception of QCD in the 1960s. The evolution of this research area has been influenced by a number of different experimental groups including the European Muon Collaboration (EMC), the Spin Muon Collaboration, and more recently HERMES, COMPASS, STAR and PHENIX [6]. The focal point of these investigations have largely centred around settling an unresolved dispute known as the spin crisis, which refers to results obtained by the EMC experiment [17] that suggested quarks accounted for 'only' a very small amount of the spin of the proton. Many of the proposed solutions to the spin crisis are based on splitting the QCD angular momentum operator up in different ways, and then arguing a particular physical interpretation of the resulting pieces. It turns out that spatial boundary terms play a prominent role in these decompositions. More specifically, during the rest of this section we will outline why the vanishing of certain spatial boundary superpotential terms is an essential assumption in this analysis.

As discussed in Section 1, in order to analyse the dynamical properties of quantised Yang–Mills theory (in this case QCD), one must choose a quantisation procedure. For the analysis in this paper we will consider the local BRST quantisation with the following Hermitian, gauge-fixed Lagrangian density as proposed by [12]:

$$\mathcal{L}_{QCD} = -\frac{1}{4}F^a_{\mu\nu}F^{\mu\nu a} + \overline{\psi}\left(\frac{i}{2}\gamma^\mu(\overrightarrow{\partial}_\mu - \overleftarrow{\partial}_\mu) + gT^a A^a_\mu \gamma^\mu - m\right)\psi + \mathcal{L}_{GF} + \mathcal{L}_{FP} \tag{3.1}$$

$$\mathcal{L}_{GF} = -(\partial^\mu B^a)A^a_\mu + \frac{\xi}{2}B^a B^a \tag{3.2}$$

$$\mathcal{L}_{FP} = -i\partial^\mu \overline{C}^a (D_\mu C)^a \tag{3.3}$$

where $C^a, \overline{C}^a$ are the Faddeev–Popov ghost fields, B^a is the auxiliary gauge fixing field, ξ is a gauge fixing parameter, and one defines: $(D_\mu C)^a := \partial_\mu C^a - gf^{abc}A^c_\mu C^b$. To determine the spin structure of QCD one must first define the energy-momentum tensor of the theory $T^{\mu\nu}_{QCD}$. As with any current density the definition of $T^{\mu\nu}$ is always ambiguous up to a sign, but for the purposes of this paper the following definition will be used:

$$T^{\mu\nu} = \frac{2}{\sqrt{-g}}\frac{\delta S}{\delta g_{\mu\nu}} \tag{3.4}$$

where: $g = \det g_{\mu\nu}$, $S = \int d^4x \sqrt{-g}\mathcal{L}$, and L has the formal functional dependence: $L=L(g_{\mu\nu},\Psi^I,\nabla_\alpha\Psi^I)$, with ∇_α the general relativistic covariant derivative[11] and $\{\Psi^I\}$ the dynamical fields in the theory (with possible internal index I). As was famously shown by Belinfante [18], this expression can always be decomposed into the following form:

$$T^{\mu\nu} = T^{\mu\nu}_c + \frac{1}{2}\partial_\rho(S^{\rho\mu\nu} + S^{\mu\nu\rho} + S^{\nu\mu\rho}) \tag{3.5}$$

where $T^{\mu\nu}_c$ is the canonical energy–momentum tensor and $S^{\rho\mu\nu}$ is the so-called spin-angular momentum density term [10]:

$$T^{\mu\nu}_c := \frac{\delta\mathcal{L}}{\delta(\partial_\mu\Psi^I)}\partial^\nu\Psi^I - g^{\mu\nu}\mathcal{L}, \qquad S^{\rho\mu\nu} := -i\frac{\delta\mathcal{L}}{\delta(\partial_\rho\Psi^I)}(s^{\mu\nu})^I{}_J\Psi^J \tag{3.6}$$

$(s^{\mu\nu})^I{}_J$ corresponds to the Lorentz generator $s^{\mu\nu}$ in the finite-dimensional representation[12] of the field Ψ^I. It is interesting to note that the question of whether the quantised Belinfante or canonical energy–momentum tensor is physically more relevant is still an on-going issue [6]. A discussion of the related subtleties between the Belinfante P^μ and canonical P^μ_c momentum operators is given in Appendix A. Applying the definition in Eq. (3.5) to QCD, one obtains the following expression for the energy–momentum tensor [12]:

$$T^{\mu\nu}_{QCD} = T^{\mu\nu}_{\text{phys}} - \left\{Q_B,\ (\partial^\mu\overline{C}^a)A^{\nu a} + (\partial^\nu\overline{C}^a)A^{\mu a} + g^{\mu\nu}\left(\frac{1}{2}\xi\overline{C}^a B^a - (\partial^\rho\overline{C}^a)A^a_\rho\right)\right\} \tag{3.7}$$

$$T^{\mu\nu}_{\text{phys}} = \frac{1}{2}\overline{\psi}\left(\frac{i}{2}\gamma^\mu(\overrightarrow{\partial^\nu} - \overleftarrow{\partial^\nu}) + gT^a A^{\nu a}\gamma^\mu\right)\psi + (\mu \leftrightarrow \nu) + F^{\mu a}_{\ \ \rho}F^{\rho\nu a} + \frac{1}{4}g^{\mu\nu}F^a_{\alpha\beta}F^{\alpha\beta a} \tag{3.8}$$

where Q_B is the BRST charge. In any field theory the current associated with Lorentz transformations is the rank-3 tensor defined by [18]:

$$M^{\mu\nu\lambda} := x^\nu T^{\mu\lambda} - x^\lambda T^{\mu\nu} \tag{3.9}$$

Using this definition, the Lorentz current in QCD can be written:

$$M^{\mu\nu\lambda}_{QCD} = x^\nu T^{\mu\lambda}_{\text{phys}} - x^\lambda T^{\mu\nu}_{\text{phys}} + ix^{[\nu}\delta_B\Big[(\partial^\mu\overline{C}^a)A^{\lambda]a} + (\partial^{\lambda]}\overline{C}^a)A^{\mu a} + g^{\mu\lambda]}\left(\frac{1}{2}\xi\overline{C}^a B^a - (\partial^\rho\overline{C}^a)A^a_\rho\right)\Big]$$

where δ_B:F→F is the BRST variation map defined for R∈F by:

$$\delta_B\mathcal{R} := [iQ_B, \mathcal{R}]_\pm \tag{3.10}$$

and ± signifies an anti-commutator or commutator depending on whether R has an odd or an even ghost number [13] respectively. The remarkable thing about this structure is that all of the ghost and gauge fixing fields are contained in a single BRST variation (coboundary) term, and this guarantees that these unphysical operators will not contribute[14] to any physical matrix element involving states in H[10]. For this reason we will not discuss this term any longer and will simply set: $T^{\mu\nu}_{QCD} \equiv T^{\mu\nu}_{\text{phys}}$. The remaining physical QCD Lorentz current then takes the form:

equation(3.11)

$$\begin{aligned}
M^{\mu\nu\lambda}_{QCD} &= x^{\nu}\Big[\frac{1}{2}\overline{\psi}\Big(\frac{i}{2}\gamma^{\mu}(\overrightarrow{\partial^{\lambda}} - \overleftarrow{\partial^{\lambda}}) + gT^{a}A^{\lambda a}\gamma^{\mu}\Big)\psi + (\mu \leftrightarrow \lambda) + F^{\mu a}_{\ \ \rho}F^{\rho\lambda a} \\
&\quad + \frac{1}{4}g^{\mu\lambda}F^{a}_{\alpha\beta}F^{\alpha\beta a}\Big] \\
&\quad - x^{\lambda}\Big[\frac{1}{2}\overline{\psi}\Big(\frac{i}{2}\gamma^{\mu}(\overrightarrow{\partial^{\nu}} - \overleftarrow{\partial^{\nu}}) + gT^{a}A^{\nu a}\gamma^{\mu}\Big)\psi + (\mu \leftrightarrow \nu) + F^{\mu a}_{\ \ \rho}F^{\rho\nu a} \\
&\quad + \frac{1}{4}g^{\mu\nu}F^{a}_{\alpha\beta}F^{\alpha\beta a}\Big] \\
&= x^{\nu}F^{\mu a}_{\ \ \rho}F^{\rho\lambda a} - x^{\lambda}F^{\mu a}_{\ \ \rho}F^{\rho\nu a} + \frac{1}{4}F^{a}_{\alpha\beta}F^{\alpha\beta a}(x^{\nu}g^{\mu\lambda} - x^{\lambda}g^{\mu\nu}) \\
&\quad + \frac{i}{4}\big[x^{\nu}\overline{\psi}(\gamma^{\mu}D^{\lambda} + \gamma^{\lambda}D^{\mu})\psi - (\nu \leftrightarrow \lambda)\big] + \text{h.c.}
\end{aligned} \tag{3.11}$$

where $D^{\mu} := \partial^{\mu} - igT^{a}A^{\mu a}$ is the covariant derivative and h.c. denotes the Hermitian conjugate. Individually, the terms in Eq. (3.11) do not have a clear interpretation. However, in the early discussion of $M^{\mu\nu\lambda}_{QCD}$ in the literature it was suggested [19] that a more meaningful expression can be obtained by factoring out total divergence terms. In this case it turns out that one can write [19]:

$$\begin{aligned}
M^{\mu\nu\lambda}_{QCD} &= \frac{i}{2}\overline{\psi}\gamma^{\mu}(x^{\nu}\partial^{\lambda} - x^{\lambda}\partial^{\nu})\psi + \text{h.c.} + \frac{1}{2}\epsilon^{\mu\nu\lambda\rho}\overline{\psi}\gamma_{\rho}\gamma^{5}\psi \\
&\quad - F^{\mu\rho a}(x^{\nu}\partial^{\lambda} - x^{\lambda}\partial^{\nu})A^{a}_{\rho} + F^{\mu\lambda a}A^{\nu a} + F^{\nu\mu a}A^{\lambda a} \\
&\quad + \frac{1}{4}F^{a}_{\alpha\beta}F^{\alpha\beta a}(x^{\nu}g^{\mu\lambda} - x^{\lambda}g^{\mu\nu}) \\
&\quad - \frac{i}{16}\partial_{\beta}\big[x^{\nu}\overline{\psi}\{\gamma^{\lambda},[\gamma^{\mu},\gamma^{\beta}]\}\psi - (\nu \leftrightarrow \lambda)\big] + \partial_{\beta}(x^{\nu}F^{\mu\beta a}A^{\lambda a} - x^{\lambda}F^{\mu\beta a}A^{\nu a})
\end{aligned} \tag{3.12}$$

If one then chooses to drop these divergence terms, the current takes the following partitioned form:

$$\begin{aligned}
M^{\mu\nu\lambda}_{QCD} &= \frac{i}{2}\overline{\psi}\gamma^{\mu}(x^{\nu}\partial^{\lambda} - x^{\lambda}\partial^{\nu})\psi + \text{h.c} + \frac{1}{2}\epsilon^{\mu\nu\lambda\rho}\overline{\psi}\gamma_{\rho}\gamma^{5}\psi \\
&\quad - F^{\mu\rho a}(x^{\nu}\partial^{\lambda} - x^{\lambda}\partial^{\nu})A^{a}_{\rho} + F^{\mu\lambda a}A^{\nu a} + F^{\nu\mu a}A^{\lambda a} \\
&\quad + \frac{1}{4}F^{a}_{\alpha\beta}F^{\alpha\beta a}(x^{\nu}g^{\mu\lambda} - x^{\lambda}g^{\mu\nu})
\end{aligned} \tag{3.13}$$

This is often referred to in the literature as the Jaffe–Manohar decomposition[6] and [20]. The fundamental difference between this expression and the expression in Eq. (3.11) is that the interaction terms between the quark and gluon fields do not feature,[15] and this makes it easier to give the individual remaining terms a physical interpretation. It also turns out that one can arrive at this same decomposition by instead using the canonical energy-momentum tensor $T^{\mu\nu}_{C}$, and in this context this is referred to as the canonical version of the Lorentz current [6]. In a similar manner to the above derivation, it is also necessary in this case to drop certain divergence terms in order to arrive at the decomposition in Eq. (3.13). By applying the definition for the angular momentum charge:

$$J^{i}_{QCD} := \frac{1}{2}\epsilon^{ijk}\int d^{3}x\, M^{0jk}_{QCD}(x) \tag{3.14}$$

the decomposition of the current in Eq. (3.13) becomes a decomposition of charges. These charges are often then individually interpreted as corresponding to physically distinct angular momentum sources [21], [22], [23], [24] and [25]. As well as the Jaffe–Manohar decomposition there also exist many other possible ways of decomposing J^{i}_{QCD} [6] and [21]. Despite the variety in structure of these different decompositions, it turns out [6] that they all differ from one another by terms of the form: $\partial_{\beta}B^{[\mu\beta][\nu\lambda]}$, which are called superpotentials [16][19]. Therefore, in order for any of these decompositions to hold one is required to drop superpotential terms, and if this procedure is justified it implies that all of the possible decompositions are physically equivalent to one another [6].

The argument in the literature [19] for dropping superpotential terms goes as follows: if two current densities $\widetilde{M}$ and M differ by a superpotential:

$$\widetilde{M}^{\mu\nu\lambda} = M^{\mu\nu\lambda} + \partial_{\beta}B^{[\mu\beta][\nu\lambda]}$$

then the corresponding charges $\widetilde{J}^{\nu\lambda} = \int d^{3}x\, \widetilde{M}^{0\nu\lambda}$ and $J^{\nu\lambda} = \int d^{3}x\, M^{0\nu\lambda}$ must be related by:

$$\widetilde{J}^{\nu\lambda} = J^{\nu\lambda} + \int d^{3}x\, \partial_{i}B^{[0i][\nu\lambda]}$$

Assuming that $B^{[0i][\nu\lambda]}(x)$ vanishes at spatial infinity, the divergence theorem implies that: $\widetilde{J}^{\nu\lambda} = J^{\nu\lambda}$, and so for computing charges the currents $\widetilde{M}^{\mu\nu\lambda}$ and $M^{\mu\nu\lambda}$ are indistinguishable. However, the results and discussions in Sections 1 and 2demonstrate that this argument is often too simplistic in the classical case, and is also not transferable to the corresponding quantised theory. Nevertheless, these arguments for dropping boundary terms have been applied to many situations, one of which being the derivation of the proton angular momentum sum rule [19] and [22]. This derivation starts by taking the full expression for $M^{\mu\nu\lambda}_{QCD}$ (Eq. (3.12)) and inserting it into Eq. (3.14), giving:

$$\begin{aligned} J^i_{QCD} = {} & \underbrace{\epsilon^{ijk}\int d^3x\left[\frac{i}{2}\overline{\psi}\gamma^0(x^j\partial^k)\psi + \text{h.c.}\right]}_{:=L^i_q} + \underbrace{\epsilon^{ijk}\int d^3x\left[\frac{1}{4}\epsilon^{0jkl}\overline{\psi}\gamma_l\gamma^5\psi\right]}_{:=S^i_q} \\ & \underbrace{-\epsilon^{ijk}\int d^3x\left[F^{0la}(x^j\partial^k)A^a_l\right]}_{:=L^i_g} + \underbrace{\epsilon^{ijk}\int d^3x\left[F^{0ka}A^{ja}\right]}_{:=S^i_g} \\ & \underbrace{-\frac{i}{16}\epsilon^{ijk}\int d^3x\,\partial_l\left[x^j\overline{\psi}\{\gamma^k,[\gamma^0,\gamma^l]\}\psi\right]}_{:=\mathcal{S}^i_1} + \underbrace{\epsilon^{ijk}\int d^3x\,\partial_l\left(x^jF^{0la}A^{ka}\right)}_{:=\mathcal{S}^i_2} \end{aligned} \tag{3.15}$$

If one then assumes that the superpotential boundary terms $\mathcal{S}^i_1$ and $\mathcal{S}^i_2$ in this expression vanish, and inserts the remaining z -components (i=3) between z-polarised proton states [17]$|p,s\rangle$, one obtains the proton angular momentum sum rule[19] and [22]:

$$\frac{1}{2} = \frac{1}{2}\Sigma + L_q + S_g + L_g \tag{3.16}$$

where the $\frac{1}{2}$ term on the left-hand side comes from the fact that: $J^3_{QCD}|p,s\rangle = \frac{1}{2}|p,s\rangle$, and each of the other terms are defined by:

$$\frac{1}{2}\Sigma = \frac{\langle p,s|S^3_q|p,s\rangle}{\langle p,s|p,s\rangle}, \qquad S_g = \frac{\langle p,s|S^3_g|p,s\rangle}{\langle p,s|p,s\rangle}, \qquad L_q = \frac{\langle p,s|L^3_q|p,s\rangle}{\langle p,s|p,s\rangle},$$

$$L_g = \frac{\langle p,s|L^3_g|p,s\rangle}{\langle p,s|p,s\rangle}$$

where a sum over quark flavour is also implicitly assumed in the definitions of Σ and L_q. Σ/S_g are then interpreted as the contributions to the z-component of the internal spin of the proton from quarks/gluons, [18] and L_q/L_g the contributions to the z-component of the orbital angular momentum of the proton from the quarks/gluons. It is clear that this derivation requires that the superpotential boundary terms either exactly vanish, or at least vanish when inserted between proton states. In the next section we will use the results of Section 2 to address whether either of these conditions is actually satisfied in this case.

SUPERPOTENTIAL BOUNDARY TERMS IN QCD

In this section we will focus on addressing the issue of angular momentum decompositions in QCD, and in particular whether the proton angular momentum sum rule holds. As discussed in Section 3, to tackle this question it is important to understand superpotential spatial boundary terms of the form: $\int d^3x\,\partial_i(x^jB^{k0i})$, where $B^{k0i}(x)$ is local. Despite the explicit x-dependence, Theorem 1 continues to hold with the replacement: $B^i \rightarrow x^jB^{k0i}$ because the function multiplication property of distributions (Eq. (1.4)) allows one to re-write the boundary operator in terms of B^{k0i} smeared with the test function $x^j\alpha\partial_if_R$, and so the local (anti-)commutativity argument continues to hold. In Section 3 it was established that in order for the derivation of the proton angular momentum sum rule to remain valid, it must be the case that either one of the following conditions is satisfied:

1. The superpotential boundary term operators $\mathcal{S}^i_1$ and $\mathcal{S}^i_2$ are exactly vanishing.
2. These operators vanish when inserted between identical z -polarised proton states $|p,s\rangle$.

Using the conditions in Eqs. (2.5) and (2.6) the first statement can be tested as follows:

$$\text{If } \exists|p\rangle \in \mathcal{H} \text{ s.t.: } \langle p|\int d^3x\,\partial_i\left(x^jB^{k0i}(x)\right)|0\rangle \neq 0 \implies \int d^3x\,\partial_i\left(x^jB^{k0i}(x)\right) \neq 0 \tag{4.1}$$

By performing an analogous calculation to the one at the end of Section 2, the matrix elements of the superpotential operator: $\int d^3x\,\partial_i(x^jB^{k0i}(x))$ can be written in the form:

$$\langle p| \int d^3x\, \partial_i \left(x^j B^{k0i}(x)\right)|0\rangle$$
$$= \begin{cases} \lim_{R\to\infty} \int d^3x\, f_R(\mathbf{x})\langle 0|B^{k0j}(0)|0\rangle, & p=0 \\ \lim_{R\to\infty} \int d^4x\, \alpha(x_0) f_R(\mathbf{x}) e^{ip_\mu x^\mu}[\langle p|B^{k0j}(0)|0\rangle \\ \quad + ip_i \langle p|x^j B^{k0i}(0)|0\rangle], & p\neq 0 \end{cases} \tag{4.2}$$

where $|p\rangle$ is some momentum eigenstate. In the case of the Jaffe–Manohar angular momentum decomposition discussed in Section 3, the superpotential boundary terms $\mathcal{S}_1^i$ and $\mathcal{S}_2^i$ are given by:

$$\mathcal{S}_1^i = -\frac{i}{16}\epsilon^{ijk} \int d^3x\, \partial_l \left(x^j \overline{\psi}\{\gamma^k, [\gamma^0, \gamma^l]\}\psi\right) \qquad \mathcal{S}_2^i = \epsilon^{ijk} \int d^3x\, \partial_l \left(x^j F^{0la} A^{ka}\right) \tag{4.3}$$

Choosing $|p\rangle = |0\rangle$, and applying Eq. (4.2), the matrix elements of these operators can then be written:

$$\langle 0|\mathcal{S}_1^i|0\rangle = \lim_{R\to\infty} \int d^3x\, \frac{1}{4} f_R(\mathbf{x}) \epsilon^{ijk} \epsilon^{0jkl} \langle 0|\overline{\psi}\gamma^l\gamma^5\psi|0\rangle$$
$$\langle 0|\mathcal{S}_2^i|0\rangle = \lim_{R\to\infty} \int d^3x\, f_R(\mathbf{x}) \epsilon^{ijk} \langle 0|F^{0ja}A^{ka}|0\rangle$$

In both cases these expressions are non-zero if the vacuum expectation values $\epsilon^{ijk}\epsilon^{0jkl}\langle 0|\overline{\psi}\gamma^l\gamma^5\psi|0\rangle$ and $\epsilon^{ijk}\langle 0|F^{0ja}A^{ka}|0\rangle$ are non-zero. It is important to note here that these expressions are non-trivial because the fields involved are solutions to the full interacting theory, and so their expectation values are non-perturbative objects. Moreover, these combinations of fields do not correspond to conserved currents, so one cannot infer their value based on conservation properties. In order to establish the value of expressions such as these one must either employ a non-perturbative technique (such as lattice gauge theory), or make use of additional symmetries to restrict their form. Calculations such as these have been performed in the literature, and there is evidence to suggest that the first of these vacuum expectation values is non-vanishing [26]. The second expression though has not to our knowledge been computed.[19] In the special case where one takes: $|\Psi\rangle = |\pi\rangle$ (the pion state) for the expectation value in the first case, then one can use the hypothesised relation:

$$\langle \pi|\overline{\psi}\gamma^l\gamma^5\psi|0\rangle = if_\pi p^l \tag{4.4}$$

where $f_\pi \neq 0$ is the pion form factor [28] and p^l is the pion's 3-momentum. Inserting this expression into Eq. (4.2) also gives a non-zero result for the $p\neq 0$ case. Applying the condition in Eq. (4.1), these results suggest that $\mathcal{S}_1^i$ is in general non-vanishing. Therefore, since $\mathcal{S}_2^i$ does not cancel $\mathcal{S}_1^i$, this casts doubt on the validity of the angular momentum operator decomposition: $J^i_{QCD} = S^i_q + L^i_q + S^i_g + L^i_g$.

To compute whether condition 2 is satisfied or not one must calculate the matrix elements of the superpotential operators $\mathcal{S}_1^i$ and $\mathcal{S}_2^i$ between the z-polarised proton states $|p,z\rangle$. Performing this calculation one obtains:

$$\langle p,s|\mathcal{S}_1^i|p,s\rangle = \lim_{R\to\infty} \int d^3x\, \frac{1}{4} f_R(\mathbf{x}) \epsilon^{ijk} \epsilon^{0jkl} \langle p,s|\overline{\psi}\gamma^l\gamma^5\psi|p,s\rangle$$
$$\langle p,s|\mathcal{S}_2^i|p,s\rangle = \lim_{R\to\infty} \int d^3x\, f_R(\mathbf{x}) \epsilon^{ijk} \langle p,s|F^{0ja}A^{ka}|p,s\rangle$$

Without applying a non-perturbative technique it is unclear whether either of these expressions are vanishing or not. However, by computing the same matrix elements for the operators S_q^i and S_g^i in Eq. (3.15), it turns out that the following exact relations hold:

$$\langle p,s|S_q^i|p,s\rangle = -\langle p,s|\mathcal{S}_1^i|p,s\rangle \tag{4.5}$$

$$\langle p,s|S_g^i|p,s\rangle = -\langle p,s|\mathcal{S}_2^i|p,s\rangle \tag{4.6}$$

This means that regardless of whether these terms vanish or not, the proton matrix elements for $\mathcal{S}_1^i$ and $\mathcal{S}_2^i$ will actually cancel the corresponding matrix elements for the 'spin' operators S_q^i and S_g^i in the angular momentum sum rule. The physical interpretation of the sum rule is therefore lost, and so L_q/L_g can no longer be interpreted as orbital angular momentum observables. It is also clear that the matrix elements Σ and S_g are not constrained since they do not contribute to the sum rule. A similar cancellation to Eq. (4.5) was also found in [29], although this approach relied on a wave-packet and form factor formulation which was later shown by [30] to not hold in general.

The analysis in this section demonstrates that neither conditions 1 nor 2 are satisfied, nor therefore the validity of the derivation of the angular momentum sum rule is undermined. Physically speaking, it is the non-perturbative structure of QCD which prevents one from forming distinct quark and gluon observables in this way. These results also provide a resolution to the spin crisis since the cancellation of the Σ term in the sum rule lifts the constraint on Σ, which means that there is no longer an a priori expectation as to what value this matrix element should take.

CONCLUSIONS

Spatial boundary term operators play an important role in quantum field theories, and in particular the issue of whether they vanish or not is a recurring theme in many of the applications of these theories. The main aim of this paper was to use an axiomatic field theory approach in order to establish a concrete condition on when these terms vanish. It turns out that a necessary and sufficient condition for this class of operators to vanish is that the operator must annihilate the vacuum state. This is a somewhat surprising result in itself because it is completely independent of how this operator acts on the full space of states. In the remainder of this paper we applied this result in order to address the issue of whether meaningful quark–gluon angular momentum operator decompositions are possible in QCD. It turns out that a common feature of these decompositions is the necessity to drop certain spatial boundary terms called superpotentials. Using the boundary term conditions established in the previous part of the paper, we analysed the superpotential terms for the specific case of the Jaffe–Manohar decomposition derived from the Belinfante energy–momentum tensor, with the conclusion that the sum of these superpotential operators is non-vanishing. In this context, this suggests that the Jaffe–Manohar angular momentum operator decomposition does not hold. An important consequence of these non-trivial boundary operators is the effect that they have on the proton angular momentum sum rule . By keeping these boundary terms explicit, we found that the sum rule is modified in a rather surprising way – the supposed gluon S_g and quark Σ spin terms are completely cancelled in the expression. This throws into doubt the physical interpretation of these terms and also provides a resolution to the proton spin crisis, since the cancellation of the Σ term in the sum rule lifts the constraint on Σ, and therefore one loses any expectation on what value it should take. Physically speaking, the boundary term conditions imply that the non-perturbative structure of QCD prevents the possibility of forming distinct quark and gluon observables in this way.

ACKNOWLEDGEMENTS

I would like to thank Thomas Gehrmann and Jürg Fröhlich for useful discussions and input. This work was supported by the Swiss National Science Foundation (SNF) under contract CRSII2_141847.

APPENDIX A.

Because of the way the Belinfante energy–momentum tensor is defined in Eq. (3.5), the Belinfante P^μ and canonical P_c^μ momentum operators are related to one another as follows:

$$P^\mu = P_c^\mu + \int d^3x\, \partial_i B^{i\mu} \tag{A.1}$$

where $B^{i\mu} = \frac{1}{2}(S^{i0\mu} + S^{0\mu i} + S^{\mu 0 i})$. So P^μ and P_c^μ differ by a spatial boundary term. Since P_c^μ is the generator of translations this means that: $[iP_c^\mu, F(y)] = \partial^\mu F(y)$, where F(y) is any local field. However, because of the relation in Eq. (A.1), it follows that:

$$\begin{aligned}
\left[iP^\mu, F(y)\right] &= \left[iP_c^\mu, F(y)\right] + i\int d^3x \left[\partial_i B^{i\mu}, F(y)\right] \\
&= \left[iP_c^\mu, F(y)\right] + i \lim_{R\to\infty} \int d^4x\, \alpha(x_0) f_R(\mathbf{x}) \left[\partial_i B^{i\mu}(x), F(y)\right] \\
&= \left[iP_c^\mu, F(y)\right] - \underbrace{i \lim_{R\to\infty} \int d^4x\, \alpha(x_0) \left(\partial_i f_R(\mathbf{x})\right) \left[B^{i\mu}(x), F(y)\right]}_{=0} \\
&= \partial^\mu F(y)
\end{aligned}$$

where the second term in the penultimate line vanishes by exactly the same local (anti-)commutativity argument as used in the proof of Theorem 1 in Section 2. So the Belinfante momentum operator P^μ is also a generator of space–time translations. Nevertheless, P^μ and P_c^μ may well give different results when applied to states in Hsince it is not necessarily the case that the spatial boundary operator: $\int d^3x\, \partial_i B^{i\mu}$ vanishes exactly. The only way to determine this definitively is to apply Theorem 1.

REFERENCES

1. S. Weinberg, The Quantum Theory of Fields, vol. 1, Cambridge University Press, 1995.
2. M. Maggiore, A Modern Introduction to Quantum Field Theory, Oxford University Press, 2005.
3. F. Mandl, G. Shaw, Quantum Field Theory, John Wiley & Sons, 2010.
4. W. Greiner, J. Reinhardt, Field Quantization, Springer-Verlag, 1996.
5. M. Srednicki, Quantum Field Theory, Cambridge University Press, 2007.
6. E. Leader, C. Lorcé, The angular momentum controversy: what's it all about and does it matter? Phys. Rep. 541 (2014) 163.
7. S. Weinberg, The Quantum Theory of Fields, vol. 2, Cambridge University Press, 1996.
8. R.F. Streater, A.S. Wightman, PCT, Spin and Statistics, and All That, W. A. Benjamin, Inc., 1964.
9. R. Haag, Local Quantum Physics, Springer-Verlag, 1996.
10. N. Nakanishi, I. Ojima, Covariant Operator Formalism of Gauge Theories and Quantum Gravity, World Scientific Publishing Co. Pte. Ltd, 1990.
11. F. Strocchi, An Introduction to Non-Perturbative Foundations of Quantum Field Theory, Oxford University Press, 2013.
12. T. Kugo, I. Ojima, Local covariant operator formalism

of non-Abelian gauge theories and quark confinement problem, Prog. Theor. Phys. Suppl. 66 (1979).

13. D. Kastler, D.W. Robinson, J.A. Swieca, Conserved currents and associated symmetries; Goldstone's theorem, Commun. Math. Phys. 2 (1966) 108.
14. R. Ferrari, L.E. Picasso, F. Strocchi, Local operators and charged states in quantum electrodynamics, Nuovo Cimento 39A (1977) 1.
15. I. Ojima, H. Hata, Observables and quark confinement in the covariant canonical formalism of Yang–Mills theory II, Z. Phys. C, Part. Fields 1 (1979) 405.
16. G. Morchio, F. Strocchi, Charge density and electric charge in quantum electrodynamics, J. Math. Phys. 44 (2003) 5569.
17. M.J. Ashman, et al., An investigation of the spin structure of the proton in deep inelastic scattering of polarised muons on polarised protons, Nucl. Phys. B 328 (1989) 1.
18. F.J. Belinfante, On the current and the density of the electric charge, the energy, the linear momentum and the angular momentum of arbitrary fields, Physica 7 (1940) 449.
19. R.L. Jaffe, A. Manohar, The g1 problem, Nucl. Phys. B 337 (1990) 509.
20. M. Wakamatsu, Is gauge-invariant complete decomposition of the nucleon spin possible? Int. J. Mod. Phys. A 29 (2014) 1430012.
21. X. Ji, Gauge-invariant decomposition of nucleon spin, Phys. Rev. Lett. 78 (1997) 610.
22. R.L. Jaffe, Gluon spin in the nucleon, Phys. Lett. B 365 (1996) 359.
23. X. Ji, J. Tang, P. Hoodbhoy, Spin structure of the nucleon in the asymptotic limit, Phys. Rev. Lett. 76 (1996) 740.
24. X. Ji, Breakup of hadron masses and the energy–momentum tensor of QCD, Phys. Rev. D 52 (1995) 271.
25. X. Ji, Lorentz symmetry and the internal structure of the nucleon, Phys. Rev. D 58 (1998) 056003.
26. J. Pasupathy, R.K. Singh, Axial vector current matrix elements and QCD sum rules, Int. J. Mod. Phys. A 21 (2006) 5099.
27. O. Pene, B. Blossier, Ph. Boucaud, A. Le Yaouanc, J.P. Leroy, J. Micheli, M. Brinet, M. Gravina, F. De Soto, Z. Liu, V. Morenas, K. Petrov, J. Rodriguez-Quintero, Vacuum expectation value of A2 from LQCD, in: Proceedings of Science, The many faces of QCD Conference, 2010, p. 010.
28. S.J. Brodsky, C.D. Roberts, R. Shrock, P.C. Tandy, Essence of the vacuum quark condensate, Phys. Rev. C 82 (2010) 022201(R).
29. G.M. Shore, B.E. White, The gauge-invariant angular momentum sum-rule for the proton, Nucl. Phys. B 581 (2000) 409.
30. B.L.G. Bakker, E. Leader, T.L. Trueman, Critique of the angular momentum sum rules and a new angular momentum sum rule, Phys. Rev. D 70 (2004) 114001.

CHAPTER 10 DECAY CONSTANTS OF PSEUDOSCALAR D-MESONS IN LATTICE QCD WITH DOMAIN-WALL FERMION

Wen-Ping Chen, Yu-Chih Chen, Ting-Wai Chiu, Han-Yi Chou, Tian-Shin Guu, Tung-Han Hsieh

We present the first study of the masses and decay constants of the pseudoscalar D mesons in two flavors lattice QCD with domain-wall fermion. The gauge ensembles are generated on the $24^3\times48$ lattice with the extent $N_s=16$ in the fifth dimension, and the plaquette gauge action at $\beta=6.10$, for three sea-quark masses with corresponding pion masses in the range 260–475 MeV. We compute the point-to-point quark propagators, and measure the time-correlation functions of the pseudoscalar and vector mesons. The inverse lattice spacing is determined by the Wilson flow, while the strange and the charm quark masses by the masses of the vector mesons $\phi(1020)$ and $J/\psi(3097)$ respectively. Using heavy meson chiral perturbation theory (HMChPT) to extrapolate to the physical pion mass, we obtain $f_D=202.3(2.2)(2.6)$ MeV and $f_{D_s}=258.7(1.1)(2.9)$ MeV.

In the Standard Model (SM), the quark and antiquark of a charged pseudoscalar meson P (with quark content $Q\bar{q}$) can decay into a charged lepton and its associated neutrino through a virtual W boson. This is the purely leptonic decay of the charged pseudoscalar meson. To the lowest order, the purely leptonic decay width can be written as

$$\Gamma(P \to l\nu) = \frac{G_F^2}{8\pi}|V_{Qq}|^2 m_l^2\left(1-\frac{m_l^2}{M_P^2}\right)^2 M_P f_P^2, \tag{1}$$

where G_F is the Fermi coupling constant, V_{Qq} is the Cabibbo–Kobayashi–Maskawa (CKM) matrix element, m_l is the mass of the lepton, M_P is the mass of the charged pseudoscalar meson, and f_P is the decay constant of the charged pseudoscalar meson, which is defined by the matrix element of the axial vector current between the QCD vacuum and the one-particle state of the charged pseudoscalar meson,

$$\langle 0|\bar{q}\gamma_\mu\gamma_5 Q(0)|P(\vec{q})\rangle = i f_P q_\mu. \tag{2}$$

According to (1), experimental measurement of the leptonic decay width gives a determination of the product $|V_{Qq}|f_P$. If the value of f_P can be obtained from another experimental measurement, then the value of $|V_{Qq}|$ can be determined, which is crucial for testing the SM via the unitarity of the CKM matrix, as a constraint for any new physics beyond the SM. On the other hand, if the value of f_P is unavailable from other experiments, then it must be determined theoretically from the first principles of QCD, before the value of $|V_{Qq}|$ can be fixed.

Theoretically, lattice QCD is a viable framework to tackle QCD nonperturbatively from the first principles of QCD, by discretizing the continuum space–time on a 4-dimensional space–time lattice [1], and computing physical observables by Monte Carlo simulation [2]. However, in practice, any lattice QCD calculation suffers from the discretization and finite volume errors, plus the systematic error due to the unphysically heavy u/d quark masses (with $M_\pi>140$ MeV). Moreover, since all quarks in QCD are excitations of Dirac fermion fields, it is vital to preserve all salient features of the Dirac fermion field on the lattice, in particular, the chiral symmetry of the massless Dirac fermion field. It is nontrivial to formulate Dirac fermion field with exact chiral symmetry at finite lattice spacing. This is realized through the domain-wall fermion (DWF) on the 5-dimensional lattice [3] and the overlap-Dirac fermion on the 4-dimensional lattice [4].

In June 2005, we determined the masses and decay constants of pseudoscalar mesons D and D_s in quenched lattice QCD with exact chiral symmetry [5], before the CLEO Collaboration announced their high-statistics measurement of f_D in July 2005. Our theoretical predictions of f_D and f_{D_s} turned out in good agreement with the experimental values from the CLEO Collaboration [6] and [7]. In 2007, we extended our study to the B-mesons [8], and determined the masses and

Citation: Wen-Ping Chen, Yu-Chih Chen, Ting-Wai Chiu, Han-Yi Chou, Tian-Shin Guu, Tung-Han Hsieh, Decay constants of pseudoscalar D-mesons in lattice QCD with domain-wall fermion, Physics Letters B, Volume 736, 7 September 2014, Pages 231-236, ISSN 0370-2693, http://dx.doi.org/10.1016/j.physletb.2014.07.025.

decay constants of B_s and B_c, as well as the lowest-lying spectra of heavy mesons with quark contents $b\bar{b}$, $c\bar{b}$, $s\bar{b}$, and $c\bar{c}$.

To remove the systematic error due to the quenched approximation, it is necessary to simulate lattice QCD with dynamical quarks. For lattice QCD with exact chiral symmetry, the challenge is how to perform the hybrid Monte Carlo (HMC) simulation [9] such that the chiral symmetry is preserved at a high precision and all topological sectors are sampled ergodically.

During 2011–2012, we demonstrated that it is feasible to perform large-scale dynamical QCD simulations with the optimal domain-wall fermion (ODWF) [10], which not only preserves the chiral symmetry to a good precision, but also samples all topological sectors ergodically. To recap, we perform HMC simulations of two flavors QCD on the $16^3\times32$ lattice (with lattice spacing a~0.1 fm), for eight sea-quark masses corresponding to the pion masses in the range 228–565 MeV. Our results of the topological susceptibility [11], as well as the mass and decay constant of the pseudoscalar meson [12], are all in good agreement with the sea-quark mass dependence predicted by the next-to-leading order (NLO) chiral perturbation theory (ChPT). This asserts that the nonperturbative chiral dynamics of the sea-quarks are well under control in our HMC simulations. In this paper, we perform HMC simulations of two flavors QCD with ODWF on the $24^3\times48$ lattice (with lattice spacing a~0.062 fm), with the purpose of studying the charm physics in lattice QCD with exact chiral symmetry.

In general, the 5-dimensional lattice Dirac operator of ODWF can be written as [13]

$$[\mathcal{D}(m)]_{xx':ss'} = (\rho_s D_w + 1)_{xx'}\delta_{ss'} + (\sigma_s D_w - 1)_{xx'} L_{ss'}, \quad (3)$$

where $\rho_s=c\omega_s+d$, $\sigma_s=c\omega_s-d$, and c, d are constants. The indices x and x′ denote the sites on the 4-dimensional space–time lattice, and s and s′ the indices in the fifth dimension, while the lattice spacing a and the Dirac and color indices have been suppressed. The weights $\{\omega_s,\ s=1,\ldots,N_s\}$ along the fifth dimension are fixed according to the formula derived in [10] such that the maximal chiral symmetry is attained. Here D_w is the standard Wilson Dirac operator plus a negative parameter $-m_0$ $(0<m_0<2)$,

$$(D_w)_{xx'} = -\frac{1}{2}\sum_\mu[(1-\gamma_\mu)U_\mu(x)\delta_{x+\hat{\mu},x'} + (1+\gamma_\mu)U_\mu^\dagger(x')\delta_{x-\hat{\mu},x'}] + (4-m_0), \quad (4)$$

where $U_\mu(x)$ denotes the link variable pointing from x to $x+\hat{\mu}$. The operator L is independent of the gauge field, and it can be written as

$$L = P_+L_+ + P_-L_-, \quad P_\pm = (1\pm\gamma_5)/2,$$

and

$$(L_+)_{ss'} = (L_-)_{s's} = \begin{cases} -m\delta_{N_s,s'}, & s=1, \\ \delta_{s-1,s'}, & 1<s\le N_s, \end{cases}$$

where N_s is the number of sites in the fifth dimension, $m\equiv rm_q$, m_q is the bare quark mass, and $r=1/[2m_0(1-dm_0)]$.

Including the action of Pauli–Villars fields (with bare mass $m_{PV}=1/r$), the partition function of ODWF in a gauge background can be written as

$$Z_{odwf} = \int [d\Psi][d\bar{\Psi}][d\Phi][d\Phi^\dagger]\exp\{-\bar{\Psi}\mathcal{D}(m)\Psi - \Phi^\dagger\mathcal{D}(1)\Phi\} \quad (5)$$

which can be integrated successively to obtain the fermion determinant of the effective 4-dimensional Dirac operator [10]

$$Z_{odwf} = \det D(m_q), \quad D(m_q) = (D_c + m_q)(1 + rD_c)^{-1}, \quad (6)$$

where

$$D_c = \frac{1}{r}\frac{1+\gamma_5 S_{opt}(H)}{1-\gamma_5 S_{opt}(H)}, \quad S_{opt}(H) = \frac{1-\prod_{s=1}^{N_s}T_s}{1+\prod_{s=1}^{N_s}T_s},$$
$$T_s = \frac{1-\omega_s H}{1+\omega_s H}, \quad H = cH_w(1+d\gamma_5 H_w)^{-1},$$
$$H_w = \gamma_5 D_w(-m_0). \quad (7)$$

Here Sopt(H)=$HR_Z(H)$, where $R_Z(H)$ is the Zolotarev optimal rational approximation of $(H^2)^{-1/2}$.

For HMC simulation of lattice QCD with ODWF, it is crucial to perform the even–odd preconditioning on the ODWF operator (3) such that the condition number of the conjugate gradient is reduced and the memory consumption is halved. Now, separating the even and the odd sites on the 4-dimensional space–time lattice, (3) can be written as

$$\mathcal{D}(m_q) = \begin{pmatrix} 4-m_0 & D_w^{EO} \\ D_w^{OE} & 4-m_0 \end{pmatrix}[c\omega(1+L)+d(1-L)] + (1-L)$$
$$= \begin{pmatrix} X & D_w^{EO}Y \\ D_w^{OE}Y & X \end{pmatrix}, \quad (8)$$

where

$$X \equiv (4-m_0)[c\omega(1+L)+d(1-L)] + (1-L),$$
$$Y \equiv c\omega(1+L)+d(1-L). \quad (9)$$

We further rewrite it in a more symmetric form by defining

$$M_5 \equiv \omega^{-1/2}YX^{-1}\omega^{1/2}$$
$$= \{(4-m_0) + \omega^{-1/2}[c(1+L)(1-L)^{-1} + d\omega^{-1}]^{-1}\omega^{-1/2}\}^{-1}, \quad (10)$$

and

$$S_1 \equiv \omega^{-1/2}YX^{-1} = M_5\omega^{-1/2}, \quad S_2 \equiv Y^{-1}\omega^{1/2}. \quad (11)$$

Then Eq. (8) becomes

$$\mathcal{D}(m_q) = S_1^{-1}\begin{pmatrix} 1 & M_5 D_w^{EO} \\ M_5 D_w^{OE} & 1 \end{pmatrix} S_2^{-1}$$
$$= S_1^{-1}\begin{pmatrix} 1 & 0 \\ M_5 D_w^{OE} & 1 \end{pmatrix}\begin{pmatrix} 1 & 0 \\ 0 & C \end{pmatrix}\begin{pmatrix} 1 & M_5 D_w^{EO} \\ 0 & 1 \end{pmatrix} S_2^{-1}. \quad (12)$$

where the Schur decomposition has been used in the last equality, with the Schur complement

$$C \equiv 1 - M_5 D_W^{\mathrm{OE}} M_5 D_W^{\mathrm{EO}}. \tag{13}$$

Since $\det \mathcal{D} = \det S_1^{-1} \cdot \det C \cdot \det S_2^{-1}$, and S_1 and S_2 do not depend on the gauge field, we can just use C for the Monte Carlo simulation. After including the Pauli–Villars fields (with $m_{PV}=1/r$), the pseudofermion action for two-flavors QCD (in the isospin symmetry limit $m_u=m_d$) can be written as

$$S_{pf} = \phi^\dagger C_1^\dagger (CC^\dagger)^{-1} C_1 \phi, \quad C_1 \equiv C(m=1), \tag{14}$$

where ϕ and $\phi^\dagger$ are complex scalar fields carrying the same quantum numbers (color, spin) of the quark fields. Including the gluon fields, the partition function for 2 flavors QCD can be written as

$$Z = \int [dU][d\phi^\dagger][d\phi] \exp(-S_g[U] - \phi^\dagger C_1^\dagger (CC^\dagger)^{-1} C_1 \phi), \tag{15}$$

where $S_g[U]$ is the lattice action for the gluon field. Here we use the plaquette gauge action

$$S_g[U] = \beta \sum_{plaq.} \left\{1 - \frac{1}{3} \mathrm{Re\,Tr}(U_p)\right\}, \quad \beta = \frac{6}{g^2}. \tag{16}$$

Further details of our HMC simulations of two flavors QCD can be found in Ref. [14]and a forthcoming long paper.

We perform the hybrid Monte Carlo simulation of two flavors QCD on the $24^3\times48$ lattice with the plaquette gauge action at $\beta=6/g^2=6.10$, for three sea-quark masses (mseaa=0.005,0.01,0.02) with the corresponding pion masses in the range 260–475 MeV. For the quark part, we use ODWF with c=1, d=0 (i.e., $H=H_w$), $N_s=16$, and $\lambda_{min}/\lambda_{max}=0.05/6.2$. For each sea-quark mass, we generate the initial 400–500 trajectories on an Nvidia GPU card (with device memory larger than 4 GB). After discarding initial 300 trajectories for thermalization, we sample one configuration every 5 trajectories, resulting 20–32 "seed" configurations for each sea-quark mass. Then we use these seed configurations as the initial configurations for independent simulations on 20–32 GPUs. Each GPU generates 200–250 trajectories independently. Then we accumulate a total of 5000 trajectories for each sea-quark mass. From the saturation of the binning error of the plaquette, as well as the evolution of the topological charge, we estimate the autocorrelation time to be around 10 trajectories. Thus we sample one configuration every 10 trajectories, and obtain ~500 configurations for each ensemble. The basic parameters of these three ensembles are summarized in Table 1.

Table 1: Basic parameters of the gauge ensembles: the sea-quark mass, the number of gauge configurations in each ensemble, the inverse lattice spacing, the residual mass, and the pion mass

Ensemble	$m_{sea}a$	N_{cfg}	a^{-1} [GeV]	$m_{res}a$	M_π [MeV]
A	0.005	535	3.194(11)(12)	0.000057(5)	259(10)(11)
B	0.010	506	3.138(16)(12)	0.000046(4)	347(7)(6)
C	0.020	501	3.044(13)(11)	0.000028(3)	474(6)(5)

In Fig. 1, we plot the histogram of the topological charge (Q_t) distribution for these three ensembles. Evidently, the probability distribution of Q_t for each ensemble behaves like a Gaussian, and it becomes more sharply peaked around $Q_t=0$ as the sea-quark mass gets smaller. Here the topological charge $Q_t=\sum_x \epsilon_{\mu\nu\lambda\sigma} \mathrm{tr}[F_{\mu\nu}(x)F_{\lambda\sigma}(x)]/(32\pi^2)$, where the matrix-valued field tensor $F_{\mu\nu}(x)$ is obtained from the four plaquettes surrounding x on the $(\hat{\mu}, \hat{\nu})$ plane. Even though the resulting topological charge is not exactly equal to an integer, the probability distribution $P(Q_t)$ suffices to demonstrate that our HMC simulation indeed samples all topological sectors ergodically. For a rigorous study of topology, we are projecting the zero modes and the low-lying eigenmodes of the overlap Dirac operator for each gauge configuration, with the same procedures as outlined in Ref. [11], and we will report our results in a forthcoming long paper.

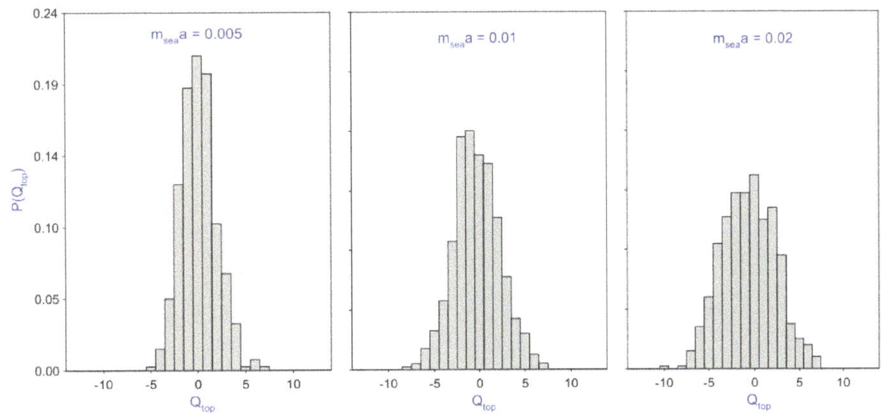

Figure 1: Histogram of topological charge distribution for three sea-quark masses, mseaa=0.005,0.01, and 0.02 respectively.

To determine the lattice scale, we use the Wilson flow [15] with the condition [16]

$$\{t^2\langle E(t)\rangle\}|_{t=t_0} = 0.3 \quad (17)$$

to obtain $\sqrt{t_0}/a$ for each gauge ensemble. Our procedures are as follows. First, we compute the Wilson flow for each gauge ensemble of the 2-flavors QCD on the $16^3\times32$ lattice at β=5.95[12], and obtain the value of $\sqrt{t_0}/a$. By linear extrapolation, we obtain $\sqrt{t_0}/a = 1.3674(67)(42)$ in the chiral limit, where the systematic error is estimated by varying the number of sea-quark masses used for the linear extrapolation. Then, using the lattice spacing a=0.1034(1)(2) fm (in the chiral limit) [12] which has been determined at β=5.95 (by heavy quark potential with Sommer parameter r_0=0.49 fm), we obtain $\sqrt{t_0} = 0.1414(7)(5)$ fm. Thus, with the input $\sqrt{t_0} = 0.1414(7)(5)$ fm, we can determine the lattice spacing of any gauge ensemble of 2-flavors QCD, with the value of $\sqrt{t_0}/a$ obtained by the Wilson flow with the condition (17). For the gauge ensembles of 2-flavors QCD on the $24^3\times48$ lattice at β=6.10, a^{-1} [GeV] = 3.194(11)(12), 3.138(16)(12), 3.044(13)(11), for mseaa=0.005,0.01, and 0.02 respectively. The inverse lattice spacing is well fitted by the linear function of mseaa, which gives a^{-1}=3.2418(35)(44) GeV in the chiral limit.

We compute the valence quark propagator with the point source at the origin, and with parameters exactly the same as those of the sea-quarks (N_s=16 and $\lambda_{min}/\lambda_{max}$=0.05/6.2). First, we solve the following linear system (with even–odd preconditioned CG),

$$\mathcal{D}(m_q)|Y\rangle = \mathcal{D}(m_{PV})B^{-1}|\text{source vector}\rangle, \quad (18)$$

where $B^{-1}_{x,s;x',s'} = \delta_{x,x'}(P_-\delta_{s,s'} + P_+\delta_{s+1,s'})$ with periodic boundary conditions in the fifth dimension. Then the solution of (18) gives the valence quark propagator

$$(D_c + m_q)^{-1}_{x,x'} = r(1 - rm_q)^{-1}\left[(BY)_{x,1;x',1} - \delta_{x,x'}\right].$$

For each gauge ensemble, we measure the time-correlation functions for pseudoscalar (P) and vector (V) mesons,

$$C_P(t) = \left\langle \sum_{\vec{x}} \text{tr}\{\gamma_5(D_c + m_Q)^{-1}_{x,0}\gamma_5(D_c + m_q)^{-1}_{0,x}\}\right\rangle_U,$$

$$C_V(t) = \left\langle \frac{1}{3}\sum_{\mu=1}^{3}\sum_{\vec{x}} \text{tr}\{\gamma_\mu(D_c + m_Q)^{-1}_{x,0}\gamma_\mu(D_c + m_q)^{-1}_{0,x}\}\right\rangle_U,$$

for the following quark contents: (m_q,m_Q)={(msea,msea),(msea,m_s),(msea,m_c),(m_s,m_s),(m_s,m_c),(m_c,m_c)}, where m_s and m_c are the bare masses of the strange quark and the charm quark.

In general, the decay constant f_P of a charged pseudoscalar meson P with quark content $Q\bar{q}$ is defined by (2). In lattice QCD with exact chiral symmetry, we can use the axial Ward identity

$$\partial_\mu(\bar{q}\gamma_\mu\gamma_5 Q) = (m_q + m_Q)\bar{q}\gamma_5 Q$$

to obtain

$$f_P = (m_q + m_Q)\frac{|\langle 0|\bar{q}\gamma_5 Q|P(\vec{0})\rangle|}{M_P^2}, \quad (19)$$

where the pseudoscalar mass M_Pa and the decay amplitude $z \equiv |\langle 0|\bar{q}\gamma_5 Q|P(\vec{0})\rangle|$ can be obtained by fitting the pseudoscalar time-correlation function $C_P(t)$ to the formula

$$\frac{z^2}{2M_Pa}\left[e^{-M_Pat} + e^{-M_Pa(T-t)}\right], \quad (20)$$

where the excited states have been neglected.

To measure the chiral symmetry breaking due to finite N_s, we compute the residual mass according to the formula [13]

$$m_{res} = \frac{\langle \text{tr}(D_c + m_q)^{-1}_{0,0}\rangle_U}{\langle \text{tr}[\gamma_5(D_c + m_q)\gamma_5(D_c + m_q)]^{-1}_{0,0}\rangle_U} - m_q, \quad (21)$$

where $(D_c+m_q)^{-1}$ denotes the valence quark propagator with m_q equal to the sea-quark mass, tr denotes the trace running over the color and Dirac indices, and the brackets $\langle\cdots\rangle_U$ denote averaging over an ensemble of gauge configurations. In Table 1, we list the residual mass of each ensemble. We see that the residual mass is at most ~1% of the bare quark mass, amounting to ~0.17 MeV, which is expected to be much smaller than other systematic errors. In Table 1, we list the pion mass which is extracted from the time-correlation function $C_P(t)$ with the valence quark masses equal to the sea-quark mass (m_q=m_Q=msea). Here the pion mass has been corrected for the finite volume effect, using the estimate within ChPT calculated up to $\mathcal{O}(M_\pi^4/(4\pi f_\pi)^4)$[17].

For each ensemble, the strange quark bare mass m_s is tuned such that the vector-meson mass M_V extracted from the time-correlation function $C_V(t)$ with the valence quark masses m_q=m_Q=m_s is equal to the mass of the vector meson ϕ(1020). The vector meson mass M_V is extracted by fitting $C_V(t)$ to a formula similar to (20), for the range $[t_1,t_2]$ in which the effective mass attaining a plateau. We estimate the statistical error of M_V using the jackknife method with the bin size (10–15 configurations) of which the statistical error saturates. To estimate the systematic errors of M_V, besides that due to the systematic error of the inverse lattice spacing (see Table 1), we also incorporate the variation of M_V based on all fittings satisfying χ^2/dof<1.1. In the following, it is understood that all masses and decay constants, and their errors are obtained with the same procedure. Similarly, the charm quark bare mass m_c is tuned such that the vector-meson mass extracted from the time-correlation function $C_V(t)$ with the valence quark masses m_q=m_Q=m_c is equal to the mass of the vector meson J/ψ(3097). The values of m_s and m_c of each ensemble are summarized in Table 2, where the error denotes the combined statistical and systematic error.

Table 2: Determination of the strange quark and the charm quark masses

Ensemble	$m_s a$	M_ϕ [MeV]	$[t_1,t_2]$	χ^2/dof	$m_c a$	$M_{J/\psi}$ [MeV]	$[t_1,t_2]$	χ^2/dof
A	0.04	1020(7)	[11,24]	0.99	0.53	3097(11)	[15,24]	0.25
B	0.04	1020(8)	[11,23]	0.11	0.55	3104(15)	[17,23]	0.58
C	0.04	1020(8)	[16,23]	0.20	0.55	3099(13)	[16,24]	0.22

Using the value of m_c in Table 2, we measure the time-correlation function of the D meson with the valence quark masses (m_q=msea, m_Q=m_c) for each ensemble. Then we fit the time-correlation function $C_D(t)$ to (20) to extract the D meson mass $M_D a$ and the decay constant $f_D a$. Similarly, using the values of m_s and m_c in Table 2, we measure the time-correlation function of the D_s meson with the valence quark masses (m_q=m_s, m_Q=m_c), and extract the mass $M_{Ds}a$ and the decay constant $f_{Ds}a$ for each ensemble. Our results of M_D, f_D, M_{Ds}, and f_{Ds} are summarized in Table 3, where the error denotes the combined statistical and systematic error.

Table 3: The mass and decay constant of D-mesons

Ensemble	M_D[MeV]	f_D[MeV]	$[t_1,t_2]$	χ^2/dof	M_{Ds}[MeV]	f_{Ds}[MeV]	$[t_1,t_2]$	χ^2/dof
A	1862(11)	216(5)	[17,24]	0.83	1969(8)	261(5)	[14,23]	0.97
B	1864(10)	220(4)	[18,23]	0.82	1968(9)	264(4)	[13,24]	0.54
C	1877(10)	227(6)	[18,22]	0.54	1962(8)	267(3)	[10,17]	0.70

We see that for all ensembles, the masses of D and D_s mesons are in good agreement with the experimental values compiled by PDG [18], M_D=1869.62±0.15 MeV, and M_{Ds}=1968.50±0.32 MeV.

For the decay constants f_D and f_{Ds}, we use HMChPT [19] to extrapolate to the physical M_π=140 MeV. In general, for the pseudoscalar meson with quark content ($c\bar{q}$), the NLO formula for N_f=2 reads

$$f_{D_q} = \frac{\kappa}{\sqrt{M_{D_q}}}\left\{1 - \frac{1+3g_c^2}{2}\left[(\xi+\xi_q)\ln\left(\frac{\xi+\xi_q}{2}\right) - \frac{\xi_q}{2}\ln\xi_q\right] + c_1\xi\right\}, \tag{22}$$

where $\xi = M_\pi^2/(4\pi f)^2$, $\xi_q = M_{qq}^2/(4\pi f)^2$, M_{qq} is the mass of the pseudoscalar meson with quark content $q\bar{q}$, g_c=0.61(7) which is determined by the experimental measurement of the coupling $g_{D^*\to D\pi}$[20], and κ and c_1 are low-energy constants. For $\bar{q}$ equal to $\bar{d}$, (22) reduces to

$$f_D = \frac{\kappa}{\sqrt{M_D}}\left[1 - \frac{3}{4}(1+3g_c^2)\xi\ln\xi + c_1\xi\right]. \tag{23}$$

Fitting the data of ensembles (A)–(C) to (23) [see the lower curve in Fig. 2(a)], we obtain κ=0.2638(28) and c_1=−1.899(206) with χ^2/dof=0.34. At the physical M_π=140 MeV, (23) gives f_D=202.3(2.2) MeV. Alternatively, performing the fit to $f_D\sqrt{M_D}$ (multiplying both sides of (23) with $\sqrt{M_D}$) [see the lower curve in Fig. 2(b)], we obtain κ=0.2673(22) and c_1=−2.012(189) with χ^2/dof=0.27. At the physical M_π=140 MeV, it gives $f_D\sqrt{M_D} = 0.2785(23)$, which in turn (with physical input M_D=1869.62 MeV) yields f_D=203.7(1.7) MeV. Comparing the results of these two HMChPT fits, we estimate the systematic error of f_D due to the chiral extrapolation to be ~1.4 MeV. Next we estimate the systematic error of f_D due to the scaling violations (i.e., the error induced through the scale setting), by performing the chiral extrapolation of $(f_D a)\sqrt{M_D a}$ in the lattice unit, using the raw data of $M_D a$ and $f_D a$ extracted from the time-correlation function. Fitting the lattice data of ensembles (A)–(C) to NLO HMChPT (see the lower curve in Fig. 3) gives κ=0.0454(2) and c_1=−0.6923(948) with χ^2/dof=0.14. At the physical M_π=140 MeV ($M_\pi a$=0.001842), it gives $f_D a\sqrt{M_D a} = 0.0477(2)$. With the inputs a^{-1}=3.2418(56) GeV (in the chiral limit) and M_D=1869.62 MeV, we obtain f_D=203.7(1.0) MeV. Thus we estimate the systematic error of f_D due to the scaling violations to be ~1.0 MeV. Since our calculation is done at one single lattice spacing, the discretization error cannot be quantified reliably. Nevertheless, we do not expect it to be much larger than the systematic errors due to the chiral extrapolation and the scaling violations, because our lattice action is free from O(a) discretization errors. Now, assuming the discretization error to be ~2 MeV, together with the systematic errors due to the scaling violations and the chiral extrapolation, we obtain

$$f_D = 202.3 \pm 2.2 \pm 2.6 \text{ MeV}, \tag{24}$$

which is in good agreement with the experimental values [21] and [22], as well as the experimental average f_D=204.6±5.0 MeV[23].

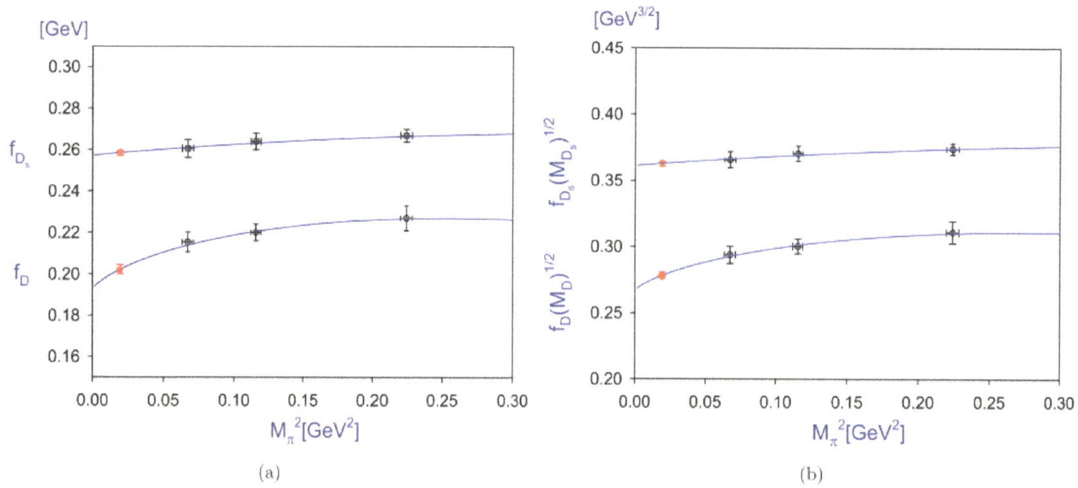

Figure 2: Fitting the data to NLO HMChPT: (a) f_D and f_{Ds}; (b) $f_D\sqrt{M_D}$ and $f_{D_s}\sqrt{M_{D_s}}$. In each figure, the solid line is the fit of the data points to HMChPT. The symbol (in red) is the extrapolated value at physical M_π=140 MeV. (For interpretation of the references to color in this figure legend, the reader is referred to the web version of this article.)

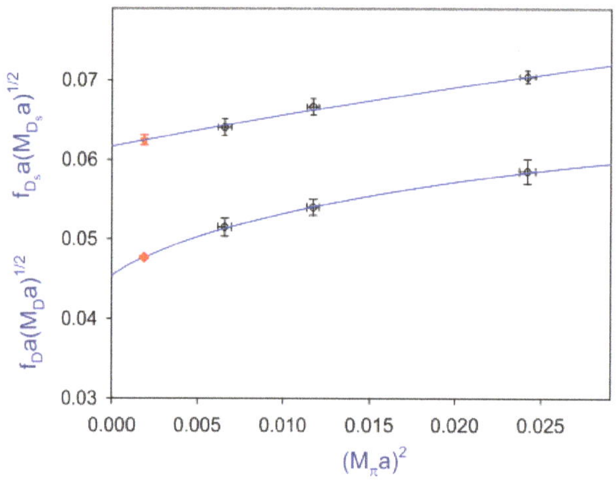

Figure 3: Fitting the lattice data of $f_Da\sqrt{M_Da}$ and $f_{D_s}a\sqrt{M_{D_s}a}$ (in the lattice unit) to NLO HMChPT. The solid line is the fit to HMChPT, and the symbol (in red) is the extrapolated value at physical M_π=140 MeV. (For interpretation of the references to color in this figure legend, the reader is referred to the web version of this article.)

Next, we turn to the decay constant of D_s meson. Fitting the data of ensembles (A)–(C) to (22) with $\bar{q}=\bar{s}$ [see the upper curve in Fig. 2(a)], we obtain κ=0.2790(12) and c_1=−0.7740(733) with χ^2/dof=0.20. At the physical M_π=140 MeV, (22) gives f_{Ds}=258.7(1.1) MeV. Alternatively, performing the fit to $f_{D_s}\sqrt{M_{D_s}}$ [see the upper curve in Fig. 2(b)], we obtain κ=0.2865(13) and c_1=−0.8949(899) with χ^2/dof=0.21. At the physical M_π=140 MeV, it gives $f_{D_s}\sqrt{M_{D_s}}=0.2785(23)$, which (with physical input M_{Ds}=1968.50 MeV) yields f_{Ds}=258.8(1.2) MeV. Comparing the results of these two HMChPT fits, we estimate the systematic error of f_{Ds} due to the chiral extrapolation to be ~0.2 MeV. Next, we estimate the systematic error of f_{Ds} due to the scaling violations, by performing the chiral extrapolation of $(f_{D_s}a)\sqrt{M_{D_s}a}$ in the lattice unit. Fitting the lattice data to NLO HMChPT (see the upper curve in Fig. 3) gives κ=0.0489(5) and c_1=0.513(185) with χ^2/dof=0.46. At the physical M_π=140 MeV ($M_\pi a$=0.001842), it gives $f_{D_s}a\sqrt{M_{D_s}a}=0.0625(6)$. With the inputs a^{-1}=3.2418(56) GeV (in the chiral limit) and M_{Ds}=1968.50 MeV, we obtain f_{Ds}=259.9(2.7) MeV. Thus we estimate the systematic error of f_{Ds} due to the scaling violations to be ~2.0 MeV. Again, assuming the discretization error to be ~2 MeV, together with the systematic errors due to the scaling violations and the chiral extrapolation, we obtain

$$f_{D_s} = 258.7 \pm 1.1 \pm 2.9 \text{ MeV}, \tag{25}$$

which is in good agreement with the experimental values [7], [24] and [25], as well as the experimental average f_{Ds}=257.5+4.6 MeV[23].

Since the ratio f_{Ds}/f_D is expected to have systematic error less than those of f_D and f_{Ds}, our results of f_D and f_{Ds} yield

$$\frac{f_{D_s}}{f_D} = 1.2788 \pm 0.0264 \tag{26}$$

in good agreement with the experimental average f_{Ds}/f_D=1.258±0.038[23].

To summarize, we perform the first study of the masses and decay constants of the pseudoscalar D -mesons in two-flavors lattice QCD with domain-wall fermion, on the $24^3\times48$ lattice with lattice spacing a~0.062 fm, for three sea-quark masses corresponding to the pion masses in the range 260–475

MeV. Our results of f_D(24), f_{Ds}(25) and their ratio (26), are all in good agreement with the experimental results, as well as with other lattice QCD results [26]. Since our calculation is done at one single lattice spacing, we cannot perform the extrapolation to the continuum limit. Nevertheless, we do not expect the combined systematic errors much larger than our estimates in (24) and (25), since the lattice spacing (a~0.062 fm) is sufficiently fine, and our lattice action is free from O(a) lattice artifacts. Likewise, since our calculation is done on a single volume, the finite volume effect cannot be estimated reliably. However, it is believed that the finite volume error of physical observables involving heavy (charm/bottom) quarks is smaller than other systematic ones. We will address these issues with calculations on different volumes as well as several lattice spacings. Moreover, to incorporate the internal quark loops of (u,d,s,c) quarks in our dynamical simulations, we are performing HMC simulations of (2+1)-flavors and (2+1+1)-flavors QCD on the $24^3 \times 48$ lattice, with the novel exact pseudofermion action for one-flavor DWF [27].

ACKNOWLEDGEMENTS

This work is supported in part by the Ministry of Science and Technology (Nos. NSC102-2112-M-002-019-MY3, NSC102-2112-M-001-011) and NTU–CQSE (No. 103R891404). We also thank NCHC for providing facilities to perform part of our calculations.

REFERENCES

1. K.G. Wilson, Phys. Rev. D 10 (1974) 2445.
2. M. Creutz, Phys. Rev. D 21 (1980) 2308.
3. D.B. Kaplan, Phys. Lett. B 288 (1992) 342; D.B. Kaplan, Nucl. Phys. B, Proc. Suppl. 30 (1993) 597.
4. H. Neuberger, Phys. Lett. B 417 (1998) 141; R. Narayanan, H. Neuberger, Nucl. Phys. B 443 (1995) 305.
5. T.W. Chiu, T.H. Hsieh, J.Y. Lee, P.H. Liu, H.J. Chang, Phys. Lett. B 624 (2005) 31.
6. M. Artuso, et al., CLEO Collaboration, Phys. Rev. Lett. 95 (2005) 251801.
7. M. Artuso, et al., CLEO Collaboration, Phys. Rev. Lett. 99 (2007) 071802.
8. T.W. Chiu, et al., TWQCD Collaboration, Phys. Lett. B 651 (2007) 171.
9. S. Duane, A.D. Kennedy, B.J. Pendleton, D. Roweth, Phys. Lett. B 195 (1987) 216.
10. T.W. Chiu, Phys. Rev. Lett. 90 (2003) 071601; T.W. Chiu, Phys. Lett. B 716 (2012) 461.
11. T.W. Chiu, T.H. Hsieh, Y.Y. Mao, TWQCD Collaboration, Phys. Lett. B 702 (2011) 131.
12. T.W. Chiu, T.H. Hsieh, Y.Y. Mao, TWQCD Collaboration, Phys. Lett. B 717 (2012) 420.
13. Y.C. Chen, T.W. Chiu, TWQCD Collaboration, Phys. Rev. D 86 (2012) 094508.
14. T.W. Chiu, TWQCD Collaboration, J. Phys. Conf. Ser. 454 (2013) 012044.
15. R. Narayanan, H. Neuberger, J. High Energy Phys. 0603 (2006) 064.
16. M. Lüscher, J. High Energy Phys. 1008 (2010) 071.
17. G. Colangelo, S. Durr, C. Haefeli, Nucl. Phys. B 721 (2005) 136.
18. J. Beringer, et al., Particle Data Group Collaboration, Phys. Rev. D 86 (2012) 010001.
19. S.R. Sharpe, Y. Zhang, Phys. Rev. D 53 (1996) 5125.
20. A. Anastassov, et al., CLEO Collaboration, Phys. Rev. D 65 (2002) 032003.
21. B.I. Eisenstein, et al., CLEO Collaboration, Phys. Rev. D 78 (2008) 052003.
22. M. Ablikim, et al., BESIII Collaboration, Phys. Rev. D 89 (2014) 051104(R).
23. J.L. Rosner, S. Stone, arXiv:1309.1924.
24. P. del Amo Sanchez, et al., BaBar Collaboration, Phys. Rev. D 82 (2010) 091103.
25. A. Zupanc, et al., Belle Collaboration, J. High Energy Phys. 1309 (2013) 139.
26. For a recent review, see, e.g., S. Aoki, Y. Aoki, C. Bernard, T. Blum, G. Colangelo, M. Della Morte, S. Durr, A.X.E. Khadra, et al., arXiv:1310.8555; A.X. El-Khadra, PoS LATTICE 2013 (2013) 001.
27. Y.C. Chen, T.W. Chiu, arXiv:1403.1683

CHAPTER 11

THE QUARK-GLUON PLASMA EQUATION OF STATE AND THE GENERALIZED UNCERTAINTY PRINCIPLE

L. I. Abou-Salem, N. M. El Naggar and I. A. Elmashad

INTRODUCTION

Essential modifications in Heisenberg's uncertainty principle are predicted near Planck scale which is called Generalized Uncertainty Principle (GUP). One of the most exciting predictions of some approaches related to quantum gravity [1], perturbative string theory [2], and black holes [3] is the minimal length concept existence. For a recent review, see [4]. These approaches seem to modify almost all mechanical Hamiltonians. Thus, quantum mechanics can be studied in the presence of a minimal length [5–8]. Also, in quantum optics, the GUP implications can be measured directly which confirm the theoretical predictions [9–11]. In particular, exact solutions of various relativistic [12–15] and nonrelativistic problems [16–22] have been obtained in the presence of a minimal length $\Delta x_0 = \hbar(\beta)^{1/2}$. The Generalized Uncertainty Principle [23, 24] was implemented on deriving the thermodynamics of ideal quark-gluon plasma of massless quark flavor [25, 26].

In other minimal length formalism [5, 6], the Heisenberg algebra associated with the momentum $\hat{p}$ and the position coordinates $\hat{x}_i$ is given by

$$\left[\hat{x}_i, \hat{p}_j\right] = i\hbar\delta_{ij}\left(1 + \beta p^2\right), \tag{1}$$

where $\beta>0$ is the minimal length parameter. The Generalized Uncertainty Principle (GUP) corresponding to it reads

$$\Delta x_i \Delta p_j \geq \delta_{ij}\left[1 + \beta\left(\Delta p\right)^2 + \beta\left\langle p\right\rangle^2\right] \tag{2}$$

which yields a minimal observable length $\Delta x_0 = \hbar(\beta)^{1/2}$. Hence, the momentum would be subject of a modification and becomes

$$p_i = p_{oi}\left(1 + \beta p_0^2\right), \tag{3}$$

where $x_i = x_{0i}$ and p_{0i} satisfy the canonical commutation relations $[x_{0i}, p_{0j}] = i\hbar\delta_{ij}$. Also, p_{0i} can be interpreted as the momentum at low energies and p_i at high energies. Since the GUP modifies the Hamiltonian, it is important to study these effects quantitatively. These effects were investigated on condensed matter, atomic systems [10–12, 27], the Liouville theorem (LT) in statistical mechanics [28], and the weak equivalence principle (WKP). Recently, another approach based on super gravity was implemented to QCD and to QGP especially [29–31].

In this paper, the effect of the GUP on QGP equation of state of massless quark flavors at a vanishing chemical potential μ is studied. We calculate the corrections to various thermodynamic quantities, like the energy density and the pressure. Then, these corrections with bag model are used to describe the quark-gluon plasma equation of state and compare it with QCD lattice results. This paper is organized as follows. In Section 2, we derive thermodynamics of QGP consisting of a noninteracting massless bosons and fermions with impact of GUP approach. The results, discussions, and conclusions are given in Sections 3 and 4.

THERMODYNAMICS OF QUARK-GLUON PLASMA WITH GUP EFFECT

In this section, we derive the thermodynamics of QGP in case of bosons and fermions taking into account the GUP impact. Then, the thermodynamical equations such as pressure and energy density of quark-gluon plasma are obtained.

In Case of Bosons.

At finite temperature T and chemical potential μ, the grand canonical partition function z_B for noninteracting massive bosons with g internal degrees of freedom is given by [32]

Citation: L. I. Abou-Salem, N. M. El Naggar, and I. A. Elmashad, "The Quark-Gluon Plasma Equation of State and the Generalized Uncertainty Principle," Advances in High Energy Physics, vol. 2015, Article ID 103576, 7 pages, 2015. doi:10.1155/2015/103576.

$$z_B = \prod_k \left[\sum_{l=0}^{\infty} \exp\left(-l\frac{E(k)-\mu}{T}\right) \right]^g \tag{4}$$

$$= \prod_k \left[1 - \exp\frac{(E(k)-\mu)}{T} \right]^{-g}, \tag{5}$$

where l is the occupation number for each quantum state with energy $E(k) = \sqrt{k^2+m^2}$ with mass m and k is the momentum of the particle. Here the infinite product is taken for all possible momentum states.

For simplicity we consider chiral limit (i.e., vanishing mass) and a vanishing chemical potential, which experiments ensure at high energy. Then the dominant excitation in the hadronic phase is a massless pion, while that in the quarkgluon plasma is massless quarks and gluons. For a particle of mass M having a distant origin and an energy comparable to the Planck scale, the momentum would be a subject of a tiny modification so that the dispersion relation would too. According to GUP approach, the dispersion relation reads

$$E^2(k) = k^2c^2\left(1+\beta k^2\right)^2 + M^2c^4, \tag{6}$$

where M and c are the mass of the particle and the speed of light as introduced by Lorentz and implemented in special relativity, respectively. For simplicity we use natural units in which $\hbar$=c=1. Hence, we have

$$E(k) = k\left(1+\beta k^2\right). \tag{7}$$

For large volume, the sum over all states of single particle can be rewritten in terms of an integral [33]

$$\sum_k \longrightarrow \frac{V}{(2\pi)^3}\int_0^\infty d^3k \longrightarrow \frac{V}{2\pi^2}\int_0^\infty \frac{k^2dk}{(1+\beta k^2)^4}. \tag{8}$$

Thus, the partition function, (4), becomes

$$\ln z_B = \frac{-Vg}{2\pi^2}\int_0^\infty k^2\frac{\ln\left[1-\exp\left(-E(k)/T\right)\right]}{(1+\beta k^2)^4}dk = \frac{-Vg}{2\pi^2} \cdot \int_0^\infty k^2\frac{\ln\left[1-\exp\left(-(k/T)\left(1+\beta k^2\right)\right)\right]}{(1+\beta k^2)^4}dk$$

$$= \frac{Vg}{2\pi^2}\left[\frac{k}{6\beta(1+\beta k^2)^3} \cdot \ln\left[1-\exp\left(-\frac{k}{T}\left(1+\beta k^2\right)\right)\right]\right]\Bigg|_0^\infty - \frac{Vg}{2\pi^2} \cdot \int_0^\infty \frac{1}{6\beta(1+\beta k^2)^3}\left[\ln\left[1 - \exp\left(-\frac{k\left(1+\beta k^2\right)}{T}\right)\right] + \frac{k\left(1+3\beta k^2\right)}{T} \cdot \frac{1}{\exp\left(k\left(1+\beta k^2\right)/T\right)-1}\right]dk, \tag{9}$$

where "Integration by Parts" was used for solving the above equation. It is obvious that the first term in (9) vanishes. Thus,

$$\ln z_B = \frac{-Vg}{2\pi^2} \cdot \int_0^\infty \frac{1}{6\beta(1+\beta k^2)^3}\left[\ln\left[1-\exp\left(-\frac{k\left(1+\beta k^2\right)}{T}\right)\right] + \frac{k\left(1+3\beta k^2\right)}{T}\frac{1}{\exp\left(k\left(1+\beta k^2\right)/T\right)-1}\right]dk. \tag{10}$$

For solving (10) let $x = (k/T)(1+\beta k^2)$ so that $dx = (1+3\beta k^2)dk/T$ and

$$\ln z_B = \frac{-Vg}{2\pi^2}\left[\int_0^\infty \frac{\ln\left[1-e^{-x}\right]T\,dx}{6\beta(1+\beta k^2)^3(1+3\beta k^2)}\right] - \frac{Vg}{2\pi^2}\left[\int_0^\infty \frac{k\,dx}{6\beta(1+\beta k^2)^3(e^x-1)}\right]. \tag{11}$$

The momentum k as a function of x variable can be approximated to the first order of β as follows:

$$k = xT - \beta k^3 = xT - \beta\left(xT-\beta k^3\right)^3 = xT - \beta x^3T^3\left(1-\frac{\beta k^3}{xT}\right)^3 \simeq xT - \beta x^3T^3. \tag{12}$$

Employing the value of k into integrant terms of (11) and approximating them to the first order of β are as follows.

For the first term,

$$\frac{T}{6\beta(1+\beta k^2)^3(1+3\beta k^2)} = \frac{T\left(1-6\beta k^2+9\beta^2k^4\right)}{6\beta} \simeq \frac{T}{6\beta} - x^2T^3 + \frac{7\beta}{2}x^4T^5. \tag{13}$$

For the second term,

$$\frac{k}{6\beta(1+\beta k^2)^3} = \frac{k-3\beta k^3}{6\beta} \simeq \frac{xT}{6\beta} - \frac{2x^3T^3}{3} + \frac{3}{2}\beta x^5T^5. \tag{14}$$

Substituting the values of terms in (13) and (14) into (11) and solving them analytically, we have

$$\ln z_B = \frac{\pi^2}{90}VgT^3 - \frac{16\pi^4}{315}g\beta T^5. \tag{15}$$

In Case of Fermions.

At finite temperature T and chemical potential μ, the grand canonical partition function z_f for noninteracting massive bosons with g internal degrees of freedom is given by [32]

$$z_f = \prod_k \left[\sum_{l=0,1} \exp\left(-l\frac{E(k)-\mu}{T}\right)\right]^g \quad (16)$$

$$= \prod_k \left[1 + \exp\frac{-(E(k)-\mu)}{T}\right]^g, \quad (17)$$

where l is the occupation number for each quantum state with energy $E(k) = \sqrt{k^2+m^2}$ with mass m and k is the momentum of the particle. Here the infinite product is taken for all possible momentum states.

For simplicity we consider chiral limit (i.e., vanishing mass) and a vanishing chemical potential, which experiments ensure at high energy. Then the dominant excitation in the hadronic phase is a massless pion, while that in the quarkgluon plasma is massless quarks and gluons. For a particle of mass M having a distant origin and an energy comparable to the Planck scale, the momentum would be a subject of a tiny modification. According to GUP approach, the dispersion relation reads

$$E^2(k) = k^2c^2\left(1+\beta k^2\right)^2 + M^2c^4, \quad (18)$$

where M and c are the mass of the particle and the speed of light as introduced by Lorentz and implemented in special relativity, respectively. For simplicity we use natural units in which $\hbar=c=1$. Hence, we have

$$E(k) = k\left(1+\beta k^2\right). \quad (19)$$

For large volume, the sum over all states of single particle can be rewritten in terms of an integral [33]

$$\sum_k \longrightarrow \frac{V}{(2\pi)^3}\int_0^\infty d^3k \longrightarrow \frac{V}{2\pi^2}\int_0^\infty \frac{k^2 dk}{\left(1+\beta k^2\right)^4}. \quad (20)$$

Thus, the partition function, (16), becomes

$$\ln z_f = \frac{Vg}{2\pi^2}\int_0^\infty \frac{k^2\ln\left(1+\exp\left(-E(k)/T\right)\right)}{\left(1+\beta k^2\right)^4 dk} = \frac{Vg}{2\pi^2} \cdot \int_0^\infty \frac{k^2\ln\left(1+\exp\left(-k\left(1+\beta k^2\right)/T\right)\right)}{\left(1+\beta k^2\right)^4 dk}$$

$$= \frac{-Vg}{2\pi^2}\left.\frac{k\ln\left(1+\exp\left(-k\left(1+\beta k^2\right)/T\right)\right)}{6\beta\left(1+\beta k^2\right)^3}\right|_0^\infty + \frac{Vg}{2\pi^2}\int_0^\infty \frac{\ln\left(1+\exp\left(-k\left(1+\beta k^2\right)/T\right)\right)dk}{6\beta\left(1+\beta k^2\right)^3}$$

$$- \frac{Vg}{2\pi^2} \cdot \int_0^\infty \frac{k\left(1+3\beta k^2\right)dk}{6\beta T\left(1+\beta k^2\right)^3\left(1+\exp\left(k\left(1+\beta k^2\right)/T\right)\right)}, \quad (21)$$

where "Integration by Parts" was used for solving the above equation. It is obvious that the first term in (21) vanishes. Thus,

$$\ln z_f = \frac{Vg}{2\pi^2}\left[\int_0^\infty \frac{\ln\left(1+\exp\left(-k\left(1+\beta k^2\right)/T\right)\right)dk}{6\beta\left(1+\beta k^2\right)^3} - \int_0^\infty \frac{k\left(1+3\beta k^2\right)dk}{6\beta T\left(1+\beta k^2\right)^3\left(1+\exp\left(k\left(1+\beta k^2\right)/T\right)\right)}\right]. \quad (22)$$

For solving (22) let $x = (k/T)(1+\beta k^2)$ so that $dx = (1+3\beta k^2)dk/T$ and

$$\ln z_f = \frac{Vg}{2\pi^2}\left[\int_0^\infty \frac{\ln\left[1+e^{-x}\right]T\,dx}{6\beta\left(1+\beta k^2\right)^3\left(1+3\beta k^2\right)} - \int_0^\infty \frac{k\left(1+3\beta k^2\right)dx}{6\beta\left(1+3\beta k^2\right)\left(e^x+1\right)}\right]. \quad (23)$$

Employing the value of k into integrant terms of (23) and approximating them to the first order of β are as follows. For the first term,

$$\frac{T}{6\beta\left(1+\beta k^2\right)^3\left(1+3\beta k^2\right)} = \frac{\left(1-3\beta k^2\right)\left(1-3\beta k^2\right)}{6\beta} \simeq \frac{T}{6\beta} - x^2T^3 + \frac{7\beta}{2}x^4T^5. \quad (24)$$

For the second term,

$$\frac{k}{6\beta\left(1+3\beta k^2\right)} = k\left(1-3\beta k^2\right) \simeq \frac{xT}{6\beta} - \frac{2}{3}x^3T^3 + \frac{3}{2}\beta x^5T^5. \quad (25)$$

Substituting the values of terms in (24) and (25) into (23) and solving them analytically, we have

$$\ln z_f = \frac{7}{8}\frac{\pi^2}{90}VgT^3 - \frac{31\pi^4}{630}Vg\beta T^5. \quad (26)$$

In case of $\beta\to 0$, the above equation is reduced to the partition function for fermions without GUP effect [32]. Now, the QGP equation of state of free massless quarks and gluons can be derived from the above equations. The total grand canonical partition function of QGP state can be given by adding the grand partition functions coming from the contribution of bosons (gluons), fermions (quarks), and vacuum [39] as follows:

$$\ln z_{QGP} = \ln z_F + \ln z_B + \ln z_V, \quad (27)$$

where $\ln z_B$, $\ln z_F$, and $\ln z_V$ are the grand canonical functions of gluons, quarks, and vacuum, respectively. In our case, the vacuum pressure can be represented with the bag constant B in case of Bag model [40]. Since the value of vacuum partition function equals $\ln z_V = -VB/T$ and $\ln z = -\Omega/T$, using these values (27) reads

$$\left.\frac{\Omega}{V}\right|_{QGP} = \left.\frac{\Omega}{V}\right|_{quarks} + \left.\frac{\Omega}{V}\right|_{gluons} + \left.\frac{\Omega}{V}\right|_{vacuum}, \quad (28)$$

where Ω is the grand canonical potential. Substituting (15) and (26) into (28), we have

$$\left.\frac{\Omega}{V}\right|_{QGP} = -\left(g_g + \frac{7}{8}g_q\right)\frac{\pi^2}{90}T^4 + \left(g_g\frac{16\pi^4}{315} + g_q\frac{31\pi^4}{630}\right)\beta T^6 + B. \quad (29)$$

Hence, the QGP equation of state reads

$$P_{QGP} = -\left.\frac{\Omega}{V}\right|_{QGP} = \left(g_g + \frac{7}{8}g_q\right)\frac{\pi^2}{90}T^4 - \left(g_g\frac{16\pi^4}{315} + g_q\frac{31\pi^4}{630}\right)\beta T^6 - B = \frac{\sigma_{SB}}{3}T^4 - \left(g_g\frac{16\pi^4}{315} + g_q\frac{31\pi^4}{630}\right)\beta T^6 - B, \quad (30)$$

where σ_{SB} is SB constant and is given by $(g_g + (7/8)g_q)\pi^2/30$. Hence, the energy density of QGP state of matter is given by

$$\varepsilon_{QGP} = \sigma_{SB}T^4 - 3\left(g_g\frac{16\pi^4}{315} + g_q\frac{31\pi^4}{630}\right)\beta T^6 + B. \quad (31)$$

RESULTS AND DISCUSSIONS

The main features of QCD lattice results show a clear N_c dependance for both energy density and pressure (i.e., they become larger as the number of degrees of freedom increases). As it is clear from the Monte Carlo lattice results [34, 35], the pressure $P(T)$ rapidly increases at $T \simeq T_c$ which may be agreed with our predictions after adding the GUP effect on QGP equation of state. The bag pressure B can be determined by comparing the obtained equation with that of QCD lattice results.

The main problem appears when one starts to adjust the behavior of the energy density, through varying the value of the bag pressure; with the QCD lattice results, one can not obtain a qualitative agreement in case of the pressure using the same bag parameter value [41]. For overcoming this problem, the bag model was modified [42]. In this technique, the thermodynamical relation between the energy density and the pressure [33] was used

$$T\frac{dP}{dT} - P(T) = \varepsilon(T). \quad (32)$$

Solving the above first-order differential equation, we obtain the QGP equation of state as follows:

$$P = \frac{\sigma_{SB}}{3}T^4 - \left(g_g\frac{16}{525} + g_q\frac{31}{1050}\right)\pi^4\beta T^6 - B + AT, \quad (33)$$

where A is a constant coming from the partial differential equation solution and can be determined from comparing the calculated QGP equation of state with the QCD lattice results.

To adjust the high temperature behavior for (T) and $\varepsilon(T)$, we will consider the suppression factor of the Stefan Boltzmann constant used in quasi particle approach [43]. In this approach, the system of interacting gluons may be treated as a noninteracting quasi particles gas with gluon quantum numbers, but with thermal mass (i.e., (T)). The modified SB constant σ reads [41]

$$\sigma = \kappa(a)\,\sigma_{SB}, \quad (34)$$

where $\kappa(a)$ is a suppression factor. For $a \to 0$, it follows $\kappa \to 1$. Also, the function (a) decreases monotonously and goes to zero at $a \to \infty$. Thus, the final form of the bag model equation of state (33) with incorporating the GUP modification is given by

$$P = \frac{\sigma}{3}T^4 - \left(g_g\frac{16}{525} + g_q\frac{31}{1050}\right)\pi^4\beta T^6 - B + AT. \quad (35)$$

With σ and B being free model parameters

$$\varepsilon = \sigma T^4 - 3\left(g_g\frac{16}{315} + g_q\frac{31}{630}\right)\pi^4\beta T^6 + B. \quad (36)$$

In case of $\beta \to 0$, the above equations (35) and (36) are reduced to the pressure and the energy density equations obtained in the modified bag model [42] (i.e., without incorporating the GUP modification).

The theoretical calculation of the pressure and the energy density of the QGP using (35) and (36) compared to the lattice results in case of $n_f = 0$ are given in Figure 1. In this case we used the following parameter values: $g_{QGP} = 16$, $g_q = 0$, $g_g = 16$, $T_c = 0.200$ GeV, and $\beta = 0.0001$ GeV^{-1} as taken in [11, 32, 43].

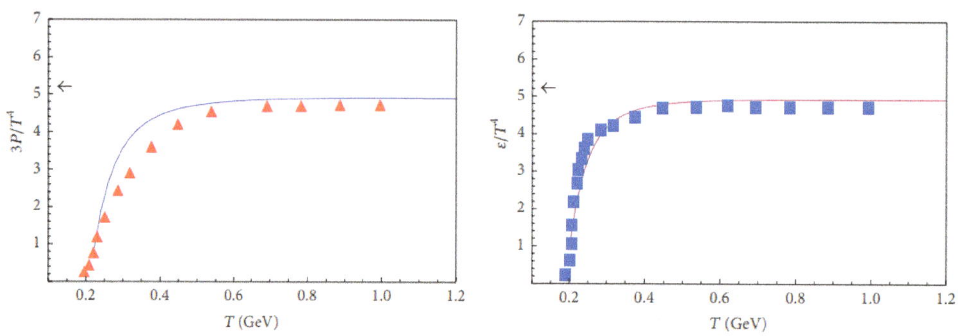

Figure 1: The symbols show the MC LR for the pressure and the energy density in the SU(3) gluodynamics [34, 35]; the line corresponds to the equation of state (35) for the pressure in the left panel and (36) for the energy density in the right panel with $\sigma = 4.20719$, $A = -2.0019T_c^3$, and $B = -1.465358T_c^4$. The arrows in the above figures correspond to $P/\varepsilon = 1/3$ (i.e., the SB limit).

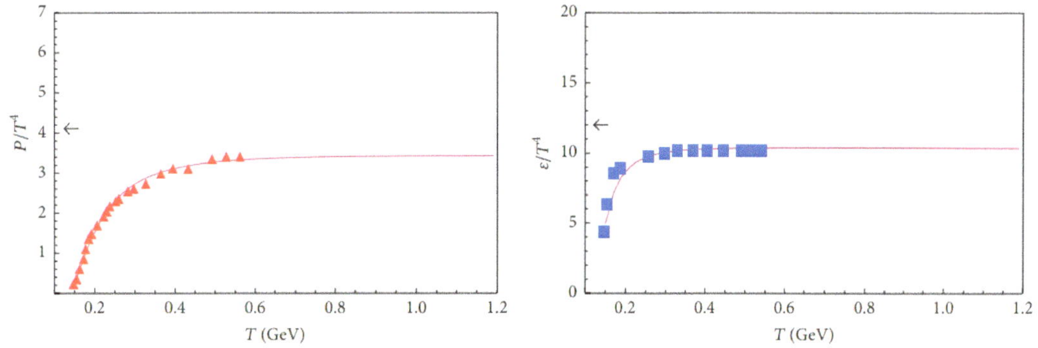

Figure 2: The symbols show the MC LR for the pressure and the energy density in case of $nf = 2$ QCD equation of state [36]; the line corresponds to the equation of state (35) for the pressure in the left panel and (36) for the energy density in the right panel with $\sigma = 10.403$, $A = -8.72895 T_c^3$, and $B = -4.9 T_c^4$. The arrows in the above figures correspond to $P/\varepsilon = 1/3$ (i.e., the SB limit).

The theoretical calculation of the pressure and the energy density of the QGP using (35) and (36) compared to the lattice results in case of $n_f = 2$ are given in Figure 2. In this case we used the following parameter values: $g_{QGP} = 37$, $g_q = 24$, $g_g = 16$, $T_c = 0.152$ GeV, and $\beta = 0.0001$ GeV^{-1} as taken in [11, 32, 43].

The theoretical calculation of the pressure and the energy density of the QGP using (35) and (36) compared to the lattice results in case of $n_f = 2+1$ are given in Figure 3. In this case we used the following parameter values: $g_{QGP} = 47.5$, $g_q = 36$, $g_g = 16$, $T_c = 0.152$ GeV, and $\beta = 0.0001$ GeV^{-1} as taken in [11, 32, 43].

Also, we can calculate the interaction measure $(\varepsilon - 3P)/T^4$ for both $n_f = 0$ and $n_f = 2+1$ and compare them with lattice results for the interaction measure in the SU(3) gluodynamics [34, 35] and in case of $n_f = 2+1$ QCD equation of state [37, 38], respectively. These results can be shown in Figure 4.

It can be shown that the inflection point of $(\varepsilon - 3P)/T^4$ is at $T = 0.1846$ GeV for $n_f = 0$ and $T = 0.202$ GeV for $n_f = 2+1$ which are different from that obtained in [42].

CONCLUSION

In the present work, the effect of the GUP on QGP of massless quark flavors at a vanishing chemical potential, μ, is studied. The equation of state and the energy density are derived for the QGP state consisting of noninteracting massless bosons and fermions with impact of GUP approach. Also, the total grand canonical partition function of QGP state is given. One can conclude that a significant effect for the GUP term exists in case of studying the thermal properties of the QGP state. The main features of QCD lattice results were quantitatively achieved in case of $n_f = 0$, $n_f = 2$, and $n_f = 2+1$ for the equation of state, the energy density, and the interaction measure. The interesting point in our results is the large value of bag pressure especially in case of $n_f = 2+1$ flavor. It nearly equals 4.46 times the value obtained in [42] without taking into account the negative sign of it, which reflects the strong correlation between quarks inside the bag which is already expected. The negative sign may be regarded as the tendency of the bag to reduce its volume. One can conclude that the modification of the QGP bag model equation of state using the GUP effect can reproduce the QCD lattice results which stands for the real data of QGP. Finally, with this study, one may be encouraged to implement GUP effect in other thermodynamic properties of QGP.

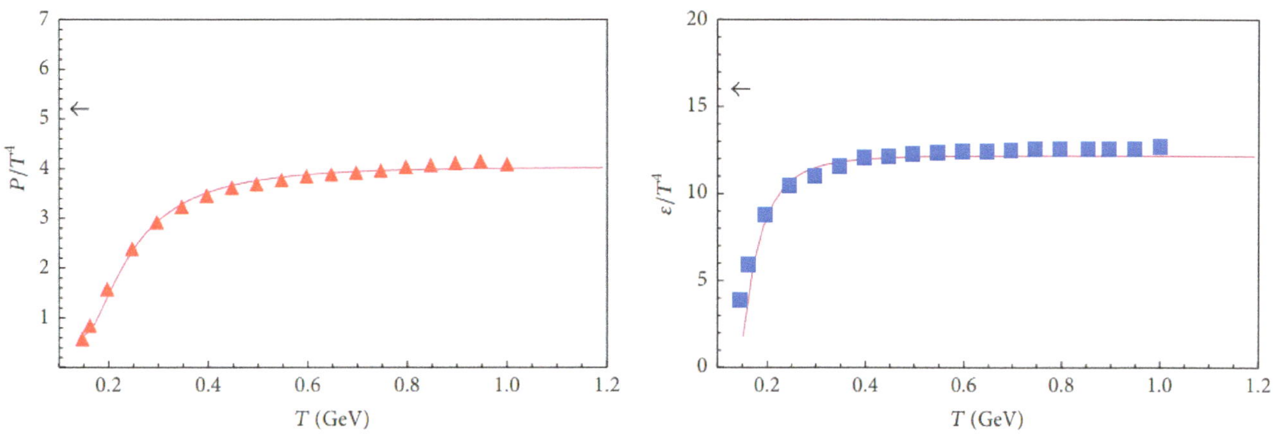

Figure 3: The symbols show the MC LR for the pressure and the energy density in case of $n_f = 2+1$ QCD equation of state [37, 38]; the line corresponds to the equation of state (35) for the pressure in the left panel and (36) for the energy density in the right panel with $\sigma = 12.22017$, $A = -13.8577 T_c^3$, and $B = -10.4353 T_c^4$. The arrows in the above figures correspond to $P/\varepsilon = 1/3$ (i.e., the SB limit).

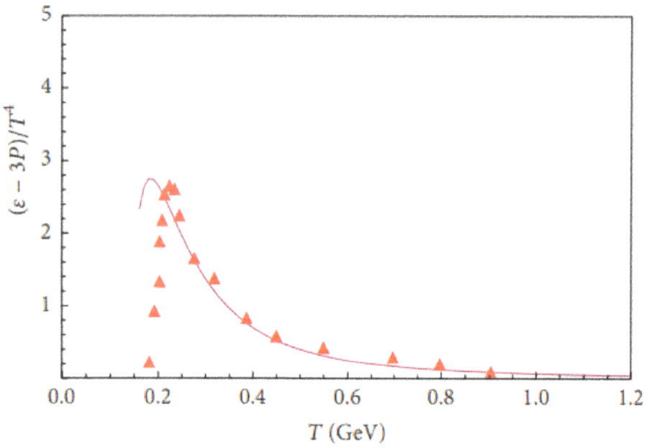

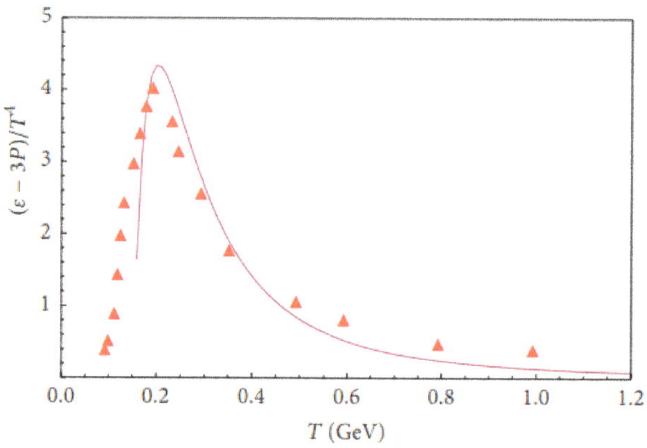

Figure 4: The symbols show the lattice results for the interaction measure in the SU(3) gluodynamics (left panel) [34, 35] and in case of n_f =2+1 QCD equation of state (right panel) [37, 38].

REFERENCES

1. L. J. Garay, "Quantum gravity and minimum length," International Journal of Modern Physics A, vol. 10, no. 2, pp. 145–166, 1995.
2. D. J. Gross and P. F. Mende, "String theory beyond the Planck scale," Nuclear Physics B, vol. 303, no. 3, pp. 407–454, 1988.
3. M. Maggiore, "A generalized uncertainty principle in quantum gravity," Physics Letters B, vol. 304, no. 1-2, pp. 65–69, 1993.
4. S. Hossenfelder, "Minimal length scale scenarios for quantum gravity," Living Reviews in Relativity, vol. 16, no. 2, 2013.
5. A. Kempf, "Uncertainty relation in quantum mechanics with quantum group symmetry," Journal of Mathematical Physics, vol. 35, no. 9, pp. 4483–4496, 1994.
6. A. Kempf, G. Mangano, and R. B. Mann, "Hilbert space representation of the minimal length uncertainty relation," Physical Review D, vol. 52, no. 2, pp. 1108–1118, 1995.
7. A. Kempf, "Non-pointlike particles in harmonic oscillators," Journal of Physics A: Mathematical and General, vol. 30, no. 6, pp. 2093–2101, 1997.
8. H. Hinrichsen and A. Kempf, "Maximal localization in the presence of minimal uncertainties in positions and in momenta," Journal of Mathematical Physics, vol. 37, no. 5, pp. 2121–2137, 1996.
9. I. Pikovski, M. R. Vanner, M. Aspelmeyer, M. S. Kim, and Č. Brukner, "Probing planck-scale physics with quantum optics," Nature Physics, vol. 8, no. 5, pp. 393–397, 2012.
10. S. Das and E. C. Vagenas, "Universality of quantum gravity corrections," Physical Review Letters, vol. 101, no. 22, Article ID 221301, 4 pages, 2008.
11. A. F. Ali, S. Das, and E. C. Vagenas, "Proposal for testing quantum gravity in the lab," Physical Review D, vol. 84, no. 4, Article ID 044013, 10 pages, 2011.
12. C. Quesne and V. M. Tkachuk, "Dirac oscillator with nonzero minimal uncertainty in position," Journal of Physics A: Mathematical and General, vol. 38, no. 8, pp. 1747–1765, 2005.
13. K. Nouicer, "An exact solution of the one-dimensional Dirac oscillator in the presence of minimal lengths," Journal of Physics A, vol. 39, no. 18, pp. 5125–5134, 2006.
14. C. Quesne and V. M. Tkachuk, "Generalized deformed commutation relations with nonzero minimal uncertainties in position and/or momentum and applications to quantum mechanics," Symmetry, Integrability and Geometry: Methods and Applications, vol. 3, article 016, 18 pages, 2007.
15. T. K. Jana and P. Roy, "Exact solution of the Klein-Gordon equation in the presence of a minimal length," Physics Letters A, vol. 373, no. 14, pp. 1239–1241, 2009.
16. L. N. Chang, D. Minic, N. Okamura, and T. Takeuchi, "Exact solution of the harmonic oscillator in arbitrary dimensions with minimal length uncertainty relations," Physical Review D, vol. 65, no. 12, Article ID 125027, 2002.
17. I. Dadić, L. Jonke, and S. Meljanac, "Harmonic oscillator with minimal length uncertainty relations and ladder operators," Physical Review D, vol. 67, no. 8, Article ID 087701, 4 pages, 2003. ·
18. K. Gemba, Z. T. Hlousek, and Z. Papp, "Algebraic solution of the harmonic oscillator with minimal length uncertainty relations," http://arxiv.org/abs/0712.2078.
19. F. Brau, "Minimal length uncertainty relation and the hydrogen atom," Journal of Physics A: Mathematical and General, vol. 32, no. 44, pp. 7691–7696, 1999.
20. T. V. Fityo, I. O. Vakarchuk, and V. M. Tkachuk, "One-dimensional Coulomb-like problem in deformed space with minimal length," Journal of Physics A: Mathematical and General, vol. 39, no. 9, p. 2143, 2006.
21. R. Akhoury and Y.-P. Yao, "Minimal length uncertainty relation and the hydrogen spectrum," Physics Letters B, vol. 572, no. 1-2, pp. 37–42, 2003.
22. S. Benczik, L. N. Chang, D. Minic, and T. Takeuchi, "Hydrogen-atom spectrum under a minimal-length hypothesis," Physical Review A—Atomic, Molecular, and Optical Physics, vol. 72, no. 1, Article ID 012104, 2005.
23. S. Das and E. C. Vagenas, "Phenomenological implications of the generalized uncertainty

principle," Canadian Journal of Physics, vol. 87, no. 3, pp. 233–240, 2009.

24. S. Das, E. C. Vagenas, and A. F. Ali, "Discreteness of space from GUP II: relativistic wave equations," Physics Letters B, vol. 690, no. 4, pp. 407–412, 2010.
25. N. M. El Naggar, L. I. Abou-Salem, I. A. Elmashad, and A. F. Ali, "A study on Quark-Gluon plasma equation of state using generalized uncertainty principle," Journal of Modern Physics, vol. 4, pp. 13–20, 2013.
26. A. Tawfik, L. I. Abou-Salem, I. Elmashad, J.-U. Nabi, and A. F. Ali, "Quantum gravity effect on the Quark-Gluon plasma," SOP Transactions on Theoretical Physics, vol. 1, no. 1, pp. 1–6, 2014.
27. A. F. Ali, S. Das, and E. C. Vagenas, "The generalized uncertainty principle and quantum gravity phenomenology," in Proceedings of the 12th Marcel Grossmann Meeting, pp. 2407–2409, Paris, France, July 2009.
28. A. F. Ali, "Minimal length in quantum gravity, equivalence principle and holographic entropy bound," Classical and Quantum Gravity, vol. 28, no. 6, Article ID 065013, 2011.
29. J. Sadeghi, B. Pourhassan, and S. Heshmatian, "Application of AdS/CFT in quark-gluon plasma," Advances in High Energy Physics, vol. 2013, Article ID 759804, 13 pages, 2013.
30. J. Sadeghi and B. Pourhassan, "STU–QCD correspondence," Canadian Journal of Physics, vol. 91, no. 12, pp. 995–1019, 2013.
31. J. Sadeghi and B. Pourhassan, "Drag force of moving quark at N=2 supergravity," Journal of High Energy Physics, vol. 2008, no. 12, article 026, 2008.
32. K. Yagi, T. Hatsuda, and Y. Miake, Quark-Gluon Plasma From Big Bang to Little Bang, Cambridge University Press, Cambridge, UK, 2005.
33. W. Greiner, L. Neise, and H. Stöcker, Thermodynamics and Statistical Mechanics, Springer, New York, NY, USA, 1997.
34. G. Boyd, J. Engels, F. Karsch, et al., "Equation of state for the SU(3) gauge theory," Physical Review Letters, vol. 75, no. 23, pp. 4169–4172, 1995.
35. G. Boyd, J. Engels, F. Karsch et al., "Thermodynamics of SU(3) lattice gauge theory," Nuclear Physics B, vol. 469, no. 3, pp. 419–444, 1996.
36. F. Karsch, E. Laermann, and A. Peikert, "The pressure in 2, 2+1 and 3 flavour QCD," Physics Letters, Section B: Nuclear, Elementary Particle and High-Energy Physics, vol. 478, no. 4, pp. 447–455, 2000.
37. S. Borsányi, Z. Fodor, C. Hoelbling, et al., "Is there still any Tc mystery in lattice QCD? Results with physical masses in the continuum limit III," Journal of High Energy Physics, vol. 2010, no. 9, article 073, 2010.
38. S. Borsanyi, G. Endrődi, Z. Fodor, et al., "The QCD equation of state with dynamical quarks," Journal of High Energy Physics, vol. 2010, article 77, 2010.
39. J. Letessier and J. Rafelski, Hadrons and Quark-Gluon Plasma, Cambridge University Press, Cambridge, UK, 2004.
40. A. Chodos, R. L. Jaffe, K. Johnson, C. B. Thorn, and V. F. Weisskopf, "New extended model of hadrons," Physical Review. D: Particles and Fields, vol. 9, no. 12, pp. 3471–3495, 1974.
41. V. V. Begun, M. I. Gorenstein, and O. A. Mogilevsky, "Modified bag models for the Quark-Gluon plasma equation of state," International Journal of Modern Physics E, vol. 20, no. 8, pp. 1805–1815, 2011.
42. V. V. Begun, M. I. Gorenstein, and O. A. Mogilevsky, "Equation of state for the quark gluon plasma with the negative bag constant," Ukrainian Journal of Physics, vol. 55, no. 9, pp. 1049–1052, 2010.
43. M. I. Gorenstein and S. N. Yang, "Gluon plasma with a medium-dependent dispersion relation," Physical Review D, vol. 52, no. 9, pp. 5206–5212, 1995.

CHAPTER 12

CALCULATING MASSES OF PENTAQUARKS COMPOSED OF BARYONS AND MESONS

M. Monemzadeh, N. Tazimi and Sh. Babaghodrat

INTRODUCTION

There are two types of hadrons, baryons and mesons. Baryons are equivalent to the bound states of three quarks, and mesons are known to be the bound states of a quark and an antiquark. However, QCD describes mesons and baryons even with a more intricate structure. There are anomalous mesons such as ggg, $q\bar{q}g$, and $q\bar{q}q\bar{q}$, as well as exotic baryons like $qqqg$, $qqqq\bar{q}$, and so forth. Pentaquarks are baryons with at least four quarks and one antiquark. In exotic pentaquarks, the antiquark has a flavor different from the other four quarks.

Exotic hadrons containing at least three valence quarks are being studied fairly extensively in modern physics. Although there are hundreds of ordinary hadrons, exotic ones have not been found stable yet. However, QCD does not reject their existence. Pentaquark θ^+, studied in photo production experiments [1, 2], is a prototype of exotic hadrons in light and strong quark sector. Theoretically, hadronic reactions contribute to θ^+ production more vividly than other types of reactions.

The quark model is commonly used to describe hadrons. In this model, mesons are described as $q\bar{q}$ and baryons as three-quark composite particles. In a more microscopic view, QCD usually serves to describe the strong interaction. According to Lipkin [3, 4] and Gignoux et al. [5], among pentaquarks, the five-quark anticharmed baryons of $P^0 = [uud\bar{c}s]$ and $\bar{p} = [udd\bar{c}s]$ or similar antibeauty baryons are the most bound.

A lot of experimental evidence on the existence of exotic hadrons has been found since 2003. Exotic hadrons' quantum numbers cannot be justified based on two- and threequark bound states. Pentaquarks of $qqqq\bar{q}$ form are examples of exotic baryon states. Conjugation quantity of C charge is not an accurate quantum number for baryons, and all combinations of total spin J and parity P can exist. However, an exotic baryon combination can be readily identified by its electric charge Q and its strangeness S. Some evidence has been reported during the last few years. For example, the pentaquark θ^{++} was proven to exist in Hermes experiment in Hamburg, Germany [6, 7].

For exotic baryons, we consider the following:

- θ^+: the existence of this exotic baryon was predicted in chiral solution model [8]. It has $S = +1$, $J^P = 1/2^+$, and I=0. It is a narrow light-mass particle of 1540 MeV. These attributes initially made θ^+ a subject of experimental observation by LEPS [9]. The most suitable hadronic decay mode to identify it is $\theta^+ \rightarrow K^0 p$.
- θ_c and θ_b: the existence of the bound exotic hadron θ_c was predicted through bound Skyrmion approach. This particle has a mass of 2650 MeV and quantum numbers $J^P = 1/2^+$ and I=0. An experiment [10] showed a positive signal at a mass of about 3.1 GeV, but it was not confirmed later [11]. In strongly bound states, the decay mode $K^+\pi^-\pi^- p$ is easy to identify. Likewise, the mass of θ_b with the same quantum numbers $J^P = 1/2^+$ and I=0 was predicted to be 5207 MeV. The possible weak decay mode is $K^+\pi^-\pi^- + \pi^+ + p$.
- θ_{cs}: it is the five-quark state with $J^P = 1/2^-$ and I=0. In a quark model which includes color-spin interaction, it can be bound and despite its strong decay, it becomes stable [12]. The mass dependent on the model parameters is predicted to be 2420 MeV. This state was traced in the Fermilab E791 experiment via $\phi\pi p$ mode [13] and $K^{*0}K^- p$ mode [14].

Lippmann-Schwinger equation for two-body bound states is solved in Section 2. In Section 3, the procedure of the study is given and pentaquark masses are calculated.

Citation: M. Monemzadeh, N. Tazimi, and Sh. Babaghodrat, "Calculating Masses of Pentaquarks Composed of Baryons and Mesons," Advances in High Energy Physics, vol. 2016, Article ID 6480926, 4 pages, 2016. doi:10.1155/2016/6480926.

LIPPMANN-SCHWINGER EQUATION FOR TWO-BODY BOUND STATES

In this part, the binding energy of the entire system (pentaquark) is calculated by numerical solution of homogeneous Lippmann-Schwinger equation for each subsystem of bound meson and baryon. Schrodinger equation for a two-body bound state with the potential V runs as the following integral equation [16, 17]:

$$\psi_b(r) = -\frac{m}{4\pi}\int_0^\infty dr' r'^2 \int_{-1}^{1} dx' \cdot \int_0^{2\pi} d\phi' \frac{\exp\left(-\sqrt{m|E_b|}\,|r-r'|\right)}{|r-r'|} V(r')\psi_b(r'), \quad (1)$$

where E_b stands for the binding energy of the two-body bound system (meson + baryon). When the interaction potential is considered as r function, (1) is of the following eigenvalue form

$$K(E_b)|\psi_b\rangle = \lambda(E_b)|\psi_b\rangle. \quad (2)$$

λ=1 is the highest positive eigenvalue. The eigenvalue equation is solved through direct method [16–19] using a regularization method [20, 21]. To discretize the integrals, Gauss-Legendre method [22] is employed. 200 grid points are considered for r and r' , whereas 100 mesh points are quite enough for x' . The kernel of integral equation is diagonized to find λ=1 in the eigenvalue spectrum. The energy corresponding to λ=1 will be the system's binding energy

NUMERICAL RESULTS AND DISCUSSION

We consider a pentaquark as a bound state of a two-particle system formed by a baryon and a meson. The hadronic molecular structure consisting of a baryon and a meson is shown in Figure 1.

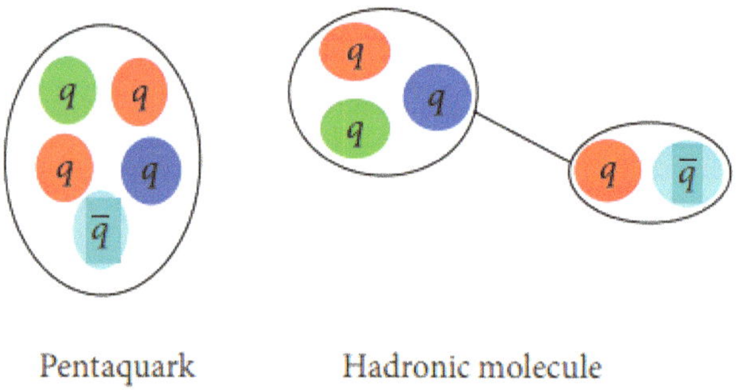

Figure 1: Pentaquark.

Interaction potential has an essential role in solving the eigenvalue equation (5). Different potentials have been introduced for meson-baryon interaction. Yukawa potential (screened Coulomb potential) is proposed as one of the appropriate ones [15, 23]. This study deals with exotic baryon states created by a meson and a nucleon. The π exchange potential is among the most prominent meson exchange forces. π is the lightest hadron that can be exchanged between a meson and a nucleon. Therefore, we consider only π exchange, and ρ and ω meson exchange will be elaborated on in our subsequent works. One-pion exchange potential (OPEP) is of the following form [15]:

$$V_\pi(r) = \begin{cases} (\vec{I}_N\cdot\vec{I}_H)\left(2S_{12}V_T(r) + 4\vec{S}_N\cdot\vec{S}_l\right)V_c(r), & r > r_0, \\ V_0, & r < r_0, \end{cases} \quad (3)$$

where for $r>r_0$

$$V_\pi(r) = (\vec{I}_N\cdot\vec{I}_H)\left(2S_{12}V_T(r) + 4\vec{S}_N\cdot\vec{S}_l\right)V_c(r) = \left(I^2 - I_N^2 - I_H^2\right)\left(2S_{12}V_T(r) + K^2 - S_N^2 - S_l^2\right)V_c(r), \quad (4)$$

Where

$$V_c(r) = \frac{g_H g_A}{2\pi f_\pi^2}\left(m_\pi^2\right)\frac{e^{-m_\pi r}}{3r}, \quad (5)$$

$$V_T(r) = \frac{g_H g_A}{2\pi f_\pi^2}\left(m_\pi^2\right)\frac{e^{-mr}}{6r}\left(\frac{3}{r^2}+\frac{3}{r}+1\right), \quad (6)$$

$$V_0 = -62.79\ (\mathrm{MeV})\ \text{or}\ -276\ (\mathrm{MeV}), \quad (7)$$

where g_A, f_π, m_π, and g_H are axial coupling constant, pion decay constant, pion mass, and heavy-meson coupling constant, respectively. I is the total isospin of meson-nucleon system and

$$S_{12} \equiv 4\left[3(\vec{S}_N\cdot\hat{r})(\vec{S}_l\cdot\hat{r}) - \vec{S}_N\cdot(\vec{S}_l)\right]. \quad (8)$$

Inserting I_N (nucleon isospin), I_H (meson isospin), S_N (nucleon spin), S_l (the lightest quark's spin in the meson), and K=S_N + S_l into the potential, we presented hadronic molecular structure of two pentaquarks in Table 2.

In this calculation we have ignored the tensor term S_{12} for studying the pentaquarks ground state [24]; hence, we used the central part of Yukawa potential. Pentaquark binding energy is defined as the energy used when breaking a pentaquark into its components, that is, meson and baryon. We adopted the following constants for the bound state of heavy pentaquarks from [15] (Table 1).

Table 1: Potential parameters used for the bound state of heavy pentaquarks [15]

Parameter	Value
g_A	1.27
f_π	131 MeV
g_H	−0.59
m_π	138 MeV
m_{Ni}	938.92 MeV
m_B	5279 MeV
m_D	1867 MeV

Table 2: Hadronic molecular structure of pentaquarks

Hadronic structure	Pentaquark (meson + nucleon)
$\theta_c(uudd\bar{c})$	$uud/udd,\ d\bar{c}/u\bar{c}$
$\theta_b(uudd\bar{b})$	$uud/udd,\ d\bar{b}/u\bar{b}$

Pentaquark masses are calculated according to

$$M\,(\text{pentaquark}) = m_{\text{meson}} + m_{\text{baryon}} + E_b. \quad (9)$$

In order to find the binding energy (E_b), first we solve Lippmann-Schwinger equation for the two-body system. In this approach, the kernel is diagnosed and the eigenvalue spectrum is identified (spin-spin interaction in the potential and spin splitting are ignored). The data required include reduced mass of mesons and nucleons, proposed binding energy, potential coefficients, r_0 = 1 or 1.5 fm, and r-cutoff = 20 fm. Table 3 shows the binding energies that we found. We adopted Yukawa potential from [15] and our calculated energies are in good agreement with [15]. We present pentaquark masses in Table 4.

RESULTS AND DISCUSSION

In this paper, we solved Lippmann-Schwinger equation for pentaquark systems. We managed to obtain the binding energy and used it to calculate the masses of these systems. The pentaquark is considered as the bound state of a baryon and a heavy meson. We used a baryon-meson picture to reduce a complicated five-body problem to one simple twobody problem. In Table 4, we have listed our numerical results for masses of pentaquark systems, and pentaquark masses are compared with the results obtained in [25–27].

Our method is appropriate for investigating tetraquark systems too. In our previous work, we investigated the tetraquark as the bound state of a heavy-light diquark and antidiquark. We used the diquark-antidiquark picture to reduce a complicated four-body problem to simple two-body problems.

We analyzed diquark-antidiquark in the framework of a two-body (pseudopoint) problem. We made use of the potential coefficients proposed by Ebert et al. We solved Lippmann-Schwinger equation numerically for charm diquark-antidiquark systems and found the eigenvalues to calculate the binding energies and masses of heavy tetraquarks with hidden charms [28].

Table 3: The calculated binding energies (MeV) for θ_c (D meson) and θ_b (B meson) for I=0. Column A: V_0 = −276 MeV and r_0 = 1 fm; column B: V_0 = −62.79 MeV and r_0 = 1.5 fm

Pentaquark	A	B	A [15]	B [15]
θ_c	113.32	8.21	113.99	8.45
θ_b	140	15.1	139.48	15.46

Table 4: Pentaquark calculated masses (MeV) for I=0

Pentaquark	Mass (A)	Mass (B)	Mass in [25–27]
θ_c	2665.92	2790.82	2650
θ_b	6104.6	6209.71	6391

ACKNOWLEDGMENTS

This work is supported by University of Kashan under Grant no. 65500.6. The authors also express their greatest gratitude to Dr. M. R. Hadizadeh for his helpful technical comments on the numerical solution of the LS integral equation and also for comments that helped them to revise the text of the paper.

REFERENCES

1. T. Nakano, D. S. Ahn, J. K. Ahn, et al., “Evidence for a narrow S=+1 baryon resonance in photoproduction from the neutron,” Physical Review Letters, vol. 91, no. 1, Article ID 012002, 2003.
2. T. Nakano, N. Muramatsu, D. S. Ahn, et al., “Evidence for the Θ+ in the γd→K+K-pn reaction by detecting K+K- pairs,” Physical Review C, vol. 79, Article ID 025210, 2009.
3. H. J. Lipkin, “New possibilities for exotic hadrons: anticharmed strange baryons,” Physics Letters B, vol. 195, no. 3, pp. 484–488, 1987.
4. H. J. Lipkin, “Multiquark physics-the great challenge for future directions in the interplays between particle and nuclear physics,” Nuclear Physics A, vol. 478, pp. 307–324, 1988.
5. C. Gignoux, B. Silvestre-Brac, and J. M. Richard, “Possibility of stable multiquark baryons,” Physics Letters B, vol. 193, no. 2-3, pp. 323–326, 1987.
6. A. Airapetianae, N. Akopovae, and Z. Akopov, “Evidence for a narrow S=1 baryon state at a mass of 1528 MeV in quasi-real photoproduction,” Physics Letters B, vol. 585, no. 3-4, pp. 213–222, 2004.
7. A. Airapetian, N. Akopov, Z. Akopov, et al., “Search for an exotic S=-2, Q=-2 baryon resonance at a mass near 1862 MeV in quasireal photoproduction,” Physical Review D, vol. 71, no. 3, Article ID 032004, 5 pages, 2005.
8. T. Nakano, D. S. Ahn, J. K. Ahn, et al., “Evidence for a narrow S=+1 baryon resonance in photoproduction from the neutron,” Physical Review Letters, vol. 91, Article ID 012002, 2003.
9. D. Diakonov, V. Petrov, and M. Polyakov, “Exotic anti-decuplet of baryons: prediction from chiral solitons,” Zeitschrift für Physik A: Hadrons and Nuclei, vol. 359, no. 3, pp. 305–314, 1997.
10. A. Aktas, V. Andreevz, T. Anthonis, et al., “Evidence for a narrow anti-charmed baryon state,” Physics Letters B, vol. 588, no. 1-2, pp. 17–28, 2004.
11. J. M. Link, P. M. Yagera, J. C. Anjos, et al., “Search for a strongly decaying neutral charmed pentaquark,” Physics Letters B, vol. 622, no. 3-4, pp. 229–238, 2005.
12. H. J. Lipkin, “New possibilities for exotic hadrons—anticharmed strange baryons,” Physics Letters B, vol. 195, no. 3, pp. 484–488, 1987.
13. E. M. Aitala, S. Amato, J. C. Anjos, et al., “Search for the Pentaquark via the Pcs-0→φπp decay,” Physical Review Letters, vol. 81, pp. 44–48, 1998.
14. E. M. Aitala, S. Amato, J. C. Anjos, et al., “Search for the pentaquark via the Pc-s0→K∗0K-p decay,” Physics Letters B, vol. 448, no. 3-4, pp. 303–310, 1999.
15. T. D. Cohen, P. M. Hohler, and R. F. Lebed, “On the existence of heavy pentaquarks: The large Nc and heavy quark limits and beyond,” Physical Review D, vol. 72, no. 7, Article ID 074010, 2005.
16. M. Monemzadeh, M. Hadizadeh, and N. Tazimi, “Identification of the mass and stability interval of strong

potential in heavy mesons," International Journal of Theoretical Physics, vol. 50, no. 3, pp. 737–743, 2011.

17. M. R. Hadizadeh, "Comment on 'Tetraquarks as diquark-antidiquark bound systems'," The European Physical Journal C, vol. 75, article 281, 2015.

18. N. Tazimi, M. Monemzadeh, and M. R. Hadizadeh, "Description of heavy quark ms mass by lippmann-schwinger equation," International Journal of Theoretical Physics, vol. 51, no. 9, pp. 2871–2877, 2012. ·

19. N. Tazimi, M. Monemzadeh, and M. R. Hadizadeh, "Heavy mesons spectroscopy," International Journal of Theoretical Physics, vol. 52, no. 7, pp. 2329–2334, 2013. ·

20. M. R. Hadizadeh and A. Khaledi-Nasab, "Heavy tetraquarks in the diquark-antidiquark picture," Physics Letters, Section B: Nuclear, Elementary Particle and High-Energy Physics, vol. 753, pp. 8–12, 2016.

21. M. R. Hadizadeh and L. Tomio, "Solution of two-body bound state problems with confining potentials," in Proceedings of the XI Hadron Physics, vol. 1296 of AIP Conference Proceedings, p. 334, São Paulo, Brazil, March 2010.

22. S. Nakamura, Applied Numerical Methods with Software, Prentice Hall, 1998.

23. Y. Yamaguchi, S. Ohkoda, S. Yasui, and A. Hosaka, "Exotic baryons from a heavy meson and a nucleon: negative parity states," Physical Review D-Particles, Fields, Gravitation and Cosmology, vol. 84, no. 1, Article ID 014032, 2011.

24. D. Ebert, R. N. Faustov, and V. O. Galkin, "Quark-antiquark potential with retardation and radiative contributions and the heavy quarkonium mass spectra," Physical Review D, vol. 62, Article ID 034014, 2000.

25. S. Cho, T. Furumoto, T. Hyodo et al., "Identifying multiquark hadrons from heavy ion collisions," Physical Review Letters, vol. 106, no. 21, Article ID 212001, 2011.

26. D. Diakonov, "Prediction of new charmed and bottom exotic pentaquarks," https://arxiv.org/abs/1003.2157.

27. M. Karliner, Pentaquark Workshop, Jlab, November 2003.

28. M. Monemzadeh, N. Tazimi, and P. Sadeghi, "Tetraquarks as diquark-antidiquark bound systems," Physics Letters, Section B: Nuclear, Elementary Particle and High-Energy Physics, vol. 741, pp. 124–127, 2015.

CHAPTER 13 A MIXED ACTION ANALYSIS OF η AND η′ MESONS

Konstantin Ottnad, Carsten Urbach, Falk Zimmermann

INTRODUCTION

Mixed action approaches, where valence and sea fermion actions are chosen differently, are used frequently in lattice QCD. They possess a number of important advantages compared to the so-called unitary case, where valence and sea quark actions are identical. In particular, it might be possible to use a valence action obeying more symmetries than the sea action in cases where the valence action cannot be used in the sea for theoretical reasons or because of too high computational costs [1]. Prominent examples are overlap [2], [3] and [4] or domain wall [5] valence quarks on a Wilson-like or staggered sea. Concrete examples can be found for instance in Refs. [6], [7] and [8].

When working in a mixed action approach, valence and sea actions need to be matched appropriately, for instance by tuning the valence quark masses such that a choice of meson masses agrees between unitary and mixed approaches. People even go one step further and try to correct for small mismatches in bare parameters used in the sea by using a partially quenched mixed action approach. The most extreme example for this approach is to use valence strange and charm on gauge configurations with only N_f=2light dynamical quark flavours. In this case sea and valence actions are not matched, but the valence parameters are tuned such as to reproduce a choice of physical observables.

This ansatz has the big advantage that the gauge configurations do not need to be re-generated. However, while apparently quite successful, it is questionable whether this procedure works for observables with a strong sea quark dependence. Due to OZI suppression there are not many examples of such observables. But their very existence makes a clear distinction between QCD and the naive quark model.

Of course, a mixed action approach has also disadvantages, most prominently the breaking of unitarity, which might for instance drive certain correlators negative [9] and [10]. Also, it is not clear a priori how big lattice artifacts one encounters in mixed formulations.

In this paper we will present results on a particular mixed action approach with so-called Osterwalder–Seiler [11] valence quarks on an N_f=2+1+1 flavour Wilson twisted mass sea [12]. This particular action combination has the advantage that exact valence quark flavour symmetry is preserved. Moreover, the respective zero modes of sea and valence quarks coincide in the chiral limit. However, $O(a^2)$ violations of flavour (and parity) stemming from the sea quarks are still reflected in the magnitude of lattice artifacts on various physical observables.

As physical example we study the η and η′ system. The large mass splitting observed among $M_{\pi}^0 \ll M_{\eta} \ll M_{\eta}'$ is thought to be due to the $U_A(1)$ anomaly, a relation established via the Witten–Veneziano formula [13], [14] and [15]. This lets one expect a significant dependence on the sea quark degrees of freedom. Speaking more technically, the corresponding correlation functions obtain significant contributions from fermionic disconnected diagrams and are, therefore, uniquely sensitive to differences between valence and sea formulations. Note that this was also discussed in the context of the validity of the fourth root trick in staggered simulations, see Refs. [16] and [17] and references therein.

After matching valence and sea actions, we compare observables extracted from unitary and valence operators. The unitary observables have been computed in Refs. [18] and [19]. We study the continuum limit with different matching conditions and find remarkably good agreement to the unitary case. However, when comparing the valence with valence plus sea strange quark mass dependence of M_{η} we find significant differences.

Citation: Konstantin Ottnad, Carsten Urbach, Falk Zimmermann, A mixed action analysis of η and η` mesons, Nuclear Physics B, Volume 896, July 2015, Pages 470-492, ISSN 0550-3213, http://dx.doi.org/10.1016/j.nuclphysb.2015.05.001.

These findings are important for future lattice QCD investigations: there are many phenomenologically interesting quantities involving flavour singlet pseudo-scalar mesons, for instance form factors of B or D_s decays to $\eta\ell\nu$. And maybe most prominently, there are anomaly related form factors of $\eta\rightarrow\gamma\gamma$, which can be used to estimate the light-by-light contribution to the hadronic part in the anomalous magnetic moment of the muon, in which we currently observe a deviation between theory and experiment at the few σ level [20], [21], [22] and [23]. The usage of Wilson twisted mass fermions described in this paper has significant advantages compared to other lattice actions due to a powerful variance reduction. And the possibility to use a mixed action will further ease those computations.

More generally, the findings here show that with a mixed action approach one can deal with fermionic disconnected diagrams, provided one applies an appropriate matching procedure. These disconnected diagrams become more and more important as they need to be treated appropriately for instance in investigations of hadron–hadron interactions. Since we show here that a mixed action approach works in the case of η, η' mesons, where the fermionic disconnected diagrams contribute significantly, we are confident that the same approach can be used for other physical observables. First accounts of this work can be found in Ref. [24]. Other studies of η and η' mesons from lattice QCD can be found in Refs. [17], [25], [26], [27], [28] and [29].

LATTICE ACTIONS

The results we will present here are obtained by evaluating correlation functions on gauge configurations provided by the European Twisted Mass Collaboration (ETMC) [30]. We use the ensembles specified in Table 1 adopting the notation from Ref. [30]. More details can be found in this reference.

Table 1: The gauge ensembles used in this study. For the labelling of the ensembles we adopted the notation in Ref. [30]. In addition to the relevant input parameters we give the lattice volume, the number of evaluated configurations N_{conf} and the block length N_b used for bootstrapping. N_b was chosen such that blocks cover at least 20 HMC trajectories of length one

Ensemble	β	$a\mu_\ell$	$a\mu_\sigma$	$a\mu_\delta$	V	N_{conf}	N_b
A40.24	1.90	0.0040	0.150	0.190	$24^3 \times 48$	1117	5
A60.24	1.90	0.0060	0.150	0.190	$24^3 \times 48$	1249	5
A80.24	1.90	0.0080	0.150	0.190	$24^3 \times 48$	2441	10
A100.24	1.90	0.0100	0.150	0.190	$24^3 \times 48$	968	5
A80.24s	1.90	0.0080	0.150	0.197	$24^3 \times 48$	2420	10
A100.24s	1.90	0.0100	0.150	0.197	$24^3 \times 48$	1196	5
B55.32	1.95	0.0055	0.135	0.170	$32^3 \times 64$	4450	5
D45.32sc	2.10	0.0045	0.0937	0.1077	$32^3 \times 64$	2220	10

The sea quark formulation is the Wilson twisted mass formulation with N_f=2+1+1dynamical quark flavours. The Dirac operator for the light quark doublet reads [12]

$$D_\ell = D_W + m_0 + i\mu_\ell\gamma_5\tau^3, \tag{1}$$

where D_W denotes the standard Wilson Dirac operator and μ_ℓ the bare light twisted mass parameter. τ^3 and in general τ^i,i=1,2,3 represent the Pauli matrices acting in flavour space. D_ℓ acts on a spinor $\chi_\ell=(u,d)^T$ and, hence, the u (d) quark has twisted mass $+\mu_\ell$ $(-\mu_\ell)$.

For the heavy unitary doublet of c and s quarks [31] the Dirac operator is given by

$$D_h = D_W + m_0 + i\mu_\sigma\gamma_5\tau^1 + \mu_\delta\tau^3. \tag{2}$$

The bare Wilson quark mass m_0 has been tuned to its critical value [30] and [32]. This guarantees automatic order O(a) improvement [33], which is one of the main advantages of the Wilson twisted mass formulation of lattice QCD.

η and η' masses have been computed in this framework in Refs. [18], [19] and [34] on the same set of gauge configurations used here (and more) – we will refer to this framework as the unitary approach. However, in order to account for – and possibly benefit from – correlations we have re-evaluated the unitary η and η' masses on exactly the same gauge configurations as used in the present study.

The splitting term in the heavy doublet (2) introduces flavour mixing between strange and charm quarks which needs to be accounted for in the analysis. However, this complication can be avoided by using a mixed action approach for the valence strange and charm quarks. Formally, we introduce so-called Osterwalder–Seiler (OS) twisted valence strange and charm quarks [11] and [35]. The Dirac operator for a single valence quark flavour q reads

$$D_q = D_W + m_0 + i\mu_q\gamma_5. \tag{3}$$

Adapting the ideas of Ref. [11] to the η, η' system, we introduce two strange and two charm quark flavours, s, s′ and c, c′, respectively. Flavours s and s′ will have quark mass with equal modulus, but opposite sign: $\mu_s=|\mu_s|=-\mu_s'$, and the same for c and c′. Formally, the lattice action is extended to include a fermionic action corresponding to the Dirac operators (3) for all valence strange and charm quark flavours, accompanied by a ghost action to exactly cancel the contributions of the additional valence quarks to the fermionic determinant. For details we refer to Ref. [11]. In this reference it was also shown that automatic O(a)-improvement stays valid in this framework and unitarity is restored in the continuum limit. In particular, flavours s and s′ (c and c′) become identical.

It is important to notice that at finite lattice spacing values correlation functions involving s and s′ ($\mu_s = -\mu_s'$) differ by lattice artifacts. For instance, the masses extracted from the correlation function of the operator

$$\mathcal{O}_K^{\mathrm{OS}} = \bar{\psi}_s i\gamma_5 \psi_d \tag{4}$$

where the fields ψ_q, $\bar{\psi}_q$ denote single quark fields in the so-called physical basis, differ from the one extracted from the operator

$$\mathcal{O}_{K^0}^{\mathrm{OS}} = \bar{\psi}_{s'} i\gamma_5 \psi_d \tag{5}$$

by $O(a^2)$ (we denote it with K^0 in remedy of the neutral pion in the light sector). Only in the continuum limit these two masses will agree again.

Valence and unitary actions need to be matched appropriately. As shown in Ref. [11], in our case the matching can be performed in principle using the relation

$$\mu_{c/s} = \mu_\sigma \pm Z_P/Z_S\, \mu_\delta\,. \tag{6}$$

However, for the strange quark mass uncertainties in Z_P/Z_S are magnified in $a\mu_s$, and thus we decided not to rely on Eq. (6). Instead, meson masses are used: in previous studies it was found that matching kaon masses determined from the operator (4) to the unitary kaon masses is best in the sense that the residual lattice artifacts in the results computed in a mixed action approach are small [36]. We will call this procedure kaon matching.

For details on how to compute the kaon mass in the unitary case we refer to Ref. [37]. We note in passing that there is no kaon mass splitting introduced by the twisted mass formalism in the unitary case for the choice of a degenerate light quark doublet $|\mu_u|=|\mu_d|=\mu_\ell$[32].

As a second matching observable for the strange quark mass we use the mass of the so-called η_s meson $M_{\eta s}$. The η_s is an artificial meson corresponding to the following interpolating operator

$$\mathcal{O}_{\eta_s}^{\mathrm{OS}} = \bar{\psi}_s i\gamma_5 \psi_{s''}\,, \tag{7}$$

for which we assume $\mu_s = \mu_s''$, unlike the s′ quark considered above, which had opposite sign. A benefit of this particular choice is the absence of disconnected diagrams in the corresponding two-point function. This procedure will be called $\eta_s$matching. For technical details, e.g. further interpolating fields and correlation functions we refer to Section 3.

For both matching procedures on each gauge ensemble one tunes the value of $a\mu_s$such that the kaon or the η_s mass agrees within errors between the mixed and the unitary formulation. The unitary values of the masses we matched to are compiled in Table 2. In order to compute the matching values for $a\mu_s$ we performed inversions on a subset of the available configurations in a range of $a\mu_s$ values and interpolated the squared OS meson masses linearly in $a\mu_s$. The matching values for $a\mu_s$ for the two matching observables and all ensembles can be found in Table 3. The values for $M_K^{\mathrm{OS}}$and $M_{\eta_s}^{\mathrm{OS}}$ at the matching points are compiled in the appendix in Table A.12. Note that in case of matching $M_{\eta s}$ we do not reach exact agreement for all ensembles within errors when recomputing $M_{\eta_s}^{\mathrm{OS}}$ from full statistics. These numerically small differences become irrelevant for the η and η′ masses themselves due to the much larger statistical uncertainties introduced by the quark disconnected diagrams.

Table 2: Values of the unitary kaon and η_s masses M_K and $M_{\eta s}$ in lattice units, which have been used to match mixed and unitary actions. In addition, we give the mass values $M_{\pi^0_{\mathrm{conn}}}$ of the connected neutral pion which becomes identical to η_s for mass degenerate light and strange quarks. This SU(3) symmetric situation is realised approximately for the A80.24s and A100.24s ensembles. The kaon mass data shown in this table has first been published in Ref. [18]

Ensemble	aM_K	$aM_{\eta s}$	$aM_{\pi^0_{\mathrm{conn}}}$
A40.24	0.25884(43)	0.30708(60)	0.2375(25)
A60.24	0.26695(52)	0.31010(65)	0.2544(26)
A80.24	0.27706(61)	0.31406(46)	0.2659(25)
A100.24	0.28807(34)	0.31575(45)	0.2883(14)
A80.24s	0.25503(33)	0.27168(49)	0.2649(16)
A100.24s	0.26490(74)	0.27455(73)	0.2841(16)
B55.32	0.22799(34)	0.26087(33)	0.2177(10)
D45.32sc	0.17570(84)	0.21126(34)	0.1494(15)

Table 3: Matching values of the OS valence strange quark masses μ_s for kaon and $M_{\eta s}$ matching. The OS valence charm quark masses μ_c^K have been determined using Eq. (8) for kaon matching only

Ensemble	$a\mu_\ell$	$a\mu_s^K$	$a\mu_s^{\eta_s}$	$a\mu_c^K$
A40.24	0.0040	0.02300(25)	0.01239(25)	0.27700(25)
A60.24	0.0060	0.02322(22)	0.01303(22)	0.27678(22)
A80.24	0.0080	0.02328(20)	0.01338(20)	0.27672(20)
A100.24	0.0100	0.02381(21)	0.01380(22)	0.27619(21)
A80.24s	0.0080	0.01884(16)	0.00883(21)	0.28116(16)
A100.24s	0.0100	0.01877(22)	0.00922(23)	0.28123(22)

B55.32	0.0055	0.01858(12)	0.01100(10)	0.25142(12)
D45.32sc	0.0045	0.01488(30)	0.01180(12)	0.17252(30)

In the following we indicate quantities determined in the OS framework with the superscript ^{OS}, while quantities determined in the unitary case have no superscript. To distinguish the two matching procedures we use the superscripts K and η_s.

As an example for the matching procedure we show in Fig. 1 $(aM_K^{OS})^2$, $(aM_{\eta_s}^{OS})^2$ and $(aM_{K^0}^{OS})^2$ as a function of the bare OS strange quark mass $a\mu_s$ for the D45.32scensemble. In Fig. 1 the aforementioned OS kaon mass splitting can be observed. In the limit $\mu_s=\mu_\ell$ this splitting corresponds to the difference between the charged pion mass and the connected only neutral pion mass. The splitting is almost independent of μ_s, decreasing slightly with increasing μ_s.

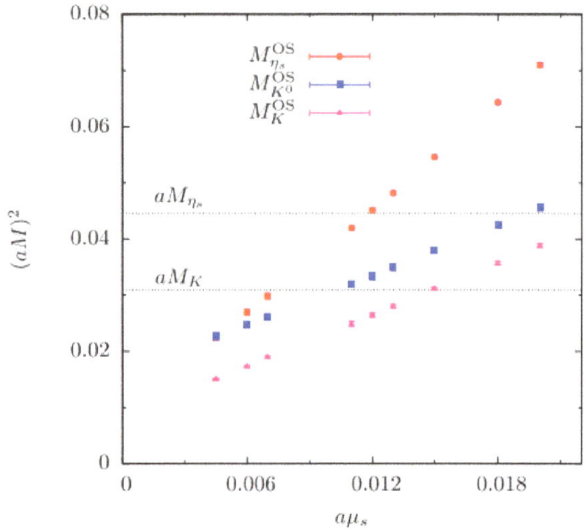

Figure 1: $(aM_K^{OS})^2$, $(aM_{\eta_s}^{OS})^2$ and $(aM_{K^0}^{OS})^2$ as a function of the bare OS strange quark mass $a\mu_s$ for the D45.32sc ensemble. Horizontal lines indicate the unitary mass values that have been used for the matching.

As expected, $M_{\eta s}$ is larger than the two kaon masses and agrees with $M_{K^0}^{OS}$ in the limit $\mu_s=\mu_\ell$. All three squared masses show a linear dependence on μ_s. The horizontal lines indicate the corresponding unitary values that have been used for computing $a\mu_s^K$ and $a\mu_s^{\eta_s}$.

For the charm quark mass the estimate from Eq. (6) is less affected by uncertainties. In order to circumvent the need for Z_P/Z_S, one can re-arrange Eq. (6) to

$$\mu_c = 2\mu_\sigma - \mu_s \,. \tag{8}$$

Because $\mu_s \ll \mu_c$ and η , η' do not depend on μ_c, we restrict ourselves to kaon matching for the μ_s value entering the charm quark mass. The corresponding values for $\mu_c \equiv \mu_c^K$ extracted in this way can be found in Table 3.

All errors are computed using a blocked bootstrap procedure to account for autocorrelation as well as all other statistical correlations in the data. The number of bootstrap samples was taken to be 1000 and the number of configurations per block N_b is given for every ensemble in Table 1. N_b itself was chosen such that the length of a block corresponds to at least 20 HMC trajectories of length one. This value turned out sufficient to compensate for autocorrelation in the observables considered in this study.

PSEUDO-SCALAR FLAVOUR-SINGLET MESONS

In order to extract η and η' states we need a set of appropriate interpolating operators. As we are going to work in the quark flavour basis our choice is

$$\mathcal{O}_\ell^p(t) = \frac{1}{\sqrt{2}}\sum_{\mathbf{x}}(\bar\psi_u i\gamma_5\psi_u(\mathbf{x},t) + \bar\psi_d i\gamma_5\psi_d(\mathbf{x},t))\,,$$

$$\mathcal{O}_s^p(t) = \sum_{\mathbf{x}} \bar\psi_s\, i\gamma_5\, \psi_s(\mathbf{x},t)\,,$$

$$\mathcal{O}_c^p(t) = \sum_{\mathbf{x}} \bar\psi_c\, i\gamma_5\, \psi_c(\mathbf{x},t)$$

in the physical basis, again denoted as $\bar\psi_q$, ψ_q. With Osterwalder–Seiler valence fermions we have to rotate the bilinears into the so-called twisted basis denoted as $\bar q$, q, see e.g. Ref. [38], in which also the Dirac operators in the previous section were written. Performing this axial rotation [11] and [12], one obtains the following operators in the so-called twisted basis

$$\mathcal{O}_\ell(t) = \frac{1}{\sqrt{2}}\sum_{\mathbf{x}}(\bar d d(\mathbf{x},t) - \bar u u(\mathbf{x},t))\,,$$

$$\mathcal{O}_s(t) = -\sum_{\mathbf{x}} \bar s s(\mathbf{x},t)\,,$$

$$\mathcal{O}_c(t) = -\sum_{\mathbf{x}} \bar c c(\mathbf{x},t)\,.$$

From these operators we build a correlation function matrix

$$\mathcal{C}(t)_{qq'} = \langle \mathcal{O}_q(t'+t)\, \mathcal{O}_{q'}^\dagger(t')\rangle\,, \quad q,q' \in \{\ell,s,c\}\,, \tag{9}$$

which allows us to obtain results for masses and amplitudes; cf. Section 3.2.

The corresponding correlation functions have fermionic connected and disconnected contributions. The case for up and down quarks is like in the unitary approach and discussed in detail in Refs. [18], [19] and [39]. Therefore, we concentrate on the disconnected contributions for strange and charm quarks. The correlation function of $O_s(t)$, for instance, has the following contributions

$$\langle \mathcal{O}_s(t)\,\mathcal{O}_s^\dagger(0)\rangle_F = -\mathrm{Tr}\{G_s^{0t}G_s^{t0}\} + \mathrm{Tr}\{G_s^{tt}\}\cdot\mathrm{Tr}\{G_s^{00}\}\,, \tag{10}$$

where $\langle . \rangle_F$ denotes the average of fermions only and

$$G_s^{xy} = (D_s^{-1})(x,y) \tag{11}$$

denotes the strange OS propagator. The first term in Eq. (10) is the connected contribution and the second the disconnected one. Note that mixed flavour correlation functions have only disconnected contributions by definition. The ground state mass extracted only from the connected piece on the r.h.s. of Eq. (10) is the mass of the artificial η_s meson, which is employed for η_s matching.

We evaluate the connected only contribution to Eq. (10) using the one-end-trick [40]. In contrast to the Wilson case, $\mathrm{Tr}\{G_s^{0t}G_s^{t0}\}$ is in general complex valued. However, the imaginary part of the corresponding trace is a pure lattice artifact. This can be shown by considering a suitable combination of connected correlation functions involving OS quarks s and s'

$$\begin{aligned}
\langle \bar\psi_s i\gamma_5\psi_s(x)\bar\psi_s i\gamma_5\psi_s(0) - \{s\to s'\}\rangle_F &= \langle \bar{s}s(x)\bar{s}s(0) - \bar{s}'s'(x)\bar{s}'s'(0)\rangle_F\\
&= -\mathrm{Tr}\{G_s^{0x}G_s^{x0}\} + \mathrm{Tr}\{G_{s'}^{0x}G_{s'}^{x0}\}\\
&= -\mathrm{Tr}\{G_s^{0x}\gamma_5(G_{s'}^{0x})^\dagger\gamma_5\} + \mathrm{Tr}\{G_{s'}^{0x}\gamma_5(G_s^{0x})^\dagger\gamma_5\}\\
&= -\mathrm{Tr}\{G_s^{0x}\gamma_5(G_{s'}^{0x})^\dagger\gamma_5\} + \mathrm{Tr}\{G_s^{0x}\gamma_5(G_{s'}^{0x})^\dagger\gamma_5\}^\dagger\\
&= -2i\,\mathrm{Im}\mathrm{Tr}\{G_s^{0x}\gamma_5(G_{s'}^{0x})^\dagger\gamma_5\}\\
&= -2i\,\mathrm{Im}\mathrm{Tr}\{G_s^{0x}G_s^{x0}\}\,,
\end{aligned}$$

where we have used the relation $D_s = \gamma_5 D_{s'}^\dagger \gamma_5$ together with the cyclic property of the trace. Since the l.h.s. of the above relation vanishes in the continuum limit, we will drop the imaginary part in our calculations.

For the disconnected contribution to Eq. (10) we need to estimate $\mathrm{Tr}\{G_s^{tt}\}$ on every gauge configuration and all t -values. $\mathrm{Tr}\{G_s^{tt}\}$ is again in general complex valued. And again, one can show that the real part is a pure lattice artifact. Similar to the case of the connected contribution this can be inferred from the following equality

$$\begin{aligned}
\langle -\bar\psi_s i\gamma_5\psi_s(x) + \bar\psi_{s'} i\gamma_5\psi_{s'}(x)\rangle_F &= \langle \bar{s}s(x) + \bar{s}'s'(x)\rangle_F\\
&= -\mathrm{Tr}\{G_s^{xx}\} - \mathrm{Tr}\{G_{s'}^{xx}\}\\
&= -\mathrm{Tr}\{G_s^{xx}\} - \mathrm{Tr}\{G_s^{xx}\}^\dagger\\
&= -2\,\mathrm{Re}\,\mathrm{Tr}\{G_s^{xx}\}\,,
\end{aligned}$$

which is zero in the continuum limit. Therefore, we will also drop the real part of the disconnected loops in the calculation. Similarly one can show that

$$\langle \bar{s}s(x) - \bar{s}'s'(x)\rangle_F = -2i\,\mathrm{Im}\,\mathrm{Tr}\{G_s^{xx}\}\,. \tag{12}$$

We remark that all of the above results hold for any further valence quark as well (e.g. the charm quark).

The full strange correlation function after subtraction of lattice artifacts is then given as

$$\langle \mathcal{O}_s(t)\,\mathcal{O}_s^\dagger(0)\rangle_F = -\mathrm{Re}\mathrm{Tr}\{G_s^{0t}G_s^{t0}\} - \mathrm{Im}\,\mathrm{Tr}\{G_s^{tt}\}\cdot\mathrm{Im}\,\mathrm{Tr}\{G_s^{00}\}\,, \tag{13}$$

and analogously for the charm. Cross flavour terms involve only disconnected diagrams and are for instance given as

$$\langle \mathcal{O}_s(t)\,\mathcal{O}_c^\dagger(0)\rangle_F = -\mathrm{Im}\,\mathrm{Tr}\{G_s^{tt}\}\cdot\mathrm{Im}\mathrm{Tr}\{G_c^{00}\}\,. \tag{14}$$

Variance reduction

The relation (12) enables us to use a very powerful variance reduction method developed originally for the disconnected contributions of the light doublet [39] also for strange and charm flavours (see also Ref. [41]). It is based on the identity (recall $\mu_s = -\mu_{s'}$)

$$D_s^{-1} - D_{s'}^{-1} = -2i\mu_s D_{s'}^{-1}\,\gamma_5\,D_s^{-1}\,.$$

Therefore, using Eq. (12) we can estimate

$$\mathrm{Im}\,\mathrm{Tr}\{G_s^{xx}\} = -\mu_s\mathrm{Tr}\{G_{s'}^{xy}\,\gamma_5\,G_s^{yx}\}\,, \tag{15}$$

and correspondingly for the charm quark. Following Ref. [39], we apply this variance reduction method also to the light doublet.

Matrix of correlation functions

By applying these results, we compute the matrix of Euclidean correlation functions in Eq. (9) and solve the generalised eigenvalue problem (GEVP) [42], [43] and [44]

$$\mathcal{C}(t)\, v^{(n)}(t, t_0) = \lambda^{(n)}(t, t_0)\, \mathcal{C}(t_0)\, v^{(n)}(t, t_0)\,, \quad (16)$$

for determining the meson masses M_η, M_η' (and possibly $M_{\eta c}$) from the principal correlators $\lambda^{(n)}(t,t_0)$, $n \in \eta, \eta'$. The effective masses are then computed by numerically solving

$$\frac{\lambda^{(n)}(t, t_0)}{\lambda^{(n)}(t+1, t_0)} = \frac{\exp^{-M_{\rm eff}^{(n)} t} + \exp^{-M_{\rm eff}^{(n)}(T-t)}}{\exp^{-M_{\rm eff}^{(n)}(t+1)} + \exp^{-M_{\rm eff}^{(n)}(T-(t+1))}}$$

for $aM_{\rm eff}^{(n)}$. The matrix C is enlarged to a 6×6 matrix by using in addition fuzzed [40] and [45] operators.

At this point we recall that in the unitary case also the mass of the η_s is not obtained from a single correlation function but rather from the ground state of a correlation function matrix involving connected correlation functions for strange and charm quarks. This minor complication arises due to the violation of flavour symmetry in the Wilson twisted mass formulation and the fact that the action can no longer be chosen flavour-diagonal for a non-degenerate doublet. Therefore, one has to consider off-diagonal connected correlation functions in addition to the ones consisting only of strange and charm quarks. However, the off-diagonal connected contributions are a pure lattice artifact and in the continuum limit the expected behaviour is restored, i.e. strange and charm sector decouple regarding the connected pieces.

Apart from meson masses also matrix elements can be extracted from the GEVP, which are needed to obtain η and η′ mixing angles. We define the mixing angles ϕ_ℓ, ϕ_s in the quark flavour basis using the pseudoscalar density matrix elements $A_{q,n} = \langle 0|O_q|n\rangle$ with $n \in \{\eta, \eta'\}$ and $q \in \{l, s\}$ as

$$\begin{pmatrix} A_{\ell,\eta} & A_{s,\eta} \\ A_{\ell,\eta'} & A_{s,\eta'} \end{pmatrix} = \begin{pmatrix} c_\ell \cos\phi_\ell & -c_s \sin\phi_s \\ c_\ell \sin\phi_\ell & c_s \cos\phi_s \end{pmatrix}, \quad (17)$$

see also Refs. [18] and [19]. From chiral perturbation theory combined with large N_C arguments $|\phi_\ell - \phi_s|/|\phi_\ell + \phi_s| \ll 1$ can be inferred according to Refs. [46], [47], [48] and [49] which is confirmed by lattice QCD [19]. Therefore, we will consider only the average mixing angle ϕ

$$\tan^2\phi \equiv -\frac{A_{l,\eta'} A_{s,\eta}}{A_{l,\eta} A_{s,\eta'}}\,. \quad (18)$$

Excited state removal

To improve the η′ (and η) mass determinations, we use a method first proposed in Ref.[50], successfully applied for the η_2 (the η′ in N_f=2 flavour QCD) in Ref. [39] and very recently to the N_f=2+1+1 case in Ref. [19]. It relies on the assumption that disconnected contributions are significant only for the η and η′ state, but negligible for higher excited states. This means, in turn, that only the connected contributions to C are affected by excited states.

Since the signal-to-noise ratio of the connected contributions is much larger than the one of the disconnected ones, we can determine the corresponding ground state at large Euclidean times and subtract the excited states at small times. This subtracted connected and the full disconnected contributions are combined in $C^{\rm sub}_{q,q'}(t)$, which is then used in the analysis. We refer to the discussion in Ref. [19] for more details.

This procedure clearly depends on the validity of the assumption. However, it can be checked in our Monte-Carlo data: if the subtracted combination of connected and disconnected contributions does not show excited states anymore, we take it as a strong hint for the validity of the assumption. This is the case for all our ensembles, and it was also the case in the unitary approach [19].

For all ensembles we observe that the mass of the η meson is unaffected by this procedure within errors. Only the error estimates get significantly smaller. This is also the case for the η′ where, however, the errors are quite significant before excited state removal.

As an example we show in Fig. 2 the effective masses of the two lowest-lying states for the A80.24s ensemble from η_s matching. The values shown in the left panel are obtained from the standard method using a 6×6 correlation function matrix build from local and fuzzed operators, whereas the right panel shows the results for a 3×3 local-only correlation function matrix with excited states removed from its connected contributions. Our fitted values are always obtained from a cosh-type fit to the respective principal correlators. The corresponding fit ranges are indicated by the bands in the plots. In general, our choice for the fit ranges $[t_1,t_2]$ and the values of t_0 for the GEVP are given in Table A.13. We remark that – since there is usually no clear plateau reached for the η′ state from the standard method – we apply a two-state fit to the corresponding principal correlator in this case. In all other cases we employed a single state fit in the plateau region.

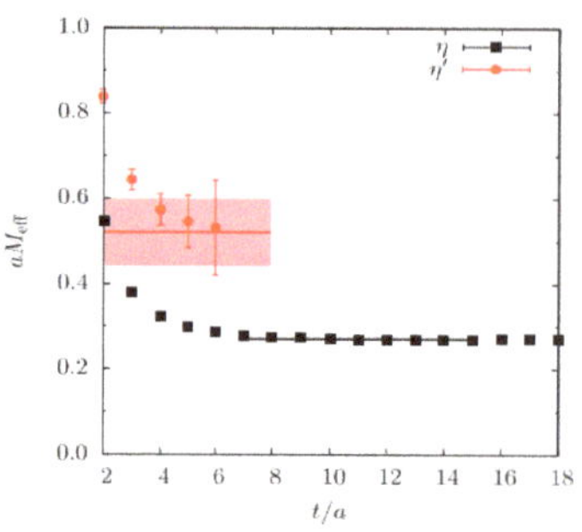

(a)

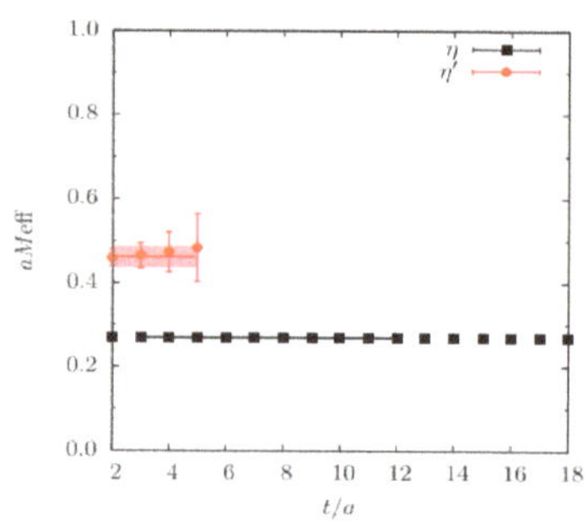

(b)

Figure 2: Effective masses $M_\eta^{\rm OS}$ and $M_{\eta'}^{\rm OS}$ for the A80.24s ensemble using η_s matching for (a) a 6 × 6 correlation matrix including local and fuzzed operators and (b) a 3 × 3 local-only correlation matrix with connected excited states subtracted. The fitted values are shown as lines with error band. The corresponding fit range is indicated by the length of the lines. For further details see text and Table A.13.

Comparing the two panels in Fig. 2 one observes that the η mass plateau is unaffected, but starts at t/a=2 in the case of removed excited states. For the η′ there is no plateau reached in the left panel before the signal is lost in noise, whereas in the right one a reasonable plateau is visible. The extracted masses still agree within errors.

RESULTS

In order to compare the mixed case with the unitary case we match the two actions as detailed in the previous sections using either the kaon or the η_s mass. Next we compute OS meson masses at these matching points. As an example we show in Fig. 3 the effective masses for the principal correlators $\lambda^{(n)}(t,t_0)$ of η and η′ as a function of t/a after removal of excited states from the connected contributions. For η and η′ a plateau in the effective masses is visible from early t/a on. The corresponding result of an exponential fit is indicated by the horizontal lines. The fit range corresponds to the extension of the lines in t/a. All OS meson masses are compiled in the appendix in Table A.7 and Table A.8.

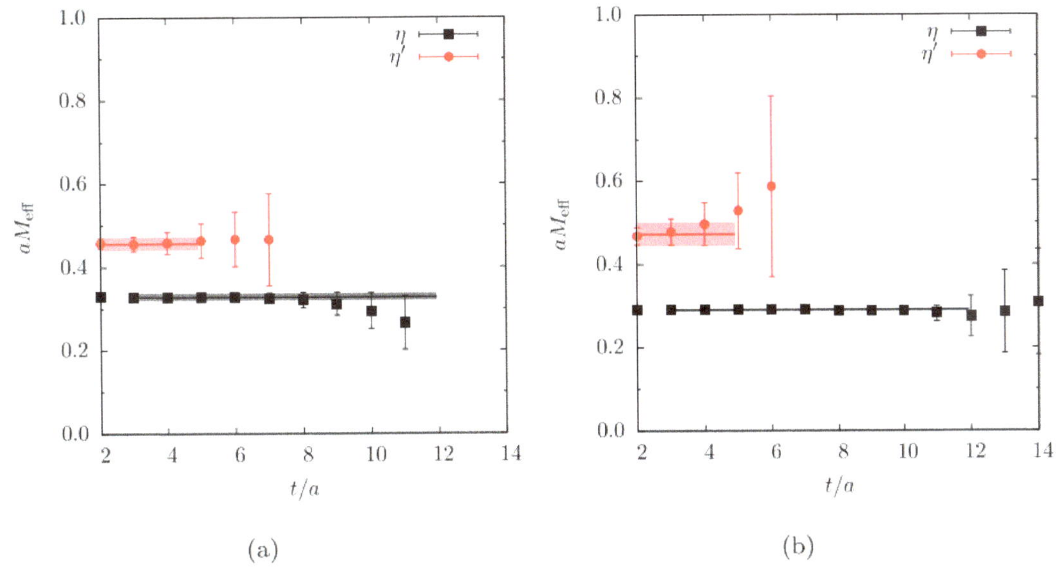

Figure 3: We show the effective masses aM_{η}^{OS} and $M_{\eta'}^{\mathrm{OS}}$ from a 3 × 3 correlation matrix after subtraction of excited states as described in the text for ensemble A 60.24. Panel (a) is for kaon matching and (b) for $\eta_s$matching.

The η_c decouples from η and η′ and the corresponding signal is lost in noise very early in t/a. Hence, we will not discuss it further here and due to the decoupling we will also not discuss the charm quark mass dependence of operators in the following.

It turns out that the choice of the matching variable makes a significant difference for the extracted value of M_{η}. Moreover, we always find $aM_{\eta}^{\eta_s} < aM_{\eta}^{K}$. On the other hand, the value of $M_{\eta'}$ is unaffected within statistical errors. We find this consistently for all the ensembles investigated; cf. Table A.7 and Table A.8.

In addition one observes $\phi^K < \phi^{\eta_s}$ by approximately 15°; cf. Table A.9. This results from a change in the overlap of mass and flavour eigenstates, leading to an increased light quark contribution to the η for kaon matching. Consequently, the light quark contribution to the η′ is reduced, while the respective strange quark contributions behave in the opposite way. Since most of the noise is introduced by the light quark disconnected diagrams in our calculation, this explains why $M_{\eta}^{\eta_s}$ in general exhibits a smaller statistical error than M_{η}^{K}, whereas the error for $M_{\eta'}^{\eta_s}$ is larger than the one for $M_{\eta'}^{K}$.

However, there is a tendency that kaon matching leads to worse plateaus than $\eta_s$matching. A particularly extreme case of this behaviour is shown in Fig. 4 for the B 55.32 ensemble. In the left panel the effective masses for the two lowest lying states from kaon matching are plotted. Clearly there is no plateau visible for the first excited state. For comparison and to guide the eye we show the situation in the unitary setup in the right panel, calculated on the same set of configurations. This is the only ensemble for which we cannot identify a plateau for the η′ safely. Therefore, we will not quote a value for the η′ mass for B55.32 and kaon matching.

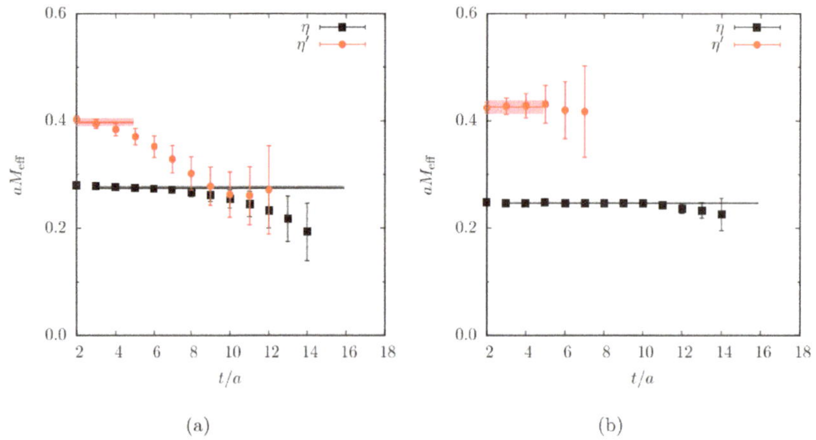

Figure 4: We show the effective masses aM_η and M_η' of the two lowest-lying states on the B55.32 ensemble for (a) kaon matching and (b) the unitary case using subtraction of excited states as described in the text.

Although the observed behaviour on B55.32 can still be interpreted as a statistical fluctuation, it might – in principle – also be caused by unitarity violation. However, it is neither possible to verify nor exclude the latter from our present data. While earlier studies [6] and [51] observed a sign flip in the scalar correlator signalling unitarity violation at least for a certain regime of valence quark masses, a similar argument cannot easily be extended to our case. The reason is that only the strange quark is treated in a mixed action approach while the light quarks are unitary. When looking at the scalar correlator made from strange quarks only, we do not observe it to become negative on any of our ensembles.

Another observation regarding the two matching methods concerns the behaviour of the ground states in the correlation functions used to build the full correlation function matrix. One expects all correlators with the same quantum numbers to asymptotically approach the same ground state mass. We observe this for the unitary case, where the η mass can be extracted from all correlators in the matrix C (diagonal and off-diagonal) at large Euclidean times. In the OS case we observe a similar behaviour for η_s matching, but for kaon matching e.g. the strange–strange correlator alone does often not reproduce the ηmass from the light–light correlator. This might signal unitarity violations for the kaon matching procedure on the one hand, and can explain the worse plateaus for this particular choice of the matching observable on the other hand.

Finally, for $M_\eta^{\eta_s}$ we observe the error to be reduced approximately by a factor of two with respect to the unitary result for all ensembles (cf. Table A.7). We attribute this to the fact that we can use the variance reduction method discussed in Section 3.1 for the OS strange quark, which is not possible for the unitary strange quark. However, the errors of $M_{\eta'}^{\eta_s}$ and ϕ^{η_s} do not show such an error reduction (cf. Table A.8 and Table A.9), presumably because the strange quark contributes little to these observables.

Light quark mass dependence

The first goal of this paper is to compare unitary to mixed action approaches and study the continuum limit of the corresponding differences. For this purpose we will study differences of quantities of the form $\Delta O = O^{OS} - O^{\text{unitary}}$. Our ensembles at different values of the lattice spacing are not at exactly identical light and strange sea quark masses. Therefore, we have to understand whether we can nevertheless study the continuum limit.

Theoretically, the answer to this question is yes: both in the unitary and in the mixed action approach we may write

$$O^{\text{lat}} = O^{\text{cont}} + \mathcal{O}(a^2\Lambda^2_{\text{QCD}})$$

and henceforth the aforementioned difference ΔO is $\mathcal{O}(a^2\Lambda^2_{\text{QCD}})$. A quark mass dependence is expected to be negligible because for the quark mass difference $\delta\mu \ll \Lambda_{\text{QCD}}$ holds. Since ΔO is always computed on identical gauge configurations there is no physical quark mass dependence that needs to be taken into account, because it cancels in the difference.

Despite this theoretical argument, let us also investigate this point numerically. We first study the light quark mass dependence of the difference between unitary and OS values of M_η and M_η'. For this purpose we focus on the A-ensembles A40, A60, A80 and A100, where we have different light quark mass values available. We denote

$$\Delta M_X = M_X^{\text{OS}} - M_X^{\text{unitary}}, \quad X = \eta, \eta'$$

the difference between unitary and OS meson masses. Analogously we define the angle difference $\Delta\phi$. $r_0\Delta M_\eta$ is shown for the A-ensembles in the left panel of Fig. 5as a function of $(r_0M_{PS})^2$. The chirally extrapolated values of r_0/a used in this study have been determined in [8] and are listed in Table 4. Filled circles represent the $\eta_s$matching results, filled boxes the corresponding kaon matching results. The differences are computed using exactly the same configurations leading to reduced statistical errors due to the correlation between unitary and OS data.

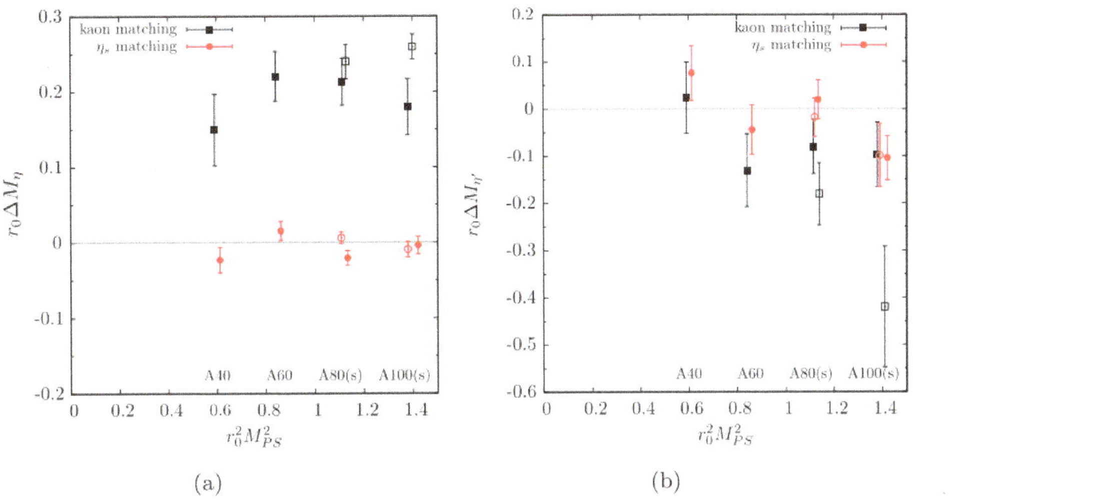

Figure 5: (a) $r_0\Delta M_\eta$ as a function of $(r_0M_{PS})^2$ for the four A ensembles A40, A60, A80(s) and A100(s). Open symbols correspond to the s-ensembles. (b) like (a), but for $r_0\Delta M_{\eta'}$. The kaon matching and s-ensemble data have been displaced horizontally for better legibility.

Table 4: The chirally extrapolated values for r_0/a at each value of β corresponding to the three different lattice spacings [8]

β	1.90	1.95	2.10
r_0/a	5.31(8)	5.77(6)	7.60(8)

For η_s matching $r_0\Delta M_\eta$ is for all four investigated ensembles compatible with zero within one sigma, while for kaon matching the difference is always positive and not compatible with zero. For both matching procedures, but in particular for η_s matching, the dependence on $(r_0M_{PS})^2$ is not significant within our statistical uncertainties.

In the right panel of Fig. 5 we show $r_0\Delta M_{\eta'}$ as a function of $(r_0M_{PS})^2$. Despite the larger uncertainties, the differences are compatible with zero for all ensembles and both matching procedures. There is a slight trend for differences with larger modulus for kaon matching. Like for the η the light quark mass dependence is not significant.

The angle difference Δϕ shows a very similar behaviour to ΔM_η, see Table A.10 and Table A.11. For η_s matching the difference is compatible with zero, while for kaon matching a value of about −15° is observed. Also here the light quark mass dependence is not significant. Besides, we find that the difference between ϕ_ℓ and ϕ_s is compatible with zero for both matching methods and compatible to the one found in the unitary setup [19], again confirming the smallness of OZI suppressed corrections.

In order to check the strange quark mass dependence of the differences ΔM and Δϕ we make use of the A80, A80s and A100, A100s ensembles. The s-ensembles differ from their non-s counterparts only by a different bare strange quark mass value. The corresponding values for the differences defined before are also displayed in Fig. 5 with open symbols. For η_s matching the differences show no dependence on the strange quark mass, whereas this cannot be concluded completely for kaon matching. In particular, we see for A100s deviations for kaon matching, but statistical errors can still account for the deviation.

Continuum limit

Next, we study the dependence on the lattice spacing. For this purpose we use the ensembles A60.24, B 55.32 and D45.32sc, which have approximately the same physical value of the pion mass, i.e. $r_0M_{PS}^{A60.24} = 0.917(14)_{\rm stat}$, $r_0M_{PS}^{B55.32} = 0.888(09)_{\rm stat}$ and $r_0M_{PS}^{D45.32sc} = 0.911(11)_{\rm stat}$, where we included the statistical error from the respective, chirally extrapolated values of r_0/a. As discussed in the previous section, we do not expect the residual differences in the light and strange quark masses to have any effect on this study.

The difference between OS and unitary results ΔM for both matching procedures is shown for η and η′ in Fig. 6 as a function of $(a/r_0)^2$ in the left and right panel, respectively. The lines represent linear fits in $(a/r_0)^2$ to our data, and the corresponding continuum extrapolated values are shown with open symbols.

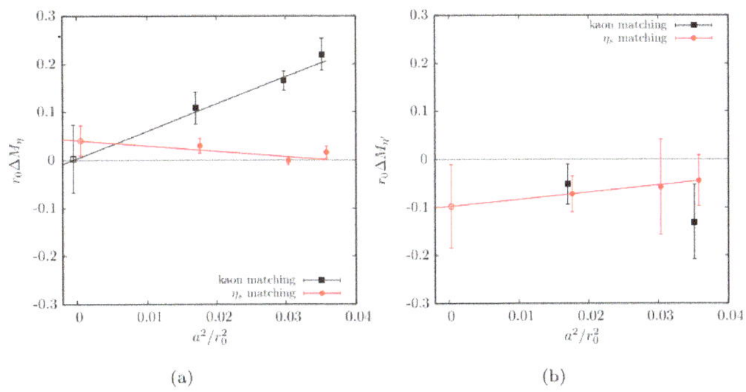

Figure 6: Continuum extrapolation of (a) $r_0\Delta M_\eta$ and (b) $r_0\Delta M_{\eta'}$ as a function of $(r_0/a)^2$. We show the continuum extrapolated results from a linear extrapolation of the three ensembles (D45.32sc, B55.32, A 60.24) in $(a/r_0)^2$ as open symbols. The continuum results are displaced horizontally for legibility.

$r_0\Delta M_\eta$ is shown in the left panel of Fig. 6. For both matching procedures we observe a linear dependence in $(a/r_0)^2$. A corresponding continuum extrapolation in $(a/r_0)^2$ leads to the expected vanishing of this difference at a=0 within errors. Kaon matching clearly exhibits larger a^2 artifacts, while η_s matching gives $r_0\Delta M_\eta$ compatible with zero for each value of the lattice spacing separately.

In the right panel of Fig. 6 we show $\Delta M_{\eta}'$, again for both matching procedures. We remark that for kaon matching it is not possible to perform a fit from our present data, because of the missing mass value on the B 55.32 ensemble which is due to a bad plateau, as discussed above. In this case, statistical errors are significantly larger. However, within their larger errors the difference for the two matching procedures seems compatible and the difference vanishes in the continuum limit for η_s matching, as indicated by the fitted line in the plot. In contrast to the η mass, it cannot be concluded that lattice artifacts for kaon matching are significantly larger than for η_s matching.

In the left panel of Fig. 7 $\Delta\phi$ is shown as a function of $(a/r_0)^2$, again for the ensembles A60.24, B 55.32 and D45.32sc. Like for ΔM_{η} we observe also for $\Delta\phi$ larger differences for kaon matching compared to η_s matching. For η_s matching the difference is in fact compatible with zero for all three ensembles separately. For both matching procedures the continuum extrapolated values are compatible with zero.

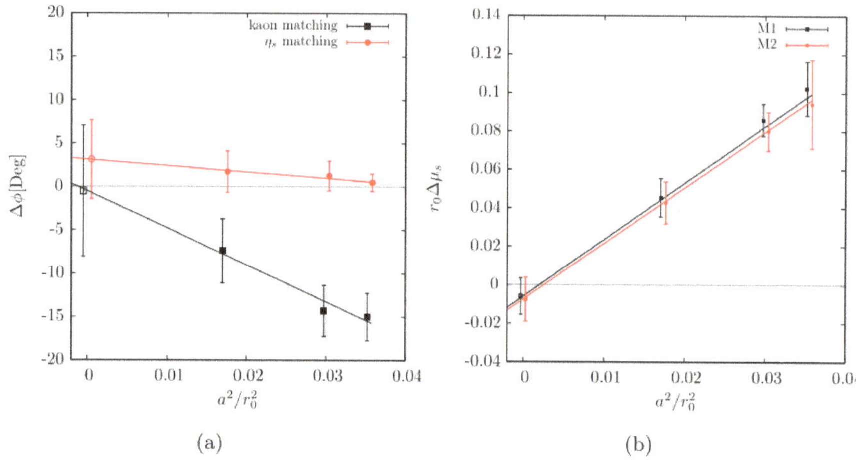

Figure 7: (a) like Fig. 6, but for the mixing angle difference $\Delta\phi$. In (b) we show $r_0\Delta\mu_s$ at 2 GeV in the $\overline{\text{MS}}$ scheme as a function of $(a/r_0)^2$ for the two methods M1 and M2 to estimate Z_P presented in Ref. [8].

Finally, we show in the right panel of Fig. 7 the quark mass difference

$$\Delta\mu_s = \frac{1}{Z_P}(\mu_s^K - \mu_s^{\eta_s}) \tag{19}$$

as a function of $(a/r_0)^2$ for the three ensembles A60.24, B 55.32 and D45.32sc at 2 GeV in the $\overline{\text{MS}}$ scheme. The renormalisation constant Z_P has been taken from Ref. [8]. The two colours correspond to the methods M1 and M2 for estimating Z_P. We refer to Ref. [8] for the details. In the continuum limit it is expected that the two matching conditions agree and the difference should vanish like a^2. This is what is confirmed by Fig. 7(b). Also, the two methods M1 and M2 give compatible results in the continuum limit, as expected.

Dependence on sea and valence strange quark mass

Next we study the dependence of the η (and in principle also the η') meson mass on the valence and sea quark mass values. As said in the introduction, the dependence on the valence and sea quark masses must be identical (at least within errors) to legitimate re-tuning in the valence quark masses only against sea strange quark mass mismatches. For this purpose we first define the dimensionless quantity

$$D_\eta^{\text{val}} = \left.\frac{d(M_\eta^{\text{OS}})^2}{d(M_K^{\text{OS}})^2}\right|_{\text{fixed sea ensemble}}, \tag{20}$$

which can be computed using the two matching points we have available for each ensemble. For estimating D_η^{val} from two μ_s values at each ensemble, we have to assume that $(M_\eta^{\text{OS}})^2$ depends linearly on $(M_K^{\text{OS}})^2$ to a good approximation. That this is the case can be seen in Fig. 8, where we show $(M_\eta^{\text{OS}})^2$ as a function of $(M_K^{\text{OS}})^2$ for the two ensembles A80.24 and A80.24s. The lines represent linear fits to our data, which describe the data well within errors.

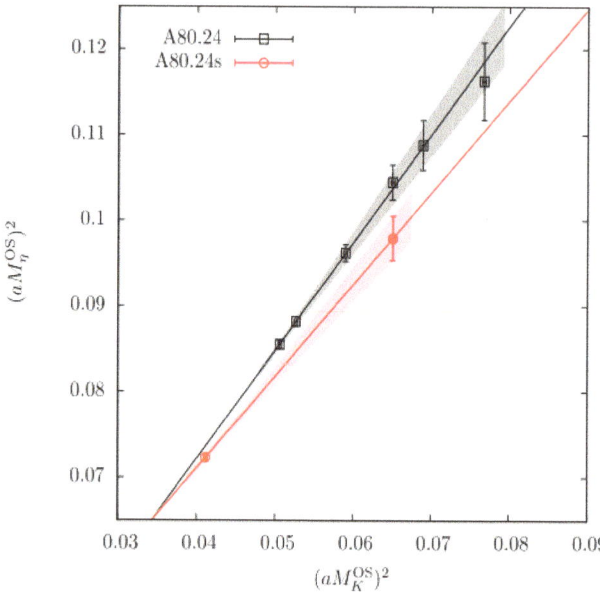

Figure 8: $(M_\eta^{\text{OS}})^2$ as a function of $(M_K^{\text{OS}})^2$ for ensembles A80.24 and A80.24s.

We expect D_η^{val} to be mostly sensitive to the valence strange quark mass if computed for several valence μ_s-values on the same ensemble. The results for D_η^{val} are compiled in Table 5 for all ensembles and η_s matching. They appear – independently of the lattice spacing, light and strange quark mass values – to be all compatible with 1. Taking the weighted average we obtain $D_\eta^{val} = 1.09(5)$, where the error is purely statistical. Concerning possible systematics we stress that there is no trend visible from the data, e.g. regarding a quark mass or lattice spacing dependence.

Table 5: The valence derivative D_η^{val} obtained in the OS case by using the mass values from the kaon and η_s matching points

Ensemble	D_η^{val}
A40.24	0.76(22)
A60.24	0.97(17)
A80.24	1.17(17)
A100.24	0.92(18)
A80.24s	1.07(11)
A100.24s	1.32(08)
B55.32	0.93(09)
D45.32sc	0.86(31)

Including the sea strange quark mass dependence, the corresponding derivative is given by

$$D_\eta = \frac{d(M_\eta)^2}{d(M_K)^2}\,.$$

For the unitary case we find D_η=1.45(10) and in the OS case $D_\eta^{OS} = 1.46(6)$, using the A100.24, A100.24s and A80.24, A80.24s ensembles and η_s matching, see Table 6. While D_η and D_η^{OS} are compatible within errors, they differ significantly from D_η^{val}. For kaon matching the relative statistical errors on D_η^{OS} turn out to be at least a factor five larger than for η_s matching. Therefore, a meaningful statement regarding the compatibility of D_η^{OS} with D_η and D_η^{val} for kaon matching is not possible from our current data.

Table 6: We list the values for D_η evaluated both for the unitary and the OS case using the two A80 and the two A100 ensembles. This derivative includes, in contrast to D_η^{val}, the valence and sea strange quark mass dependence. The values for D_η^{OS} are for η_s matching

Ensemble	D_η	D_η^{OS}
A80/A80s	1.54(13)	1.37(07)
A100/A100s	1.34(15)	1.67(11)

We take this as an indication that M_η is indeed a quantity with a significant sea strange quark mass dependence. Therefore, correcting for mismatches of the sea strange quark mass value in the valence sector only is not enough for M_η.

In principle, the difference between D_η and D_η^{val} that we found for a single lattice spacing could also be a lattice artifact. We do not think this is the case for two reasons: first, in Ref. [18] the value of D_η was used to correct a mismatch in the strange quark mass tuning for all three lattice spacings available. And we did not observe large cut-off effects introduced by this procedure. Second, also D_η^{OS} is merely independent of the lattice spacing. In fact, as the D_X are computed from differences and the leading lattice artifacts are independent of the quark mass, it is expected that these quantities are not plagued by large cut-off effects.

Unfortunately, the statistical uncertainty on the η′ meson masses is too large to allow for a meaningful investigation of D_η'. Within errors this quantity is always zero, irrespective of whether the valence or the full strange quark mass dependence is considered. Moreover, M_η' has a larger light than strange quark contribution. It would, therefore, be interesting to perform the same study for the valence light quark mass instead of the valence strange.

Finally, we remark that it is in principle possible to calculate the difference between D_η and D_η^{val} from chiral perturbation theory [52], [53] and [54]. At leading order the corresponding prediction is 4/3 for both derivatives, implying that the difference is an NLO effect.

SUMMARY AND DISCUSSION

In this paper we have studied η and η′ mesons in a mixed action approach and in comparison to the unitary results. The mixed action was so-called Osterwalder–Seiler fermions on a twisted mass sea with N_f=2+1+1 dynamical quark flavours.

We have found that indeed the difference between mixed and unitary results vanishes as the continuum limit is approached. The rate is as expected of $O(a^2)$[11] for all quantities and matching procedures investigated in this paper.

For the η mass we find a significant dependence of the size of the cutoff effects on the matching procedure. Lattice artifacts in the difference to the unitary result are compatible with zero when the two actions are matched using the η_s meson, while they are of normal size when the kaon is used as a matching variable. The same is true for the mixing angle and the strange quark mass.

For M_η' we do not observe a strong dependence on the matching procedure. This can have two reasons: first the error of M_η' is large making precise statements difficult. Second, the η′ receives a strong contribution from sea quarks, because it is mainly the singlet state. The sea quark contributions are unaffected by different choices of the valence strange quark mass. Hence, this finding might reflect the physical properties of the η′ meson.

This shows that the mixed action approach can also be applied in practice for flavour singlet quantities and, more generally, for observables involving fermionic disconnected diagrams. In case of η_s matching we find even reduced statistical errors for the η mass which might turn out to be an important advantage of the mixed approach. Thus, we will use the mixed action to investigate more complicated problems like $\eta\to\gamma\gamma$ form factors or Kπ scattering for I=1/2 in the future.

Another important result of this paper is that for the η meson it is not sufficient to re-tune the valence quark masses to correct for small mismatches in the simulation runs. Our data shows that the valence strange quark mass dependence of the η differs significantly from the dependence on the sea plus valence strange quark mass. And this difference is not vanishing as the continuum limit is approached. For the η′ we cannot make such a statement due to too large statistical errors, but we expect a similar result once M_{η}' can be determined with higher accuracy.

The latter finding seems to contradict previous studies where the valence quark masses have been re-tuned to their physical values in the same mixed action approach for instance to determine non-singlet pseudo-scalar decay constants or quark masses (see for instance Refs. [8], [55] and [56]). However, these investigations were concerned with observables for which the quark mass dependence is expected to be mainly governed by the valence quarks. For the η and η′ mesons studied here this is not the case as OZI violating contributions are anomalously large. However, with high enough accuracy the effect seen here should also show up in other physical quantities, but on the current level of precision it is likely to be a negligible systematic uncertainty.

ACKNOWLEDGEMENTS

We thank the members of ETMC for the most enjoyable collaboration. The computer time for this project was made available to us by the John von Neumann Institute for Computing (NIC) on the JUDGE, Jugene and Juqueen systems in Jülich. In particular we thank U.-G. Meißner for granting us access on JUDGE. We thank C. Michael and S. Simula for useful comments. We thank R. Frezzotti for useful discussions and very helpful comments on the draft of this paper. This project was funded by the DFG as a project in the SFB/TR 16 and in part by the Sino–German CRC110. Two of the authors (K.O. and C.U.) were supported by the Bonn–Cologne Graduate School (BCGS) of Physics and Astronomy. The open source software packages tmLQCD [57], Lemon [58] and R [59] have been used.

APPENDIX A. DATA TABLES

In this appendix we have compiled all data in tables for convenience.

Table A.7: Results for aM_{η} for the unitary and the mixed action approach. For the latter we show the values corresponding to η_s and kaon matching

Ensemble	aM_{η}	$aM_{\eta}^{\eta_s}$	aM_{η}^{K}
A40.24	0.2837(47)	0.2793(35)	0.312(12)
A60.24	0.2870(28)	0.2899(16)	0.3285(74)
A80.24	0.3009(20)	0.2970(09)	0.3410(67)
A100.24	0.3074(23)	0.3067(13)	0.3412(75)
A80.24s	0.2678(13)	0.2690(07)	0.3130(42)
A100.24s	0.2759(17)	0.2741(10)	0.3247(30)
B55.32	0.2467(12)	0.2465(07)	0.2753(34)
D45.32sc	0.1890(34)	0.1929(39)	0.2032(72)

Table A.8: Same as Table A.7, but for aM_{η}'

Ensemble	aM_{η}'	$aM_{\eta'}^{\eta_s}$	$aM_{\eta'}^{K}$
A40.24	0.443(27)	0.457(28)	0.448(15)
A60.24	0.482(27)	0.474(27)	0.458(15)
A80.24	0.481(26)	0.485(29)	0.466(17)
A100.24	0.461(24)	0.441(21)	0.442(13)
A80.24s	0.465(23)	0.461(25)	0.431(13)
A100.24s	0.542(42)	0.523(40)	0.463(20)
B55.32	0.425(12)	0.415(12)	bad plateau
D45.32sc	0.278(12)	0.269(12)	0.271(09)

Table A.9: Same as Table A.7, but for the mixing angle ϕ

Ensemble	ϕ	ϕ^{η}_{s}	ϕ^{K}
A40.24	47.1(2.0)	47.3(2.2)	30.6(4.1)
A60.24	49.2(1.4)	49.8(1.6)	34.2(3.7)
A80.24	49.9(1.1)	51.1(1.1)	38.0(3.5)
A100.24	49.6(1.3)	50.3(1.3)	31.9(4.1)
A80.24s	51.3(0.8)	52.6(0.5)	36.2(3.2)
A100.24s	53.8(0.7)	55.1(0.6)	44.2(2.7)
B55.32	48.2(0.8)	49.5(1.3)	34.0(2.9)
D45.32sc	45.6(2.5)	47.3(4.4)	38.2(5.8)

Table A.10: Values for $a\Delta M_{\eta}$, $a\Delta M_{\eta}'$ and $\Delta\phi$ [deg] as defined in the main text for kaon matching procedure

Ensemble	$a\Delta M_{\eta}$	$a\Delta M_{\eta}'$	$\Delta\phi$ [deg]
A40.24	+0.0281(89)	+0.004(14)	−16.5(2.6)
A60.24	+0.0415(61)	−0.024(15)	−15.0(2.7)
A80.24	+0.0401(58)	−0.015(11)	−11.9(2.6)
A100.24	+0.0338(69)	−0.018(13)	−17.7(3.3)
A80.24s	+0.0452(43)	−0.034(12)	−15.1(2.8)
A100.24s	+0.0489(31)	−0.079(24)	−9.5(2.7)
B55.32	+0.0286(35)	N/A	−14.3(2.9)
D45.32sc	+0.0142(44)	−0.007(06)	−7.4(3.7)

Table A.11: Same as Table A.10, but for η_s matching procedure

Ensemble	$a\Delta M_{\eta}$	$a\Delta M_{\eta}'$	$\Delta\phi$ [deg]
A40.24	−0.0044(32)	+0.014(11)	+0.2(1.2)
A60.24	+0.0029(24)	−0.008(10)	+0.5(1.0)
A80.24	−0.0039(18)	+0.004(08)	−1.2(0.6)

A100.24	−0.0007(22)	−0.020(09)	+0.7(1.0)
A80.24s	+0.0011(15)	−0.004(08)	+1.3(0.9)
A100.24s	−0.0018(20)	−0.019(13)	+1.3(0.8)
B55.32	−0.0002(14)	−0.010(17)	+1.3(1.5)
D45.32sc	+0.0039(20)	−0.010(05)	+1.8(2.4)

Table A.12: OS kaon and η_s mass values for both matching procedures

Ensemble	aM_K^K	$aM_K^{\eta_s}$	$aM_{\eta_s}^K$	$aM_{\eta_s}^{\eta_s}$
A40.24	0.2583(16)	0.2031(17)	0.3824(13)	0.3046(18)
A60.24	0.2669(09)	0.2161(09)	0.3838(10)	0.3123(14)
A80.24	0.2770(06)	0.2297(06)	0.3860(08)	0.3126(12)
A100.24	0.2880(08)	0.2422(09)	0.3923(11)	0.3178(16)
A80.24s	0.2551(07)	0.2027(07)	0.3533(09)	0.2720(12)
A100.24s	0.2649(07)	0.2173(07)	0.3538(12)	0.2695(18)
B55.32	0.2280(04)	0.1893(04)	0.3221(04)	0.2640(06)
D45.32sc	0.1758(10)	0.1618(10)	0.2319(04)	0.2096(05)

Table A.13: List of fit parameters. $t_1^{\eta_{l,s}}$ and $t_2^{\eta_{l,s}}$ define the fit intervals for the ground state of the connected correlation function in the light (strange) sector, which is required for the subtraction of excited states from the full correlator, before solving the GEVP for the resulting correlation function matrix. The t_0 values for the GEVP used to determine the η , η′ states are given by $t_0^{\eta,\eta'}$ whereas $t_1^{\eta,\eta'}$, $t_2^{\eta,\eta'}$ denote the respective fit ranges to the principal correlators. In the last row we give the parameters for the GEVP applied to the full 6 × 6 correlation function matrix of the A 80.24 ensemble for the η_s matching case, as shown in the left panel of Fig. 2

Ensemble	$t_1^{\eta_\ell}$	$t_2^{\eta_\ell}$	$t_1^{\eta_s}$	$t_2^{\eta_s}$	t_0^{η}	t_1^{η}	t_2^{η}	$t_0^{\eta'}$	$t_1^{\eta'}$	$t_2^{\eta'}$
A-ensembles	12	22	12	22	1	3	12	1	2	5
B55.32	15	25	15	25	1	3	16	1	2	5
D45.32sc	18	30	18	30	1	3	16	1	2	5
A80.24$^{6\times6}$					2	7	15	1	2	8

REFERENCES

1. O. Bär, G. Rupak, N. Shoresh, Simulations with different lattice Dirac operators for valence and sea quarks, Phys. Rev. D 67 (2003) 114505, http://dx.doi.org/10.1103/PhysRevD.67.114505.
2. H. Neuberger, Exactly massless quarks on the lattice, Phys. Lett. B 417 (1998) 141–144, http://dx.doi.org/10.1016/S0370-2693(97)01368-3.
3. H. Neuberger, More about exactly massless quarks on the lattice, Phys. Lett. B 427 (1998) 353–355, http://dx.doi.org/10.1016/S0370-2693(98)00355-4.
4. O. Bar, M. Golterman, Y. Shamir, Flavor symmetry breaking in lattice QCD with a mixed action, Phys. Rev. D 83 (2011) 054501, http://dx.doi.org/10.1103/PhysRevD.83.054501.
5. D.B. Kaplan, A method for simulating chiral fermions on the lattice, Phys. Lett. B 288 (1992) 342–347, http://dx.doi.org/10.1016/0370-2693(92)91112-M.
6. K. Bowler, B. Joo, R. Kenway, C. Maynard, R. Tweedie, Lattice QCD with mixed actions, J. High Energy Phys. 0508 (2005) 003, http://dx.doi.org/10.1088/1126-6708/2005/08/003.
7. K. Cichy, V. Drach, E. Garcia-Ramos, G. Herdoiza, K. Jansen, Overlap valence quarks on a twisted mass sea: a case study for mixed action lattice QCD, Nucl. Phys. B 869 (2013) 131–163, http://dx.doi.org/10.1016/j.nuclphysb.2012.12.011.
8. N. Carrasco, et al., Up, down, strange and charm quark masses with Nf = 2 + 1 + 1 twisted mass lattice QCD, Nucl. Phys. B 887 (2014) 19–68, http://dx.doi.org/10.1016/j.nuclphysb.2014.07.025.

9. W.A. Bardeen, A. Duncan, E. Eichten, N. Isgur, H. Thacker, Chiral loops and ghost states in the quenched scalar propagator, Phys. Rev. D 65 (2001) 014509, http://dx.doi.org/10.1103/PhysRevD.65.014509.

10. W.A. Bardeen, E. Eichten, H. Thacker, Chiral Lagrangian parameters for scalar and pseudoscalar mesons, Phys. Rev. D 69 (2004) 054502, http://dx.doi.org/10.1103/PhysRevD.69.054502.

11. R. Frezzotti, G.C. Rossi, Chirally improving Wilson fermions. II: four-quark operators, J. High Energy Phys. 10 (2004) 070, http://dx.doi.org/10.1088/1126-6708/2004/10/070.

12. R. Frezzotti, P.A. Grassi, S. Sint, P. Weisz, Lattice QCD with a chirally twisted mass term, J. High Energy Phys. 08 (2001) 058.

13. E. Witten, Current algebra theorems for the U(1) Goldstone boson, Nucl. Phys. B 156 (1979) 269, http://dx.doi.org/10.1016/0550-3213(79)90031-2.

14. G. Veneziano, U(1) without instantons, Nucl. Phys. B 159 (1979) 213–224, http://dx.doi.org/10.1016/ 0550-3213(79)90332-8.

15. G. Veneziano, Goldstone mechanism from gluon dynamics, Phys. Lett. B 95 (1980) 90, http://dx.doi.org/10.1016/0370-2693(80)90406-2.

16. E.B. Gregory, A.C. Irving, C.M. Richards, C. McNeile, Pseudoscalar flavor-singlets and staggered fermions, PoS LAT2006 (2006) 176.

17. E.B. Gregory, A.C. Irving, C.M. Richards, C. McNeile, A study of the η and η mesons with improved staggered fermions, Phys. Rev. D 86 (2012) 014504, http://dx.doi.org/10.1103/PhysRevD.86.014504.

18. K. Ottnad, et al., η and η mesons from Nf=2+1+1 twisted mass lattice QCD, J. High Energy Phys. 1211 (2012) 048, http://dx.doi.org/10.1007/JHEP11(2012)048.

19. C. Michael, K. Ottnad, C. Urbach, η and η mixing from lattice QCD, Phys. Rev. Lett. 111 (2013) 181602, http://dx.doi.org/10.1103/PhysRevLett.111.181602.

20. G. Bennett, et al., Final report of the muon E821 anomalous magnetic moment measurement at BNL, Phys. Rev. D 73 (2006) 072003, http://dx.doi.org/10.1103/PhysRevD.73.072003.

21. T. Aoyama, M. Hayakawa, T. Kinoshita, M. Nio, Complete tenth-order QED contribution to the muon g-2, Phys. Rev. Lett. 109 (2012) 111808, http://dx.doi.org/10.1103/PhysRevLett.109.111808.

22. M. Davier, A. Hoecker, B. Malaescu, Z. Zhang, Reevaluation of the hadronic contributions to the muon g-2 and to α(MZ), Eur. Phys. J. C 71 (2011) 1515, http://dx.doi.org/10.1140/epjc/s10052-012-1874-8, http://dx.doi.org/ 10.1140/epjc/s10052-010-1515-z.

23. K. Hagiwara, R. Liao, A.D. Martin, D. Nomura, T. Teubner, (g − 2)μ and α(M2 Z) re-evaluated using new precise data, J. Phys. G 38 (2011) 085003, http://dx.doi.org/10.1088/0954-3899/38/8/085003.

24. K. Cichy, V. Drach, E.G. Ramos, K. Jansen, C. Michael, et al., Properties of pseudoscalar flavour-singlet mesons from 2 + 1 + 1 twisted mass lattice QCD, PoS LATTICE2012 (2012) 151.

25. N. Christ, C. Dawson, T. Izubuchi, C. Jung, Q. Liu, et al., The η and η mesons from lattice QCD, Phys. Rev. Lett. 105 (2010) 241601, http://dx.doi.org/10.1103/PhysRevLett.105.241601.

26. T. Kaneko, et al., Flavor-singlet mesons in N(f) = 2 + 1 QCD with dynamical overlap quarks, PoS LAT2009 (2009) 107.

27. J.J. Dudek, R.G. Edwards, B. Joo, M.J. Peardon, D.G. Richards, et al., Isoscalar meson spectroscopy from lattice QCD, Phys. Rev. D 83 (2011) 111502, http://dx.doi.org/10.1103/PhysRevD.83.111502.

28. J.J. Dudek, R.G. Edwards, P. Guo, C.E. Thomas, Toward the excited isoscalar meson spectrum from lattice QCD, Phys. Rev. D 88 (9) (2013) 094505, http://dx.doi.org/10.1103/PhysRevD.88.094505.

29. G.S. Bali, S. Collins, S. Dürr, I. Kanamori, Ds → η, η semileptonic decay form factors with disconnected quark loop contributions, Phys. Rev. D 91 (1) (2015) 014503, http://dx.doi.org/10.1103/PhysRevD.91.014503, arXiv: 1406.5449

30. R. Baron, et al., Light hadrons from lattice QCD with light (u, d), strange and charm dynamical quarks, J. High Energy Phys. 06 (2010) 111, http://dx.doi.org/10.1007/JHEP06(2010)111.

31. R. Frezzotti, G.C. Rossi, Twisted-mass lattice QCD with mass non-degenerate quarks, Nucl. Phys. B, Proc. Suppl. 128 (2004) 193–202.

32. T. Chiarappa, F. Farchioni, K. Jansen, I. Montvay, E. Scholz, et al., Numerical simulation of QCD with u, d, s and c quarks in the twisted-mass Wilson formulation, Eur. Phys. J. C 50 (2007) 373–383, http://dx.doi.org/10.1140/ epjc/s10052-006-0204-4.

33. R. Frezzotti, G.C. Rossi, Chirally improving Wilson fermions. I: O(a) improvement, J. High Energy Phys. 08 (2004) 007.

34. K. Ottnad, Properties of pseudoscalar flavor singlet mesons from lattice QCD, Ph.D. thesis, University of Bonn, Bonn, February 2014.

35. B. Blossier, et al., Light quark masses and pseudoscalar decay constants from Nf = 2 lattice QCD with twisted mass fermions, J. High Energy Phys. 04 (2008) 020, http://dx.doi.org/10.1088/1126-6708/2008/04/020.

36. R. Frezzotti, G. Martinelli, M. Papinutto, G. Rossi, Reducing cutoff effects in maximally twisted lattice QCD close to the chiral limit, J. High Energy Phys. 0604 (2006) 038, http://dx.doi.org/10.1088/1126-6708/2006/04/038.

37. R. Baron, et al., Computing K and D meson masses with Nf = 2 + 1 + 1 twisted mass lattice QCD, Comput. Phys. Commun. 182 (2011) 299–316, http://dx.doi.org/10.1016/j.cpc.2010.10.004.

38. A. Shindler, Twisted mass lattice QCD, Phys. Rep. 461 (2008) 37–110, http://dx.doi.org/10.1016/j.physrep.2008.03.001.

39. K. Jansen, C. Michael, C. Urbach, The eta-prime meson from lattice QCD, Eur. Phys. J. C 58 (2008) 261–269, http://dx.doi.org/10.1140/epjc/s10052-008-0764-6.

40. P. Boucaud, et al., Dynamical twisted mass fermions with light quarks: simulation and analysis details, Comput. Phys. Commun. 179 (2008) 695–715, http://dx.doi.org/10.1016/j.cpc.2008.06.013.
41. C. Alexandrou, M. Constantinou, V. Drach, K. Hadjiyiannakou, K. Jansen, et al., Evaluation of disconnected quark loops for hadron structure using GPUs, Comput. Phys. Commun. 185 (2014) 1370–1382, http://dx.doi.org/ 10.1016/j.cpc.2014.01.009.
42. C. Michael, I. Teasdale, Extracting glueball masses from lattice QCD, Nucl. Phys. B 215 (1983) 433, http://dx.doi.org/10.1016/0550-3213(83)90674-0.
43. M. Lüscher, U. Wolff, How to calculate the elastic scattering matrix in two-dimensional quantum field theories by numerical simulation, Nucl. Phys. B 339 (1990) 222–252, http://dx.doi.org/10.1016/0550-3213(90)90540-T.
44. B. Blossier, M. Della Morte, G. von Hippel, T. Mendes, R. Sommer, On the generalized eigenvalue method for energies and matrix elements in lattice field theory, J. High Energy Phys. 0904 (2009) 094, http://dx.doi.org/10.1088/1126-6708/2009/04/094.
45. P. Lacock, A. McKerrell, C. Michael, I.M. Stopher, P.W. Stephenson, Efficient hadronic operators in lattice gauge theory, Phys. Rev. D 51 (1995) 6403–6410.
46. R. Kaiser, H. Leutwyler, Pseudoscalar decay constants at large N(c), arXiv:hep-ph/9806336, 1998.
47. R. Kaiser, H. Leutwyler, Large N(c) in chiral perturbation theory, Eur. Phys. J. C 17 (2000) 623–649, http:// dx.doi.org/10.1007/s100520000499.
48. T. Feldmann, P. Kroll, B. Stech, Mixing and decay constants of pseudoscalar mesons: the sequel, Phys. Lett. B 449 (1999) 339–346, http://dx.doi.org/10.1016/S0370-2693(99)00085-4.
49. T. Feldmann, P. Kroll, B. Stech, Mixing and decay constants of pseudoscalar mesons, Phys. Rev. D 58 (1998) 114006, http://dx.doi.org/10.1103/PhysRevD.58.114006.
50. H. Neff, N. Eicker, T. Lippert, J.W. Negele, K. Schilling, On the low fermionic eigenmode dominance in QCD on the lattice, Phys. Rev. D 64 (2001) 114509, http://dx.doi.org/10.1103/PhysRevD.64.114509.
51. S. Prelovsek, C. Dawson, T. Izubuchi, K. Orginos, A. Soni, Scalar meson in dynamical and partially quenched two-flavor QCD: lattice results and chiral loops, Phys. Rev. D 70 (2004) 094503, http://dx.doi.org/10.1103/PhysRevD.70.094503.
52. C.W. Bernard, M.F. Golterman, Chiral perturbation theory for the quenched approximation of QCD, Phys. Rev. D 46 (1992) 853–857, http://dx.doi.org/10.1103/PhysRevD.46.853.
53. S.R. Sharpe, N. Shoresh, Physical results from unphysical simulations, Phys. Rev. D 62 (2000) 094503, http://dx.doi.org/10.1103/PhysRevD.62.094503.
54. J. Bijnens, N. Danielsson, T.A. Lahde, Three-flavor partially quenched chiral perturbation theory at NNLO for meson masses and decay constants, Phys. Rev. D 73 (2006) 074509, http://dx.doi.org/10.1103/PhysRevD.73.074509.
55. N. Carrasco, P. Dimopoulos, R. Frezzotti, P. Lami, V. Lubicz, et al., Leptonic decay constants fK, fD, and fDs with Nf = 2 + 1 + 1 twisted-mass lattice QCD, Phys. Rev. D 91 (5) (2015) 054507, http://dx.doi.org/10.1103/PhysRevD.91.054507, arXiv:1411.7908
56. B. Blossier, et al., Pseudoscalar decay constants of kaon and D-mesons from Nf = 2 twisted mass lattice QCD, J. High Energy Phys. 0907 (2009) 043, http://dx.doi.org/10.1088/1126-6708/2009/07/043.
57. K. Jansen, C. Urbach, tmLQCD: a program suite to simulate Wilson twisted mass lattice QCD, Comput. Phys. Commun. 180 (2009) 2717–2738, http://dx.doi.org/10.1016/j.cpc.2009.05.016.
58. A. Deuzeman, S. Reker, C. Urbach, Lemon: an MPI parallel I/O library for data encapsulation using LIME, Comput. Phys. Commun. 183 (2012) 1321–1335, http://dx.doi.org/10.1016/j.cpc.2012.01.016, arXiv:1106.4177
59. R Development Core Team, R: A Language and Environment for Statistical Computing, R Foundation for Statistical Computing, Vienna, Austria, ISBN 3-900051-07-0, 2005, http://www.R-project.org.

CHAPTER 14 INTRINSIC HEAVY QUARK CONTENT OF THE NUCLEON

S. J. Brodsky, A. Kusina, F. Lyonnet, I. Schienbein, H. Spiesberger, and R. Vogt

INTRODUCTION

The existence of a nonperturbative intrinsic heavy quark component in the nucleon is a rigorous prediction of Quantum Chromodynamics (QCD). An unambiguous experimental confirmation is still missing and would represent a major discovery. The goal of this paper is to summarize our current understanding of this subject with a particular focus on the potential of a high energy and high luminosity fixed target experiment using the LHC beams (AFTER@LHC) [1–4] to search for intrinsic charm.

Production processes sensitive to the intrinsic heavy quark distributions of protons and nuclei are among the most interesting hadronic physics topics that can be investigated with AFTER@LHC. In contrast to the familiar extrinsic contributions which arise from gluon splitting in perturbative QCD, the intrinsic heavy quarks have multiple connections to the valence quarks of the proton and thus are sensitive to its nonperturbative structure. For example, if the gluongluon scattering box diagram, $gg \to Q\bar{Q} \to gg$ (the analog of QED light-by-light scattering), is inserted into the proton self-energy, the cut of this amplitude generates five-quark Fock states of the proton $|uudQ\bar{Q}\rangle$; see Figure 1.

Intrinsic strange, charm, and bottom quarks are thus a fundamental property of the wavefunctions of hadronic bound states [5–8]. While the extrinsic contributions to the heavy quark parton distribution functions (PDFs) are most important at low x and depend logarithmically on the heavy quark mass M_Q, the intrinsic heavy quark contributions are dominant at high x and depend on $1/M_Q^2$. Because the extrinsic heavy quarks are generated by gluon splitting, their PDFs are always softer than those of the parent gluon by a factor of $(1 - x)$. In contrast, the high x intrinsic heavy quark contributions are kinematically dominated by the regime where the $|uudQ\bar{Q}\rangle$ state is minimally off shell, corresponding to equal rapidities of the constituent quarks. The resulting momentum and spin distributions of the intrinsic Q and $\bar{Q}$ can be distinct, for example, $s(x) \neq \bar{s}(x)$, since the comoving $uudQ\bar{Q}$ quarks are sensitive to the global quantum numbers of the protons.

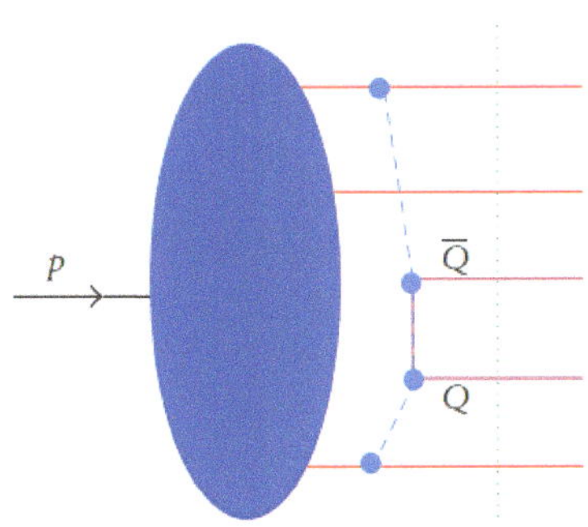

Figure 1: Five-quark Fock state $|uudQ\bar{Q}\rangle$ of the proton and the origin of the intrinsic sea.

A finite intrinsic charm contribution to the nucleon has been extracted from lattice QCD. An analysis by the MILC collaboration [9] yields a probability for the charm matrix element $\langle N|c\bar{c}|N\rangle$ in the range of 5-6%, consistent with a four-loop perturbative QCD calculation [10].

While the first experimental evidence of intrinsic heavy quarks came from the EMC measurement of the large x charm structure function [11], a variety of other charm hadrons and charmonium measurements are consistent with the existence of intrinsic charm. Open charm observables in hadroproduction include forward Λ_c production at the ISR [12]. Similarly, the

---

Citation: S. J. Brodsky, A. Kusina, F. Lyonnet, I. Schienbein, H. Spiesberger, and R. Vogt, Intrinsic Heavy Quark Content of the Nucleon, vol. 2015, Article ID 231547, 12 pages, 2015. doi:10.1155/2015/231547.

coalescence of comoving b, u, and d quarks from the $|uud\bar{b}b\rangle$ intrinsic bottom Fock state in the proton can explain the high x_F production of the $\Lambda(udb)$ baryon, as observed at the ISR [12]. and asymmetries between leading and nonleading charms ($\bar{D}$ mesons which share valence quarks with the projectile and D mesons which do not, resp.) measured as functions of x_F and p_T in fixed target experiments, WA89 and WA82 at CERN; E791 and SELEX at Fermilab; see [13–15] and references therein. Previous fixed target J/ψ measurements also give indications of important intrinsic charm contributions, particularly from the nuclear mass, or A, dependence, as measured by NA3 at CERN as well as E772 and, later, E866 at Fermilab; see, for example, [16]. Indeed, the A dependence, proportional to A^{α}, is quite different than the $\alpha \sim 1$ expected from extrinsic-type production [17]. At large x_F, there are indications of $A^{2/3}$ dependence, consistent with a nuclear surface-type interaction instead of the volume dependence of pQCD. In addition, the NA3 collaboration measured double J/ψ production at forward x_F in πA interactions, difficult to explain without an intrinsic charm mechanism [18]. All of these observables can be studied with higher energies and luminosities at AFTER@LHC, making precision measurements possible for the first time.

In addition to the typical observables for intrinsic heavy quarks, these intrinsic heavy quarks also contribute to a number of more exotic observables and inclusive and diffractive Higgs production $pp \to ppH$, in which the Higgs boson carries a significant fraction of the projectile proton momentum [19, 20]. There are also important implications for intrinsic charm and bottom quarks in Standard Model physics, as in the weak decays of the B-meson [21] and a novel solution to the $J/\psi \to \rho\pi$ problem [22]. AFTER@LHC could also shed light on these topics.

The rest of this paper is organized as follows. In Section 2, we give an overview of the theoretical models predicting the x-shape (but not the normalization) of the intrinsic charm and bottom parton distribution functions. In Section 3, we discuss the constraints on the normalization of the intrinsic charm (IC) obtained in global analyses of PDFs. Section 4 is devoted to the intrinsic bottom (IB) content of the nucleon, for which there are currently no quantitative constraints. In Section 5 we review collider observables sensitive to an intrinsic charm or bottom PDF. Finally, in Section 6 we present our conclusions.

THEORETICAL MODELS

The QCD wavefunction of a hadron can be represented as a superposition of quark and gluon Fock states. For example, at fixed light-front time, a hadron wavefunction can be expanded as a sum over the complete basis of free quark and gluon states: $|\Psi_h\rangle = \sum_m |m\rangle\, \psi_{m/h}(x_i, k_{T,i})$ where the color-singlet states, $|m\rangle$, represent the fluctuations in the hadron wavefunction with the Fock components $|q_1q_2q_3\rangle$, $|q_1q_2q_3g\rangle$, $|q_1q_2q_3c\bar{c}\rangle$, and so forth. The boost-invariant light-front wavefunctions, $\psi_{m/h}(x_i, k_{T,i})$, are functions of the relative momentum coordinates $x_i = k_i^+/P^+$ and $k_{T,i}$ where k_i denotes the parton momenta and P the hadron momentum. Momentum conservation demands $\sum_{i=1}^n x_i = 1$ and $\sum_{i=1}^n \vec{k}_{T,i} = 0$ where n is the number of partons in state $|m\rangle$. For example, as predicted by Brodsky and collaborators, in the BHPS model intrinsic charm fluctuations [5, 23] can be liberated by a soft interaction which breaks the coherence of the Fock state [24] provided the system is probed during the characteristic time that such fluctuations exist.

Microscopically, the intrinsic heavy quark Fock component in the proton wavefunction, $|uudc\bar{c}\rangle$, is generated by virtual interactions such as $gg \to Q\bar{Q}$ where the gluons couple to two or more valence quarks. The probability for $c\bar{c}$ fluctuations to exist in a hadron is higher twist since it scales as $1/m_c^2$ relative to the extrinsic, EC, leading-twist production by photon-gluon fusion [18].

The dominant Fock state configurations are not far off shell and thus have minimal invariant mass, $M^2 = \sum_i^n \hat{m}_i^2/x_i$, , where $\hat{m}_i^2 = m_i^2 + \langle \vec{k}_{T,i}^2\rangle$ is the square of the average transverse mass of parton i.The general form of the Fock state wavefunction for a hadron with mass m_h appropriate to any frame at fixed light-front time is

$$\Psi\left(x_i, \vec{k}_{\perp i}\right) = \frac{\Gamma\left(x_i, \vec{k}_{\perp i}\right)}{m_h^2 - M^2}, \tag{1}$$

where Γ is a vertex function, expected to be a slowly varying, decreasing function of $m_h^2 - M^2$. The particle distributions are then controlled by the light-front energy denominator and phase space. This form for the higher Fock components is applicable to an arbitrary number of light and heavy partons. Intrinsic $c\bar{c}$ Fock components with minimum invariant mass correspond to configurations with equal rapidity constituents. Thus, unlike extrinsic heavy quarks generated from a single parton, intrinsic heavy quarks carry a larger fraction of the parent momentum than the light quarks in the state [5, 23].

The parton distributions reflect the underlying shape of the Fock state wavefunction. Assuming that it is sufficient to use $\langle k_T^2\rangle$ for the transverse momentum, the probability distribution as a function of x in a general n-particle intrinsic $c\bar{c}$ Fock state is

$$\frac{dP_{\rm IC}}{dx_i \cdots dx_n} = N_n \frac{\delta\left(1 - \sum_{i=1}^n x_i\right)}{\left(m_h^2 - \sum_{i=1}^n \left(\hat{m}_i^2/x_i\right)\right)^2}, \tag{2}$$

where N_n normalizes the n-particle Fock state probability. At LO in the heavy quark limit, $\hat{m}_c, \hat{m}_{\bar{c}} \gg m_h, \hat{m}_q$,

$$\frac{dP_{\rm IC}}{dx_i \cdots dx_n} = N_n \frac{x_c^2 x_{\bar{c}}^2}{\left(x_c + x_{\bar{c}}\right)^2} \delta\left(1 - \sum_{i=1}^{n} x_i\right), \tag{3}$$

leading to

$$\begin{aligned} F_{2c}^{\rm ICLO}(x) &= \frac{8}{9} x c(x) \\ &= \frac{8}{9} x \int dx_1 \cdots dx_{\bar{c}} \frac{dP_{\rm IC}}{dx_i \cdots dx_{\bar{c}} dx_c}. \end{aligned} \tag{4}$$

There are many applications of intrinsic charm in charm hadron production. See, for example, [13–16, 18] for more details.

Paiva et al. have also calculated an intrinsic charm component of the nucleon sea within the context of the meson cloud model [25]. They assumed that the nucleon can fluctuate into $\bar{D}\Lambda_c$. The $\bar{c}$ distribution in the nucleon is then

$$x\bar{c}_N(x) = \int_x^1 dy f_{\bar{D}}(y) \frac{x}{y} \bar{c}_{\bar{D}}\left(\frac{x}{y}\right), \tag{5}$$

Where

$$f_{\bar{D}}(y) = \frac{g^2_{\bar{D}N\Lambda_c}}{16\pi^2} y \int_{-\infty}^{t_{\max}} dt \frac{\left[-t + \left(m_{\Lambda_c} - m_N\right)^2\right]}{\left[t - m^2_{\bar{D}}\right]^2} F^2(t), \tag{6}$$

with $F(t)$ being a form factor at the $DN\Lambda$ vertex and $t_{\max} = m^2_N y - m^2_{\Lambda_c} y/(1-y)$. In this case they chose a monopole form factor with $\Lambda_m = 1.2$ GeV. The coupling constant was assumed to be $g_{\bar{D}N\Lambda_c} = -3.795$. From heavy quark effective theories [26], the $\bar{c}$ distribution in the $\bar{D}$ is expected to be hard because, in the bound state, the c exchanges momenta much less than m_c. They make the extreme assumption that the entire $\bar{D}$ momentum is carried by the charm quark, $\bar{c}_{\bar{D}} = (x - y)$.

Next, Steffens et al. investigated all the charm structure function data with two variants of intrinsic charm [27]. The first was that of (4), called IC1 in their paper, while the second was a meson cloud model, IC2. In the second approach, the $\bar{c}$ distribution is obtained from the light-front distribution of $\bar{D}^0$ mesons in the nucleon:

$$\bar{c}^{\mathrm{IC2}}(x) \approx f_{\bar{D}}(x) = \frac{1}{16\pi^2}\int_0^\infty dk^2_\perp \cdot \frac{g^2\left(x, k^2_\perp\right)}{x(1-x)\left(s_{\bar{D}\Lambda_c} - m^2_N\right)^2} \cdot \frac{k^2_\perp + \left(m_{\Lambda_c} - (1-x)m_N\right)^2}{1-x}. \tag{7}$$

A hard charm momentum distribution was assumed in the $\bar{D}$, similar to that of [25]. The vertex function $g^2(x, k^2_\perp)$ is parameterized as $g^2 = g^2_0(\Lambda^2 + m^2_N)/(\Lambda^2 + s_{\bar{D}\Lambda_c})$ where $s_{\bar{D}\Lambda_c}$ is the square of the center-of-mass energy of the $\bar{D}\Lambda_c$ system and g^2_0 the coupling constant at $s_{\bar{D}\Lambda_c} = m^2_N$. For an intrinsic charm probability of 1%,$\Lambda \approx 2.2$ GeV. The charm distribution is then

$$c^{\mathrm{IC2}}(x) \approx \frac{3}{2} f_{\Lambda_c}\left(\frac{3x}{2}\right), \tag{8}$$

where the charm distribution in the Λ_c is assumed to be $c_{\Lambda c} \sim \delta(x - 2/3)$ and $f_{\Lambda_c}(x) = f_{\bar{D}}(1-x)$.

Pumplin [28] considered a model where a point scalar particle of mass m_0 couples with strength g to N scalar particles with mass $m_1, m_2, \ldots,$. The probability density is then

$$dP = \frac{g^2}{\left(16\pi^2\right)^{N-1}(N-2)!} \prod_{j=1}^N dx_j \delta\left(1 - \sum_{j=1}^N x_j\right) \int_{s_0}^\infty ds \cdot \frac{\left(s - s_0\right)^{N-2}}{\left(s - m^2_0\right)^2} |F(s)|^2, \tag{9}$$

where $s_0 = \sum_{j=1}^N (m^2_j/x_j)$. The form factor (s) suppresses higher mass state contributions. If the quark transverse momenta are neglected, with m_c much greater than all other mass scales, and $(s) = 1$, then the BHPS model is recovered. Two types of form factors were studied, an exponential $|F(s)|^2 = \exp[-(s - m^2_0)/\Lambda^2]$ and a power law $|F(s)|^2 = 1/(s+\Lambda^2)^n$ where the cutoff Λ is varied between 2 and 10 GeV. Hobbs et al. employed a meson cloud type approach but specified the spin and parity of all lowest mass charm meson-baryon combinations from the 5-particle $|uudc\bar{c}\rangle$ Fock states of the proton [29]. They pointed out that treating quarks as scalar point-like particles, as in, for example, [28], does not conserve spin and parity. They calculated the appropriate meson-baryon splitting functions for the mesonbaryon combinations and found that the production of charm mesons would be almost entirely through D^* mesons. To study the phenomenological distributions of charm mesons and baryons in this approach, they studied exponential and confining vertex functions, $\propto \exp[-(s - m^2_D)/\Lambda^2]$ and $(s - m^2_D)\exp[-(s - m^2_D)/\Lambda^2]$, respectively. They used these results to compare to the Λ_c distribution from the ISR [30] and the $\Lambda_c/\bar{\Lambda}_c$ asymmetry from SELEX [31]. See [29] for details.

GLOBAL ANALYSES OF PDFS WITH INTRINSIC CHARM

In the standard approach employed by almost all global analyses of PDFs, the heavy quark distributions are generated radiatively, according to DGLAP evolution equations [32–34], starting with a perturbatively calculable boundary condition [35, 36] at a scale of the order of the heavy quark mass. In other words, there are no free fit parameters associated with the heavy quark distribution and it is entirely related to the gluon distribution function at the scale of the boundary condition. As a consequence, also the PDF uncertainties for the heavy quark and the gluon PDFs are strongly correlated as has been discussed in the context of inclusive Higgs production at the Tevatron and the LHC [37]. However, a purely perturbative treatment might not be adequate in particular for the charm quark with a mass $m_c \simeq 1.3$ GeV which is not much bigger than typical hadronic scales but also for the bottom quark with a mass $m_b \simeq 4.5$ GeV. Indeed, as discussed above, light-front models predict a nonperturbative ("intrinsic") heavy quark component in the proton wavefunction [5, 23]. Motivated by the theoretical predictions of the BHPS light-front model, analyses of the charm distribution in the proton going beyond the common assumption of purely radiatively generated charm date back almost as far as the BHPS predictions themselves. For definiteness, in the following we refer to the radiatively generated charm by $c_0(x, Q)$ and to the intrinsic charm by $c_1(x, Q)$. The full charm parton distribution is then given by the sum $(x, Q) = c_0(x, Q) + c_1(x, Q)$. Strictly speaking, this decomposition is defined at the initial scale $Q_0 \simeq m_c$ of the DGLAP evolution but holds to a good approximation at any scale since the intrinsic component $c1$ is governed (to a very good approximation) by a standalone nonsinglet evolution equation [38]. A similar decomposition is understood for the bottom quark which will be discussed in Section 4.

The BHPS model of the $|uudc\bar{c}\rangle$ Fock state predicts a simple form for $F_2(x)$:

$$F_{2c}^{IC}(x) = \left(\frac{8}{9}x\right)\frac{1}{2}N_5 x^2\left[\frac{1}{3}(1-x)\left(1+10x+x^2\right) + 2x(1+x)\ln x\right]. \tag{10}$$

If there is a 1% intrinsic charm contribution to the proton PDF, $N_5 = 36$. Hoffmann and Moore incorporated mass effects and introduced next-to-leading order corrections as well as scale evolution [39]. They compared their result to the EMC F_{2c} data from muon scattering on iron at high x and Q^2 with the intrinsic charm contribution added to the leading order calculation of F_{2c} by photon-gluon fusion.

A complete next-to-leading order analysis of both the "extrinsic" radiatively generated charm component and the intrinsic component was later carried out by Harris et al. [7]. The EMC data with $\bar{\nu} = \overline{Q^2/2m_p x} = 53, 95,$', and 168 GeV were fit by a sum of the extrinsic and intrinsic components [7]. The normalization of the two components was left as free parameters:

$$F_{2c}\left(x,\mu^2,m_c^2\right) = \epsilon F_{2c}^{\gamma p}\left(x,\mu^2,m_c^2\right) + \delta F_{2c}^{IC}\left(x,\mu^2,m_c^2\right), \tag{11}$$

with the scale $\mu = \sqrt{m_{c\bar{c}}^2 + Q^2}$. The parameter ϵ, typically larger than unity, was considered to be an estimate of the NNLO contribution to the extrinsic contribution. Since a 1% normalization of the IC component was assumed in (11), the fitted value of δ is the fraction of this normalization. Given the quality of the data, no statement could be made about the intrinsic charm content of the proton when $\bar{\nu} = 53$ and 95 GeV. However, with $\bar{\nu} = 168$ GeV an intrinsic charm contribution of $(0.86 \pm 0.60)\%$ was indicated. These results were consistent with those of the original analysis by Hoffmann and Moore [39].

The BHPS light-front model assumes that $c_1(x) = \bar{c}_1(x)$. . Meson cloud models, introduced later, treat the 5-particle Fock state as a combination of (predominantly) $\overline{D}^0\Lambda_c^+$. In this case, of course, $c_1(x) \neq \bar{c}_1(x)$ with the $\bar{c}$ quark in the $\overline{D}^0$ carrying more momentum than the c quark in the charm baryon. An analysis by Steffens et al. in the context of the meson cloud model and using a hybrid scheme to interpolate between massless evolution at high Q^2 and "extrinsic" production at low Q^2 found a limit of ~0.4% [27]. Regardless of whether or not the models predict $\bar{c}_1(x) - c_1(x) > 0$, intrinsic charm should provide the dominant contribution to the charm density in the proton at large x [28].

For some time, no other analyses of the charm structure function were made. The EMC data remain the only measurement of the charm structure function in the relevant (x, Q^2) regime and are the only DIS data cited as evidence for intrinsic charm. The HERA data on F_{2c} were at too low x to address the issue.

The first global analyses of the proton PDFs with an intrinsic charm contribution included were performed by members of the CTEQ collaboration [40, 41]. In addition to the BHPS and meson cloud approaches, they also allowed for a "sea-like" contribution with the same shape as the radiatively generated charm distribution. They characterized the magnitude of the intrinsic charm component ($c_1(x, Q^2)$) by the first moment of the charm distribution at the input scale $Q_0 = m_c = 1.3$ GeV (Note that at $Q_0 = m_c$ the radiatively generated charm component ($c_0(x, Q^2)$) vanishes at NLO in the $\overline{\text{MS}}$ scheme so that $c(x, Q_0^2) = c_1(x, Q_0^2)$.):

$$c_1\left(N=1,Q_0^2\right) = \int_0^1 dx\, c_1\left(x,Q_0^2\right) = 0.01, \tag{12}$$

which translates into a momentum fraction

$$\langle x\rangle_{c_1+\bar{c}_1} = \int_0^1 dx\, x\left[c_1\left(x,Q_0^2\right) + \bar{c}_1\left(x,Q_0^2\right)\right] = 0.0057. \tag{13}$$

They found that the global analyses of hard-scattering data provided no evidence for or against the existence of intrinsic charm up to $\langle x\rangle_{c_1+\bar{c}_1} = 0.0057$; that is, the quality of the fit is insensitive to $\langle x\rangle_{c_1+\bar{c}_1}$ in this interval. They also found that the allowed range was greatest for the sea-like IC expected since this shape is rather easily interchangeable with other sea quark components while the other, harder, charm distributions are not [40]. In addition, they concluded that the enhancement due to IC relative to analyses without it persisted up to scales of ~100 GeV and could have an influence on charm-initiated processes at the LHC, as is discussed later. The CTEQ6.6C proton PDFs were generated as a result of this analysis [41].

There are two recent updates to the global analyses, reaching different conclusions about the importance of intrinsic charm. The first, by Dulat et al. [42], follows the previous work in the context of the CTEQ collaboration [40, 41].The second, by Jimenez-Delgado et al. [43], included more lower energy data than the previous global analyses.

The result of Dulat et al. [42] was based on the CT10 NNLO parton densities. Here the strong coupling, (Q^2), the evolution equations, and the matrix elements are calculated at NNLO. Only the inclusive jet data still required NLO expressions. Their analysis included DIS data from BCDMS, NMC, CDHSW, and CCFR; SIDIS data from NuTeV and CCFR; the combined DIS and F_{2c} data from HERA; Drell-Yan production; the W charge asymmetry and Z_0 rapidity from CDF and D0; and the inclusive jet measurements from CDF and D0; see [42] for a complete list.

Two models of IC were considered: the BHPS light-front model and the sea-like IC introduced in [40]. They found a broader possible probability range for IC in this analysis, $\langle x\rangle_{IC} = \langle x\rangle_{c_1+\bar{c}_1}(Q_0^2) \lesssim 0.025$ for BHPS and $\langle x\rangle_{IC} \lesssim 0.015$ for the sea-like IC, summarized in Figure 2. This finding differs from the previous work which found a larger upper limit on IC for the sea-like model. They believe that the difference is caused by the improved treatment of the charm quark mass in the later study [42].

In addition to the global fit, they also tested the sensitivity of their result to individual experiments by introducing a penalty factor, $T_2(i)$, for each experiment i.This penalty factor is designed to increase more rapidly than the χ_i^2 for that experiment when χ_i^2 goes beyond the 90% confidence level. The penalty factor employs an equivalent Gaussian variable Sn which measures the goodness of fit for each individual data set. Values of $S_n \leq |1|$ are considered good fits, $S_n > 3$ is considered to be a poor fit, and values of $S_n < -3$ are better fits than expected from usual statistical analyses. Using the Sn dependence on $\langle x \rangle_{IC}$, they determined which of the data sets used in the global analyses are most sensitive to intrinsic charm. The upper limit on the BHPS value of $\langle x \rangle_{IC}$ comes from the CCFR structure function data while the HERA combined charm data sets the upper limit on IC from the sealike model [42].

They also studied the sensitivity of their sea-like result to the charm quark mass and found that if the charm quark mass was raised from 1.3 GeV, as in the CT10 fits, to 1.67 GeV, then the minimum χ^2 for the global analyses would support $\langle x \rangle_{IC} = 0.01$ rather than 0 although the global χ^2 is worse for the larger charm mass [42]. Finally, they showed how W and Z production at the LHC might be affected by a nonzero IC contribution.

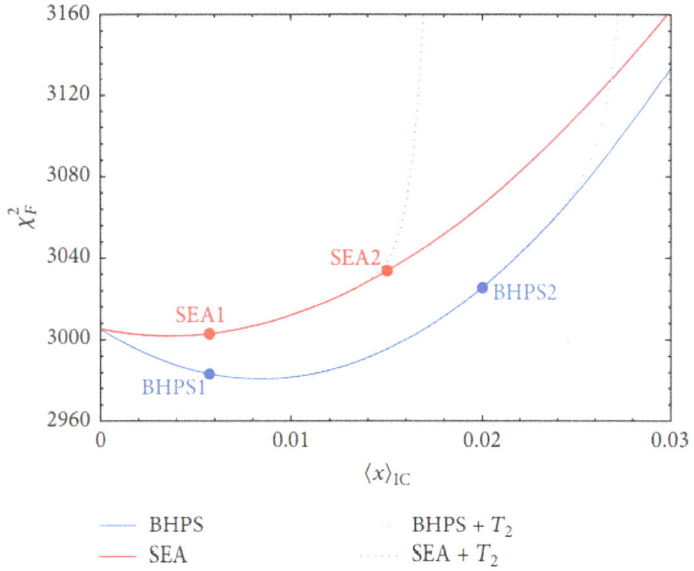

Figure 2: The global chi-square function versus charm momentum fraction $\langle x \rangle_{IC}$. The two curves are determined from fits with many values of $\langle x \rangle_{IC}$. Two exemplary fits for each IC model are shown as dots. Blue dots denote the BHPS model; the dots have $\langle x \rangle_{IC} = 0.57\%$ and 2%, which are denoted as BHPS1 and BHPS2. Red denotes SEA model; the dots have $\langle x \rangle_{IC} = 0.57\%$ and 1.5%, which are denoted as SEA1 and SEA2. Additionally the dotted lines show global chisquare function with additional penalty, $T_2(i)$, used to set the upper limits on the allowed IC component (figure taken from [42]).

In the most recent study, Jimenez-Delgado et al. [43] included the full range of high energy scattering data by using looser kinematic cuts $Q^2 \geq 1$ GeV^2 and $W^2 \geq 3.5$ GeV^2 . In particular, they included the lower energy SLAC fixed target data which did not pass the more stringent standard DIS cuts on the (Q^2 , W^2) plane applied in the previous work [40–42]. The EMC F_{2c} data, cited as the strongest evidence for intrinsic charm in DIS, are used as a consistency check. The low energy, high x, fixed target data lie precisely in the region where IC is expected to be most important. Thus including these data could enhance the sensitivity of the global fit to IC. Note, however, that some of these newly added data are on heavier targets than the deuteron and thus target mass corrections, nuclear corrections for $A > 2$, and higher-twist effects need to be taken into account [43].

They followed the framework of the JR14 [44] global fit which decomposed F_2 into light and heavy components. The charm component is itself separated into the "extrinsic" and intrinsic charm components. The fixed-flavor number scheme is used to compute the extrinsic contribution. In this scheme, the charm quark mass enters the PDF evolution only indirectly through the running of α_s [43]. They employed a charm quark mass of 1.3 GeV, as did Dulat et al. [42]. They used all three intrinsic charm models previously considered: BHPS, the meson cloud model (this time including pseudoscalar and vector mesons as well as spin 1/2 and spin 3/2 charm baryons—the CTEQ analyses only included the scalar $\overline{D}\Lambda c$ fluctuation), and the sea-like component [43]. The IC contribution was evolved up to NLO.

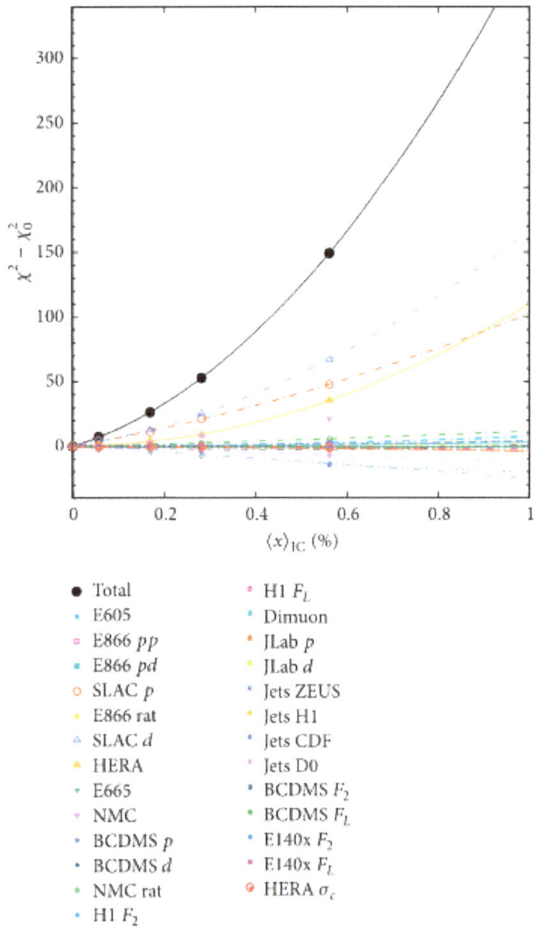

Figure 3: Contributions to the total χ^2 (black circles), relative to the value χ_0^2 for no IC, of various data sets as a function of the momentum fraction $\langle x \rangle_{IC}$ (figure taken from [43]).

They found that the total χ^2 is minimized for $\langle x \rangle_{IC} = 0$ with $\langle x \rangle_{IC} < 0.1\%$ at the 5σ level. When a hadron suppression factor to suppress charm contributions near threshold is applied, they find a minimum χ^2 at $\langle x \rangle_{IC} = (0.15 \pm 0.09)\%$ for the full data set. The SLAC F_2 (large x), NMC cross sections (medium x), and HERA F_{2c} (low x) display the greatest sensitivity to IC; see Figure 3 for details. However, fits without the SLAC data still give a low IC contribution [43]. The difference between their results and previous results is in part due to the very different tolerance criteria, $\Delta\chi^2 = 1$ for their fit and $\Delta\chi^2 = 100$ for Dulat et al. [42]. Increasing the tolerance to $\Delta\chi^2 = 100$ would also accommodate $\langle x \rangle_{IC} = 1\%$ at the 1σ level [43]. (For a critical discussion of the analysis in [44] and in particular of the tolerance criterion $\Delta\chi^2 = 1$ see [45, 46].)

When checked against the EMC F_{2c} data, a clear preference for IC is found, as expected, for the highest x data. (The EMC μ-p data have not been included in some global analyses because of a possible conflict with HERA e-p data at very low x; however, the relative suppression of the low x and low Q^2 EMC data could be accounted for by nuclear shadowing, suppressed evolution, higher twist, and other effects.)

Given that the two most recent analyses set significantly different limits on IC, it is important to collect further large x data, particularly on F_{2c} to try and place greater confidence on the limit of IC in the nucleon. This would be an important measurement at the future electron-ion collider.

PREDICTIONS FOR INTRINSIC BOTTOM

In contrast to the case of intrinsic charm, there is currently no global analysis available that investigates the possibility of an intrinsic bottom (IB) content of the nucleon. The main reason for this is the lack of experimental data that could constrain it. The BHPS light-front model [5] predicts the existence of IB with an x-shape very similar to the one of IC given in (10) but with a normalization which is parametrically suppressed by the ratio m_c^2/m_b^2. This fact, together with the observation that the IB PDF is governed (to an excellent approximation) by an independent nonsinglet evolution equation [38], can be used to investigate IB in a flexible way without the need of a dedicated global analysis. Such a study has been done in [38] where a set of decoupled IB (and IC) PDFs has been provided and used together with the CTEQ6.6 PDFs [41] to estimate the impact of the IB on new physics searches at the LHC. The advantage of this approach is that the provided IB (IC) PDF can be used with any standard set of PDFs and the normalization of the intrinsic component can be freely adjusted. This is especially useful for studies of possible IB effects, as, in that case, there are no experimental limits on what amount of IB is allowed.

In the following we show some of the results found in [38]. In this work, the boundary condition for the IB distribution was modeled using the IC distributions in the CTEQ analyses [40, 41] scaled down by the mass factor m_c^2/m_b^2. The result of such an intrinsic bottom distribution $b_1(x, Q^2)$, with normalization $\int_0^1 dx\, b_1(x, m_c^2) = 0.01 \times m_c^2/m_b^2$, is shown in Figure 4, where the ratio of the intrinsic (b_1) and the radiatively generated (b_0) component of the bottom PDF is plotted. As always in the light-front models the intrinsic component is mostly present at large x values. We can see that for low scales $Q \sim 10$ GeV the modification of the bottom PDF, $\kappa_b = 1 + b_1/b_0$, can reach $\kappa_b = 2.5$. However, it decreases rapidly with the rising scale. Since $b1$ evolves independently of the other PDFs the change in the normalization of the IB component in Figure 4 can be done by simply rescaling the curves in the figure. If we allowed for a $0.035 \times m_c^2/m_b^2$ normalization of the IB the modification of the bottom PDF would be given by $\kappa_b = 1 + b_1/b_0 \times 3.5$, which for high x and $Q \sim 10$ GeV would result in an enhancement of the bottom PDF by a factor ~6.25. However, at a scale of around 100 GeV and x below 0.2–0.3, even with the higher IB normalization, the effect is becoming negligible.

In Figure 5 we show the sum of the intrinsic bottom PDF b_1 and the dynamically generated PDF b_0 from CTEQ6.6 for different normalizations of the IB component, namely, 0.01 and $0.035 \times m_c^2/m_b^2$. We compare this sum to the asymmetric uncertainties (The asymmetric errors are computed following [47, 48].) of the CTEQ6.6 PDF set (a). In the same figure is also shown the ratio of the same PDFs to the central value of CTEQ6.6 (b). As can be seen, the IB curve with the $0.035 \times m_c^2/m_b^2$ normalization clearly lies outside the uncertainty band whereas the one with the smaller normalization is marginally outside the band (up to $x \lesssim 0.6$).

If we are looking for new physics with couplings proportional to the mass, the suppression of IB compared to the IC would be partly compensated by the square of the coupling. For a more detailed study of the relevant partonparton luminosities please see [38].

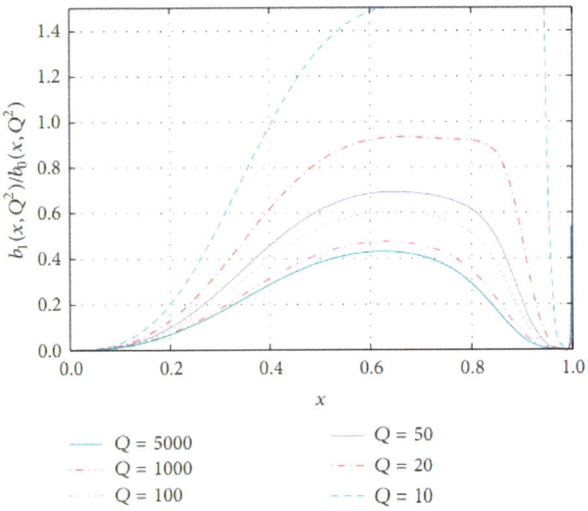

Figure 4: Ratio of intrinsic (b_1) and dynamically generated (b_0) bottom PDFs for various Q scales. The perturbative bottom PDF from CTEQ6.6c0 [41] is used; the normalization of the IB is taken to be such that $\int_0^1 dx\, b_1(x, m_c^2) = 0.01 \times m_c^2/m_b^2$ (figure taken from [38]).

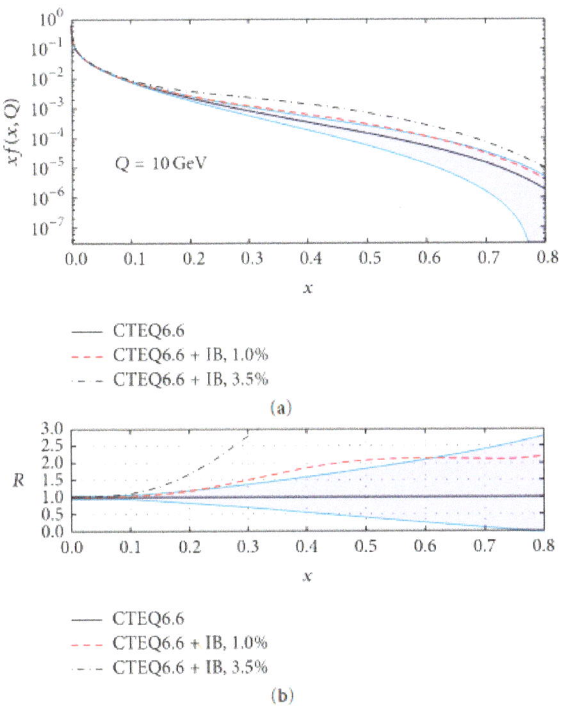

Figure 5: CTEQ6.6 + b_1 for different normalizations of the intrinsic bottom quark PDF at the scale Q = 10 GeV, compared to the asymmetric PDF errors from the same set (a). Also shown is the ratio of the same PDF sets to the central value of CTEQ6.6 (b).

COLLIDER OBSERVABLES

Several collider observables receive large contributions from heavy quark initiated subprocesses and are hence potentially sensitive to an intrinsic charm content in the nucleon. In order to expect optimal effects the heavy quark PDF should be probed at large $x \gtrsim 0.2$ (for light-front models) and not too large factorization scales. This kinematic region is best accessible at lower energies in the center-of-mass system (cms) and/or large rapidities. Therefore, a fixed target experiment like AFTER@ LHC [1–4] operating at cms energy $\sqrt{s}$ = 115 GeV with a high luminosity is ideally suited for searches of IC effects. In the following we review some of the collider processes which have been studied in the literature in this respect.

Open Heavy Flavor Production.

Inclusive charm hadron (D^0 , D^+, D^{*+}, Λ_c,...) production in hadronic collisions was advocated in [49] as a laboratory to probe IC inside the colliding hadrons. In this analysis, predictions for the differential cross section in dependence on the transverse momentum p_T were obtained in the general-mass variable-flavor-number scheme (GM-VFNS) [50–52] at next-to-leading order (NLO). In this scheme, the charm quark is an active parton and the differential cross sections of inclusive charm meson production depend heavily on the PDF of the charm quark. The sensitivity of these cross sections to IC was studied for the Tevatron at cms energy of 1960 GeV and the Relativistic Heavy Ion Collider (RHIC) at cms energies of 200 GeV (RHIC200) and 500 GeV (RHIC500). The different IC models from the CTEQ6.5c global analysis [40] were employed together with the fragmentation functions for charm mesons from [53]. While the effects at the Tevatron were found to be very moderate and likely not testable, large enhancements were found at RHIC200 reaching values of ~3 at pT = 20 GeV. Unfortunately, the measurements at RHIC200 are limited by the luminosity. At RHIC500 the cross section is increased by about a factor 3.6. However, the sensitivity to IC for the lightfront models is greatly reduced.

More recently, the GM-VFNS was applied to obtain predictions for the production of inclusive D mesons at the LHC for cms energy of 7 TeV (LHC7) [54]. It was found that the production cross sections at large rapidities $y \gtrsim 4$ are sensitive to an IC component. These predictions can be tested by measurements at forward rapidities with the LHCb detector.

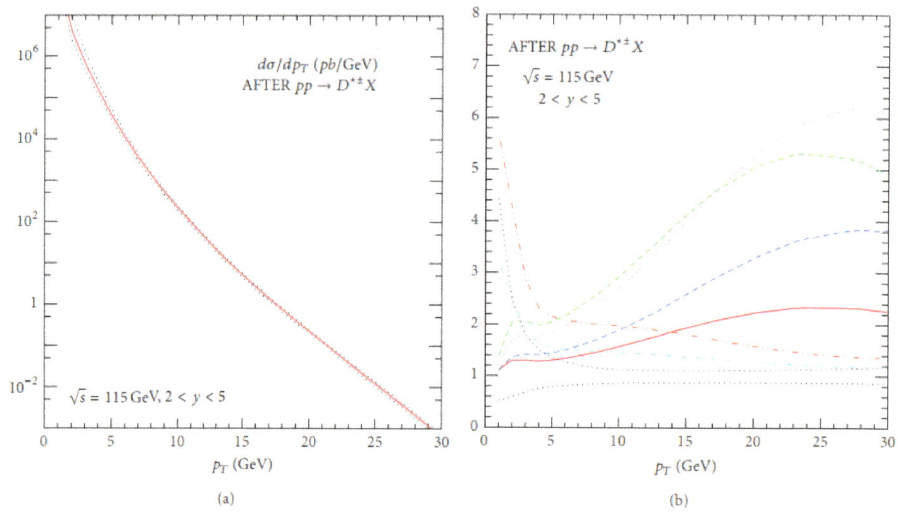

Figure 6: NLO predictions for inclusive D^* meson production at AFTER@LHC versus the transverse momentum of the D meson. (a) Differential cross section on an absolute scale without intrinsic charm. (b) Ratio with respect to to the central prediction of (a). Shown are results using the IC parametrizations from [40] for $n = 1$ (red, solid line), 2 (violet, dotted line), 3 (blue, dashed line), 4 (green, long dashed line), 5 (cyan, dot-dashed line), and 6 (orange, double-dot-dashed line). In both figures, the black dotted lines have been obtained by varying the renormalization scale around the central choice ($\mu_R = m_T$) to $\mu_R = m_T/2$ (upper line) and $\mu_R = 2m_T$ (lower line).

The ideal experiment to search for the effects of IC would be a high luminosity fixed target experiment such as AFTER@LHC operating at cms energy of 115 GeV. In Figure 6 we show results for inclusive D^* meson production as a function of the transverse momentum of the D^* meson integrated over the rapidity range $2 < y < 5$ (in the laboratory frame) in essentially the same setup as in [49] to which we refer for details. The only difference is that, following [55], the default choice for the renormalization and factorization scales is $\mu_R = m_T$, $\mu_F = \mu_F' = m_T/2$, where $m_T = \sqrt{p_T^2 + m^2}$ is the transverse mass. The theoretical predictions are shown on an absolute scale in Figure 6(a) and as a ratio with respect to the default results in Figure 6(b). In both figures, the black dotted lines have been obtained by varying the renormalization scale around the central choice to $\mu R = m_T/2$ (upper line) and $\mu_R = 2m_T$ (lower line). In Figure 6(b) we repeat the calculation of the central prediction in turn with PDF sets CTEQ6.5Cn for $n = 1,\ldots, 6$ and normalize the outcome to the default prediction with zero IC of Figure 6(a). We observe that the ratios for $n = 1, 2, 3, 4$ corresponding to the BHPS ($n = 1, 2$) or meson cloud ($n = 3, 4$) models become very large at large p_T. Indeed, the default cross section can be increased by more than a factor 5 at $p_T = 20$ GeV in scenarios with maximally allowed intrinsic charm ($n = 2, 4$). Even for the IC sets with smaller normalization ($n = 1, 3$) corresponding to $\langle x\rangle_{c_1+\bar{c}_1} = 0.57\%$ and $\langle x\rangle_{c_3+\bar{c}_3} = 0.96\%$ the cross section would be enhanced by a factor larger than 2 (red solid line) or 3 (blue dashed line) at $p_T = 20$ GeV. It is also interesting to note that the phenomenological models for a sea-like IC ($n = 5, 6$) lead to a significant enhancement of the cross section at small $p_T \sim m_c$ which would be probed at AFTER@LHC as well.

Production of a Photon in Association with a Charm Quark.

Another process with a wide range of phenomenological applications in pp, pA, and AA collisions [56–58] which is very sensitive to the heavy quark PDF is the associated production of a photon with a heavy quark. A dedicated study of this process at the LHC operating at $\sqrt{s} = 8$ TeV (LHC8) was performed in [59, 60] where it was demonstrated that the existence of IC in the proton can be visible at large transverse momenta of the photons and heavy quark jets at rapidities $1.5 < |y_\gamma| < 2.4$, $|y_c| < 2.4$. Indeed, for the BHPS model the cross section can be enhanced by a factor of 2-3 for $p_T^\gamma > 200$ GeV (see Figure 5 in [60]). This comes with the penalty that the cross section falls rapidly with increasing transverse momentum so that this measurement will be limited by statistics.

Again, as for open heavy flavor production, the lower cms energy together with the high luminosity makes a fixed target experiment like AFTER@LHC the ideal place to discover IC using $\gamma+c$ production. This can be seen in Figure 7, where the differential cross section is enhanced by a factor 5 at $p_T^\gamma = 20$ GeV (b) with a not too small cross section (a).

Vector Boson Production.

Dulat et al. [42] studied the sensitivity of $W^\pm$ and Z^0 production to the presence of IC. Vector boson production at the LHC is an interesting testing ground for IC because they are produced at relatively large x and $Z^0 \to l^+l^-$ is a rather clean final state. They did a NNLO calculation of W and Z production including IC based on their global fits at $\sqrt{s} = 8$ and 14 TeV. They also studied the ratio $d\sigma_{W^+ + W^-}(y)/d\sigma_{Z^0}(y)$ relative to the result with no IC. Neither of these calculations showed an effect larger than the uncertainties due to the CT10 sets themselves. However, when the Z^0 p_T distribution with IC was compared to that without, they saw a factor of two enhancement at $p_T \sim 500$ GeV for $\sqrt{s} = 8$ TeV in the range $|\eta| < 2.1$. The corresponding enhancement at 14 TeV was smaller at the same p_T because the x value reached is reduced at the higher energy [42].

We show a simple test case here for W and Z production to NLO at $\sqrt{s} = 7$ TeV. We use only the BHPS IC parameterization for the five-particle Fock state, shown in (10). We assume a 1% normalization and no Q^2 evolution to maximize the possible effect at forward rapidity. The p_T- integrated rapidity distribution is shown in Figure 8, as is the ratio of the result with IC to that without as a function of rapidity. The rapidity distributions without IC are given by the solid curves while the dashed curves are the calculations with the BHPS IC contribution to the charm parton density. With BHPS IC, one expects enhancement only at forward rapidity. The enhancement from IC appears for $|y| > 2.5$. Note that if the sea-like IC would be used instead, the enhancement would be small but finite over all rapidity.

The W^+ cross section is largest and most forward peaked, because of the $u\bar{d}$ contribution. The contribution from the $c\bar{d}$ part is a very small addition since the u valence contribution is large and peaks at large x, making the y distribution larger at $|y| \sim 2$ than at $y = 0$. Indeed, it gives the smallest IC contribution. The W^- distribution should have the largest possible contribution from IC because both $d\bar{u}$ and $d\bar{c}$ peak at low x and because the d valence distribution peaks at lower x so that the W^- rapidity distribution has a maximum at $y = 0$. At $|y| \sim 4$, the IC enhancement is ~40%. Finally, the Z^0 distribution, with a plateau over $|y| < 1.5$, also has a very small IC contribution because the charm enhancement only comes through $c\bar{c}$.

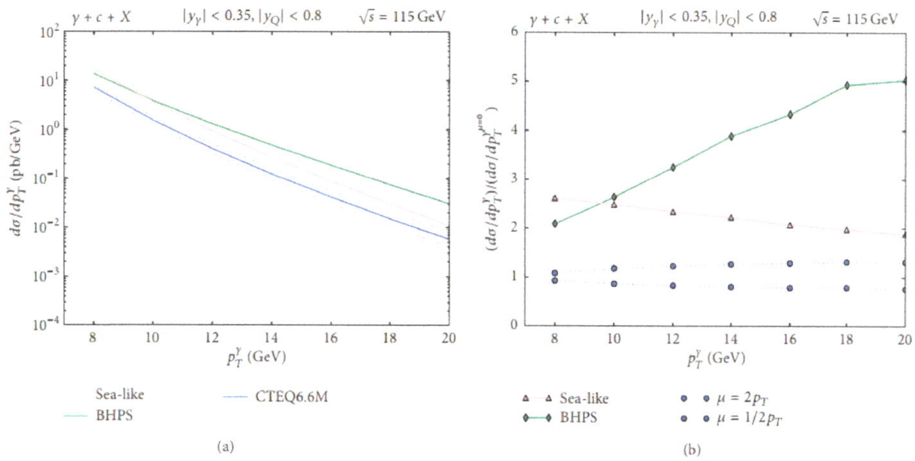

Figure 7: (a) NLO predictions for the production of a prompt photon in association with a charm quark jet in pp collisions at AFTER@LHC versus the transverse momentum of the photon. Shown are results for a BHPS and a sea-like intrinsic charm using the CTEQ6.6c PDFs. For comparison, the predictions without an IC using the CTEQ6.6M PDFs are shown as well together with the uncertainty band obtained by varying the central factorization scale $\mu F = p_T^\gamma$ by a factor 2 up and down (blue, dotted curves). (b) Depicting the ratio of the curves in (a) with respect to the central prediction without intrinsic charm.

Such IC enhancements are only visible outside the midrapidity acceptance of the collider detector coverage of CMS and ATLAS. However, LHCb or ALICE covers this forward rapidity range with muons and could detect forward Z^0. They could also look at the lepton rapidity asymmetry $(W^+ - W^-)/(W^+ + W^-)$ at forward rapidity. The statistical accuracy of the measurement would need to be high to distinguish an IC enhancement from the no IC result, especially since the 1% BHPS IC is likely an upper limit on this enhancement. Note that the higher energy of LHC Run 2 will reduce the potential enhancement even though it would increase the rates.

CONCLUSIONS

The existence of nonperturbative intrinsic charm and bottom components is a fundamental prediction of QCD. In this paper, we have reviewed the current status of our understanding of this intrinsic heavy quark content of the nucleon which yet remains to be confirmed experimentally. In particular, after introducing theoretical models predicting the intrinsic heavy quark distributions we have turned to a summary of the available information on intrinsic charm coming from global analyses of parton distribution functions. There are no global analyses of intrinsic bottom available and we have described how IB can be modeled in order to explore its impact on collider observables keeping in mind that bottom quark initiated subprocesses play an important role in certain electroweak observables and in models for physics beyond the Standard Model. We then have turned to a discussion of collider processes where IC could be discovered. Generally, the effects of IC are larger at colliders with a lower center-of-mass energy and for hard processes with moderate factorization scales. Therefore, a high luminosity fixed target experiment like AFTER@LHC operating at a center-of-mass energy $\sqrt{s}$ = 115 GeV would be ideally suited to discover or constrain IC.

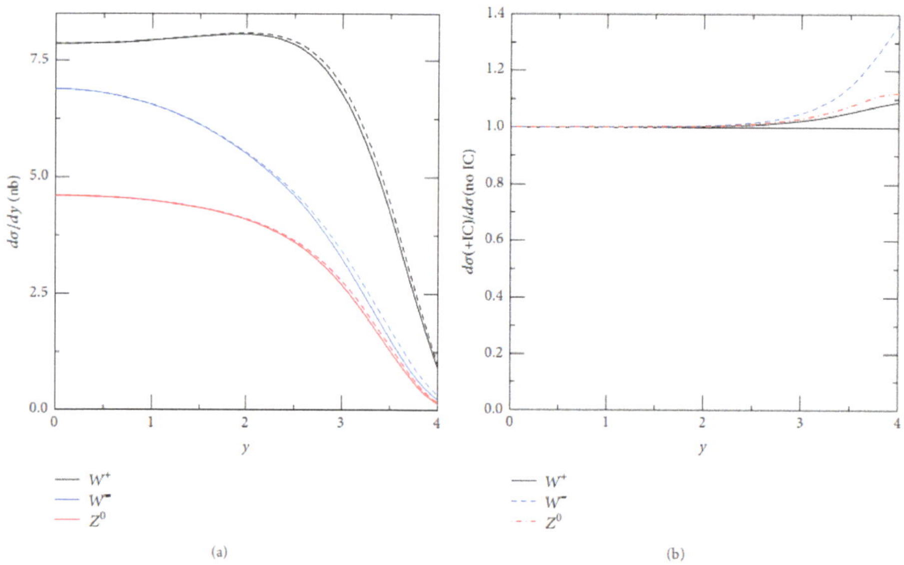

Figure 8: The W^+ (black), W^- (blue), and Z^0 (red) rapidity distributions (a). The solid curves are the results without IC while the dashed curves include 1% BHPS IC. The ratios of the dashed curves to the solid curves showing the enhancement of the rapidity distributions due to IC for W^+ (solid black), W^- (blue dashed), and Z^0 (red dot dashed) are shown in (b).

ACKNOWLEDGMENTS

The authors are grateful to T. Stavreva for providing Figure 7. The work of S. J. Brodsky was supported by the Department of Energy Contract no. DE-AC02-76SF00515. The work of R. Vogt was performed under the auspices of the US Department of Energy by Lawrence Livermore National Laboratory under Contract DE-AC52-07NA27344.

REFERENCES

1. S. J. Brodsky, F. Fleuret, C. Hadjidakis, and J. P. Lansberg, "Physics opportunities of a fixed-target experiment using LHC beams," Physics Reports, vol. 522, no. 4, pp. 239–255, 2013. ·
2. J. P. Lansberg, S. J. Brodsky, F. Fleuret, and C. Hadjidakis, "Quarkonium physics at a fixed-target experiment using the LHC Beams," Few-Body Systems, vol. 53, no. 1-2, pp. 11–25, 2012.
3. J. Lansberg, R. Arnaldi, S. Brodsky, et al., "AFTER@ LHC: a precision machine to study the interface between particle and nuclear physics," EPJ Web of Conferences, vol. 66, Article ID 11023, 4 pages, 2014.
4. A. Rakotozafindrabe, M. Anselmino, R. Arnaldi, et al., "Studying the high x frontier with afixed-target experiment at the LHC," in Proceedings of the 21st International Workshop on Deep-Inelastic Scattering and Related Subject (DIS '13), p. 250, Marseille, France, April 2013.
5. S. J. Brodsky, P. Hoyer, C. Peterson, and N. Sakai, "The intrinsic charm of the proton," Physics Letters B, vol. 93, no. 4, pp. 451–455, 1980.
6. S. J. Brodsky, J. C. Collins, S. D. Ellis, J. F. Gunion, and A. H. Mueller, "Intrinsic chevrolets at the SSC," in Proceedings of the 1984 Summer Study on the SSC, Snowmass, Colo, USA, June-July 1984, Edited by R. Donaldson and J. G. Morfin (AIP, New York, NY, USA, 1985).
7. B. Harris, J. Smith, and R. Vogt, "Reanalysis of the EMC charm production data with extrinsic and intrinsic charm at NLO," Nuclear Physics B, vol. 461, pp. 181–196, 1996.
8. M. Franz, M. V. Polyakov, and K. Goeke, "Heavy quark mass expansion and intrinsic charm in light hadrons," Physical Review D, vol. 62, Article ID 074024, 2000.
9. W. Freeman and D. Toussaint, "Intrinsic strangeness and charm of the nucleon using improved staggered fermions," Physical Review D, vol. 88, no. 5, Article ID 054503, 18 pages, 2013.
10. A. Kryjevski, "Heavy quark q-q matrix elements in the nucleon from perturbative QCD," Physical Review D, vol. 70, Article ID 094028, 2004.
11. J. J. Aubert, G. Bassompierre, K. H. Becks, et al., "Production of charmed particles in 250 GeV μ+-iron interactions," Nuclear Physics B, vol. 213, no. 1, pp. 31–64, 1983.
12. G. Bari, M. Basile, G. Bruni, et al., "The Λb0 beauty baryon production in proton-proton interactions at s=62 GeV: a second observation," Il Nuovo Cimento A, vol. 104, no. 12, pp. 1787–1800, 1991.
13. R. Vogt, S. J. Brodsky, and P. Hoyer, "Systematics of charm production in hadronic collisions," Nuclear Physics B, vol. 383, no. 3, pp. 643–684, 1992.
14. R. Vogt and S. J. Brodsky, "Charmed hadron asymmetries in the intrinsic charm coalescence model," Nuclear Physics B, vol. 478, no. 1-2, pp. 311–332, 1996.
15. T. Gutierrez and R. Vogt, "Leading charm in hadron-nucleus interactions in the intrinsic charm model," Nuclear Physics B, vol. 539, pp. 189–214, 1999.
16. R. Vogt, S. J. Brodsky, and P. Hoyer, "Systematics of J/ψ production in nuclear collisions," Nuclear Physics B, vol. 360, no. 1, pp. 67–96, 1991.
17. P. Hoyer, M. Vanttinen, and U. Sukhatme, "Violation of factorization in charm hadroproduction," Physics Letters B, vol. 246, pp. 217–220, 1990.

18. R. Vogt and S. J. Brodsky, "Intrinsic charm contribution to double quarkonium hadroproduction," Physics Letters B, vol. 349, no. 4, pp. 569–575, 1995.
19. S. J. Brodsky, B. Kopeliovich, I. Schmidt, and J. Soffer, "Diffractive Higgs production from intrinsic heavy flavors in the proton," Physical Review D, vol. 73, no. 11, Article ID 113005, 13 pages, 2006.
20. S. J. Brodsky, A. S. Goldhaber, B. Z. Kopeliovich, and I. Schmidt, "Higgs hadroproduction at large Feynman x," Nuclear Physics B, vol. 807, no. 1-2, pp. 334–347, 2009.
21. S. J. Brodsky and S. Gardner, "Evading the CKM hierarchy: intrinsic charm in B decays," Physical Review D, vol. 65, Article ID 054016, 2002.
22. S. Brodsky, G. de Teramond, and M. Karliner, "Puzzles in hadronic physics and novel quantum chromodynamics phenomenology," Annual Review of Nuclear and Particle Science, vol. 62, pp. 1–35, 2012.
23. S. J. Brodsky, C. Peterson, and N. Sakai, "Intrinsic heavy-quark states," Physical Review D, vol. 23, no. 11, p. 2745, 1981.
24. S. J. Brodsky, P. Hoyer, A. H. Mueller, and W.-K. Tang, "New QCD production mechanisms for hard processes at large x," Nuclear Physics B, vol. 369, no. 3, pp. 519–542, 1992.
25. S. Paiva, M. Nielsen, F. S. Navarra, F. O. Duraes, and L. L. Barz, "Virtual meson cloud of the nucleon and intrinsic strangeness and charm," Modern Physics Letters A, vol. 13, no. 34, pp. 2715–2724, 1998.
26. M. Neubert, "Heavy-quark symmetry," Physics Reports, vol. 245, no. 5-6, pp. 259–395, 1994.
27. F. M. Steffens, W. Melnitchouk, and A. W. Thomas, "Charm in the nucleon," The European Physical Journal C—Particles and Fields, vol. 11, no. 4, pp. 673–683, 1999.
28. J. Pumplin, "Light-cone models for intrinsic charm and bottom," Physical Review D, vol. 73, Article ID 114015, 2006.
29. T. Hobbs, J. Londergan, and W. Melnitchouk, "Phenomenology of nonperturbative charm in the nucleon," Physical Review D, vol. 89, Article ID 074008, 2014.
30. P. Chauvat, R. Cousins, K. Hayes, et al., "Production of Λc with large XF at the isr," Physics Letters B, vol. 199, no. 2, pp. 304–310, 1987.
31. F. Garcia, G. Alkhazov, A. G. Atamantchouk, et al., "Hadronic production of Λ_c from 600 GeV/cπ^-, $\Sigma^-$and p beams," Physics Letters B, vol. 528, no. 1-2, pp. 49–57, 2002.
32. G. Altarelli and G. Parisi, "Asymptotic freedom in parton language," Nuclear Physics B, vol. 126, pp. 298–318, 1977.
33. V. N. Gribov and L. N. Lipatov, "Deep inelastic ep scattering in perturbation theory," Soviet Journal of Nuclear Physics, vol. 15, p. 438, 1972.
34. Y. L. Dokshitzer, "Calculation of the structure functions for deep inelastic scattering and e^+e^-annihilation by perturbation theory in quantum chromodynamics," Soviet Physics—JETP, vol. 46, pp. 641–653, 1977.
35. J. C. Collins and W.-K. Tung, "Calculating heavy quark distributions," Nuclear Physics B, vol. 278, no. 4, pp. 934–950, 1986.
36. M. Buza, Y. Matiounine, J. Smith, and W. L. van Neerven, "Charm electroproduction viewed in the variable-flavour number scheme versus fixed-order perturbation theory," The European Physical Journal C, vol. 1, no. 1-2, pp. 301–320, 1998.
37. A. Belyaev, J. Pumplin, W.-K. Tung, and C. P. Yuan, "Uncertainties of the inclusive Higgs production cross section at the tevatron and the LHC," Journal of High Energy Physics, vol. 2006, no. 1, article 069, 2006.
38. F. Lyonnet, A. Kusina, T.Ježo et al., "On the intrinsic bottom content of the nucleon and its impact on heavy new physics at the LHC," Journal of High Energy Physics, http://arxiv.org/abs/1504.05156.
39. E. Hoffmann and R. Moore, "Sub-leading contributions to the intrinsic charm of the nucleon," Zeitschrift für Physik C, vol. 20, no. 1, pp. 71–82, 1983.
40. J. Pumplin, H. L. Lai, and W. K. Tung, "Charm parton content of the nucleon," Physical Review D, vol. 75, Article ID 054029, 2007.
41. P. M. Nadolsky, H.-L. Lai, Q.-H. Cao, et al., "Implications of CTEQ global analysis for collider observables," Physical Review D, vol. 78, no. 1, Article ID 013004, 21 pages, 2008. ·
42. S. Dulat, T.-J. Hou, J. Gao et al., "Intrinsic charm parton distribution functions from CTEQ-TEA global analysis," Physical Review D, vol. 89, no. 7, Article ID 073004, 18 pages, 2014.
43. P. Jimenez-Delgado, T. J. Hobbs, J. T. Londergan, and W. Melnitchouk, "New limits on intrinsic charm in the nucleon from global analysis of parton distributions," Physical Review Letters, vol. 114, Article ID 082002, 2015.
44. P. Jimenez-Delgado and E. Reya, "Delineating parton distributions and the strong coupling," Physical Review D, vol. 89, no. 7, Article ID 074049, 17 pages, 2014.
45. S. J. Brodsky and S. Gardner, "Comment on 'New limits on intrinsic charm in the nucleon from global analysis of parton distributions'," http://arxiv.org/abs/1504.00969.
46. P. Jimenez-Delgado, T. J. Hobbs, J. T. Londergan, and W. Melnitchouk, "Reply to comment on 'new limits on intrinsic charm in the nucleon from global analysis of parton distributions'," http://arxiv.org/abs/1504.06304.
47. D. Stump, J. Huston, J. Pumplin, et al., "Inclusive jet production, parton distributions, and the search for new physics," Journal of High Energy Physics, vol. 2003, no. 10, article 046, 2003.
48. P. M. Nadolsky and Z. Sullivan, "PDF uncertainties in WH production at Tevatron," http://arxiv.org/abs/hep-ph/0110378.
49. B. A. Kniehl, G. Kramer, I. Schienbein, and H. Spiesberger, "Open charm hadroproduction and the charm content of the proton," Physical Review D, vol. 79, no. 9, Article ID 094009, 10 pages, 2009.
50. B. A. Kniehl, G. Kramer, I. Schienbein, and H. Spiesberger, "Inclusive D*± production in pp- collisions

with massive charm quarks," Physical Review D, vol. 71, Article ID 014018, 2005. ·

51. B. A. Kniehl, G. Kramer, I. Schienbein, and H. Spiesberger, "Collinear subtractions in hadroproduction of heavy quarks," The European Physical Journal C—Particles and Fields, vol. 41, no. 2, pp. 199–212, 2005.
52. B. A. Kniehl, G. Kramer, I. Schienbein, and H. Spiesberger, "Hadroproduction of D and B mesons in a massive VFNS," in Proceedings of the 13th International Workshop on Deep Inelastic Scattering (DIS '05), Madison, Wis, USA, April 2005.
53. T. Kneesch, B. Kniehl, G. Kramer, and I. Schienbein, "Charmed-meson fragmentation functions with finite-mass corrections," Nuclear Physics B, vol. 799, no. 1-2, pp. 34–59, 2008.
54. B. Kniehl, G. Kramer, I. Schienbein, and H. Spiesberger, "Inclusive charmed-meson production at the CERN LHC," The European Physical Journal C, vol. 72, article 2082, 2012.
55. B. Kniehl, G. Kramer, I. Schienbein, and H. Spiesberger, "Inclusive B-meson production at small p_T in the general-mass variable-flavor-number scheme," The European Physical Journal C, vol. 75, p. 140, 2015.
56. T. P. Stavreva and J. F. Owens, "Direct photon production in association with a heavy quark at hadron colliders," Physical Review D, vol. 79, Article ID 054017, 2009.
57. T. Stavreva, I. Schienbein, F. Arleo et al., "Probing gluon and heavy-quark nuclear PDFs with γ + Q production in pA collisions," Journal of High Energy Physics, vol. 2011, no. 1, article 152, 2011.
58. T. Stavreva, F. Arleo, and I. Schienbein, "Prompt photon in association with a heavy-quark jet in Pb-Pb collisions at the LHC," Journal of High Energy Physics, vol. 2013, no. 2, article 072, 2013.
59. V. Bednyakov, M. Demichev, G. Lykasov, T. Stavreva, and M. Stockton, "Searching for intrinsic charm in the proton at the LHC," EPJ Web of Conferences, vol. 60, Article ID 20047, 3 pages, 2013.
60. V. A. Bednyakov, M. A. Demichev, G. I. Lykasov, T. Stavreva, and M. Stockton, "Searching for intrinsic charm in the proton at the LHC," Physics Letters, Section B: Nuclear, Elementary Particle and High-Energy Physics, vol. 728, no. 1, pp. 602–606, 2014.

CHAPTER 15 MESON PRODUCTION IN HEAVY-ION COLLISION

Md. Nasim, Vipul Bairathi, Mukesh Kumar Sharma, Bedangadas Mohanty, and Anju Bhasin

INTRODUCTION

According to quantum chromodynamics (QCD) [1–4], at very high temperature (T) and/or at high density, a deconfined phase of quarks and gluons is expected to be present, while at low T and low density the quarks and gluons are known to be confined inside hadrons. The heavy-ion collisions (A + A) provide a unique opportunity to study QCD matter in the laboratory experiments. The medium created in the heavy-ion collision is very hot and dense and also extremely short-lived (~5–10 fm/c). In experiments, we are only able to detect the freely streaming final state particles emerging from the collisions. Using the information carried by these particles as probes, we try to understand the properties of the medium created in the collision.

The ϕ vector meson, which is the lightest bound state of s and $\bar{s}$ quarks, is considered as a good probe for the study of QCD matter formed in heavy-ion collisions. It was discovered at Brookhaven National Laboratory in 1962 through the reaction $K+p \rightarrow \Lambda+K+\bar{K}$ as shown in Figure 1 [5].

It has a mass of 1.019445 ± 0.000020 GeV/c^2 which is comparable to the masses of the lightest baryons such as proton (0.938 GeV/c^2) and Λ (1.115 GeV/c^2). The interaction cross-section (σ) of the ϕ meson with nonstrange hadrons is expected to have a small value [6]. The data on coherent ϕ photo-production shows that $\sigma_{\phi N} \sim$ 10 mb [7]. This is about a factor of 3 times lower than $\sigma_{\rho N}$ and $\sigma_{\pi N}$, about a factor of 4 times lower than $\sigma_{\Lambda N}$ and σ_{NN}, and about a factor of 2 times lower than σ_{KN}. Therefore its production is expected to be less affected by the later stage hadronic interactions in the evolution of the system formed in heavy-ion collisions. A hydrodynamical inspired study of transverse momentum (p_T) distribution of ϕ meson seems to suggest that it freezes out early compared to other hadrons [8]. The life time of the ϕ meson is ~42 fm/c. Because of longer life time the ϕ meson will mostly decay outside the fireball and therefore its daughters will not have much time to rescatter in the hadronic phase. Therefore, properties of ϕ meson are primarily controlled by the conditions in the early partonic phase and those can be considered as a clean probe to investigate the properties of matter created in heavy-ion collisions.

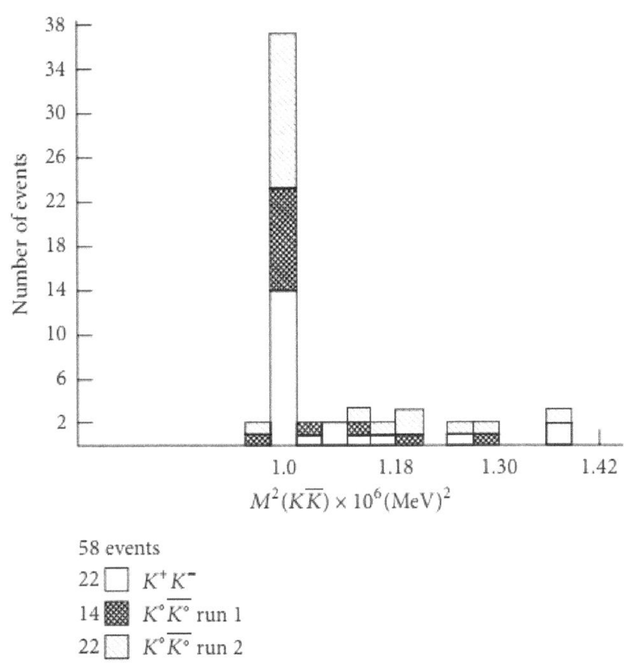

Figure 1: Number of events versus square of invariant mass of $K\bar{K}$ pairs from the reaction $K+p \rightarrow \Lambda+K+\bar{K}$ in bubble chamber experiments at Brookhaven National Laboratory (BNL) [5].

Citation: Md. Nasim, Vipul Bairathi, Mukesh Kumar Sharma, Bedangadas Mohanty, and Anju Bhasin, "Meson Production in Heavy-Ion Collision," Advances in High Energy Physics, vol. 2015, Article ID 197930, 16 pages, 2015. doi:10.1155/2015/197930.

Strange particle production is one of the observables that is expected to deliver detailed information on the reaction dynamics of relativistic nucleus-nucleus collisions [9]. In experiments at the CERN SPS accelerator it was found that the ratio of the number of produced kaons to that of pions is higher by a factor of about two compared to that in proton-proton reactions at the same energy [10– 13]. In the past, several possible reasons for this strangeness enhancement have been discussed. Firstly, if nucleus-nucleus reactions proceed through a deconfined stage, then strangequark production should be enhanced relative to a no QGP scenario [14]. Alternative ideas of Canonical suppression of strangeness in small systems (proton-proton) as a source of strangeness enhancement in high energy nucleus-nucleus collisions have been proposed [15]. This led to a lot of ambiguity in understanding the true physical origin of the observed enhancement in the strange particle production in high energy heavy-ion collisions. The ($s\bar{s}$) meson due to its zero net strangeness is not subjected to Canonical suppression effects. Therefore measurement of ϕ meson yield in both nucleus-nucleus and proton-proton would provide the true answer for observed strangeness enhancement.

Experimentally measured results on V_2 (= $\langle \cos 2(\varphi - \Psi)\rangle$, a measure of the azimuthal angle (φ) anisotropy of the produced particles, and Ψ is the reaction plane angle) of identified hadrons as a function of p_T show that at low p_T (<2 GeV/c) elliptic flow follows a pattern ordered by mass of the hadron (the V_2 values are smaller for heavier hadrons than that of lighter hadrons). At the intermediate p_T (2 GeV/c to 6 GeV/c) all mesons and all baryons form two different groups [35]. When V_2 and p_T are scaled by the number of constituent quarks (nq) of the hadrons, the magnitude of the scaled V_2 is observed to be the same for all the hadrons at the intermediate p_T.This observation is known as number-of-constituent quark scaling (NCQ scaling). This effect has been interpreted as collectivity being developed at the partonic stage of the evolution of the system in heavyion collision [36, 37]. Since ϕ meson has mass (1.0194 GeV/c^2) comparable to the masses of the lightest baryons such as proton and at the same time it is a meson, the study of ϕ meson V_2 would be more appropriate to understand the mass type and/or particle type (baryon-meson) dependence of $V_2(p_T)$.

In this review we have compiled all the available experimental measurements on ϕ meson production in high energy heavy-ion collisions as a function of p_T, azimuthal angle (φ), and rapidity (y). This paper is organised in the following manner. In Section 2, measurement of ϕ meson invariant yield has been presented from SPS to LHC energy. Section 3 describes the compilation of the azimuthal anisotropy measurements in ϕ meson production from all the available experimental results. Finally, the summary and conclusion have been discussed in Section 4.

INVARIANT YIELD OF ϕ MESON

Invariant Transverse Momentum Spectra.

We have compiled the data on the invariant pT spectra of the ϕ meson measured in p+p, d + A, and A + A systems for different collision centralities at various centre-of-mass energies ($\sqrt{s_{NN}}$ = 17.3 GeV–7 TeV) [16–22] and those are shown in Figure 2. Only the statistical errors are indicated as error bars. Measurement of ϕ meson invariant yield in p+p collisions at $\sqrt{s}$=7 TeV by the ATLAS collaboration [38] is consistent with that from the ALICE experiment and hence is not shown in Figure 2. The dashed black lines in Figure 2 are fits to the experimental data using an exponential function of the form

$$\frac{1}{2\pi p_T}\frac{d^2N}{dydp_T} = \frac{dN/dy}{2\pi T\left(m_0+T\right)}\exp\left[-\frac{\sqrt{m_0^2+p_T^2}-m_0}{T}\right]. \quad (1)$$

The blue solid lines in Figure 2 are the fits to the data with Levy function of the form given by

$$\frac{1}{2\pi p_T}\frac{d^2N}{dydp_T} = \frac{dN}{dy}\frac{(n-1)(n-2)}{2\pi nT\left(nT+m_0(n-2)\right)} \times\left(1+\frac{\sqrt{p_T^2+m_0^2}-m_0}{nT}\right)^{-n}. \quad (2)$$

T is known as the inverse slope parameter, dN/dy is the ϕ meson yield per unit rapidity, m_0 is the rest mass of ϕ meson, and n is the Levy function parameter. Levy function is similar in shape to an exponential function at low p_T and has a power-law-like shape at higher p_T. In fact, the exponential function is the limit of the Levy function as n approaches infinity. From Figure 2, it can be seen that the exponential and Levy functions both fit the central collision data equally well.

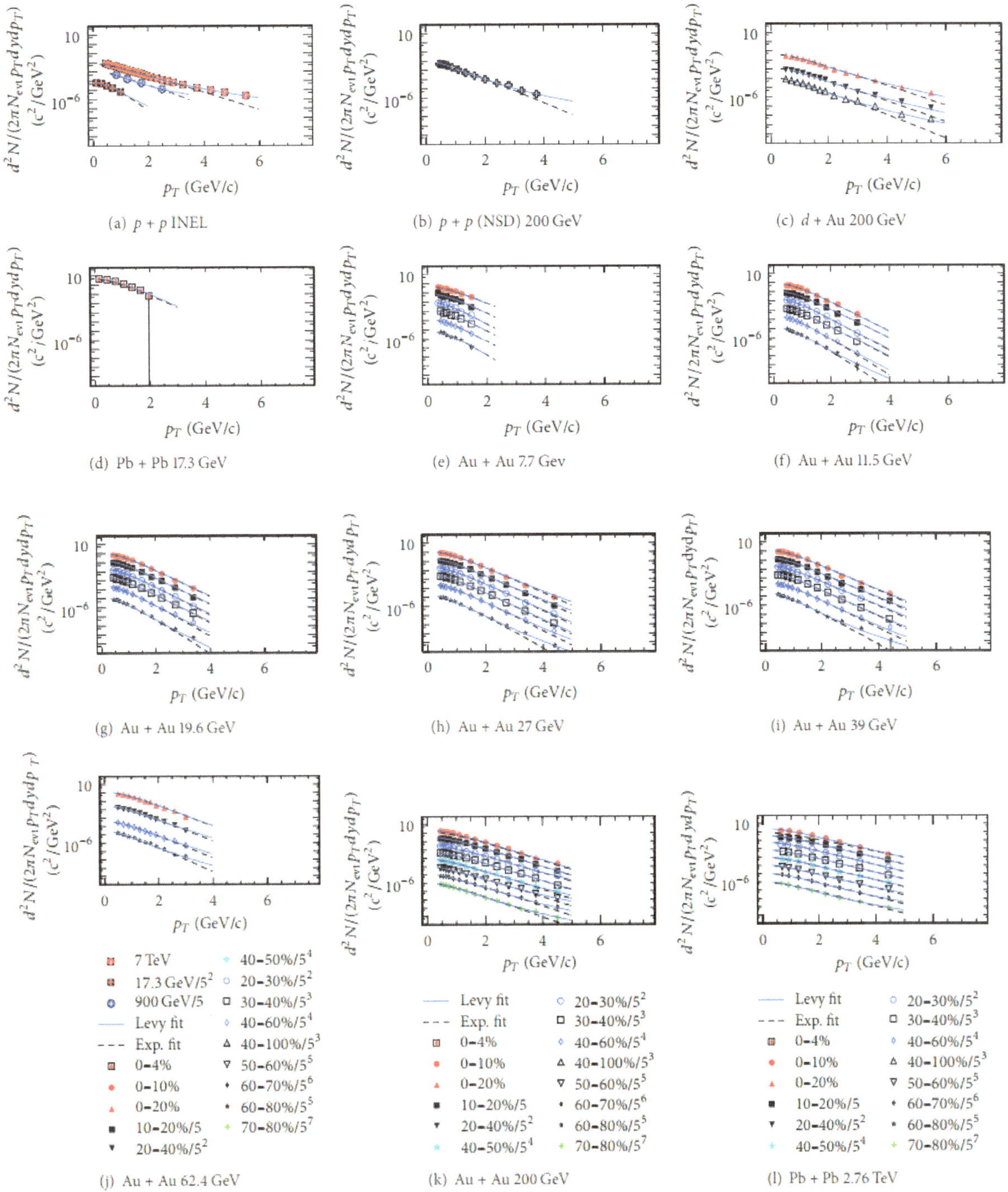

Figure 2: The invariant yield of ϕ mesons as a function of p_T measured for different system and different centralities at various centre-of-mass energies [16–21]. The black dashed (blue solid) line represents an exponential (Levy) function fit to the data.

However, with decreasing centrality, the exponential fits diverge from the data at higher transverse momentum and the Levy function fits the data better. The χ^2 /ndf values are larger for exponential function fits in peripheral collisions compared to Levy function fits (see Tables 1 and 2). This indicates a change in shape of the p_T spectra (deviations from exponential distribution and more towards a power law distribution) at high p_T for peripheral collisions. Tsallis function also describes the measured identified spectra equally well as Levy, which is shown in [39, 40]. Like Levy, Tsallis function describes both the low pT exponential and the high pT power law behaviors. The Tsallis function has two parameters while number of parameters for Levy is three. The exponential function fails to explain data at high p_T for p+p and d + Au collisions whereas Levy function describes data for all p_T. This evolution in the shape of the spectra from exponential-like in central collisions to more power-law-like in peripheral collisions reflects the increasing contribution from pQCD (hard) processes to ϕ meson production in more peripheral collisions at higher p_T. Particle production at low p_T is expected to be due to nonperturbative soft processes and with sufficient interactions the system could be thermalized, and that is why both exponential and Levy functions fit the data for all centralities at lowp_T. All the Levy and exponential fit parameters for A + A collisions are given in Tables 1 and 2. One can see that the values of parameter n are large in the case central A + A collisions where both Levy and exponential functions fit the data well. The dN/dy values increase from peripheral to central collisions, indicating increasing production of ϕ meson with increase in collision centrality.

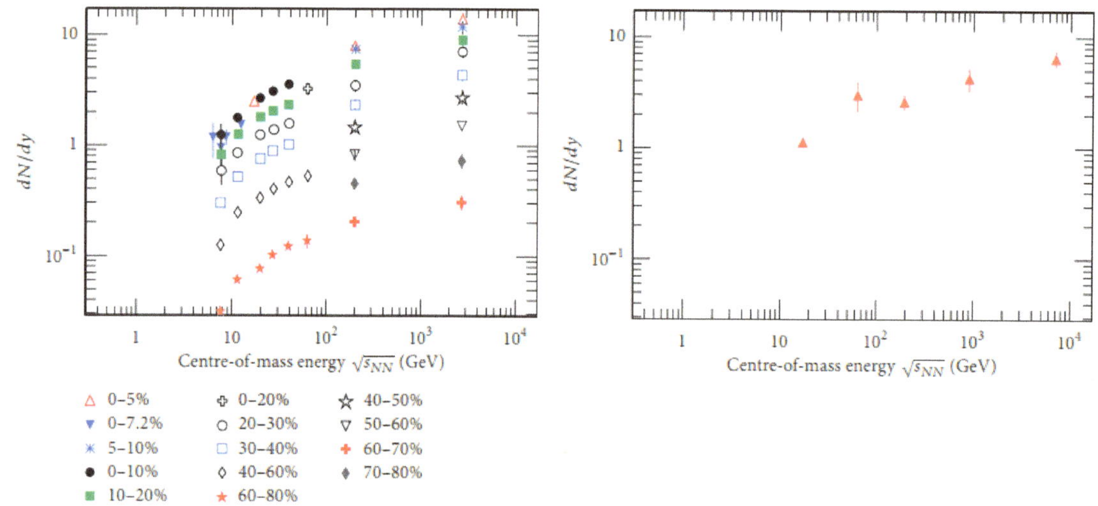

(a) A + A

(b) $p + p$ $(dN/dy) \times 200$

Figure 3: The ϕ meson midrapidity yield (dN/dy) as a function of $\sqrt{s_{NN}}$ for A + A [16–18, 21] and p+p collisions [18–20, 22, 23]. For RHIC BES energies ($\sqrt{s_{NN}}$ = 7.7–39 GeV) only statistical errors are shown whereas for other energies systematic errors are added in quadrature with statistical errors.

ϕ Meson Yield per Unit Rapidity.

In Figure 3 we present all available measurements of p_T integrated ϕ meson yield (dN/dy) at midrapidity as a function of centre-of-mass energy in both nucleus-nucleus [16–18, 21] and protonproton collisions [18–20, 22, 23]. For A + A collisions, different centralities are shown by different marker styles in Figure 3(a). The measured midrapidity yield increases with centrality and for the same centrality it increases with the collision energy for both A + A and p+p collisions. The rate of increases with $\sqrt{s_{NN}}$ is higher in A + A collisions compared to p+p collisions. We have observed that the measured midrapidity yield per participant (N_{part}) pair, $(dN/dy)/(0.5N_{\text{part}})$, increases nonlinearly with centrality and for the same N_{part} $(dN/dy)/(0.5N\text{part})$ increases with the collision energy of the A + A collisions. The former suggests that particle production does not scale with N_{part} and the latter is expected because of the increase of energy available to produce the ϕ mesons.

Strangeness Enhancement.

The ratio of strange hadron production normalised to $\langle N_{\text{part}} \rangle$ in nucleus-nucleus collisions relative to corresponding results from p+p collisions at 200 GeV [24] is shown in the left upper panel of Figure 4. The results are plotted as a function of $\langle N_{\text{part}} \rangle$. K^-, Λ, and Ξ are found to exhibit an enhancement (value > 1) that increases with the number of strange valence quarks. Furthermore, the observed enhancement in these open-strange hadrons increases with collision centrality, reaching a maximum for the most central collisions. However, the enhancement of ϕ meson production from Cu + Cu and Au + Au collisions shows a deviation in ordering in terms of the number of strange constituent quarks. Such deviation is also observed in central Pb + Pb collisions at SPS energy (as shown in the right bottom panel of Figure 4).The difference in the ordering does not seem to be a baryon-meson effect, since K^- and Λ have similar enhancement, or a mass effect, since Λ and ϕ have similar mass but different enhancement factors.

In heavy-ion collisions, the production of ϕ mesons is not Canonically suppressed due to its $s\bar{s}$ structure. The p+p collisions at RHIC are at an energy which is ~25 times higher than energies where violations of the Okubo-Zweig-Iizuka (OZI) rule were reported [41, 42].The observed enhancement of ϕ meson production then is a clear indication for the formation of a dense partonic medium being responsible for the strangeness enhancement in Au + Au collisions at 200 GeV. Furthermore, ϕ mesons do not follow the strange quark ordering as expected in the Canonical picture for the production of other strange hadrons. The observed enhancement in ϕ meson production being related to medium density is further supported by the energy dependence shown in the lower panel of Figure 4. The ϕ meson production relative to p+p collisions is larger at higher beam energy, a trend opposite to that predicted in Canonical models for other strange hadrons.

Table 1: Results from Levy fits to the transverse mass distributions of the ϕ meson. All values are for midrapidity ($|y| < 0.5$)

	Centrality	χ^2/ndf	T (GeV)	n	dN/dy
Pb + Pb (2.76 TeV)	0–10%	4.282/5	0.5005 ± 0.0128	$1.121e^{06} \pm 1.414e^{06}$	12.41 ± 0.796
	10–20%	4.793/5	0.5131 ± 0.0132	$4.969e^{05} \pm 2.6e^{05}$	8.978 ± 0.553
	20–30%	3.976/5	0.4926 ± 0.0429	162 ± 681.2	6.870 ± 0.415
	30–40%	1.796/5	0.4897 ± 0.0454	129.6 ± 436.7	4.190 ± 0.262
	40–50%	2.25/5	0.4401 ± 0.0424	28.34 ± 21	2.605 ± 0.162
	50–60%	3.177/5	0.3946 ± 0.042	15.41 ± 6.68	1.457 ± 0.094
	60–70%	0.551/5	0.3801 ± 0.0432	14.68 ± 6.23	0.7157 ± 0.050
	70–80%	0.910/5	0.3401 ± 0.04351	10.41 ± 3.32	0.297 ± 0.022
Au + Au (200 GeV)	0–10%	9.4/11	0.3572 ± 0.002331	102.9 ± 116	7.421 ± 0.106
	10–20%	19.2/11	0.3529 ± 0.002514	93.2 ± 101	5.142 ± 0.108
	20–30%	15.2/11	0.3591 ± 0.002343	41.6 ± 5.6	3.442 ± 0.071
	30–40%	16.2/11	0.3595 ± 0.002448	38.9 ± 20	2.189 ± 0.045
	40–50%	21.4/11	0.3153 ± 0.003802	22.7 ± 4.3	1.392 ± 0.033
	50–60%	6.9/11	0.2905 ± 0.003404	0.0138 ± 1.9	0.806 ± 0.023
	60–70%	7.4/11	0.2916 ± 0.002911	0.0186 ± 3.6	0.419 ± 0.014
	70–80%	5.5/11	0.2430 ± 0.002543	0.0130 ± 2.3	0.202 ± 0.009
Au + Au (62.4 GeV)	0–20%	9.2/8	0.3211 ± 0.00624	$9.993e^{06} \pm 6.39e^{06}$	3.693 ± 0.353
	20–40%	8.8/8	0.3217 ± 0.00353	$4.24e^{08} \pm 7.41e^{06}$	1.590 ± 0.142
	40–60%	14.5/8	0.2910 ± 0.00644	$9.409e^{07} \pm e^{10}$	0.580 ± 0.071
	60–80%	6.78/6	0.2681 ± 0.001426	21.45 ± 17.89	0.151 ± 0.020
Au + Au (39 GeV)	0–10%	1.645/9	0.2995 ± 0.01611	$6.619e^{10} \pm 8.931e^{05}$	3.402 ± 0.812
	10–20%	1.046/9	0.3131 ± 0.03514	$1.108e^{07} \pm 1.414e^{04}$	2.216 ± 0.278
	20–30%	1.55/8	0.2996 ± 0.01301	$1.532e^{08} \pm 6.651e^{05}$	1.597 ± 0.150
	30–40%	1.047/9	0.2957 ± 0.04906	945 ± 356	1.019 ± 0.0773
	40–60%	1.6383/9	0.2363 ± 0.03561	24.47 ± 14.96	0.456 ± 0.057
	60–80%	1.705/9	0.2110 ± 0.03410	20.2 ± 10.72	0.128 ± 0.018
Au + Au (27 GeV)	0–10%	3.719/9	0.2861 ± 0.007709	89.17 ± 78.22	3.051 ± 0.178
	10–20%	10.05/9	0.2851 ± 0.0039	56.62 ± 38.42	2.004 ± 0.0278
	20–30%	12.85/9	0.2747 ± 0.01754	46.3 ± 32.98	1.345 ± 0.082
	30–40%	14.06/9	0.2574 ± 0.01741	40.36 ± 30.44	0.846 ± 0.053
	40–60%	2.573/9	0.2114 ± 0.01696	19.3 ± 6.155	0.404 ± 0.026
	60–80%	15.946/9	0.1901 ± 0.01368	23.34 ± 7.753	0.107 ± 0.007
Au + Au (19.6 GeV)	0–10%	10.28/8	0.2803 ± 0.00466	87.17 ± 69.63	2.603 ± 0.051
	10–20%	10.96/8	0.2676 ± 0.01202	59.38 ± 47.27	1.786 ± 0.037
	20–30%	9.75/8	0.2367 ± 0.01038	32.32 ± 13.11	1.263 ± 0.029
	30–40%	8.804/8	0.2397 ± 0.01144	33.25 ± 26.33	0.759 ± 0.018
	40–60%	14.31/8	0.1947 ± 0.00865	15.19 ± 2.743	0.336 ± 0.007
	60–80%	7.346/8	0.1913 ± 0.00553	15.76 ± 2.849	0.080 ± 0.001
Pb + Pb (17.3 GeV)	0–4%	1.329/2	0.1746 ± 0.01043	$1000 \pm 1.175e^{05}$	0.00227 ± 0.0013
Au + Au (11.5 GeV)	0–10%	6.694/7	0.2662 ± 0.009276	100.73 ± 75.39	1.733 ± 0.112
	10–20%	8.171/7	0.2653 ± 0.01187	60.29 ± 42.09	1.121 ± 0.076
	20–30%	4.599/7	0.2282 ± 0.02887	37.3 ± 32.05	0.772 ± 0.054
	30–40%	10.18/7	0.2353 ± 0.03014	34.36 ± 30.39	0.467 ± 0.034
	40–60%	2.898/7	0.1846 ± 0.02255	18.9 ± 10.09	0.205 ± 0.015
	60–80%	1.034/7	0.1438 ± 0.01981	11.31 ± 3.681	0.056 ± 0.005
Au+Au (7.7 GeV)	0–10%	1.5978/4	0.3082 ± 0.03636	90.33 ± 79.97	1.21 ± 0.098
	10–20%	2.8/4	0.2419 ± 0.02615	64.29 ± 40.33	0.719 ± 0.061
	20–30%	2.946/4	0.1639 ± 0.07128	50.19 ± 35.39	0.518 ± 0.091
	30–40%	1.299/4	0.2039 ± 0.02201	63.3 ± 39.39	0.275 ± 0.024
	40–60%	1.833/4	0.1562 ± 0.04774	7.10 ± 6.67	0.139 ± 0.015
	60–80%	2.816/4	0.1423 ± 0.0842	15.02 ± 52.03	0.033 ± 0.007

Table 2: Results from exponential fits to the transverse mass distributions of the ϕ meson. All values are for midrapidity ($|y| < 0.5$)

	Centrality	χ^2/ndf	T (GeV)	dN/dy
Pb + Pb (2.76 TeV)	0–10%	4.28/6	0.5005 ± 0.0128	12.41 ± 0.795
	10–20%	4.793/6	0.5131 ± 0.013	8.973 ± 0.502
	20–30%	4.032/6	0.5024 ± 0.0127	4.032 ± 0.412
	30–40%	1.88/6	0.5027 ± 0.0130	4.171 ± 0.013
	40–50%	4.036/6	0.4961 ± 0.0137	2.550 ± 0.0162
	50–60%	8.356/6	0.4901 ± 0.0146	1.408 ± 0.014
	60–70%	5.937/6	0.4891 ± 0.0156	0.679 ± 0.047
	70–80%	10.12/6	0.4815 ± 0.0182	0.271 ± 0.024
Au + Au (200 GeV)	0–10%	11.2/12	0.3562 ± 0.002431	7.440 ± 0.106
	10–20%	9.7/12	0.3522 ± 0.002614	5.371 ± 0.108
	20–30%	26.7/12	0.3731 ± 0.002413	3.435 ± 0.071
	30–40%	21.1/12	0.3873 ± 0.002548	2.291 ± 0.045
	40–50%	26.4/12	0.3671 ± 0.00307	1.342 ± 0.033
	50–60%	70/12	0.3605 ± 0.003604	0.727 ± 0.023
	60–70%	54.4/12	0.3516 ± 0.003911	0.380 ± 0.014
	70–80%	31.7/12	0.3330 ± 0.004543	0.170 ± 0.009
Au + Au (62.4 GeV)	0–20%	8.4/9	0.328 ± 0.00624	3.523 ± 0.353
	20–40%	8.4/9	0.324 ± 0.00353	1.590 ± 0.140
	40–60%	14.5/9	0.308 ± 0.00644	0.584 ± 0.072
	60–80%	13.3/9	0.2791 ± 0.01426	0.152 ± 0.022
Au + Au (39 GeV)	0–10%	10.36/10	0.3015 ± 0.0052	3.549 ± 0.070
	10–20%	15.54/10	0.3019 ± 0.00522	2.347 ± 0.045
	20–30%	3.946/9	0.3004 ± 0.005711	1.618 ± 0.030
	30–40%	12.57/10	0.2826 ± 0.00398	1.067 ± 0.019
	40–60%	20.62/10	0.267 ± 0.00362	0.498 ± 0.006
	60–80%	21.08/10	0.2371 ± 0.00501	0.124 ± 0.002

Au + Au (27 GeV)	0–10%	26.13/10	0.29 ± 0.00245	3.46 ± 0.036
	10–20%	32.3/10	0.2873 ± 0.00227	2.152 ± 0.023
	20–30%	24.93/10	0.2766 ± 0.00236	1.471 ± 0.016
	30–40%	13.57/10	0.2645 ± 0.00225	0.926 ± 0.010
	40–60%	25.19/10	0.2601 ± 0.00217	0.422 ± 0.004
	60–80%	71.08/10	0.2296 ± 0.00250	0.100 ± 0.001
Au + Au (19.6 GeV)	0–10%	28.87/9	0.2805 ± 0.003265	2.609 ± 0.038
	10–20%	15.28/9	0.2788 ± 0.003722	1.817 ± 0.026
	20–30%	13.63/9	0.2608 ± 0.003246	1.187 ± 0.017
	30–40%	7.811/9	0.2652 ± 0.003917	0.763 ± 0.011
	40–60%	17.78/9	0.2404 ± 0.003352	0.345 ± 0.004
	60–80%	10.06/9	0.2142 ± 0.003558	0.072 ± 0.001
Pb + Pb (17.3 GeV)	0–4%	1.337/3	0.1748 ± 0.01043	0.0022 ± 0.0012
Au + Au (11.5 GeV)	0–10%	11.5/8	0.2621 ± 0.006674	1.754 ± 0.047
	10–20%	9.638/8	0.265 ± 0.006731	1.281 ± 0.033
	20–30%	6.885/8	0.2359 ± 0.006009	0.843 ± 0.022
	30–40%	1.946/8	0.2453 ± 0.005765	0.506 ± 0.013
	40–60%	2.862/8	0.2071 ± 0.00545	0.235 ± 0.005
	60–80%	11.12/8	0.1947 ± 0.00888	0.051 ± 0.001
Au + Au (7.7 GeV)	0–10%	3.36/5	0.3082 ± 0.03087	1.207 ± 0.069
	10–20%	10.46/5	0.2553 ± 0.01811	0.783 ± 0.036
	20–30%	9.767/5	0.2207 ± 0.0141	0.471 ± 0.023
	30–40%	4.177/5	0.2375 ± 0.01926	0.329 ± 0.017
	40–60%	0.4763/5	0.2195 ± 0.01467	0.147 ± 0.007
	60–80%	3.958/5	0.1992 ± 0.02282	0.035 ± 0.002

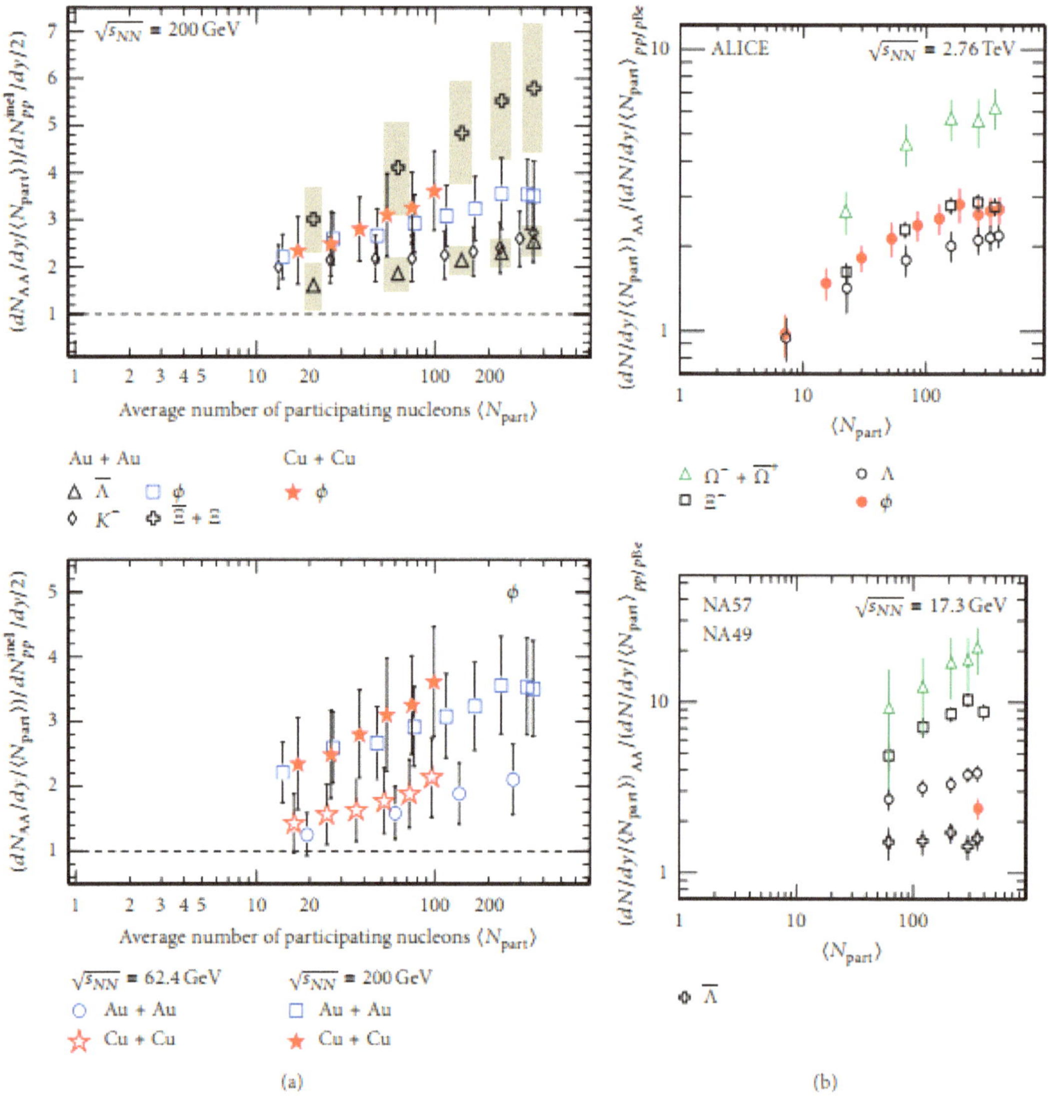

Figure 4: (a) The ratio of the yields of K^-, ϕ, $\overline{\Lambda}$, and $\Xi + \overline{\Xi}$ normalised to $\langle N_{part}\rangle$ nucleus-nucleus collisions and to corresponding yields in proton-proton collisions as a function of $\langle N_{part}\rangle$ at 62.4 and 200 GeV [24]. Error bars are quadrature sum of statistical and systematic uncertainties. (b) The ratio of ⟨*N*part⟩ normalised yield of ϕ, Λ, $\overline{\Lambda}$, Ξ^-, and $\Omega^- + \overline{\Omega}^+$ in Pb + Pb collisions to the corresponding yield in *p*+*p* (*p* + Be) collisions at 17.3 GeV (NA57 & NA49) [25, 26] and 2.76 TeV (ALICE) [21]. Only statistical uncertainties are shown.

The right upper panel of Figure 4 shows the enhancement in Pb + Pb with respect to p+p reference yields for ϕ, Λ, Ξ^-, and $\Omega^- + \overline{\Omega}^+$ at $\sqrt{s_{NN}}$ = 2.76 TeV [21]. The ϕ, Ξ, and Ω yield in p+p collisions at $\sqrt{s}$ = 2.76 TeV have been estimated by interpolating between the measured yields at $\sqrt{s}$ = 0.9 TeV and $\sqrt{s}$=7 TeV. The reference Λ yield in p+p collisions at $\sqrt{sNN}$ = 2.76 TeV is estimated by extrapolating from the measured yield in (inelastic) p+p collisions available up to $\sqrt{s}$ = 0.9 TeV. Details can be found in [21]. Enhancement factor increases linearly with N_{part} until $N_{part} \approx 100$; then the enhancement values seem to be saturated for higher values of N_{part}. Unlike SPS and RHIC, the order of ϕ enhancement is the same as Ξ^- at LHC energy. We have observed that the ϕ enhancement at central collisions increases from SPS to RHIC energy but the enhancement factor is comparable, within errors, to the values at RHIC and LHC.

These findings tell us that the observed ϕ meson enhancement is not due to the Canonical suppression effects. Therefore this enhancement is very likely due to the formation of a deconfined medium. Since other strange hadrons also emerge from the same system, their enhancement is most likely also due to formation of deconfined matter or quark-gluon plasma (QGP) in heavy-ion collisions.

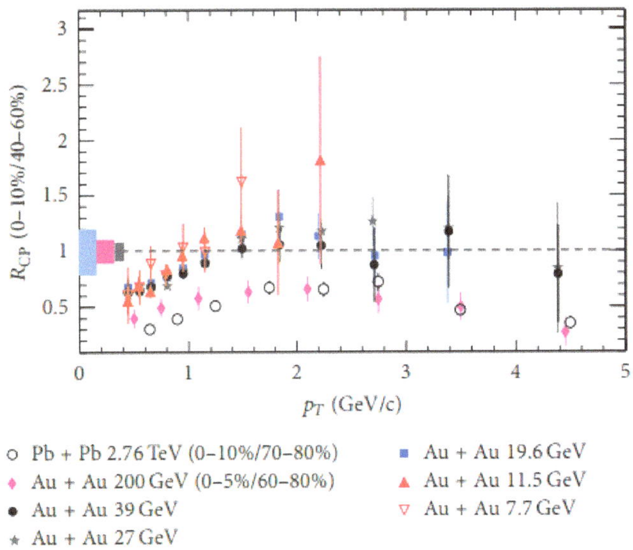

Figure 5: The ϕ mesonR_{CP} as a function of p_T in Au + Au [17,18] and Pb + Pb [21] collisions at various beam energies. Error bars are only statistical uncertainties. Bands represent normalisation error from N_{bin} which is approximately 20% for $\sqrt{sNN}$ = 7.7–39 GeV, ~10% for 200 GeV, and ~7% for 2.76 TeV.

Nuclear Modification Factor.

In order to understand parton energy loss in the medium created in high energy heavyion collisions for different centralities in A + A collisions, the nuclear modification factor (R_{CP}) is measured which is defined as follows:

$$R_{CP} = \frac{\text{Yield}_{\text{central}}}{\text{Yield}_{\text{peripheral}}} \times \frac{\langle N_{\text{bin}} \rangle_{\text{peripheral}}}{\langle N_{\text{bin}} \rangle_{\text{central}}}, \tag{3}$$

where $\langle N_{bin} \rangle$ is the average number of binary collisions for the corresponding centrality. The value of Nbin was calculated from the Monte Carlo Glauber simulation [43]. R_{CP} value is equal to one when the nucleus-nucleus collisions are simply the superposition of nucleon-nucleon collisions. Therefore deviation of R_{CP} from the unity would imply contribution from the nuclear medium effects, specifically jet-quenching [44]. Nuclear modification factors (R_{CP}) of ϕ mesons at midrapidity in Au + Au collisions at $\sqrt{s_{NN}}$ = 7.7, 11.5, 19.6, 27, 39, and 200 GeV [17, 18] and in Pb + Pb collisions at $\sqrt{s_{NN}}$ = 2.76 TeV [21] are shown in Figure 5. We can see that the R_{CP} of ϕ mesons goes below unity at 200 GeV and 2.76 TeV in nucleus-nucleus collisions. The most feasible explanation of this observation to date is due to the energy loss of the partons traversing the high density QCD medium. This implies that a deconfined medium of quarks and gluons was formed at 200 GeV and 2.76 TeV [18, 21]. For $\sqrt{s_{NN}} \leq 39$ GeV, ϕ meson R_{CP} is greater than or equal to unity at the intermediate p_T, which indicates that at low energy the parton energy loss contribution to R_{CP} measurement could be less important. In order to confirm that the $R_{CP} < 1$ is due to parton energy loss or jet-quenching phenomenon, it is important to study R_{CP} in p + A or d + A collisions. Nuclear modifications in such systems are expected to be effected by the Cronin effect [45] and not by QGP effect. Due to the Cronin effect the value of R_{CP} at high p_T is expected to be greater than one.

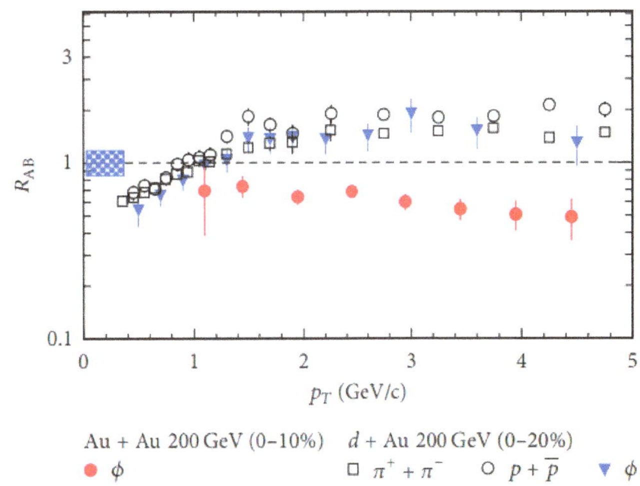

Figure 6: The nuclear modification factor R_{AB} as a function of p_T in Au + Au and d + Au [18, 27] collisions at $\sqrt{s_{NN}}$ = 200 GeV. Rectangular bands show the uncertainties associated with estimation of number of binary collisions. Error bars are quadrature sum of statistical and systematic uncertainties for ϕ in d + Au and only statistical for three other cases.

Figure 6 presents the p_T dependence of the nuclear modification factorR_{AB} in Au + Au and d + Au collisions at$\sqrt{s_{NN}}$ = 200 GeV [18, 27]. The definition of R_{AB} is the ratio of the yields of the hadron produced in the nucleus (A) + nucleus (B) collisions to the corresponding yields in the inelastic p+p collisions normalised by Nbin. The R_{AB} of ϕ mesons for d + Au collisions show a similar enhancement trend as those for $\pi^+ + \pi^-$ and $p + p$ at the intermediate p_T. This enhancement in d + Au collisions was attributed to be due to the Cronin effect [45]. The Cronin enhancement may result either from momentum broadening due to multiple soft [46] (or semihard [47, 48]) scattering in the initial state or from final state interactions as suggested in the recombination model. These mechanisms lead to different particle type and/or mass dependence in the nuclear modification factors as a function of pT. Current experimental measurements on ϕ meson R_{AB} in d + Au do not seem to have the precision to differentiate between particle type dependence types [49, 50]. On the other hand, the R_{AB} in Au + Au (i.e., R_{AA}) at 200 GeV is lower than that in d + Au at 200 GeV and is less than unity [27]. These features are consistent with the scenario of energy loss of the partons in a QGP medium formed in central Au + Au collisions.

Mean Transverse Mass.

Figure 7 shows the difference in mean transverse mass ($m_T = \sqrt{p_T^2 + m_0^2}$) and rest mass ($m_0$), that is, $\langle m_T \rangle - m_0$ for ϕ meson, as a function of centre-of-mass energy for p+p [18, 20, 22], Au + Au [17, 18], and Pb + Pb [16, 21] collisions. Due to limited statistics, result for p+p at $\sqrt{s}$ = 900 GeV is not shown. The data points in Figure 7 are connected by the lines to guide the eye of the reader. One can see that $\langle m_T \rangle - m_0$ increases monotonically with $\sqrt{sNN}$ in p+p collisions whereas the corresponding data in A + A collisions changes slope twice as a function of center-of-mass energy. In A + A collisions, $\langle m_T \rangle - m_0$ first increases with $\sqrt{s_{NN}}$ and then stays independent of energy from approximately 17 GeV to 39 GeV, followed by again an increase with $\sqrt{s_{NN}}$. For a thermodynamic system, the $\langle m_T \rangle - m_0$ can be interpreted as a measure of temperature of the system, and $dN/dy \propto \ln(\sqrt{s_{NN}})$ may represent its entropy. In such a scenario, this observation could reflect the characteristic signature of a first order phase transition [51]. Then the constant value of $\langle m_T \rangle - m_0$ for ϕ meson from 17 GeV to 39 GeV could be interpreted as a formation of a mixed phase of a QGP and hadrons during the evolution of the heavy-ion system.

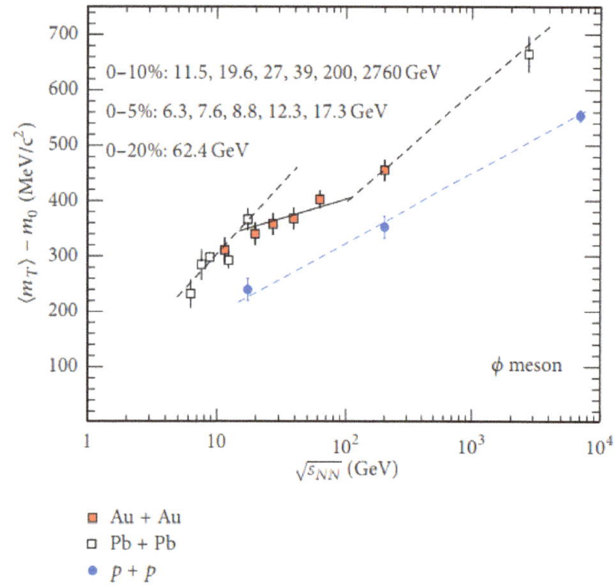

Figure 7: $\langle m_T \rangle - m_0$ as a function centre-of-mass energies in central A + A and $p + p$ collisions. Only statistical errors are shown. The dashed and solid lines are the straight lines connected to the data to guide the eye of the reader.

Particle Ratios

ϕ/K^-.

The mechanism for ϕ meson production in high energy collisions has remained an open issue. In an environment with many strange quarks, ϕ mesons can be produced readily through coalescence, bypassing the OZI rule [52]. On the other hand, a naive interpretation of ϕ meson production in heavy-ion collisions would be the ϕ production via kaon coalescence. In the latter case one could expect an increasing trend of ϕ/K ratio as function of collision centrality and centre-of-mass energy. Models that include hadronic rescatterings such as UrQMD [53, 54] have predicted an increase of the ϕ/K^- ratio at midrapidity as a function of centrality [18]. Therefore, the ratio of ϕ meson yield to that of the kaons can be used to shed light on ϕ meson production mechanism. Figure 8 shows the ϕ/K^- ratio as a function of number of participants for different centre-ofmass energies [18]. The UrQMD model prediction for ϕ/K^- in Au + Au 200 GeV collisions is shown by red dashed line. However, this prediction was disproved from experimental data. It is clear from Figure 8 that ϕ/K^- is independent of centrality and also centre-of-mass energy. In addition, if ϕ production is dominantly from $K\bar{K}$ coalescence, one expects the width of the rapidity distribution of ϕ mesons to be related to those for charged kaons as $1/\sigma_\phi^2 = 1/\sigma_{K^-}^2 + 1/\sigma_{K^+}^2$. Measurements at SPS energies show a clear deviation of the data from the above expectation [55]. Finally, if ϕ production is dominantly from $K\bar{K}$ coalescence it would be reflected in elliptic flow (V_2) measurements. We observe at intermediate p_T that the V_2 of ϕ mesons and kaons are comparable (discussed in Section 3). All these measurements effectively rule out kaon coalescence as the dominant production mechanism for the ϕ meson for this energy region.

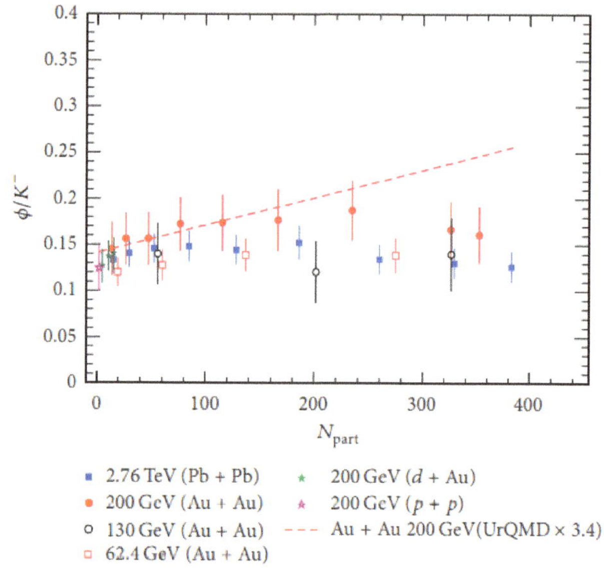

Figure 8: ϕ/K^- ratio as a function of number of participants in Au + Au [18] and Pb + Pb [21] collision at various beam energies.

Ω/ϕ.

The production mechanism of multistrange hadrons (e.g., ϕ and Ω) is predicted to be very sensitive to the early phase of nuclear collisions [56], because both ϕ and Ω freeze out early, have low hadronic interaction cross-section, and are purely made of strange and antistrange quarks. Therefore the ratio $(\Omega)/(\phi)$ is expected to reflect the information of strange quark dynamics in the early stage of the system created in the nucleus-nucleus collision [57].

Figure 9 shows the baryon-to-meson ratio in strangeness sector, $N(\Omega^- + \overline{\Omega}^+)/2N(\phi)$, as a function of p_T in Au + Au collisions at $\sqrt{s_{NN}}$ = 11.5 GeV to 2760 GeV [17, 18, 21]. The dashed lines are the results from the recombination model calculation by Hwa and Yang for $\sqrt{s_{NN}}$ = 200 GeV [57]. In this model the ϕ and Ω yields in the measured p_T region are mostly from the recombination of thermal strange quarks, which were assumed to follow an exponential p_T distribution.

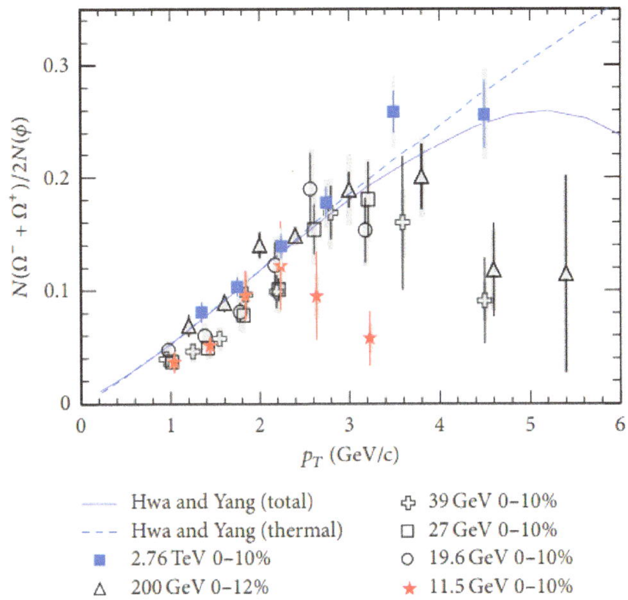

Figure 9: The baryon-to-meson ratio, $N(\Omega^- + \overline{\Omega}^+)/2N(\phi)$, as a function of p_T in midrapidity ($|y| < 0.5$) from central A + A collisions at $\sqrt{s_{NN}}$ = 11.5, 19.6, 27, 39, 200, and 2760 GeV [17, 18, 21]. Gray bands denote systematical errors.

The thermal s quark distribution was determined by fitting the low p_T data of kaon production. The contribution from hard parton scattering was assumed to be negligible unless p_T is large. Details of this recombination model have been given in [57]. We can see from Figure 9 that, in central A + A collisions at $\sqrt{s_{NN}} \geq 19.6$ GeV, the ratios of $N(\Omega^- + \Omega^+)/2N(\phi)$ in the intermediate p_T range are explained by the recombination model with thermal strange quarks and show a similar trend. The model agrees well with the trend of the data up to $p_T \sim$ 4 GeV/c which covers ~95% of the total yields for the ϕ and Ω. The observations imply that the production of ϕ and Ω in central Au + Au collisions is predominantly through the recombination of thermal s quarks for $\sqrt{s_{NN}} \geq 19.6$ GeV. But at $\sqrt{s_{NN}}$ = 11.5 GeV, the ratio at the highest measured p_T shows a deviation from the trend of other energies. This may indicate a change in Ω and/or ϕ production mechanism at $\sqrt{s_{NN}}$ = 11.5 GeV.

AZIMUTHAL ANISOTROPY IN ϕ MESON PRODUCTION

In noncentral nucleus-nucleus collisions, the initial spatial anisotropy is transformed into a final state momentum anisotropy in the produced particle distributions because of pressure gradient developed due to the interactions among the systems constituents [58–62]. The elliptic flow (V_2) [63– 65] is a measure of the second order azimuthal anisotropy of the produced particles in the momentum space. It can be used as probe for the properties of the medium created in the heavy-ion collisions. Because of its self-quenching nature, it carries information from the early phase. Although elliptic flow is an early time phenomenon, its magnitude might still be affected by the later stage hadronic interactions. Since the hadronic interaction cross-section of ϕ meson is smaller than the other hadrons [6] and freezes out relatively early [8], its V_2 remain almost unaffected by the later stage interaction. Therefore ϕ meson V_2 can be considered as good and clean probe for early system created in nucleus-nucleus collisions. Further the ϕ mesons seem to be formed by coalescence of strange quarks and antiquarks in a deconfined medium of quarks and gluons; hence the measurement of collectivity in ϕ mesons would reflect the collectivity in the partonic phase. In addition, its mass is comparable to the masses of the lightest baryon (p and Λ); therefore comparison of ϕ meson V_2 with that of proton and Λ will be helpful to distinguish the mass effect and/or baryon-meson effect in $V_2(p_T)$.

Differential ϕ Meson V2.

Elliptic flow of ϕ meson as a function of p_T measured at midrapidity [28–31] is shown in Figure 10. The shape of ϕ $V_2(p_T)$ is similar for $\sqrt{s_{NN}}$ = 19.6 GeV to 2760 GeV. But at 7.7 and 11.5 GeV, the ϕ V_2 values at the highest measured p_T bins are observed to be smaller than other energies. Various model studies predicted that ϕ meson V_2 will be small for a system with hadronic interactions [66, 67]. Small interaction cross-section of ϕ meson in hadronic phase suggests that ϕ meson V_2 mostly reflects collectivity from the partonic phase; hence small ϕ meson V_2 indicates less contribution to the collectivity from partonic phase. So the large ϕ meson V_2 at $\sqrt{s_{NN}} \geq 15$ GeV indicates the formation of partonic matter and small V_2 at $\sqrt{s_{NN}}$ ≤ 11.5 could indicate dominance of hadron interactions.

Number-of-Constituent Quark Scaling.

In Figure 11, the V2 scaled by number-of-constituent quarks (n_q) as a function $(m_T - m_0)/n_q$ for identified hadrons in Au + Au collisions at $\sqrt{s_{NN}}$ = 7.7–200 GeV are presented. We can see from Figure 11 that for $\sqrt{s_{NN}}$ = 19.6–200 GeV the V_2 values follow a universal scaling for all the measured hadrons. This observation is known as the NCQ scaling. The observed NCQ scaling at RHIC can be explained by considering particle production mechanism via the quark recombination model and can be considered as a good signature of partonic collectivity [36, 37]. Therefore, such a scaling should vanish for a purely hadronic system if formed in the heavy-ion collisions at the lower energies. At the same time the study of NCQ scaling of identified hadrons from UrQMD model shows that the pure hadronic medium can also reproduce such scaling in V2 [68–70]. This is due to modification of initially developed V_2 by later stage hadronic interactions and the production mechanism as implemented in the model [69]. Hence to avoid these ambiguities, the V_2 of those particles which do not interact with hadronic interaction will be the clean and good probe for early dynamics in heavy-ion collisions. Due to small hadronic interaction cross-section, ϕ mesons V_2 are almost unaffected by later stage interaction and it will have negligible value if ϕ mesons are not produced via s and $\overline{s}$ quark coalescence [66, 67]. Therefore, NCQ scaling of ϕ mesons V_2 can be considered as the key observables for the partonic collectivity in heavy-ion collisions. As we can see from Figure 11, at $\sqrt{s_{NN}} = 7.7$ and 11.5 GeV, the ϕ meson V_2 deviates from the trend of the other hadrons at the highest measured p_T values by 1.8σ and 2.3σ, respectively. This could be the effect for a system, where hadronic interactions are more important.

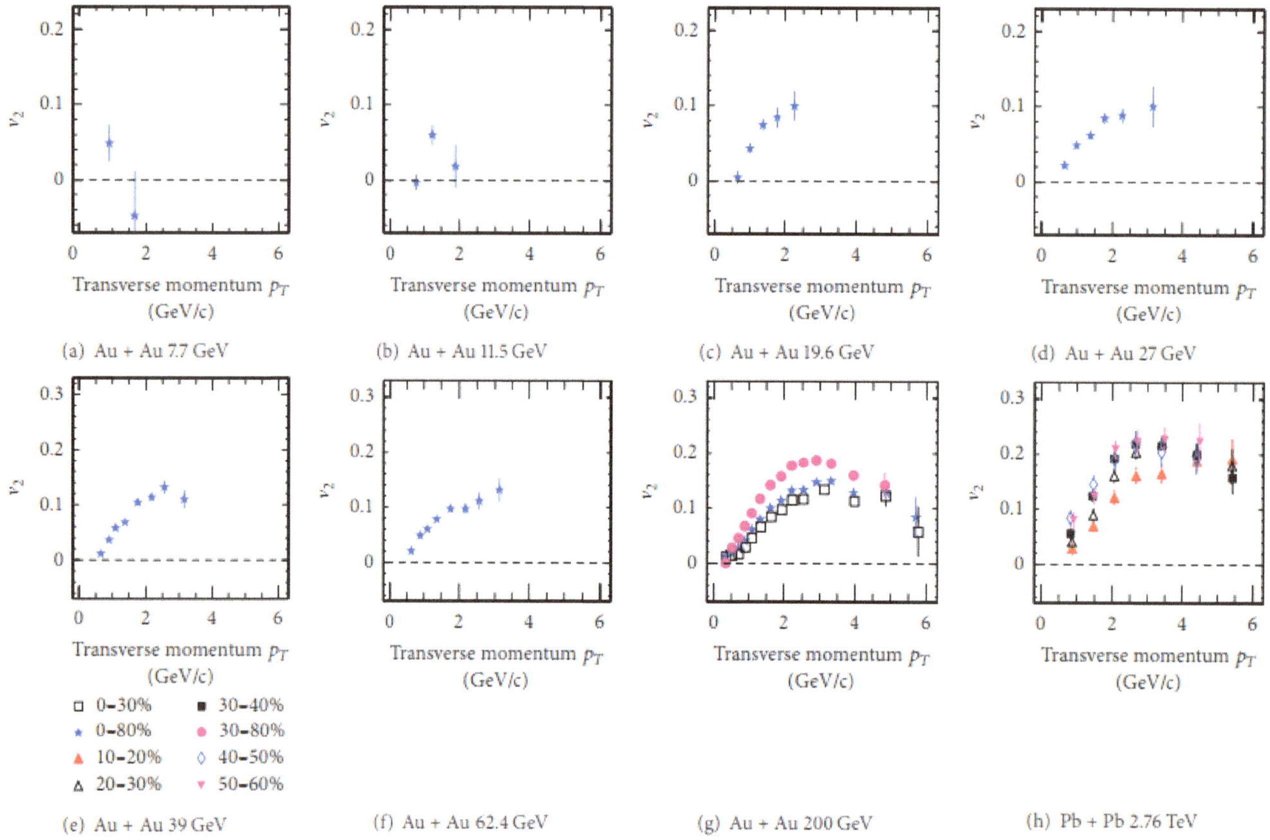

Figure 10: The ϕ meson $V_2(p_T)$ at midrapidity in Au+Au collisions at $\sqrt{s_{NN}} = 7.7$–62.4 GeV for 0–80% centrality [28] and at $\sqrt{s_{NN}} = 200$ GeV for 0–80%, 0–30%, and 30–80% centralities [29, 30] and in Pb + Pb collisions at $\sqrt{s_{NN}} = 2.76$ TeV [31] for different collisions centralities. The vertical lines are statistical uncertainties.

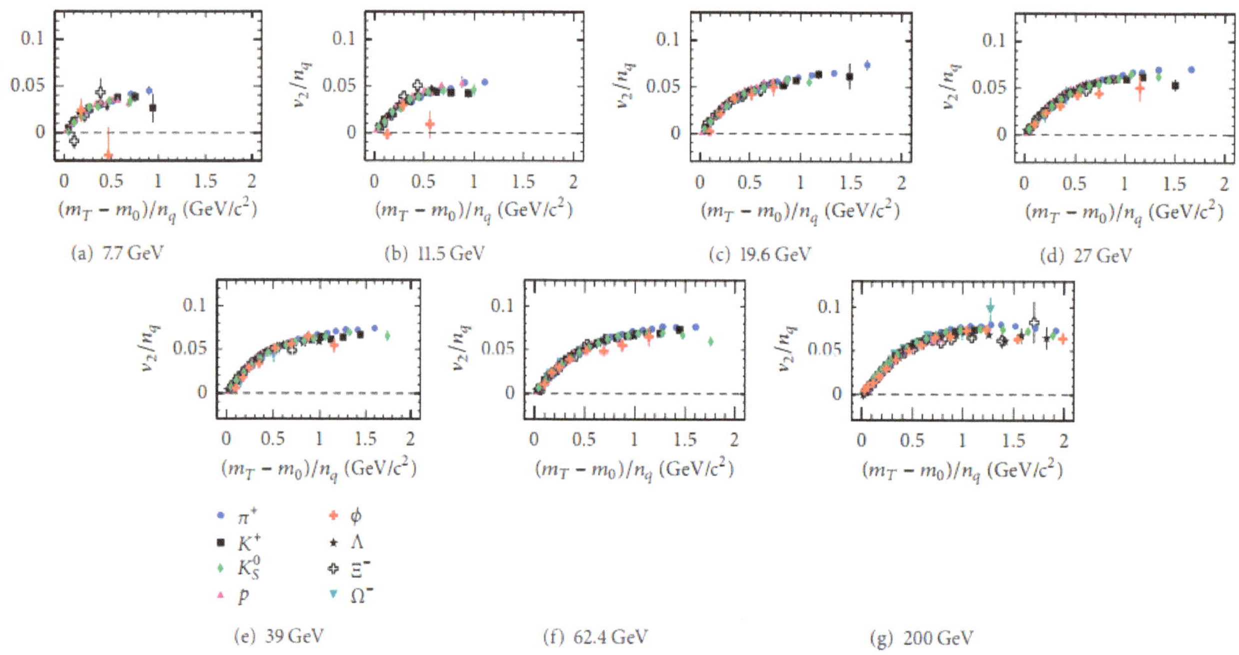

Figure 11: The NCQ-scaled elliptic flow, V_2/n_q, versus $(m_T - m_0)/n_q$ for 0–80% central Au + Au collisions for selected identified particles [28–30]. Only statistical error bars are shown.

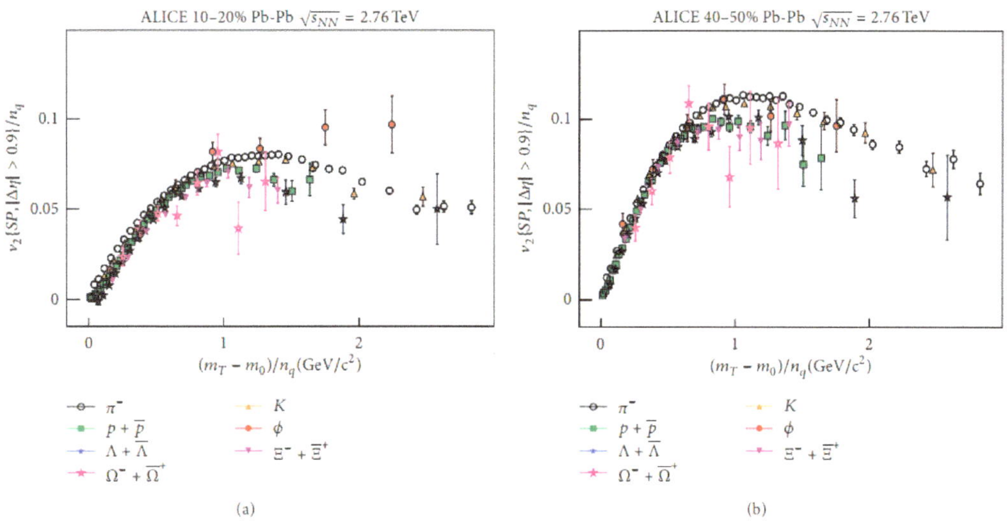

Figure 12: The NCQ-scaled elliptic flow, V_2/n_q, versus $(m_T - m_0)/n_q$ for 10–20% and 40–50% central Pb + Pb collisions for selected identified particles [31]. Only statistical error bars are shown. The figure has been taken from the presentation at Quark Matter 2014 by ALICE collaboration.

Figure 12 presents the $(m_T - m_0)/n_q$ dependence of V_2/n_q for 10–20% and 40–50% central Pb + Pb collisions for selected identified particles [31]. It can be seen that, at higher value of $(m_T - m_0)/nq$, the scaling is not good compared to that observed at RHIC energies. There are deviations at the level of ±20% with respect to the reference ratio as shown in [31].This larger deviation at LHC energy could be related to observed large radial flow at LHC compared to RHIC.

p_T Integrated ϕ Meson V_2.

The p_T integrated elliptic flow ($\langle V_2\rangle$) can be calculated as

$$\langle v_2 \rangle = \frac{\int v_2(p_T)(dN/dp_T)\,dp_T}{\int (dN/dp_T)\,dp_T}. \quad (4)$$

Figure 13 shows p_T integrated ϕ meson (red star) and proton (blue circle) V_2 as a function of centre-of-mass energy for 0–80% centrality [32]. One can see that for both particle species the $\langle V_2\rangle$ increases with increasing beam energy. ϕ meson $\langle V_2\rangle$ from a multiphase transport (AMPT) model for three different scenarios is shown by shaded bands. Green band corresponds to AMPT default model which includes only hadronic interaction whereas black and yellow bands correspond to AMPT with string melting scenario with parton-parton cross-sections of 3 mb and 10 mb, respectively. In contrast to observations from the data, the $\langle V_2\rangle$ values from model remain constant for all the energies. This is because they have been obtained for a fixed parton-parton interaction cross-section. The $\langle V_2\rangle$ of ϕ mesons for $\sqrt{s_{NN}} \geq 19.6$ GeV can be explained by the AMPT with string melting (SM) version, by varying the parton-parton cross-section. On the other hand, both the AMPT-SM and the AMPT default models overpredict data at $\sqrt{s_{NN}} = 11.5$ GeV. The comparison to AMPT model results indicates negligible contribution of the partonic interactions to the final measured collectivity for $\sqrt{s_{NN}} = 11.5$ GeV. For $\sqrt{s_{NN}} > 19.6$ GeV, proton and ϕ meson show similar magnitude of $\langle V_2\rangle$. The proton is a baryon and ϕ is a meson; in addition they are composed of different quark flavours, yet they have similar $\langle V_2\rangle$; this is a strong indication of large fraction of the collectivity being developed in the partonic phase. However at $\sqrt{s_{NN}} \leq 19.6$ GeV, ϕ meson $\langle V_2\rangle$ values show deviation from that for proton and at $\sqrt{s_{NN}} = 11.5$ GeV ϕ meson $\langle V_2\rangle$ becomes small (~1.5%). This tells us that due to the lack of enough partonic interactions at lower beam energies a larger $\langle V_2\rangle$ could not be generated for ϕ mesons. The contribution to ϕ $\langle V_2\rangle$ from hadronic interactions is small because of small ϕ-hadron interaction cross-sections. However the observed higher collectivity of protons compared to ϕ mesons at the lower beam energies could be due to the protons having larger hadronic interaction cross-section.

Hadronic Rescattering Effect on V_2.

Recent phenomenological calculation based on ideal hydrodynamical model together with the later stage hadron cascade (hydro + JAM) shows that the mass ordering of V_2 could be broken between that of ϕ meson and that of proton at low p_T ($p_T < 1.5$ GeV/c) [71]. This is because of late stage hadronic rescattering effects on proton V_2. The model calculation was done by considering low hadronic interaction cross-section for ϕ meson and larger hadronic interaction cross-section for proton. The ratio between ϕ V_2 and proton V_2 is shown in Figure 14 for minimum bias Au + Au collisions at $\sqrt{s_{NN}} = 200$ GeV. The data from the STAR experiment are shown by solid red square and blue solid circle [29, 30]. Solid red square and blue solid circle correspond to 0–30% and 30–80% centralities, respectively. The ratios are larger than unity at low p_T region ($p_T < 0.7$ GeV/c) for 0–30% centrality although mass of the ϕ meson (1.019 GeV/c^2) is greater than mass of the proton (0.938 GeV/c^2).This is qualitatively consistent with the model calculation using hydro + JAM shown by red bands.Therefore this observation is consistent with the physical scenario of larger effect of hadronic rescattering on proton V_2 which reduces its value, as predicted in the theoretical model [67, 71]. Due to small hadronic interaction cross-section ϕ meson V_2 remains unaffected by later stage hadronic rescattering.

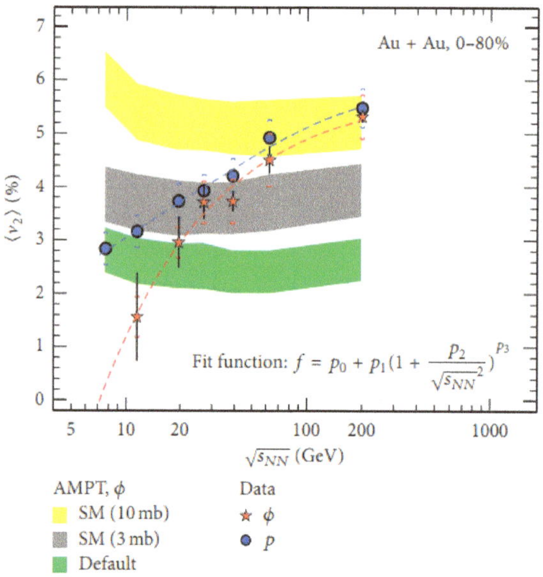

Figure 13: The p_T integrated proton and ϕ meson V_2 for various centre-of-mass energies for 0–80% centrality in Au + Au collisions [32]. Vertical lines are the statistical error and systematic errors are shown by cap symbol. For lower RHIC energies, STAR preliminary p_T spectra were used for ϕ and proton $\langle V_2 \rangle$ calculation [17, 33, 34]. The red and blue lines are the fit to the ϕ and proton V_2 by empirical function just to guide the eye of the reader.

SUMMARY

We have presented a review on the experimentally measured data on ϕ production (specifically the transverse momentum distributions and azimuthal anisotropy measurements) in high energy heavy-ion collisions. The differential (p_T, φ, y) measurements of ϕ meson production have been compared from heavy-ion collisions at the SPS, RHIC, and LHC energies. Transverse momentum spectra of ϕ meson for different centralities, different energies, and different collision systems are presented. The shape of the transverse momentum distribution changes from exponential to Levy functional form as one goes from central to peripheral collisions at a given beam energy. This indicates an increasing contribution of hard processes in the peripheral collisions.

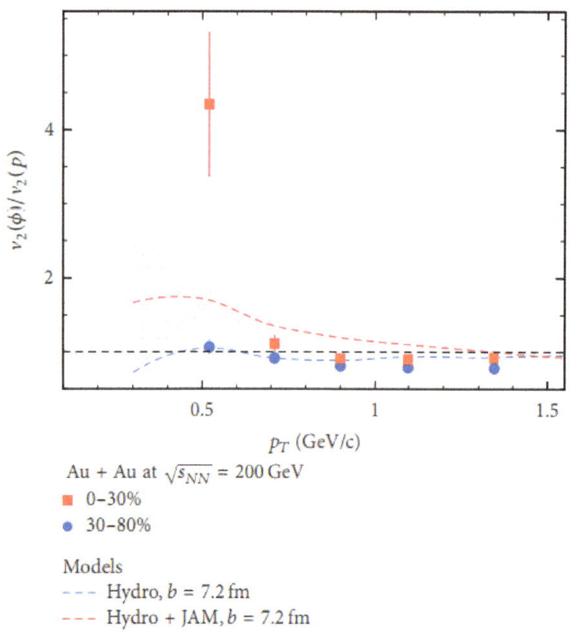

Figure 14: Ratio between ϕ and p V_2 for 0–30% and 30–80% centrality in Au + Au collisions $\sqrt{s_{NN}}$ = 200 GeV [29, 30].

The centrality and energy dependence of the enhancement in ϕ meson production support the physical origin to be due to the enhanced production of (s)-quarks in a dense partonic medium formed in high energy heavy-ion collisions. We have discussed beam energy dependence of the nuclear modification factors of ϕ meson. The values of nuclear modification factors are less than unity for beam energies of 200 GeV and 2.76 TeV, indicating formation of a dense medium with color degrees of freedom. The nuclear modification factor values at the intermediate p_T are observed to be equal to or higher than unity at $\sqrt{s_{NN}} \leq 39$ GeV. This indicates that parton energy loss effect became less important and suggests dominance of hadronic interactions at the lower beam energies. The ratio $N(\phi)/N(K^-)$ is observed to be almost constant as a function of centrality and centre-ofmass energy, disfavouring ϕ meson production through kaon coalescence. The ratio of $N(\Omega^- + \overline{\Omega}^+)/2N(\phi)$ versus p_T shows a similar trend for $\sqrt{s_{NN}}$ $\geq$ 19.6 GeV, but at $\sqrt{s_{NN}}$ = 11.5 GeV the ratio at the highest measured p_T shows a deviation from the trend at other higher energies. This may suggest a change in Ω and/or ϕ production mechanism and strange quark dynamics in general at $\sqrt{s_{NN}}$ = 11.5 GeV.

The measurement of ϕ meson V_2 as a function of p_T and collision centrality are discussed. We observe that ϕ meson $V_2(p_T)$ has similar values for $\sqrt{s_{NN}} \geq 19.6$ GeV and NCQ scaling also holds for $\sqrt{s_{NN}} \geq 19.6$ GeV. But at $\sqrt{s_{NN}}$ = 7.7 and 11.5 GeV, the ϕ meson V_2 shows deviation from the other hadrons at the highest measured p_T values by 1.8σ and 2.3σ, respectively. Since the V_2 of ϕ mesons mostly reflect collectivity from partonic phase, therefore the small ϕ V_2 observed at $\sqrt{s_{NN}}$ = 7.7 and 11.5 GeV indicates a smaller contribution to the collectivity from partonic phase. We find that the ϕ $\langle V_2 \rangle$ can be explained by AMPT model with partonic interactions having cross-section values between 3 mb and 10 mb for $\sqrt{s_{NN}} \geq 19.6$ GeV, but the model with and without partonic interactions overpredicts the data at $\sqrt{s_{NN}}$ = 7.7 and 11.5 GeV. Also at $\sqrt{s_{NN}} > 19.6$ GeV, proton and ϕ meson show similar magnitude of $\langle V_2 \rangle$; however, at $\sqrt{s_{NN}} \leq 19.6$ GeV, ϕ meson $\langle V_2 \rangle$ shows deviation from corresponding proton values. At $\sqrt{s_{NN}} = 11.5$ GeV ϕ meson $\langle V_2 \rangle$ value is small and is about 1.5%. This further emphasises our conclusion that at lower beam energies the hadronic interactions are dominating. In addition, we observe that the mass ordering between ϕ and proton V_2 breaks down in the lower momentum range at $\sqrt{s_{NN}} =$ 200 GeV. This could be because of the larger effect of hadronic rescattering on proton V_2, which reduces the proton V_2 values.

The main conclusions of the review are the following.

(a) The coalescence of K^+ and K^- is not the dominant production mechanism for ϕ meson in high energy heavyion collisions. (b) The study of Ω/ϕ and comparison to quark recombination model calculations indicate that ϕ mesons are produced via coalescence of thermalized s quarks for $\sqrt{s_{NN}} \geq$ 19.6 GeV. (c) The observed ϕ meson enhancement (unaffected by Canonical suppression effects) in heavy-ion collisions suggests that strangeness enhancement is due to the formation of a dense partonic medium. (d) The nuclear modification factor measurements for ϕ mesons and the measurements of ϕ meson V2 indicate the formation of partonic media in heavy-ion collisions at $\sqrt{s_{NN}} \geq 19.6$ GeV, while for $\sqrt{s_{NN}} \leq 11.5$ GeV hadronic interactions dominate. (e) Finally ϕ meson production provided us with a benchmark to study the rescattering effect. The comparison of ϕ and proton V_2 shows that mass ordering in $V_2(p_T)$ could be broken due to effect of later stage rescattering effects on the proton distributions.

ACKNOWLEDGMENTS

Financial assistance from the SwarnaJayanti Fellowship of the Department of Science and Technology. Government of India, is gratefully acknowledged. Md. Nasim is supported by DOE Grant of Department of Physics and Astronomy, UCLA, USA.

REFERENCES

1. F. Karsch, "Lattice results on QCD thermodynamics," Nuclear Physics A, vol. 698, no. 1–4, pp. 199–208, 2002.
2. R. V. Gavai and S. Gupta, "On the critical end point of QCD," Physical Review D, vol. 71, no. 11, Article ID 114014, 21 pages, 2005.
3. K. G. Wilson, "Confinement of quarks," Physical Review D, vol. 10, p. 2445, 1974. ·
4. S. Bethke, "Experimental tests of asymptotic freedom," Progress in Particle and Nuclear Physics, vol. 58, no. 2, pp. 351–386, 2007.
5. L. Bertanza, V. Brisson, P. L. Connolly, et al., "Possible Resonances in the Ξπ and KK- Systems," Physical Review Letters, vol. 9, no. 4, pp. 180–183, 1962.
6. A. Shor, "φ-meson production as a probe of the Quark-Gluon plasma," Physical Review Letters, vol. 54, no. 11, pp. 1122–1125, 1985.
7. A. Sibirtsev, H.-W. Hammer, U.-G. Meißner, and A. W. Thomas, "φ-meson photoproduction from nuclei," The European Physical Journal, vol. 29, no. 2, pp. 209–220, 2006.
8. J. Adams, M. M. Aggarwal, Z. Ahammed, et al., "Experimental and theoretical challenges in the search for the quark-gluon plasma: The STAR Collaboration's critical assessment of the evidence from RHIC collisions," Nuclear Physics A, vol. 757, no. 1-2, pp. 102–183, 2005.
9. P. Koch, B. Müllerb, and J. Rafelskic, "Strangeness in relativistic heavy ion collisions," Physics Reports, vol. 142, no. 4, pp. 167–262, 1986.
10. T. Alber, H. Appelshäuser, J. Bächler, et al., "Strange particle production in nuclear collisions at 200 GeV per nucleon," Zeitschrift für Physik C, vol. 64, no. 2, pp. 195–207, 1994.
11. P. G. Jones, "Hadron yields and hadron spectra from the NA49 experiment," Nuclear Physics A, vol. 610, p. 188c, 1996.
12. F. Siklér, J. Bächlerf, D. Barna, et al., "Hadron production in nuclear collisions from the NA49 experiment at 158 GeV/c. A," Nuclear Physics A, vol. 611, no. 1–4, pp. 45–54, 1996. ·
13. C. Hohne, "Strangeness production in nuclear collisions—recent results from experiment NA49," Nuclear Physics A, vol. 611, pp. 485–488, 1999.
14. J. Rafelski and B. Müller, "Strangeness production in the quark-gluon plasma," Physical Review Letters, vol. 48, no. 16, pp. 1066–1069, 1982.
15. K. Redlich and A. Tounsi, "Strangeness enhancement and energy dependence in heavy ion collisions," European Physical Journal C, vol. 24, no. 4, pp. 589–594, 2002. ·
16. C. Alt, T. Anticic, B. Baatar, et al., "Energy dependence of ϕ meson production in central Pb + Pb collisions at sNN=6 to 17 GeV," Physical Review C, vol. 78, no. 4, 15 pages, 2008.
17. M. Nasim, "Probing the QCD phase diagram with measurements of ϕ-meson production and elliptic flow in heavy-ion collisions at STAR," Journal of Physics: Conference Series, vol. 509, no. 1, Article ID 10.1088/1742-6596/509/1/012070, 2014.
18. B. I. Abelev, et al., "Measurements of ϕ meson production in relativistic heavy-ion collisions at the BNL Relativistic Heavy Ion Collider (RHIC)," Physical Review C, vol. 49, Article ID 064903, 2009.
19. B. Abelev, A. Abrahantes Quintana, D. Adamová, et al., "Strange particle production in proton–proton collisions at s = 0.9 TeV with ALICE at the LHC," The European Physical Journal C, vol. 71, p. 1594, 2011.
20. B. Abelev, J. Adam, D. Adamová, et al., "Production of K*(892)° and ϕ(1020) in pp collisions at s=7 TeV," The European Physical Journal C, vol. 72, article 2183, 2012.
21. B. B. Abelev, J. Adam, D. Adamova, et al., "K*(892)0 and Φ(1020) production in Pb-Pb collisions at sNN=2.76 TeV," In press, http://arxiv.org/abs/1404.0495.
22. S. V. Afanasiev, T. Anticicu, J. Bächler, et al., "Production of φ-mesons in p + p, p + Pb and central Pb + Pb collisions at E_{beam} = 158 A GeV," Physics Letters B, vol. 491, no. 1-2, pp. 59–66, 2000.
23. T. Akesson, "Inclusive vector-meson production in the central region of pp collisions at s = 63 GeV," Nuclear Physics B, vol. 203, no. 1, pp. 27–99, 1982.
24. B. I. Abelev, M. M. Aggarwal, Z. Ahammed, et al., "Energy and system size dependence of ϕ meson

production in Cu + Cu and Au + Au collisions," Physics Letters B, vol. 673, pp. 183–191, 2009.

25. F. Antinori, P. Bacon, A. Badalà, et al., "Enhancement of hyperon production at central rapidity in 158 AGeV/c Pb–Pb collisions," Journal of Physics G, vol. 32, no. 4, article 427, 2006.

26. S. V. Afanasev, T. Anticic, J. Bächler, et al., "Production of φ-mesons in p + p, p + Pb and central Pb + Pb collisions at E_{beam} = 158A GeV," Physics Letters B, vol. 491, p. 59, 2000.

27. A. Adare, S. Afanasiev, C. Aidala, et al., "Nuclear modification factors of ϕ mesons in d+Au, Cu+Cu, and Au+Au collisions at sNN=200," Physical Review C, vol. 83, no. 2, Article ID 024909, 10 pages, 2011.

28. L. Adamczyk, J. K. Adkins, G. Agakishiev, et al., "Elliptic flow of identified hadrons in Au+Au collisions at sNN=7.7–62.4 GeV," Physical Review C, vol. 88, Article ID 014902, 2013.

29. M. Nasim and The STAR Collaboration, "Systematic investigation of partonic collectivity through centrality dependence of elliptic flow of multi-strange hadrons in Au+Au collisions at sNN=200 GeV," Nuclear Physics A, vol. 904-905, pp. 413c–416c, 2013.

30. R. Haque and M. K. Sharma, STAR presentation at QM2014.

31. B. Abelev, "Elliptic flow of identified hadrons in Pb-Pb collisions at sNN=2.76 TeV," http://arxiv.org/abs/1405.4632.

32. M. Nasim, "Using ϕ-meson elliptic flow to map the strength of the partonic interaction," Physical Review C, vol. 89, Article ID 034909, 2014.

33. L. Kumar, "STAR results from the RHIC beam energy scan-I," Nuclear Physics A, vol. 904-905, pp. 256c–263c, 2013.

34. S. Das and The STAR Collaboration, "Study of freeze-out dynamics in STAR at RHIC beam energy scan program," Journal of Physics: Conference Series, vol. 509, conference 1, Article ID 012066, 2014.

35. J. Adams, C. Adler, M. M. Aggarwal, et al., "Particle-type dependence of azimuthal anisotropy and nuclear modification of particle production in Au+Au collisions at sNN=200 GeV," Physical Review Letters, vol. 92, no. 5, Article ID 052302, 6 pages, 2004.

36. D. Molnar and S. A. Voloshin, "Elliptic flow at large transverse momenta from quark coalescence," Physical Review Letters, vol. 91, no. 9, Article ID 092301, 4 pages, 2003.

37. R. J. Fries, B. Müller, C. Nonaka, and S. A. Bass, "Hadronization in heavy-ion collisions: recombination and fragmentation of partons," Physical Review Letters, vol. 90, no. 20, Article ID 202303, 2003.

38. G. Aad, T. Abajyan, B. Abbott, et al., "The differential production cross section of the ϕ (1020) meson in s=7 TeV pp collisions measured with the ATLAS detector," The European Physical Journal C, vol. 74, p. 2895, 2014.

39. V. Khachatryan, A. M. Sirunyan, A. Tumasyan, et al., "Transverse-momentum and pseudorapidity distributions of charged hadrons in pp collisions at s=7 TeV," Physical Review Letters, vol. 105, Article ID 022002, 2010.

40. S. Chatrchyan, V. Khachatryan, A. M. Sirunyan, et al., "Study of the inclusive production of charged pions, kaons, and protons in pp collisions at s = 0.9, 2.76, and 7 TeV," The European Physical Journal C, vol. 72, p. 2164, 2012.

41. V. Blobel, H. Fesefeldt, H. Franz, et al., "Test of the Zweig selection rule in Φ production by pp collisions," Physics Letters B, vol. 59, no. 1, pp. 88–92, 1975.

42. A. Sibirtsev, J. Haidenbauer, and U.-G. Meissner, "Aspects of phi-meson production in proton-proton collisions," The European Physical Journal A, vol. 27, no. 3, pp. 263–268, 2006.

43. M. L. Miller, K. Reygers, S. J. Sanders, and P. Steinberg, "Glauber modeling in high-energy nuclear collisions," Annual Review of Nuclear and Particle Science, vol. 57, pp. 205–243, 2007.

44. X. N. Wang and M. Gyulassy, "Gluon shadowing and jet quenching in A+A collisions at s =200A GeV," Physical Review Letters, vol. 68, no. 10, pp. 1480–1483, 1992.

45. J. W. Cronin, H. J. Frisch, M. J. Shochet, J. P. Boymond, P. A. Piroué, and R. L. Sumner, "Production of hadrons at large transverse momentum at 200, 300, and 400 GeV," Physical Review D, vol. 11, p. 3105, 1975.

46. M. Lev and B. Petersson, "Nuclear effects at large transverse momentum in a QCD parton model," Zeitschrift für Physik C Particles and Fields, vol. 21, no. 1, pp. 155–161, 1983.

47. A. Accardi and M. Gyulassy, "Cronin effect vs. geometrical shadowing in d + Au collisions at RHIC," Physics Letters B: Nuclear, Elementary Particle and High-Energy Physics, vol. 586, no. 3-4, pp. 244–253, 2004.

48. G. Papp, P. Lévai, and G. Fai, "Saturating Cronin effect in ultrarelativistic proton-nucleus collisions," Physical Review C, vol. 61, no. 2, Article ID 021902(R), 5 pages, 1999.

49. R. C. Hwa, C. B. Yang, and R. J. Fries, "Forward production in d+Au collisions by parton recombination," Physical Review C, vol. 71, no. 2, Article ID 024902, 9 pages, 2005.

50. B. I. Abelev, J. Adams, M. M. Aggarwal, et al., "Rapidity and species dependence of particle production at large transverse momentum for d+Au collisions at sNN=200 GeV," Physical Review C, vol. 76, Article ID 054903, 2007.

51. L. van Hove, "Multiplicity dependence of p_t spectrum as a possible signal for a phase transition in hadronic collisions," Physics Letters B, vol. 118, no. 1–3, pp. 138–140, 1982.

52. N. Isgur and H. B. Thacker, "Origin of the Okubo-Zweig-Iizuka rule in QCD," Physical Review D, vol. 64, no. 9, Article ID 094507, 13 pages, 2001.

53. S. A. Bass, M. Belkacem, M. Bleicher, et al., "Microscopic models for ultrarelativistic heavy ion collisions," Progress in Particle and Nuclear Physics, vol. 41, pp. 255–369, 1998.

54. M. Bleicher, E. Zabrodin, C. Spieles, et al., "Relativistic hadron-hadron collisions in the ultra-relativistic quantum molecular dynamics model," Journal of Physics G: Nuclear and Particle Physics, vol. 25, no. 9, article 1859, 1999.

55. C. Alt, T. Anticic, B. Baatar, et al., "System-size dependence of strangeness production in nucleus-nucleus collisions at sNN=17.3 GeV," Physical Review Letters, vol. 94, no. 5, Article ID 052301, 5 pages, 2005.

56. J. Adams, C. Adler, M. M. Aggarwal, et al., "Multistrange Baryon Production in Au-Au Collisions at sNN= 130 GeV," Physical Review Letters, vol. 92, no. 18, Article ID 182301, 6 pages, 2004.

57. R. C. Hwa and C. B. Yang, "Production of strange particles at intermediate p_T in central Au+Au collisions at high energies," Physical Review C, vol. 75, Article ID 054904, p. 054904, 2007.

58. P. F. Kolb and U. Heinz, "Emission angle dependent HBT at RHIC and beyond," Nuclear Physics A, vol. 715, pp. 653c–656c, 2003.

59. D. Teaney, J. Lauret, and E. V. Shuryak, "Flow at the SPS and RHIC as a quark-gluon plasma signature," Physical Review Letters, vol. 86, no. 21, pp. 4783–4786, 2001.

60. P. F. Kolb and U. Heinz, "Hydrodynamic description of ultrarelativistic heavy-ion collisions," http://arxiv.org/abs/nucl-th/0305084.

61. P. F. Kolba, P. Huovinend, U. Heinza, and H. Heiselberge, "Elliptic flow at SPS and RHIC: from kinetic transport to hydrodynamics," Physics Letters B, vol. 500, no. 3-4, pp. 232–240, 2001.

62. H. Sorge, "Elliptical flow: a signature for early pressure in ultrarelativistic nucleus-nucleus collisions," Physical Review Letters, vol. 78, pp. 2309–2312, 1997.

63. J. Y. Ollitrault, "Anisotropy as a signature of transverse collective flow," Physical Review D, vol. 46, p. 229, 1992.

64. P. Huovinen, P. F. Kolb, U. Heinz, et al., "Radial and elliptic flow at RHIC: further predictions," Physics Letters B, vol. 503, pp. 58–64, 2001.

65. A. M. Poskanzer and S. A. Voloshin, "Methods for analyzing anisotropic flow in relativistic nuclear collisions," Physical Review C, vol. 58, no. 3, pp. 1671–1678, 1998.

66. B. Mohanty and N. Xu, "Probe of the QCD phase diagram with phi-mesons in high-energy nuclear collisions," Journal of Physics G, vol. 36, no. 6, Article ID 064022, 2009.

67. M. Nasim, B. Mohanty, and N. Xu, "Elliptic flow of ϕ mesons as a sensitive probe for the onset of the deconfinement transition in high energy heavy-ion collisions," Physical Review C, vol. 87, no. 1, Article ID 014903, 4 pages, 2013.

68. P. Bhaduri and S. Chattopadhyay, "Differential elliptic flow of identified hadrons and constituent quark number scaling at the GSI Facility for Antiproton and Ion Research (FAIR)," Physical Review C, vol. 81, Article ID 034906, 2010.

69. K. J. Wu, F. Liu, and N. Xu, "The number-of-quark scaling of v2 in sNN = 9.2 GeV Au+Au collisions," Nuclear Physics A, vol. 834, no. 1, pp. 303c–305c, 2010.

70. M. Bleicher and X. Zhu, "Is constituent quark scaling a unique sign of parton recombination?" European Physical Journal C, vol. 49, no. 1, pp. 303–308, 2007.

71. T. Hirano, U. Heinz, D. Kharzeev, R. Lacey, and Y. Nara, "Mass ordering of differential elliptic flow and its violation for ϕ mesons," Physical Review C, vol. 77, Article ID 044909, 2008.

CHAPTER 16

STUDIES OF BACKWARD PARTICLE PRODUCTION WITH A FIXED-TARGET EXPERIMENT USING THE LHC BEAMS

Federico Alberto Ceccopieri

INTRODUCTION

In hadronic collisions a portion of the produced particle spectrum is characterised by hadrons carrying a sizeable fraction of the available centre-of-mass energy, the so-called leading particle effect. It is phenomenologically observed that for such hadrons their valence-parton composition is almost or totally conserved with respect to the one of initial-state hadrons [1]. In pp collisions, for example, protons, neutrons, and lambdas show a significant leading particle effect. For such processes, the production cross section peaks at very small transverse momenta with respect to the collision axis, a regime where perturbative techniques cannot be applied, giving insight into nonperturbative aspects of QCD dynamics in high energy collisions.

Quite interestingly, the leading particle effect has been observed in Semi-Inclusive Deep Inelastic Scattering (SIDIS). At variance with the hadronic processes mentioned above, such a process naturally involves a large momentum transfer. The presence of a hard scale enables the derivation of a dedicated factorisation theorem [2–5] which ensures that QCD factorisation holds for backward particle production in DIS. The relevant cross sections can then be factorised into perturbatively calculable short distance cross sections and new distributions, fracture functions, which simultaneously encode information both on the interacting parton and on the spectator fragmentation into the observed hadron. Despite of being nonperturbative in nature, their scale dependence can be calculated within perturbative QCD [6]. The factorisation theorem [2–5] guarantees that fracture functions are universal distributions, at least in the context of SIDIS.

Detailed experimental studies of hard diffraction at HERA have shown to support the hypothesis of QCD factorisation and evolution inherent the fracture function formalism. Furthermore, they led to a quite accurate knowledge of diffractive parton distributions [7–11], a special case of fracture functions in the very backward kinematic limit. For particles other than protons, proton-to-neutron fracture functions have been extracted from a pQCD analysis of forward neutron production in DIS in [12]. A set of proton-to-lambda fracture functions has been obtained by performing a combined pQCD fit to a variety of semi-inclusive DIS lambda production data in [13].

As theoretically anticipated in [2–4, 14, 15] and experimentally observed in hard diffraction in $p\bar{p}$ collisions at Tevatron [16, 17], QCD factorisation is violated for fracture functions in hadronic collisions. On general grounds, it might be expected, in fact, that the dynamics of target-remnants hadronisation is affected by the coloured environment resulting from the scattering in a rather different way with respect to the Deep Inelastic Scattering case.

Nonetheless, the tools mentioned above allow us to investigate quantitatively particle production mechanisms in the very backward and forward regions, to test the concept of factorisation at the heart of QCD, and to study the dependencies of factorisation breaking upon the species and the kinematics of the selected final state particle.

This physics program could be successfully carried on at a Fixed-Target Experiment using the LHC beams [18]. Novel experimental techniques are, in fact, available to extract beam-halo protons or heavy ions from LHC beams without affecting LHC performances. Such a resulting beam would be then impinged on a high-density and/or long-length fixed target, guaranteeing high luminosities. Furthermore, most importantly for the physics program to be discussed here, the entire backward hemisphere (in the centre-of-mass system of the collision) would be accessible with standard experimental techniques allowing high precision studies of the target

Citation: Federico Alberto Ceccopieri, "Studies of Backward Particle Production with a Fixed-Target Experiment Using the LHC Beams," Advances in High Energy Physics, vol. 2015, Article ID 652062, 9 pages, 2015. doi:10.1155/2015/652062.

fragmentation. Although measurements of particle production in the very forward region (close to the beam axis) might be challenging experimentally due to the high particle densities and large energy flow, the installation of dedicated detectors, like forward neutron calorimeters and/or proton taggers, could further broaden the physics program outlined above giving access to the beam fragmentation region.

The paper is organised as follows. In Section 2 we first give a brief theoretical introduction on the fracture functions formalism and to higher order corrections to the semi-inclusive Drell-Yan process. In Section 3 we outline different analyses which could be performed at AFTER@LHC with special focus on single hard diffraction. In Section 4 we summarise our results.

COLLINEAR FACTORISATION FORMULA

Fracture functions, originally introduced in DIS, do depend on a large momentum transfer. Therefore, in order to use them in hadronic collisions, a hard process must be selected. We consider here the semi-inclusive version of the Drell-Yan process:

$$H_1(P_1) + H_2(P_2) \longrightarrow H(h) + \gamma^*(q) + X, \tag{1}$$

in which one hadron H is measured in the final state together with a Drell-Yan pair. In such a process the high invariant mass of the lepton pair, $q^2 = Q^2$, allows the applicability of perturbative QCD, while the detected hadron H can be used, without any phase space restriction, as a local probe to investigate particle production mechanisms.

The associated production of a particle and a Drell-Yan pair in terms of partonic degrees of freedom starts at $O(\alpha s)$. One of the contributing diagrams is depicted in Figure 1.

Assuming that the hadronic cross sections admit a factorisation in terms of long distance nonperturbative distributions and short distance perturbative calculable matrix elements for the partonic process $i(p_1) + j(p_2) \longrightarrow l(k) + \gamma^*(q)$, predictions based on perturbative QCD are obtained convoluting the relevant partonic subprocess cross sections, $d\hat{\sigma}^{ij\rightarrow l\gamma^*}$, with parton distribution functions, f_{i/H_1} and f_{j/H_2}, and fragmentation functions, $D^{H/l}$.

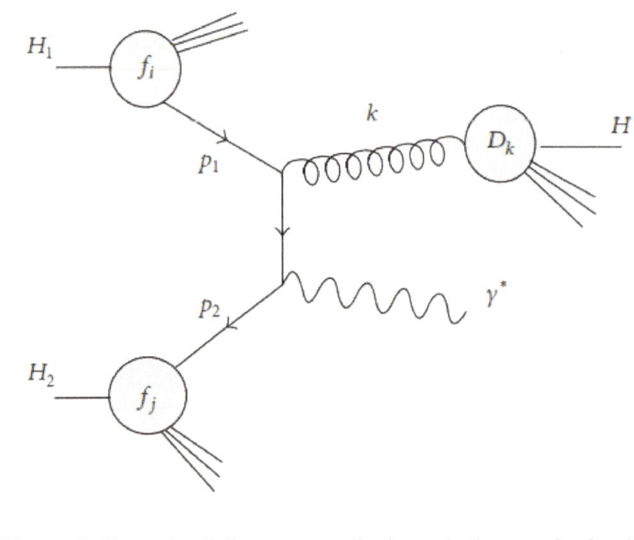

Figure 1: Example of diagram contributing to hadron production in the central fragmentation region to order $O(\alpha_s)$ in (2).

The hadronic cross section, at centre of mass energy squared $s = (P_1 + P_2)^2$, can be symbolically written as [19, 20]

$$\frac{d\sigma^{H,C,(1)}}{dQ^2 dz} \propto \sum_{i,j,l} \int \frac{dx_1}{x_1} \int \frac{dx_2}{x_2} \int \frac{d\rho}{\rho} f_i^{[1]}(x_1) f_j^{[2]}(x_2) \cdot D^{H/l}\left(\frac{z}{\rho}\right) \frac{d\hat{\sigma}^{ij\rightarrow l\gamma^*}}{dQ^2 d\rho}, \tag{2}$$

where the convolutions are over the momentum fractions of the incoming and outgoing partons. The partonic indices i, j, and l in the sum run on the available partonic subprocesses. The superscripts label the incoming hadrons and the presence of crossed terms is understood. This type of factorised hadronic cross section is expected to hold for hadrons produced at sufficiently high transverse momentum and it is widely and successfully used to compute cross sections for large momentum transfer processes in hadronic collisions. The Lorentz-invariant variable z in (2) is defined by

$$z = \frac{2h \cdot (P_1 + P_2)}{s} \equiv \frac{2E_H^*}{\sqrt{s}}. \tag{3}$$

In the hadronic centre-of-mass frame, where the second identity holds, z is just the observed hadron energy, $E^*_{H'}$, scaled down by the beam energy $\sqrt{s}/2$. The variable ρ, appearing in (2), is its partonic equivalent. Within this production mechanism, the observed hadron H is generated by the fragmentation of the final state parton l, and for this reason we address it as central. The amplitudes squared [21], however, are singular when the transverse momentum of the final state parton vanishes. In such configurations, the parent parton l of the observed hadron H is collinear either to the incoming parton i or j. As these phase space regions are approached, perturbation theory loses its predictivity. This class of collinear singularities escape the usual renormalisation procedure which amounts to reabsorb collinear divergences into a redefinition of bare parton distribution and fragmentation functions. Such singularities are likely to appear in every fixed order calculation in the same kinematical limits spoiling the convergence of the perturbative series.

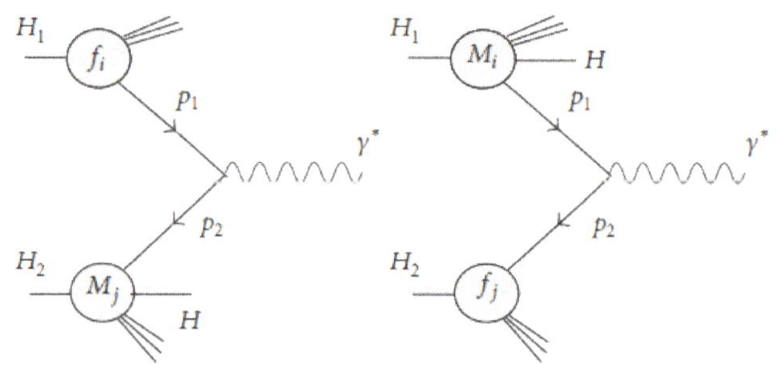

Figure 2: Pictorial representation of the parton model formula (4) for the associated production of a Drell-Yan pair and a particle in the target fragmentation regions.

In [19, 20] a generalised procedure for the factorisation of such additional collinear singularities is proposed. The latter is the same as the one proposed in Deep Inelastic Scattering [22] where the same singularities pattern is also found, confirming the universality of collinear radiation between different hard processes. Such a generalised collinear factorisation makes use of fracture functions. These distributions obey DGLAP-type evolution equations which contain an additional inhomogeneous term resulting from the subtraction of collinear singularities in the target fragmentation region [6, 22]. Such equations allow us to resume the corresponding large logarithmic corrections to all orders in perturbation theory. Bare fracture functions, $M_i^{H/H_1}(x,z)$, describe the hadronization of the spectators system in hadron-induced reactions. They express the conditional probability to find a parton i initiating the hard scattering while an hadron H is produced with fractional momentum z in the target fragmentation region of the incoming hadron H_1.

The use of fracture functions allows for particles production already to $O(\alpha_s^0)$, since the hadron H can be nonpertubatively produced by a fracture function itself. Therefore, the lowest order parton model formula can be symbolically written as

$$\frac{d\sigma^{H,T,(0)}}{dQ^2dz} \propto \sum_{i,j} \int \frac{dx_1}{x_1} \int \frac{dx_2}{x_2} \cdot \left[M_i^{[1]}(x_1,z) f_j^{[2]}(x_2) + M_i^{[2]}(x_2,z) f_j^{[1]}(x_1)\right] \cdot \frac{d\hat{\sigma}^{ij\to\gamma^*}}{dQ^2} \tag{4}$$

and it is sketched in Figure 2. The superscripts in (4) indicate from which incoming hadron, H_1 or H_2, the outgoing hadron H is produced through a fracture function. In order to complete the calculation to $O(\alpha_s)$ accuracy we should consider higher order corrections to (4). Since in this case the hadron H is already produced by fracture functions, final state parton radiation should be integrated and the resulting contribution added to virtual corrections. One of the contributing diagrams is depicted in Figure 3.

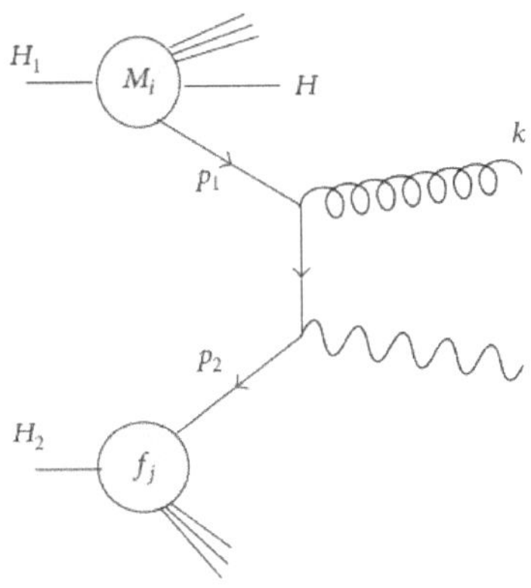

Figure 3: Example of diagram contributing to $O(\alpha_s)$ corrections in the target fragmentation region (5).

The general structure of these terms is

$$\frac{d\sigma^{H,T,(1)}}{dQ^2dz} \propto \sum_{i,j} \int \frac{dx_1}{x_1} \int \frac{dx_2}{x_2} \cdot \left[M_i^{[1]}(x_1,z) f_j^{[2]}(x_2) + M_i^{[2]}(x_2,z) f_j^{[1]}(x_1)\right] \cdot \frac{d\hat{\sigma}^{ij\to(l)\gamma^*}}{dQ^2}. \tag{5}$$

We refer to them as to the target fragmentation contributions. Their calculation is, apart from minor differences in kinematics, completely analogous to the one of the inclusive Drell-Yan case. The factorisation procedure, first elaborated in [22] in the context of SIDIS, amounts to substitute in (4) the bare fracture and parton distributions functions with their renormalised version [19, 20]. Renormalised parton distributions and fracture functions homogeneous terms do cancel, as in the inclusive Drell-Yan case, all singularities present in (5). The additional singularities in (2) are cancelled by the combination of parton distributions and fracture functions inhomogeneous renormalisation terms. Adding all the various contributions, the resulting hadron-pt integrated cross section, up to order $O(\alpha_s)$, is then infrared finite [19, 20] and can be symbolically written as

$$\frac{d\sigma^H}{dQ^2dz} = \frac{\sigma_0}{N_c s}\sum_{i,j}\left[M_i^{[1]}\otimes f_j^{[2]} + M_i^{[2]}\otimes f_j^{[1]}\right]\left(1+\frac{\alpha_s}{2\pi}C^{ij}\right) + \frac{\sigma_0}{N_c s}\frac{\alpha_s}{2\pi}\sum_{i,j,l} f_i^{[1]}\otimes f_j^{[2]}\otimes D^{H/l}\otimes K_l^{ij}, \tag{6}$$

where $\sigma_0 = 4\pi\alpha_{em}^2/3Q^2$ and N_c is the number of colors. We refer to the previous equation as to the collinear factorisation formula for the process under study. The next-toleading order coefficients C^{ij} and K_l^{ij} have been calculated in [20], making the whole calculation ready for numerical implementation.

We stress, however, that our ability to consistently subtract collinear singularities in such a semi-inclusive process is a necessary but not sufficient condition for factorisation to hold in hadronic collisions. The one-loop calculation outlined above in fact does involve only the so-called active partons. It completely ignores multiple soft parton exchanges between active and spectators partons, whose effects should be accounted for in any proof of QCD factorisation. Therefore, there is no guarantee that fracture functions extracted from SIDIS can be successfully used to describe forward or backward particle production in hadronic collisions. Reversing the argument, such a comparison may instead offer new insights into nonperturbative aspects of QCD and to the breaking of factorisation.

SINGLE HARD DIFFRACTION AT AFTER@LHC

As an application of the formalism presented in the previous sections we will consider single hard diffractive production of a Drell-Yan pair:

$$p_1(P_1) + p_2(P_2) \longrightarrow p(P) + \gamma^*(q) + X, \tag{7}$$

where we have indicated in parentheses the four momenta of the relevant particles. We present in the following cross sections differential in the virtual photon variables. The subsequent decay of the virtual photon into a lepton pair can be easily included so that realistic cuts on leptons rapidity and transverse momentum can then be applied. We consider the AFTER@LHC kinematic setting in which a 7 TeV proton beam collides on a fixed target proton leading to a centre-ofmass energy of $\sqrt{s}$ = 115 GeV. We consider the projectile proton p_1 moving in the positive z direction and p_2 at rest in the laboratory. The diffractively produced proton p has in general almost the incoming projectile proton p_1 energy and very small transverse momentum as measured with respect to the collision axis.The detection of such fast protons will in general require the installation of forward proton taggers. The lepton pair instead will be measured by the main AFTER@LHC detector. This kinematical configuration is pictorially represented in the right plot of Figure 2.

Diffractive processes have been intensively analysed in DIS at HERA *ep* collider, revealing their leading twist nature. From scaling violations of the diffractive structure functions [7,10,11] and dijet production in the final state [8, 9] quite precise diffractive parton distributions functions (dPDFs) have been extracted from HERA data, which parametrise the parton content of the color singlet exchanged in the t-channel. The comparison of QCD predictions for single diffractive hard processes based on diffractive parton distributions measured at HERA against data measured at Tevatron [16, 17] ($\sqrt{s}$ = 1.96 TeV), adopting a factorised ansatz as in (4), has indeed revealed that these processes are, not unexpectedly [14, 15], significantly suppressed in hadronic collisions. This conclusion persists even after the inclusion of higher order QCD corrections [23]. Complementing these results with the forthcoming ones from LHC at higher centre-of-mass energies ($\sqrt{s}$ = 13 TeV) and the ones from AFTER@LHC at $\sqrt{s}$ = 115 GeV will give insight into the energy dependence of the so-called rapidity gap survival (RGS) probability in a wide range in $\sqrt{s}$. Since the theoretical computation of the RGS factor is highly model dependent, we decided not to include it in our predictions. Therefore, the latter must be considered as upper bounds.

Diffractive parton distributions f_i^D are in general protonto-proton fracture functions Mi. They depend upon the final state proton fractional energy loss, $x_{IP} = 1 - z$ with z given in (3), the fractional momentum of the interacting parton with respect to the pomeron momentum, $\beta = x/x_{IP}$ and the virtuality Q^2 . In general fracture functions may depend also upon the invariant momentum transfer $t = (P - P_1)^2$ at the proton vertex [24]. In all diffractive structure functions measurements at HERA, out of which dPDFs are determined, t is integrated up to some $t_{max} \ll Q^2$. In this case dPDFs obey ordinary DGLAP evolution equations [25] as their extended, t-dependent version [24]. In the present paper we use dPDFs form [7] which are defined by $|t| < 1$ GeV^2 . Since they are extracted from large rapidity gap data where the proton is not directly measured, they contain a contribution (23%) from the so-called proton dissociation contribution. In order to use dPDFs in the present context we first note that

$$M_i\left(x_1, z, Q^2\right) = x_{IP}^{-1} f_i^D\left(\beta, x_{IP}, Q^2\right). \tag{8}$$

The extra factor x_{IP}^{-1} comes from the Jacobian of the change $x_1 = \beta x_{IP}$. We then rearrange (4) in terms of new variables obtaining

$$\frac{d\sigma^D}{dQ^2 dx_{IP}} = \frac{\sigma_0}{N_c s}\int_{\tau/x_{IP}}^{1}\frac{d\beta}{\beta}\sum_{q,\bar{q}} e_q^2 x_{IP}^{-1} f_q^D\left(\beta, x_{IP}, \mu_F^2\right) f_{\bar{q}}\left(\frac{\tau}{\beta x_{IP}}, \mu_F^2\right), \tag{9}$$

with $\tau = Q^2/s$. For simplicity we consider here leading order formulas but the extension to higher order is straightforward. In (9) we use parton distribution functions from [26]. We show explicitly the dependence of fracture and parton distributions functions upon the factorisation scale, μ_F^2. Predictions are obtained with this scale set to $\mu_F^2 = Q^2$. Theoretical errors associated with higher order corrections are instead estimated varying such scale in the range $\mu_F^2 = 1/4Q^2$ and $\mu_F^2 = 4Q^2$.

In Figure 4we present predictions for the x_{IP} distribution. In left plot we consider a Drell-Yan pair of mass $Q^2 = 100$ GeV^2 . The distribution shrinks as lower x_{IP}-values are approached whereas, from hard diffraction at HERA, it is well known that diffractive cross sections rise as an inverse power of x_{IP}. Such an effect therefore is then attributed to phase space threshold effects. The Drell-Yan invariant mass constraint can be rewritten in the diffractive case as $Q^2 = \beta x_{IP} x_2 s$, which can be cast (for $\beta \to 1$ and $x_2 \to 1$) in upper bound on the invariant mass $Q^2 < x_{IP} s$ at fixed x_{IP} and s. This hypothesis is further supported in the right plot of Figure 4, where differential distributions are presented for three values of Q^2 . The lowest values of x_{IP} are then accessed only by lowering the invariant mass of the pair.

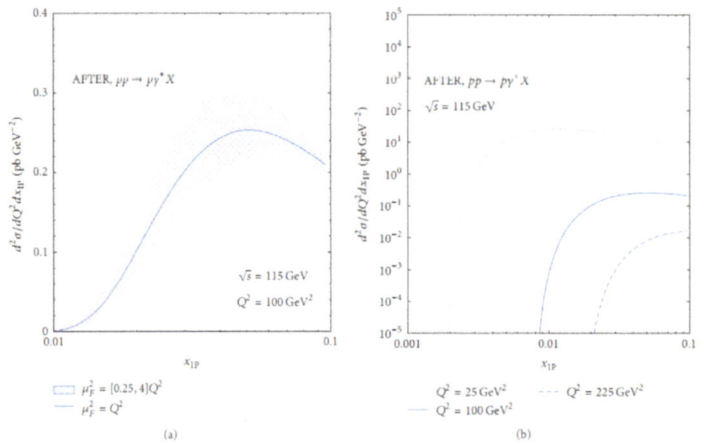

Figure 4: (a) Double differential cross sections for the production of a Drell-Yan pair of mass $Q^2 = 100$ GeV^2 . Blue error bands represent theoretical errors estimation, as described in the text. (b) Double differential cross sections for three different invariant masses.

We note that, even considering the maximum value of $x_{IP} = 0.1$, single diffractive production of $W^{\pm}$ and Z is beyond the kinematic reach at AFTER@LHC. In the left panel of Figure 5we present the prediction for the Q^2 distribution at a fixed value of xIP = 0.05. The cross section, as expected, is fast falling as an inverse power of Q^2 . The Q^2 distribution is particularly instructive since it allows us to study the possible dependence of the RGS factor on Q^2 and therefore to get an insight into the underlying dynamics. In the right panel of Figure 5 we present the Q^2 -differential cross section multiplied by a factor Q^4 , which compensates the fast fall off of the electromagnetic cross section so that all the Q^2 dependence is accounted for by that of fracture and parton distributions. These curves and the corresponding slopes, however, can not be readily interpreted as genuine results of QCD evolution of fracture and parton distributions functions because of the threshold effect mentioned above appearing at such moderate values of $\sqrt{s}$.

By changing variable from β to the virtual photon centreof-mass rapidity, y^{CM},

$$\beta = \frac{\sqrt{\tau}}{x_{IP}} e^{y^{CM}},$$

$$x_2 = \sqrt{\tau} e^{-y^{CM}}, \qquad (10)$$

equation (9) can be further manipulated to give the threedifferential cross section:

$$\frac{d\sigma^D}{dQ^2 dx_{IP} dy^{CM}} = \frac{\sigma_0}{N_c s} \sum_{q,\bar{q}} e_q^2 x_{IP}^{-1} f_q^D\left(\beta, x_{IP}, \mu_F^2\right) f_{\bar{q}}\left(x_2, \mu_F^2\right). \qquad (11)$$

The rapidity range for diffractive Drell-Yan production reads

$$\ln\sqrt{\tau} < y^{CM} < \ln\frac{\sqrt{\tau}}{x_{IP}} \qquad (12)$$

which, as expected, turns out to be asymmetric given the kinematic constraint $x_1 < x_{IP}$. The rapidity range for the inclusive Drell-Yan case is recovered simply setting $x_{IP} = 1$. The rapidity distribution is particularly sensitive to the shape of the diffractive parton distributions. This distribution will be useful to investigate any possible kinematic dependence of the RGS factor. In the left panel of Figure 6 we present the centre-of-mass rapidity distribution at fixed $Q^2 = 100$ GeV^2 and $x_{IP} = 0.05$. In this frame the distribution is shifted at negative values of y^{CM}. Therefore, on average, the parton originating from the target proton carries more momentum than the one originating from the pomeron.

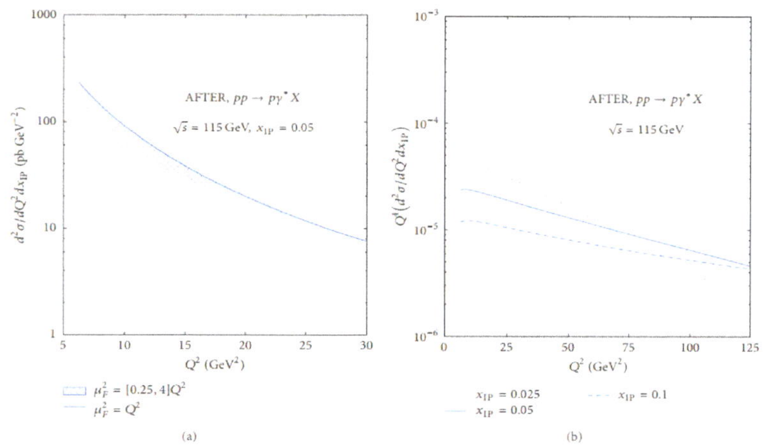

Figure 5: (a) Double differential cross sections for the production of a Drell-Yan pair at $x_{IP} = 0.05$. Blue error bands represent theoretical errors estimation, as described in the text. (b) Double differential cross sections times Q^4 for three different x_{IP}-values.

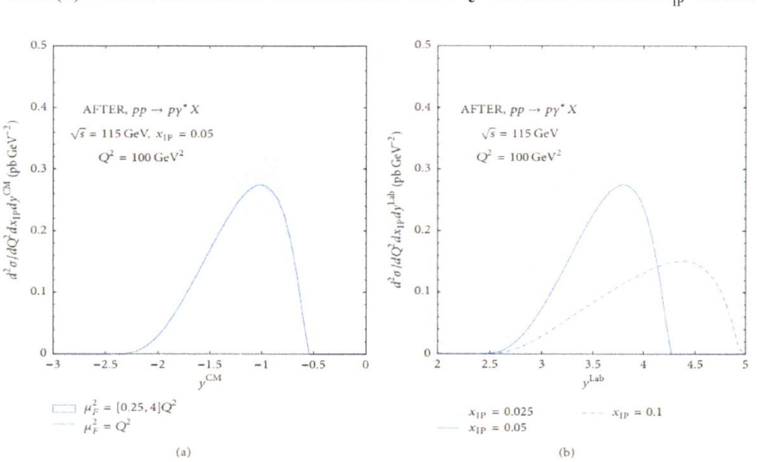

Figure 6: (a) Triple differential cross sections for the production of a Drell-Yan pair at $x_{IP} = 0.05$ and of mass $Q^2 = 100$ GeV^2 . Blue error bands represent theoretical errors estimation, as described in the text. (b) Triple differential cross sections for three different x_{IP}-values.

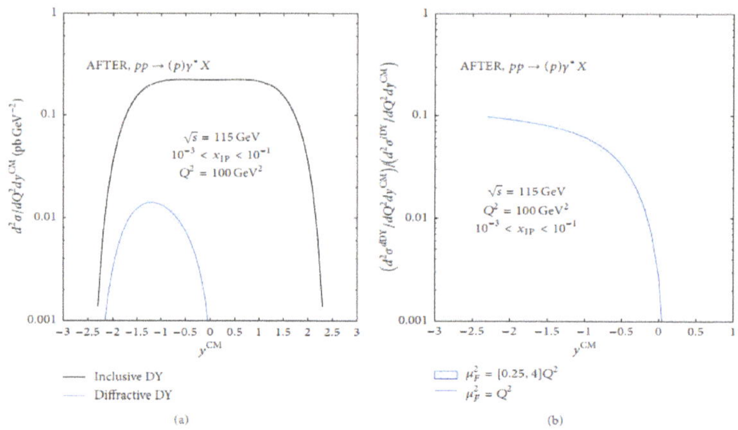

Figure 7: (a) Rapidity distributions for inclusive and diffractive Drell-Yan of mass $Q^2 = 100\ \mathrm{GeV}^2$. Blue error bands represent theoretical errors estimation, as described in the text. (b) Diffractive to inclusive Drell-Yan rapidity distributions ratio.

Since the rapidity is additive under boost along the collision axis we may easily boost the y^{CM} to the laboratory frame by using

$$y^{\mathrm{Lab}} = \frac{1+c}{1-c} + y^{\mathrm{CM}} \quad \text{with } c = \sqrt{1 - \frac{4m_p^2}{s}} \tag{13}$$

with m_p the proton mass. In the AFTER@LHC kinematics this implies a rapidity shift $\Delta y = y^{\mathrm{Lab}} - y^{\mathrm{CM}} = 4.8$. The rapidity distributions in the laboratory frame for a DrellYan pair of mass $Q^2 = 100\ \mathrm{GeV}^2$ and for three different x_{IP} values are presented in the right panel of Figure 6. One may notice from the plot that for increasing x_{IP} the DrellYan pair spans a wider rapidity range and the corresponding spectrum is increasingly more forward. It might be useful to discuss the single diffractive Drell-Yan pair production in conjunction with the analogous inclusive process. Such a comparison is presented for the centre-of-mass rapidity distributions in the left plot of Figure 7 for a common DrellYan pair of mass squared $Q^2 = 100\ \mathrm{GeV}^2$ and, for the diffractive case, integrated in the range $10^{-3} < x_{\mathrm{IP}} < 10^{-1}$. The rapidity distribution in the single diffractive case is strongly asymmetric whereas in the inclusive case it is symmetric around $y^{\mathrm{CM}} = 0$. This effect is primarily due to the different kinematics of the two processes and to the different fractional momentum distributions between parton and fracture distributions. In the right plot of Figure 7 the ratio between the two previous distributions is presented. The ratio gives direct information on the suppression factor between the single diffractive to the inclusive process, assuming a factorised expression for the former (11). Such a ratio might be convenient from the experimental side since many lepton detection systematics will cancel. On the theoretical one it is expected to be more stable against the inclusion of higher order corrections. In the present case, for example, the factorisation scale is simultaneously varied both on the numerator and denominator resulting in a reduced theoretical error band with respect to the one obtained for absolute cross sections.

We wish to end this section with a brief overview of other possible applications of the proposed formalism. A completely analogous program can be performed for the associated production of forward neutron and a Drell-Yan pair, $p{+}p \to n{+}\gamma^* {+}X$. The production of forward neutron in DIS at HERA has shown a leading twist nature. From scaling violations of the semi-inclusive neutron structure functions a set of proton-to-neutron fracture functions has been extracted from data in [12] which can be used to predict forward neutron rate in hadronic collisions. As in the case of hard diffraction, both physics programs would highly benefit from the installation of a dedicated instrumentation for the measurements of fast neutrons and protons quite close to the beam axis. Measurements in the forward region, although problematic experimentally, give in fact direct access to the study of the beam fragmentation region.

As a third application we consider hyperon production associated with a Drell-Yan pair, $p{+}p \to V{+}\gamma^* + X$, where V generically indicates either a Λ^0 or $\overline{\Lambda}^0$ hyperon. At very low transverse momentum, the Λ^0 longitudinal momentum spectrum should show a significant leading particle effect, which can be predicted, assuming factorisation as in $d\sigma^{H,T}$, by the proton-to-lambda fracture functions set obtained from a fit to SIDIS data in [13]. On the other hand, the $\overline{\Lambda}^0$ spectrum in the same kinematical conditions should instead show almost no leading particle effect, giving access to the proton-to-$\overline{\Lambda}^0$ fracture functions. We note, in general, that the particle-to-antiparticle fracture function is indeed an interesting and almost unknown distribution. On the other hand, if one considers Λ_0 or $\overline{\Lambda}^0$ at sufficiently large transverse momentum, their combined analysis, described by $d\sigma^{H,}$, should allow for the investigation of parton hadronisation into hyperons in the QCD vacuum as parametrised by fragmentation functions.

As a last application we consider the associated production of one particle and a Drell-Yan pair in the context of multiparton interactions.The latter process has already been used to investigate the contamination of the so-called underlying event [27] to jet observable and has been successfully used to study underlying event properties [28]. If the detected hadron is measured at sufficiently large transverse momentum, the latter constitutes a natural infrared regulator for the partonic matrix elements. In this kinematical conditions we also expect a rather small contribution from fracture functions. Therefore, the central term, $d\sigma^{H,(1)}$, can be used to estimate the single parton scattering contribution to the process. The latter might be considered as the baseline to study the double (or multiple) parton scattering contributions to the same final state, where, for example, the primary scatterer produces a Drell-Yan pair while the secondary one produces the detected hadron H.

CONCLUSIONS

We have briefly reviewed a perturbative approach to single particle production associated with a Drell-Yan pair in hadronic collisions. On the theoretical side we have shown that the introduction of fracture functions allows a consistent factorisation of new class of collinear singularities arising in this type of processes. The factorisation procedure coincides with the one used in DIS confirming, as expected, the universal structure of collinear singularities and supports the proposed collinear factorisation formula. On the phenomenological side we have outlined some areas in which the formalism can be fully tested. In particular, focusing on the AFTER@LHC kinematical range, we have presented numerical predictions for the single diffractive production of virtual photons. The study of such a process might improve our understanding of nonperturbative aspects of QCD and it allows one to explore in detail the nature of factorisation breaking at intermediate energies.

REFERENCES

1. M. Basile, G. Cara Romeo, L. Cifarelli, et al., "The 'leading'-particle effect in hadron physics," Nuovo Cimento A, vol. 66, no. 2, pp. 129–163, 1981.
2. J. C. Collins, "Proof of factorization for diffractive hard scattering," Physical Review D, vol. 57, no. 5, pp. 3051–3056, 1998.
3. J. C. Collins, "Erratum: proof of factorization for diffractive hard scattering [Phys. Rev. D 57, 3051 (1998)]," Physical Review D, vol. 61, no. 1, Article ID 019902, 1 pages, 1999.
4. J. C. Collins, "Factorization in hard diffraction," Journal of Physics G: Nuclear and Particle Physics, vol. 28, no. 5, pp. 1069–1077, 2002.
5. M. Grazzini, L. Trentadue, and G. Veneziano, "Fracture functions from cut vertices," Nuclear Physics B, vol. 519, no. 1-2, pp. 394–404, 1998.
6. L. Trentadue and G. Veneziano, "Fracture functions. An improved description of inclusive hard processes in QCD," Physics Letters B, vol. 323, no. 2, pp. 201–211, 1994.
7. A. Aktas, V. Andreev, T. Anthonis, et al., "Measurement and QCD analysis of the diffractive deep-inelastic scattering cross section at HERA," The European Physical Journal C—Particles and Fields, vol. 48, no. 3, pp. 715–748, 2006.
8. H1 Collaboration, "Dijet cross sections and parton densities in diffractive DIS at HERA," Journal of High Energy Physics, vol. 2007, no. 10, article 042, 2007.
9. ZEUS Collaboration, S. Chekanov, M. Derrick et al., "A QCD analysis of ZEUS diffractive data," Nuclear Physics B, vol. 831, no. 1-2, pp. 1–25, 2010.
10. F. A. Ceccopieri and L. Favart, "Revisiting QCD fits in diffractive DIS," http://arxiv.org/abs/1205.6356.
11. F. A. Ceccopieri and L. Favart, "QCD fits in diffractive DIS revisited," http://arxiv.org/abs/1110.4829.
12. F. A. Ceccopieri, "QCD analysis of forward neutron production in DIS," The European Physical Journal C, vol. 74, article 3029, 2014.
13. F. A. Ceccopieri and D. Mancusi, "QCD analysis of Lambda hyperon production in DIS target-fragmentation region," The European Physical Journal C, vol. 73, article 2435, 2013.
14. A. Berera and D. E. Soper, "Diffractive jet production in a simple model with applications to DESY HERA," Physical Review D, vol. 50, article 4328, 1994.
15. J. C. Collins, L. Frankfurt, and M. Strikman, "Diffractive hard scattering with a coherent pomeron," Physics Letters B, vol. 307, no. 1-2, pp. 161–168, 1993.
16. CDF Collaboration, "Diffractive dijets with a leading antiproton in p-p collisions at s=1800 GeV," Physical Review Letters, vol. 84, p. 5043, 2000.
17. T. Aaltonen, A. González, S. Amerio, et al., "Diffractive W and Z production at the Fermilab Tevatron," Physical Review D, vol. 82, no. 11, Article ID 112004, 10 pages, 2010.
18. S. J. Brodsky, F. Fleuret, C. Hadjidakis, and J. P. Lansberg, "Physics opportunities of a fixed-target experiment using LHC beams," Physics Reports, vol. 522, no. 4, pp. 239–255, 2013.
19. F. A. Ceccopieri and L. Trentadue, "NLO semi-inclusive Drell-Yan cross-section in quantum chromodynamics as a factorization analyzer," Physics Letters B, vol. 668, no. 4, pp. 319–323, 2008.
20. F. A. Ceccopieri, "Associated production of one particle and a Drell-Yan pair in hadronic collisions," Physics Letters B, vol. 703, no. 4, pp. 491–497, 2011.
21. G. Altarelli, R. K. Ellis, and G. Martinelli, "Large perturbative corrections to the Drell-Yan process in QCD," Nuclear Physics B, vol. 157, no. 3, pp. 461–497, 1979.
22. D. Graudenz, "One-particle inclusive processes in deeply inelastic lepton-nucleon scattering," Nuclear Physics B, vol. 432, no. 1-2, pp. 351–373, 1994.
23. M. Klasen and G. Kramer, "Survival probability for diffractive dijet production in pp- collisions from next-to-leading order calculations," Physical Review D, vol. 80, Article ID 074006, 2009.
24. G. Camici, M. Grazzini, and L. Trentadue, "Fracture functions and Jet Calculus," Physics Letters, Section B, vol. 439, no. 3-4, pp. 382–388, 1998.
25. F. A. Ceccopieri and L. Trentadue, "A new fracture function approach to QCD initial state radiation," Physics Letters B, vol. 655, no. 1-2, pp. 15–25, 2007.
26. A. D. Martin, R. G. Roberts, W. J. Stirling, and R. S. Thorne, "NNLO global parton analysis," Physics Letters B, vol. 531, no. 3-4, pp. 216–224, 2002.
27. T. Affolder, H. Akimoto, A. Akopian, et al., "Charged jet evolution and the underlying event in proton-antiproton collisions at 1.8 TeV," Physical Review D, vol. 65, no. 9, Article ID 092002, 22 pages, 2002.
28. T. Aaltonen, J. Adelman, B. Álvarez González, et al., "Studying the underlying event in Drell-Yan and high transverse momentum jet production at the Tevatron," Physical Review D, vol. 82, Article ID 034001, 2010.

CHAPTER 17

QUARKONIUM PRODUCTION AND PROPOSAL OF THE NEW EXPERIMENTS ON FIXED TARGET AT THE LHC

A. B. Kurepin and N. S. Topilskaya

INTRODUCTION

The existence of the Quark-Gluon Plasma (QGP) is predicted by lattice QCD at high temperature and large energy density. Quarkonia are important probes to study the properties of the deconfined matter. The dissociation of heavy-quark resonances by color screening in a deconfined medium was suggested by Matsui and Satz as a possible signal of the Quark-Gluon Plasma formation in ultrarelativistic heavy ion collisions [1].

Quarkonium productions have been previously studied at the CERN SPS by NA38 [2], NA50 [3–7], and NA60 [8, 9] experiments, at the FNAL [10], and by fixed target p-A experiments at the HERA-B [11]. In 1997, the NA50 experiment has observed an "anomalous" suppression of the J/ψ production in central Pb-Pb collisions [3]. However, the cold nuclear matter (CNM) effects and feed-down production of J/ψ from higher charmonium states, which affect the extraction of hot and dense matter effects, should be carefully measured. In spite of many experimental results already obtained on the J/ψ production in p-A collisions, they are still not well understood, especially at small Bjorken x_B and large Feynman x_F, where the CNM effects are expected to be large. Quarkonium productions have been studied in the Brookhaven National Laboratory at RHIC by PHENIX [12–16] and STAR experiments [17–19]. It was shown that the J/ψ suppression measured by PHENIX experiment at $\sqrt{s}$ = 200 GeV is of the same order as the suppression at the SPS energies for Pb-Pb collisions [13, 20]. In order to get better agreement between theoretical description of the RHIC and the SPS experimental results the models that include the regeneration of J/ψ were developed [21–24]. The quarkonium productions have been measured at significantly higher energies at the LHC by ALICE [25, 26], CMS [27, 28], ATLAS [29, 30], and LHCb [31, 32] experiments. At the LHC energies the contribution of B-decay should be taken into account [27, 29, 31–33]. The high statistics measurements at the LHC could investigate the properties of matter at high energy density and temperature. However, the quarkonium study at energies below the LHC energy is also very important in order to investigate the mechanism of production, the medium effects, and conditions for the Quark-Gluon Plasma formation. If it would be possible to use the proton and ion beams at the LHC with fixed targets, the data in the energy interval between maximum energy of the SPS ($\sqrt{s}$ ~ 29 GeV) and the nominal RHIC energy ($\sqrt{s}$ = 200 GeV) in pA and A-A collisions could be obtained. For 0.45 TeV proton beam, the energy in N-N centre-of-mass system is $\sqrt{s_{NN}}$ = 29.1 GeV, for 7 TeV proton beam $\sqrt{s_{NN}}$ = 114.6 GeV and for 2.75 TeV Pb beam $\sqrt{s_{NN}}$ = 71.8 GeV. The high statistics data on quarkonium production at these energies will give the possibility of clarifying the mechanism of charmonium production to investigate the importance of recombination process, since the probability of recombination decreases with decreasing the energy of collisions.

CHARMONIUM PRODUCTION AT THE CERN SPS

The productions of J/ψ and (2S) have been studied at the CERN SPS by the NA50 experiment. Data have been taken for Pb-Pb collisions at 158 GeV per nucleon [3, 7] and for p-A collisions at 450 [4, 5] and 400 GeV [6]. The nuclear suppression of the J/ψ production in proton-nucleus reactions and an anomalous suppression in central lead-lead collisions were observed [3, 7]. In the NA60 experiment, the J/ψ production was measured in In-In collisions at 158 GeV per nucleon [8] and in p-A collisions at 400 and 158 GeV [9]. The NA60 results in p-A collisions at 400 GeV confirm the NA50 values of absorption cross section at the same energy. On the other hand, the NA60 158 GeV p-A data give higher values of absorption. Rather

Citation: A. B. Kurepin and N. S. Topilskaya, "Quarkonium Production and Proposal of the New Experiments on Fixed Target at the LHC," Advances in High Energy Physics, vol. 2015, Article ID 760840, 13 pages, 2015. doi:10.1155/2015/760840.

strong energy dependence of the absorption cross section was observed. However, the older NA3 J/ψ results at 200 GeV [34] are in partial contradiction with these observations and give values of absorption cross section similar to those obtained at the higher energy. This difference has not yet been satisfactorily explained by theoretical models. Therefore, it is important to measure the charmonium production cross sections in p-p, p-A, and A-A collisions at the same energy and in the same kinematic domain.

The heavy-quark pair can interact with the nuclear matter and eventually break up. It is the effect of nuclear absorption. The cold nuclear matter (CNM) effect in suppression of the J/ψ production includes not only nuclear absorption, but also the shadowing effect, that is, the modification of the parton distribution function in nuclear matter compared to the free nucleon. If shadowing is taken into account, the amount of the anomalous J/ψ suppression is reduced. Also, it is necessary to consider the energy loss of partons in nuclear matter. The charmonium can also interact with surrounding comovers and lose energy or even break up.

For In-In collisions at 158 GeV per nucleon after adjustment for the effects of cold nuclear matter, the additional suppression becomes quite small. But the value of the anomalous suppression in the most central Pb-Pb collisions is of the order of 20–30%. So, to extract the effects of hot and dense nuclear matter in the J/ψ production at the SPS energies, the CNM effects and feed-down production from higher charmonium states must be taken into account.

CHARMONIUM PRODUCTION AT THE RELATIVISTIC HEAVY ION COLLIDER RHIC

At the RHIC, the production of J/ψ was measured in p-p, d-Au, Au-Au, and Cu-Cu collisions by the PHENIX experiment at $\sqrt{s_{NN}}$ = 200 GeV energy in the N-N centreof-mass [12–16]. To quantify the suppression of the J/ψ production in heavy ion collisions, the nuclear modification factor R_{AA} was introduced:

$$R_{AA} \equiv \frac{dN_{AA}/dy}{\left(dN_{pp}/dy \cdot \langle N_{coll}\rangle\right)}, \quad (1)$$

which is the ratio of the J/ψ yield in A-A collisions, normalized to the number of binary collisions, N_{coll}, to the J/ψ yield in p-p collisions. The nuclear modification factor has been measured as a function of several parameters: the centrality of collisions, the multiplicity, the transverse momentum, and the number of participants (N_{part}).

The suppression of the J/ψ production in Au-Au collisions is considerably stronger at forward rapidity range $1.2 < |y| < 2.2$ than at midrapidity $|y| < 0.35$. The PHENIX data for Au-Au and d-Au collisions were analyzed simultaneously for estimation of the CNM contribution. The nuclear modification factor for cold nuclear matter R_{AA}(CNM)was obtained in Au-Au collisions at the measured rapidity ranges [20]. The ratio R_{AA}/R_{AA}(CNM) estimates the value of additional suppression of the J/ψ production in the hot and dense nuclear matter produced in relativistic heavy ions collisions.The R_{AA}/R_{AA}(CNM)ratio at $\sqrt{s_{NN}}$ = 200 GeV is similar to the forward rapidity and midrapidity and reaches about 50% for the most central events.

Comparison of the R_{AA}/R_{AA}(CNM) ratio estimated for NA50, NA60, and PHENIX data versus $dN/d\eta$ at pseudorapidity η=0, which is proportional to the energy density, shows that at RHIC energy anomalous suppression of J/ψ for colliding nuclei is the same as at SPS energy [20] (Figure 1).

The nuclear modification factor for cold matter R_{AA}(CNM) was obtained under the assumption that in p-p, p-A, and d-Au collisions at SPS and RHIC energies the hot and dense nuclear matter is not formed. There are some indications that the hot matter effects could present in d-Au collisions at the RHIC energies [35]. These effects may contribute to p-p and p-A collisions at higher energies.

The production of J/ψ in p-p, Au-Au, and Cu-Cu collisions was measured also by STAR experiment at the RHIC at midrapidity $|y| < 1$ for low $p_T < 5$ GeV/c [17] and for high $pT > 5$ GeV/c [18, 19]. The value of R_{AA} in Au-Au collisions shows strong suppression at low transverse momentum but the suppression decreases significantly with increasing momentum. The J/ψ suppression in AuAu collisions at low p_T < 5 GeV/c and at midrapidity is compatible to that measured by PHENIX. It was found that the results are consistent with models that include color screening and regeneration [36, 37]. The data from PHENIX show a significant suppression in midcentral and central Cu-Cu collisions [15]. The STAR Cu-Cu data [19] exhibit no suppression, but the precision is limited by the available statistics.

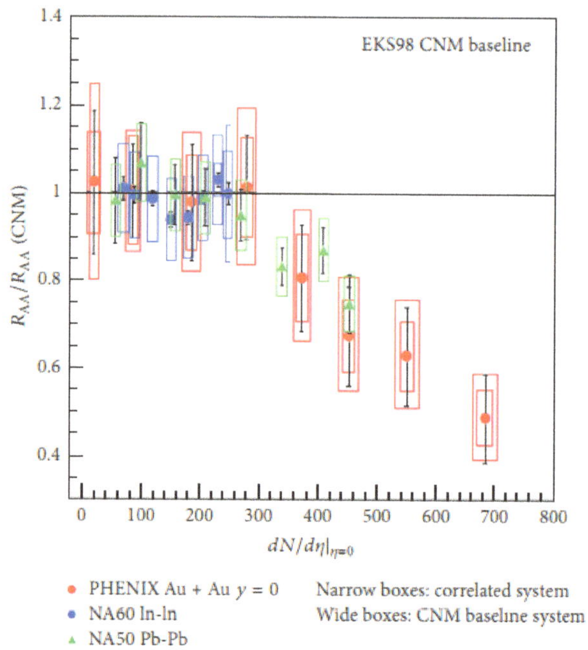

Figure 1: Comparison of the R_{AA}/R_{AA}(CNM) ratio obtained at the SPS and PHENIX versus $dN/d\eta$ at η=0.

The production of J/ψ in asymmetric Cu-Au heavy ion collisions was measured also at the RHIC at $\sqrt{s_{NN}}$ = 200 GeV for both forward (Cu-going direction) and backward (Augoing direction) rapidity [38]. The suppression in the Augoing direction is found to be consistent with the suppression measured in Au-Au collisions. In the Cu-going direction, the J/ψ suppression is stronger. The difference may be due to the CNM effects which are different at forward and backward rapidity.

Recently at the RHIC, the J/ψ production in U-U collisions was measured at $\sqrt{s_{NN}}$ = 193 GeV [39, 40]. The nuclear modification factor for J/ψ in U-U collisions for forward rapidity is very close to the Au-Au data with a hint of a slightly weaker suppression in central U-U collisions.

The RHIC experimental data could be described by theoretical models based on the regeneration of J/ψ. Additional J/ψ mesons are expected to be produced from deconfined charm quarks by kinetic recombination in the QGP [21, 22, 36, 37] or by statistical hadronization at the phase boundary [23, 24].

Bottomonium productions were also studied at the RHIC in p-p, d-Au, and Au-Au collisions [16, 39].The suppression of total Y(1S + 2S + 3S) yield measured by PHENIX is consistent with measurements made by STAR. The regeneration of bottomonia is expected to be small as compared to the J/ψ. But the energy resolution at RHIC was insufficient to separate the contributions of individual states.

The nuclear modification factor R_{AA} for the J/ψ production measured in Au-Au collisions at lower energies 62.4 and 39 GeV reveals approximately the same J/ψ suppression as at 200 GeV but has large statistical errors due to low luminosity and large systematical errors because of lack of p-p collisions at the same energies [41]. At the RHIC, the luminosity strongly decreases with decreasing the energy of collisions. The investigations at lower energy with high statistics are very important for understanding the mechanism of charmonium production, for investigation the contribution of cold and hot nuclear matter effects. In addition, the contribution of recombination process decreases with decreasing the energy of collision.

The fixed target experiment at high luminosity LHC beams could provide measurements for p-p, p-A, and AA collisions with the same equipment for different nuclei targets. The energy of collisions would be lower than the nominal RHIC energy (200 GeV). The charmonium production could be investigated with high statistics and low systematic errors and may provide reference from p-p data at these energies for RHIC.

QUARKONIUM PRODUCTION AT THE CERN LARGE HADRON COLLIDER LHC

At the LHC at CERN quarkonium productions were measured at the energy more than ten times higher than at RHIC. Five quarkonium states from two families are under study, charmonia:/ψ and ψ(2S) and bottomonia: Y(1S), Y(2S), and Y(3S). In four experiments: ALICE [25, 26, 42–46], CMS [27, 28, 47–49], ATLAS [29, 30], and LHCb [31, 32, 50, 51], the quarkonium productions were measured in different energy, rapidity, and transverse momentum ranges. In all experiments p-p, p-Pb, and Pb-Pb collisions were measured, except for LHCb where there was no Pb-Pb program.

The quarkonium production in p-p collisions at $\sqrt{s_{NN}}$ = 2.76, 7, and 8 TeV is useful for investigating the production mechanism, for comparison with different QCD model calculations and as a reference for understanding any additional effects in p-A and A-A collisions. The study of quarkonium production in p-Pb collisions at $\sqrt{s_{NN}}$ = 5.02 TeV is important to distinguish the effects of Quark-Gluon Plasma from cold nuclear matter and to provide input to understanding of nucleus-nucleus collisions. The information about properties of hot and dense medium created in the collisions could be obtained from the measurements of Pb-Pb collisions at $\sqrt{s_{NN}}$ = 2.76 TeV.

Charmonium Production at the LHC.

At the LHC the charmonium J/ψ and (2S) productions were measured in p-p collisions at 2.76, 7, and 8 TeV and in Pb-Pb collisions at 2.76 TeV. Measurement of the J/ψ production in p-p collision at the same energy as in Pb-Pb collision provides the reference for extracting the nuclear modification factor R_{AA}. There is a good agreement for p-p collisions between the data obtained by ALICE [25], CMS [27], ATLAS [29], and LHCb [31] experiments in the same kinematic domains [52]. Comparison of charmonium J/ψ and ψ(2S) production cross sections at forward rapidity in p-p collisions at 7 TeV [42], obtained by ALICE and LHCb experiments, is shown in Figure 2.

The mechanism of charmonium production was investigated by measuring the cross sections in p-p collisions as functions of energy, transverse momentum, and rapidity. Cross sections are compared to the theoretical models [42]. Unfortunately, none of the models, including NLO QCD, are able to describe simultaneously different aspects of quarkonium production. By increasing the energy of collisions, the mean transverse momentum and production cross section of J/ψ become larger. For forward rapidity, the cross section is smaller than at midrapidity. The contribution of Bdecay to J/ψ production cross section was measured. This contribution depends on rapidity, increases with growing of J/ψ transverse momentum, and is of the order of 10% for transverse momentum about 1.5 GeV/c [27, 29, 33].

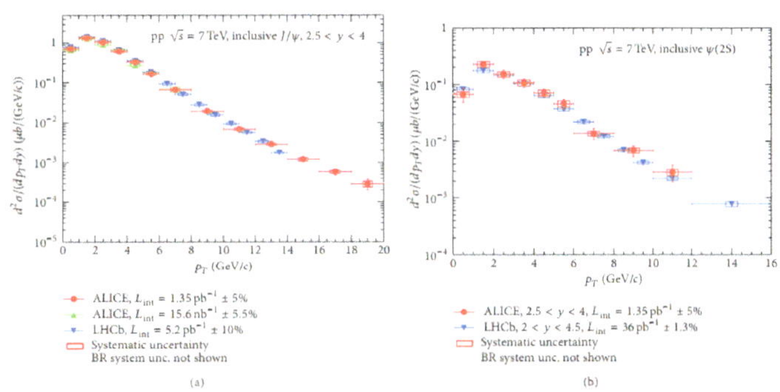

Figure 2: Comparison of charmonium production cross sections in p-p collisions at 7 TeV versus transverse momentum for ALICE and LHCb data for J/ψ (a) and (2S) (b).

ALICE [26, 43], CMS [28], and ATLAS [30] experiments have measured charmonium productions in Pb-Pb collisions at $\sqrt{s_{NN}}$ = 2.76 TeV. In the ALICE experiment, J/ψ mesons were detected in rapidity range $|y| < 0.9$ (for J/ψ decay into two electrons) and $2.5 < y < 4$ (for muon channel) with transverse momentum values from about zero up to 8 GeV/c. In ATLAS and CMS experiments, the J/ψ production was measured in the rapidity range $|y| < 2.4$, but the range of transverse momentum values depends on the rapidity. In ATLAS experiment, only the data for J/ψ mesons production with large transverse momentum $p_T > 6.5$ GeV/c were obtained. The comparison of PHENIX and ALICE nuclear modification factors R_{AA} for inclusive J/ψproduction [13, 26] is shown in Figure 3.

Smaller suppression in the ALICE measurements of the inclusive J/ψ production in Pb-Pb collisions at $\sqrt{s_{NN}}$ = 2.76 TeV [26, 43] compared to PHENIX results in Au-Au collisions at $\sqrt{s_{NN}}$ = 0.2 TeV [13] was found. These measurements are compatible with a regeneration mechanism. This additional hot nuclear matter effect works in an opposite direction to the suppression by color screening. At the LHC energy, the charm quark density produced in the collisions is larger than at SPS and RHIC energies. The probability of $c\bar{c}$ recombination increases with increasing the energy of collision and at the LHC energies may become dominant. In particular, this mechanism predicts an increase of R_{AA} from forward rapidity to midrapidity, where the density of charm quarks is higher. Moreover, in order to recombine, two charm quarks need to be close enough in phase space, so the effect will be larger at lowp_T of J/ψin agreement with experimental data [36, 37].

At the LHC, charmonium productions were measured in p-Pb collisions at the energy $\sqrt{s_{NN}}$ = 5.02 TeV by ALICE [44, 45] and LHCb [50] experiments. The inclusive J/ψ production has been studied by ALICE [44]. The measurement is performed down to zero transverse momentum in the center of mass rapidity domains $2.03 < y < 3.53$ and $-4.46 < y < -2.96$ in muon channel. In p-Pb collisions, the J/ψ production was also measured in electron channel in rapidity domain $-1.37 < y < 0.43$ [44]. Since p-p collisions data at $\sqrt{s_{NN}}$ = 5.02 TeV for determination of R_{pPb} are not available, the reference $\sigma_{pp}(J/\psi)$ has been obtained by an interpolation procedure, based on p-p collisions results at $\sqrt{s_{NN}}$ = 2.76 TeV and 7 TeV obtained by the ALICE experiment. At forward rapidity, corresponding to the proton beam direction, a suppression of the J/ψ yield with respect to binary-scaled p-p collisions is observed, but in the backward region no suppression is found. The experimental results and comparison with theoretical predictions [44] are shown in Figure 4.

The Color Glass Condensate (CGC) model [53] could not describe the data. Theoretical calculations based on nuclear shadowing [54] as well as on the models including, in addition, a contribution from partonic energy loss [55, 56] are in better agreement with the experimental results (Figure 4).

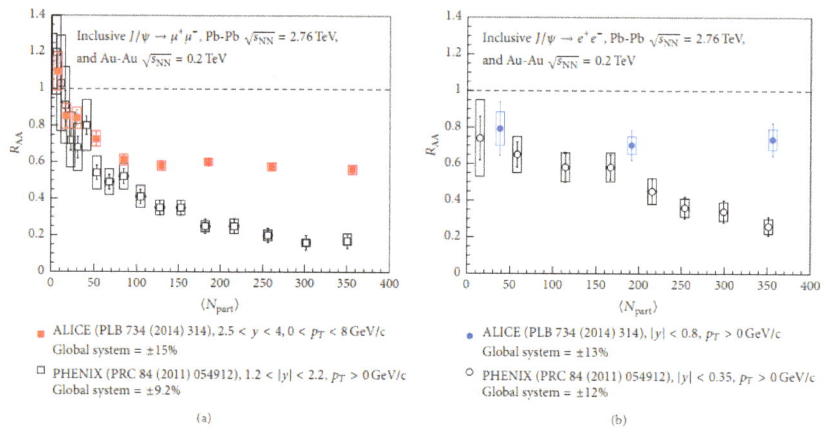

Figure 3: Comparison of R_{AA} results, obtained by PHENIX and ALICE experiments. The two panels show the data at forward rapidity (a) and midrapidity (b).

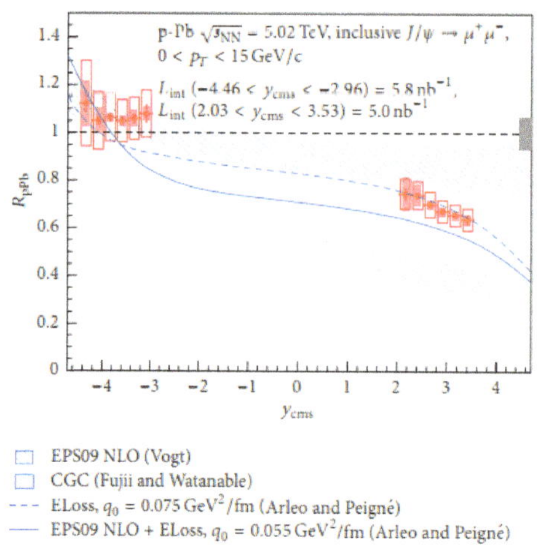

Figure 4: Comparison of ALICE R_{pPb} results for inclusive J/ψ with theoretical predictions.

The ALICE results are in agreement with results for inclusive J/ψ mesons presented by LHCb collaboration [57]. Nuclear modification factors are determined separately for prompt J/ψ mesons and for J/ψ from B-hadron decays by LHCb experiment. The suppression of prompt J/ψ mesons in p-Pb collisions with respect to p-p collisions at large rapidity is observed, while the production from B-hadron decays is less suppressed. The (2S) production was measured in p-Pb collisions by ALICE [45].

The nuclear modification factor for inclusive (2S) is evaluated. A significantly larger suppression of the (2S) compared to the inclusive J/ψ was obtained. Theoretical models predictions, which include parton shadowing and coherent energy loss mechanism, reproduce J/ψ suppression but could not describe (2S) data [56]. The models underestimate the (2S) suppression. A comparison with theoretical predictions is shown in Figure 5. Additional effects should be considered for interpretation of the results.

Bottomonium Production at the LHC.

In p-p collisions at $\sqrt{s_{NN}}$ = 7 TeV, the inclusive production cross sections of Υ(1S) and Υ(2S) have been measured as a function of p_T and rapidity by ALICE [42]. There is a good agreement of ALICE data with measurements from LHCb experiment [31, 58] in the similar p_T and rapidity ranges. The data are complemented to CMS measurements at midrapidity [47, 59]. The results could be described by NLO NRQCD calculations.

The CMS Collaboration at the LHC has observed the sequential suppression of Υ(1S), Υ(2S), and Υ(3S) bottomonium states in Pb-Pb collisions at $\sqrt{s_{NN}}$ = 2.76 TeV [48, 49]. The Υ(1S) yield is suppressed by approximately a factor of two with respect to the expectation from p-p collisions obtained by scaling with the number of binary nucleonnucleon collisions. The Υ(2S) and the Υ(3S) are strongly suppressed. Due to the lower production cross section of $b\bar{b}$ pairs compared to $c\bar{c}$ pairs, the regeneration of Υ(1S) is expected to be smaller than that of J/ψ [60]. On the other hand, the CNM effects can also modify the production of bottomonia in nucleus-nucleus collisions.

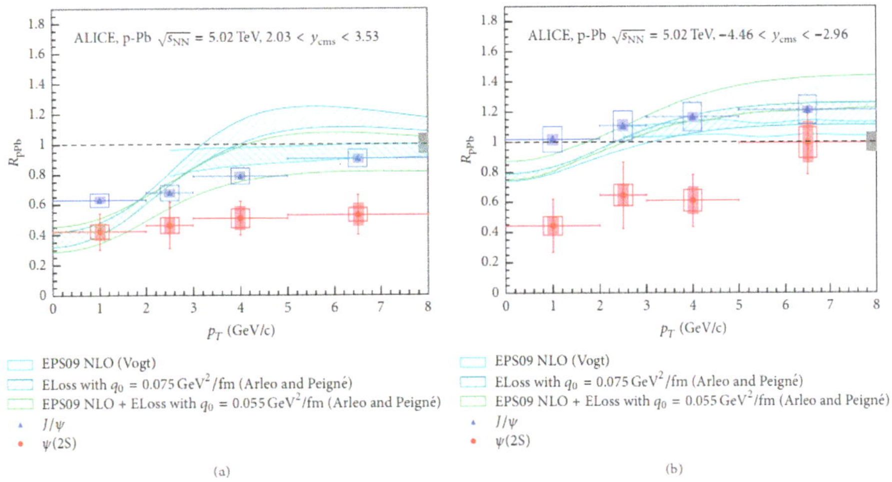

Figure 5: Comparison of ALICE R_{pPb} results for J/ψ and (2S) at forward (a) and backward (b) rapidity with theoretical predictions.

The inclusive Υ(1S) production in Pb-Pb collisions was measured by ALICE at forward rapidity in muon channel down to zero transverse momentum [61]. A strong suppression was observed with respect to p-p collisions scaled by the number of nucleon-nucleon collisions. The ALICE results are compared with CMS data ($|y|$ < 2.4) [48]. The observed suppression is stronger at forward rapidity than at midrapidity. The transport model [60] predicts a nearly constant R_{AA} as a function of rapidity and could not reproduce ALICE and CMS data. The transport model includes both suppression and regeneration effects. CNM effects were estimated by using the effective absorption cross section. The transport model clearly underestimates the observed suppression, but the shape of the centrality dependence is quite well reproduced. Another transport model [62] also includes suppression and small regeneration component and CNM effect with shadowing. The model reproduces the CMS data but underestimates the ALICE data at forward rapidity. Both transport models [60, 62] predict stronger suppression for central events in agreement with centrality dependence of ALICE data. The comparison of the data with transport model calculations [60] is shown in Figure 6.

The inclusive Υ(1S) and Υ(2S) productions were measured also by ALICE [46] and LHCb [57] in p-Pb collisions at $\sqrt{s_{NN}}$ = 5.02 TeV. At forward rapidity suppression of the inclusive Υ(1S) yield in p-Pb collisions with respect to the yield from p-p collisions scaled by the number of binary nucleon-nucleon collisions is observed. But at backward rapidity the suppression is less (Figure 7). The results are compared with several theoretical model calculations [54, 55] including partonic energy loss effects with and without nuclear shadowing. Only models with energy loss plus shadowing could describe the data at forward rapidity but these models underestimate the suppression at backward rapidity.

The production of charmonia and bottomonia is the object of intense theoretical and experimental investigations. Their production mechanism in p-p collisions is described by models based on Quantum Chromodynamics (QCD) and gives reference for comparison with p-A and A-A data. At the LHC in Pb-Pb collisions, the evidence for additional J/ψ production from regeneration at low p_T and strong J/ψ suppression at large transverse momentum was obtained. For p-Pb collisions theoretical models including nuclear suppression, parton shadowing, and coherent energy loss effects could reproduce J/ψ production but fail to describe additional suppression of (2S) and underestimate the observed Υ(1S) suppression at forward rapidity.

At the LHC and RHIC the collective and hot nuclear matter effects may also be present in p-Pb at 5.02 TeV and in d-Au collisions at 200 GeV [35]. Therefore, the measurement of nuclear effects at lower energies and assuming the absence of the QGP formation in p-p and p-A collisions could give a reference for the study of hot nuclear matter effects. Moreover, the contribution of recombination process is small at low energies. The RHIC heavy flavor program with the energy scan can perform this investigation but, unfortunately, the luminosity at RHIC strongly decreases with decreasing the energy of collision.

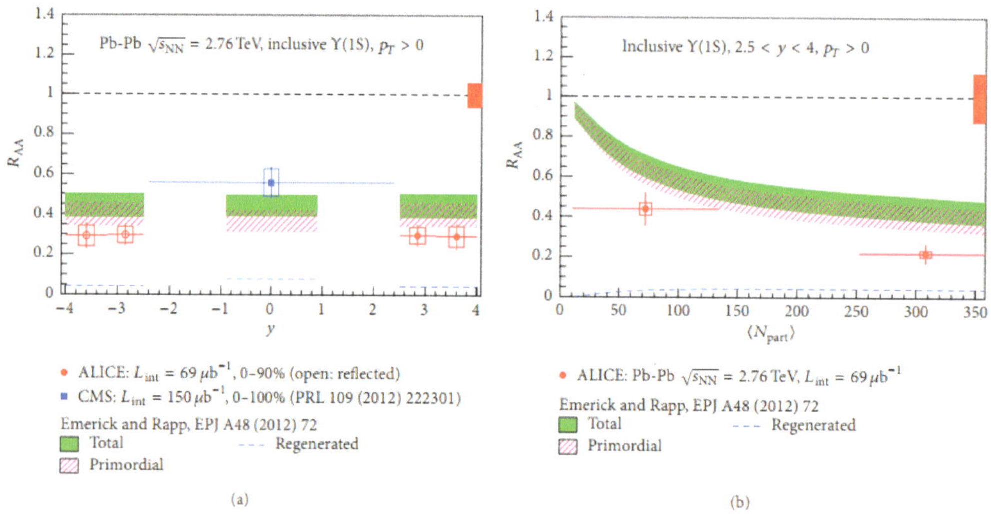

Figure 6: Comparison with theoretical predictions ALICE and CMS R_{PbPb} results for Υ(1S) versus rapidity (a) and ALICE results versus N_{part} (b).

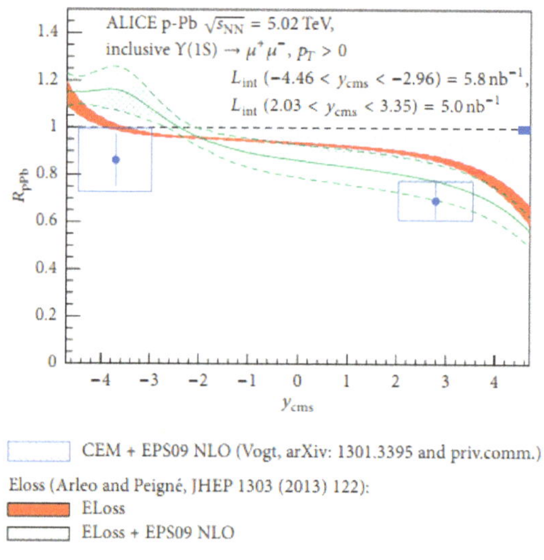

Figure 7: Comparison of ALICE R_{pPb} results for Υ(1S) versus rapidity with theoretical predictions.

QUARKONIUM PRODUCTION AT FIXED TARGET AT THE LHC BEAMS

The energy intervals between the SPS, the RHIC, and the LHC are very important to study the mechanism of quarkonium production and suppression and for the investigation of the medium effects and conditions of the Quark-Gluon Plasma formation. Investigation of the cold nuclear matter effect and understanding the properties of matter require systematic and high statistics measurements of quarkonium production also in the low energy region. To clarify the CNM effects it will be possible to study the mechanism of quarkonium production and suppression with high statistics at low energies up to 35 GeV/per nucleon in the CBM experiment at FAIR [63] and in the MPD experiment at NICA collider in Dubna [64]. At the CERN SPS, the fixed target experiment CHIC (Charm in Heavy Ion Collisions) for charmonium study at energy up to $\sqrt{s_{NN}}$ ~20 GeV is under preparation [65]. The Beam Energy Scan (BES-I) program at RHIC was performed by STAR and PHENIX collaborations. At STAR there is an ongoing fixed target program, with data already taken in the gold target test during 14.5 GeV Au-Au run in 2014 with $\sqrt{s_{NN}} = 3.9$ GeV [66].

If the LHC proton and ion beams would be used with fixed targets, the energy below the nominal RHIC energy (200 GeV) in p-A and A-A collisions could be investigated. For 7 TeV proton beam, the energy in N-N c.m. is $\sqrt{s_{NN}} = 114.6$ GeV, and for Pb beam at 2.75 TeV it is $\sqrt{s_{NN}}$ = 71.8 GeV. By using the LHC beams with reduced energy, the data could be taken in the range 30–100 GeV. QCD phase diagram may have interesting features probed by 10–100 GeV beams on fixed target. Search for signature of the phase transition and location of the critical point is the main goal of energy scan. Moreover, this is a possibility of investigating the mechanism of charmonium, J/ψ, and (2S) productions. It will be possible to separate the mechanism of hard production and then suppression by hadronic dissociation in QGP from secondary production with recombination, since the probability of recombination could decrease with decreasing energy of collision like in thermal model [67].

The existing system to inject the gas target for the luminosity measurement (SMOG) at LHCb experiment could be used for fixed target physics [68]. There are some testing measurements of p-Ne and Pb-Ne collisions, but, for the moment, there are no precise measurements of the gas density and pressure.

The main parameter for quarkonium production measurement in the fixed target experiment is the acceptance. In order to study the feasibility of using the fixed target at the LHC for charmonium production, the geometrical acceptances for J/ψ production on fixed target by means of AliRoot-FAST simulations were obtained. The detailed description of the calculation was given in [69]. For the model of quarkonium production the phenomenological Color Octet Model (COM) was used [70, 71]. For prompt J/ψ production the rapidity and transverse momentum distributions are obtained, respectively, as a parameterization of the COM predictions and by extrapolating to LHC energies the J/ψ transverse momentum (p_T) distribution, measured at midrapidity by the CDF experiment at energy $\sqrt{s_{NN}}$ = 1.8 TeV. The geometrical acceptances of the J/ψ production for PHENIX at RHIC and fixed target experiments NA50 at SPS and HERA-B were also calculated for comparison [69].

Geometrical Acceptances for Fixed Target Pb-Pb Collisions at the Energy T = 2.75 TeV per Nucleon, $\sqrt{s_{NN}}$ = 71.8GeV.

The geometrical acceptance for the J/ψ production at ALICE dimuon spectrometer was calculated by the beam axis at the interaction point z=0, at the points z = +50 cm (in the direction to the dimuon spectrometer) and z = −50 cm (outside the ITS) and at point z = +350 cm. The transverse momentum distribution for J/ψ was generated using p_T spectra in the CDF form, the same as for HERA and PHENIX, consistent with COM:

$$\frac{dN}{dp_T} \sim p_T \left[1 + \left(\frac{35\pi p_T}{256\langle p_T\rangle}\right)^2\right]^{-6}. \quad (2)$$

For rapidity distribution the Gaussian spectra were used. The parameters of distribution were energy scaled. For Pb-Pb collisions p_T spectra with the value $\langle pT\rangle$ = 1.4 and Gaussian rapidity distribution with mean values ycm = 0 and σ = 1.1 were used. The acceptances at z = +50 cm and z = −50 cm are approximately the same, but at point z = +350 cm the geometrical acceptance is much less. The results were shown in [69]. In Figure 8, the Pb-Pb results for z=0 and z = +50 cm are plotted. The J/ψ are accepted in the rapidity range −2.5 $< \eta <$ −4.0 (−2.97 $<\eta<$−4.09) for J/ψ production at the point z = 0 (z = +50 cm). The geometrical acceptances, I, are equal to (12.0±0.2) % for J/ψ production at z=0 point and (8.0±0.2) % at z = +50 cm.

Geometrical Acceptances for Fixed Target p-p and p-A Collisions at the Energy T=7 TeV, Corresponding to $\sqrt{s_{NN}}$ = 114.6GeV.

The J/ψ productions in p-p and p-Pb collisions are generated using p_T spectra in the same form as for Pb-Pb collision, but with the energy scaled parameter:

$$\frac{dN}{dp_T} \sim p_T \left[1 + \left(\frac{35\pi p_T}{256\langle p_T\rangle}\right)^2\right]^{-6}, \quad (3)$$

where $\langle p_T\rangle$ = 1.6. For rapidity distribution the Gaussian spectra were used with mean value being equal to y_{cm} = 0 and σ = 1.25. Geometrical acceptances for J/ψ production were calculated at points z=0, z = +50 cm, and z = −50 cm (outside the ITS) and at point z = +350 cm.The results were shown in [69]. The acceptances at z = +50 cm and z = −50 cm are roughly the same, but at point z = +350 cm the geometrical acceptance is much less. In Figure 9, the results for z=0 and z = +50 cm are plotted.

The geometrical acceptances, I, are equal to (8.5 ± 0.2) % for J/ψ production at z=0 point and (6.0 ± 0.3) % at z = +50 cm. We have calculated also the geometrical acceptances for J/ψ with the cut on transverse momentum of the single muon p_T > 1 GeV/c. The calculated geometrical acceptances for fixed target measurements are of the same order and even larger than geometrical acceptances for colliding nuclei in ALICE [69].

Luminosity, Cross Sections, and Counting Rates for p-p, pA, and Pb-Pb Collisions.

The target in the form of thin ribbon could be placed around the main orbit of the LHC, as it was already used for the experiment on collider with a fixed target at HERA-B [72]. The life time of the beam is determined by the beam-beam and beam-gas interactions. Therefore, after some time the particles will leave the main orbit and interact with the target ribbon. So for fixed target measurements only halo of the beam will be used. Therefore, no deterioration of the main beam will be introduced. The experiments at other interaction points will not feel any presence of the fixed target. Since the target ribbon should not interfere during the beam formation and acceleration process it should be lifted in the working position after the tuning of the beam.

In the ALICE measurements in 2011 in p-p run it was $1.2\cdot10^{11}$ protons per bunch, 1380 bunches and life time of 14.5 hours. From these parameters one can estimate the particle loss of $1.1\cdot10^{13}$ protons during one hour ($3.1\cdot10^{9}$ protons/s) and mean luminosity about $3\cdot10^{30}$ cm^{-2}s^{-1} for 500-micron lead ribbon. For Pb beam it was $1\cdot10^{8}$ ions per bunch, 358 bunches and life time of 6.5 hours. The particle loss is about $5.1\cdot10^{9}$ ions during one hour ($1.4\cdot10^{6}$ ions/s). The mean luminosity is about $1.7\cdot10^{27}$ cm^{-2}s^{-1} for 500-micron lead ribbon. The luminosity estimates are shown in Table 1. The value of the nucleon-nucleon charmonium total production cross section shown in Table 1 for 14 TeV was calculated by CEM (Color Evaporation Model) [73] with MRST HO PDF. The cross sections for lower energies were obtained by interpolation of proton-proton cross sections measured at RHIC at $\sqrt{s_{NN}} = 200$ GeV and at SPS in the NA51 experiment at 450 GeV per nucleon ($\sqrt{s_{NN}} = 29.1$ GeV), and were extracted from the data of the NA50 experiment at $\sqrt{s_{NN}} = 27.4$ GeV for proton-lead collisions.

The high statistics results could be obtained by fixed target experiments on J/ψ production in p-Pb and Pb-Pb collisions with the counting rate values presented in Table 1. For PbPb collisions, luminosity is smaller but the production cross section is larger. The measurement of(2S) production is also feasible with better statistical accuracy than at RHIC collider. The fixed target experiment AFTER using the LHC beams extracted by a bent crystal was proposed [74].The experiment AFTER has a wide physical program and gives possibility of using different targets with high thickness, so it gets higher luminosity (20 times more for 1 cm target versus 500 μm). But the experiment AFTER demands a lot of space for installation and has high cost.

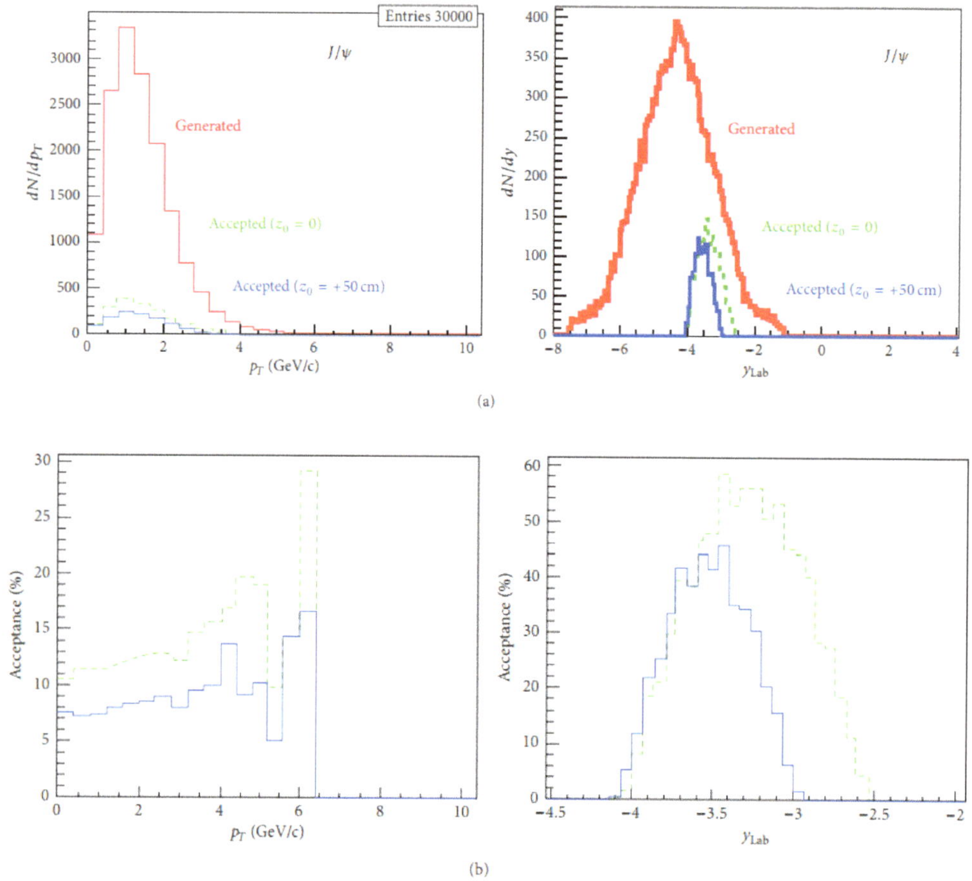

Figure 8: Transverse momentum and rapidity distributions for J/ψ events, produced at $z = +50$ cm (solid lines) and z=0 (dashed lines) in fixed target Pb-Pb collisions at $\sqrt{s} = 71.8$ GeV per nucleon pair (a) and corresponding acceptances (b).

Table 1: Luminosity, cross sections ($x_F > 0$), and counting rates

System	$\sqrt{s}$, TeV	σ_{NN}, μb	σ_{pA}, μb ($A^{0.92}\sigma_{NN}$)	I, %	IB σ_{pA}, μb	L, cm^{-2}s^{-1}	Rate, h^{-1}
pp	14	54.1	54.1	4.7	0.150	$3 \cdot 10^{30}$	1620
pp_{RHIC}	0.200	2.7	2.7	3.6	0.0057	$1 \cdot 10^{31}$	205
pPb_{NA50}	0.0274	0.19	25.7	14.0	0.212	$7 \cdot 10^{29}$	535
pPb_{fixed}	0.1146	0.65	80.2	6.0	0.310	$3 \cdot 10^{30*}$	3360
pPb_{fixed}	0.0718	0.55	74.6	8.0	0.349	$3 \cdot 10^{30*}$	3780
$PbPb_{fixed}$	0.0718	0.55	11970	8.0	42.9	$1.7 \cdot 10^{27**}$	292

$^*pPb_{fixed}$, 500 μ wire, $3.1 \cdot 10^9$ protons/s.
$^{**}PbPb_{fixed}$, 500 μ wire, $1.4 \cdot 10^6$ ions/s.

Fixed target experiment with the target in the form of thin ribbon may be placed at the existing experimental installation (e.g., LHCb). The target could be lifted in the working position with the aid of rotation system only after beam tuning. This experiment will use only halo of the beam (the target could be used as extra collimator).This fixed target experiment with the target in the form of thin ribbon looks like a first stage of more complicated experiment AFTER.

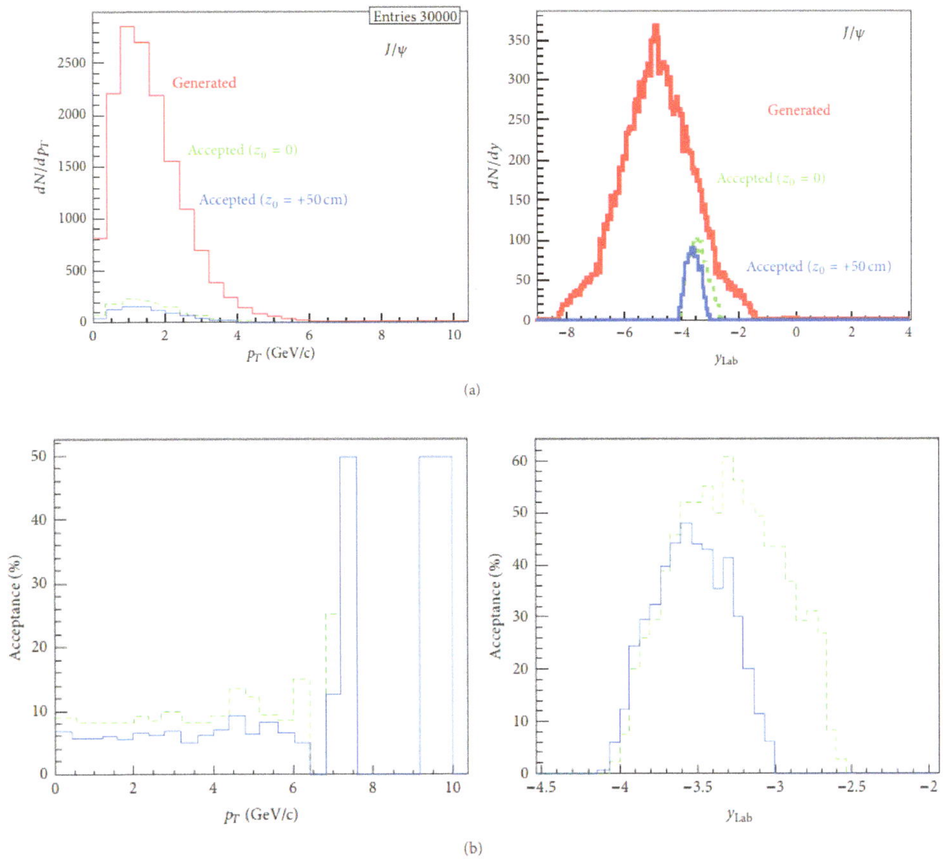

Figure 9: Transverse momentum and rapidity distributions for J/ψ events, produced at z = +50 cm (solid line) and at z=0 (dashed line) in fixed target p-A collisions at $\sqrt{s_{NN}}$ = 114.6 GeV per nucleon pair (a) and corresponding acceptances (b).

CONCLUSION

The fixed target experiment at LHC will give possibility for precise quarkonium studies at energies below nominal RHIC energy (200 GeV). It has advantage of high luminosity compared to collider experiments. The use of fixed target at LHC could provide in a short time the data for different targets and maybe for different projectile nuclei with high statistics. By using the LHC beams the data could be taken in the range $\sqrt{s_{NN}}$ = 29–115 GeV. Search for signature of the phase transition and location of the critical point is the main goal of energy scan. Moreover, this is a possibility of investigating the mechanism of charmonium, J/ψ, and (2S) productions to separate the possibilities of hard production and then suppression by hadronic dissociation in QGP or secondary production with recombination. Additional J/ψ mesons are expected to be produced from deconfined charm quarks by kinetic recombination in the QGP or by statistical hadronization at the phase boundary. The probability of recombination decreases with decreasing energy of collision. Therefore, the important information about mechanism of charmonium production and possible QGP formation could be obtained. The experiment with the fixed target in the form of thin ribbon looks like a first stage of more complicated experiment AFTER on the LHC beams.

REFERENCES

1. T. Matsui and H. Satz, "J/ψ suppression by quark-gluon plasma formation," Physics Letters B, vol. 178, no. 4, pp. 416–422, 1986.
2. M. C. Abreu, J. Astruc, C. Baglin, et al., "J/ψ and ψ′ production in p, O and S induced reactions at SPS energies," Physics Letters B, vol. 466, no. 2–4, pp. 408–414, 1999.
3. M. C. Abreu, B. Alessandro, C. Alexa, et al., "Anomalous J/ψ suppression in Pb-Pb interactions at 158 GeV/c per nucleon," Physics Letters B, vol. 410, no. 2–4, pp. 337–343, 1997.
4. B. Alessandro, C. Alexa, R. Arnaldi, et al., "A new measurement of J/ψ suppression in Pb-Pb collisions at 158 GeV per nucleon," The European Physical Journal C, vol. 39, pp. 335–345, 2005. ·
5. B. Alessandro, C. Alexa, R. Arnaldi, et al., "Charmonia and Drell-Yan production in proton-nucleus collisions at the CERN SPS," Physics Letters B, vol. 553, pp. 167–178, 2003.
6. B. Alessandro, C. Alexa, R. Arnaldi, et al., "J/ψ and ψ′ production and their normal nuclear absorption in proton-nucleus collisions at 400 GeV," The European Physical Journal C, vol. 48, no. 2, pp. 329–341, 2006.
7. B. Alessandro, C. Alexa, R. Arnaldi, et al., "ψ′ production in Pb–Pb collisions at 158 GeV/nucleon," The European Physical Journal C, vol. 49, no. 2, pp. 559–567, 2007.
8. R. Arnaldi and The NA60 Collaboration, "J/ψ production in p-A and A-A collisions at fixed target experiments," Nuclear Physics A, vol. 830, no. 1–4, pp. 345c–352c, 2009.

9. B. Alessandro, C. Alexa, R. Arnaldi, et al., "J/ψ production in proton-nucleus collisions at 158 and 400 GeV," Physics Letters B, vol. 706, no. 4-5, pp. 263–267, 2012.
10. I. Abt, M. Adams, M. Agari, et al., "Kinematic distributions and nuclear effects of J/ψ production in 920 GeV fixed-target proton-nucleus collisions," The European Physical Journal C, vol. 60, no. 4, pp. 525–542, 2009.
11. M. J. Leitch, W. M. Lee, M. E. Beddo, et al., "Measurement of differences between J/ψ and ψ′ suppression in p–A collisions," Physical Review Letters, vol. 84, no. 15, pp. 3256–3260, 2000.
12. A. Adare, S. Afanasiev, C. Aidala, et al., "J/ψ Production versus centrality, transverse momentum, and rapidity in Au + Au collisions at sNN=200 GeV," Physical Review Letters, vol. 98, Article ID 232301, 2007.
13. A. Adare, S. Afanasiev, C. Aidala, et al., "J/ψ suppression at forward rapidity in Au + Au collisions at sNN=200 GeV," Physical Review C, vol. 84, Article ID 054912, 2011.
14. A. Adare, S. Afanasiev, C. Aidala, et al., "Ground and excited state charmonium production in p+pcollisions at s=200 GeV," Physical Review D, vol. 85, no. 9, Article ID 092004, 27 pages, 2012.
15. A. Adare, S. Afanasiev, C. Aidala, et al., "J/ψ production in sNN=200 GeV Cu + Cu collisions," Physical Review Letters, vol. 101, Article ID 122301, 2008.
16. A. Adare, S. Afanasiev, C. Aidala, et al., "Cold nuclear matter effects on J/ψ yields as a function of rapidity and nuclear geometry in d+A collisions at sNN=200 GeV," Physical Review Letters, vol. 107, Article ID 142301, 2011.
17. L. Adamczyk, J. K. Adkins, G. Agakishiev, et al., "J/ψ production at low pT in Au+Au and Cu+Cucollisions at sNN=200 GeV with the STAR detector," Physical Review C, vol. 90, no. 2, Article ID 24906, 13 pages, 2014.
18. J. Adams, C. Adler, M. M. Aggarwal, et al., "J/ψ production at high transverse momenta in p+p and Au + Au collisions at sNN=200 GeV," Physics Letters B, vol. 722, pp. 55–62, 2013.
19. B. I. Abelev, M. Aggarwal, Z. Ahammed, and etaL, "J/ψ production at high transverse momenta in p+pand Cu + Cu collisions at sNN=200 GeV," Physical Review C, vol. 80, Article ID 041902(R), 2009.
20. H. Brambilla, S. Eidelman, B. K. Heltsley, et al., "Heavy quarkonium: progress, puzzles, and opportunities," The European Physical Journal C, vol. 71, article 1534, 2011.
21. R. L. Thews, "Quarkonium formation in statistical and kinetic models," The European Physical Journal C, vol. 43, no. 1–4, pp. 97–102, 2005.
22. R. L. Thews and M. L. Mangano, "Momentum spectra of charmonium produced in a quark-gluon plasma," Physical Review C, vol. 73, no. 1, Article ID 014904, 2006.
23. A. Andronic, P. Braun-Munzinger, K. Redlich, and J. Stachel, "Charmonium and open charm production in nuclear collisions at SPS/FAIR energies and the possible influence of a hot hadronic medium," Physics Letters B, vol. 659, no. 1-2, pp. 149–155, 2008.
24. P. Braun-Munzinger and J. Stachel, "(Non)thermal aspects of charmonium production and a new look at J/ψ suppression," Physics Letters B, vol. 490, no. 3-4, pp. 196–202, 2000.
25. K. Aamodt, A. Abrahantes Quintana, D. Adamová, et al., "Erratum to 'Rapidity and transverse momentum dependence of inclusive J/ψ production in pp collisions at s=7 TeV', Physics Letters B, vol. 704, no. 5, p. 442, 2011," Physics Letters B, vol. 718, no. 2, pp. 692–698, 2012.
26. B. Abelev, J. Adam, D. Adamova, et al., "Centrality, rapidity and transverse momentum dependence of J/ψ suppression in Pb-Pb collisions at sNN=2.76," Physics Letters B, vol. 734, pp. 314–327, 2014.
27. S. Chatrchyan, V. Khachatryan, A. M. Sirunyan, et al., "J/ψ and ψ (2S) production in pp collisions at s=7 Tev," Journal of High Energy Physics, vol. 2012, article 11, 2012.
28. S. Chatrchyan, V. Khachatryan, A. M. Sirunyan, et al., "Suppression of non-prompt J/ψ, prompt J/ψ, and Y (1S) in PbPb collisions at sNN=2.76 TeV," Journal of High Energy Physics, vol. 2012, no. 5, article 63, 2012.
29. G. Aad, B. Abbott, J. Abdallah, et al., "Measurement of the differential cross-sections of inclusive, prompt and non-prompt J/ψ production in proton–proton collisions at s=7 TeV," Nuclear Physics B, vol. 850, no. 3, pp. 387–444, 2011.
30. G. Aad, B. Abbott, J. Abdallah, et al., "Measurement of the centrality dependence of J/ψ yields and observation of Z production in lead-lead collisions with the ATLAS detector at the LHC," Physics Letters B, vol. 697, no. 4, pp. 294–312, 2011.
31. R. Aaij, B. Adeva, M. Adinolfi, et al., "Measurement of J/ψ production in pp collisions at s=7 TeV," The European Physical Journal C, vol. 71, article 1645, 2011.
32. R. Aaij, C. A. Beteta, A. Adametz, et al., "Measurement of J/ψ production in pp collisions at s=2.76 TeV," Journal of High Energy Physics, vol. 2013, no. 02, article 041, 2013.
33. B. Abelev, J. Adam, D. Adamova, et al., "Measurement of prompt J/ψ and beauty hadron production cross sections at mid-rapidity in pp collisions at s=7 TeV," Journal of High Energy Physics, vol. 2012, no. 11, article 065, 2012.
34. J. Badier, J. Boucrot, J. Bourotte, et al., "Experimental J/ψ hadronic production from 150 to 280 GeV/c," Zeitschrift für Physik C, vol. 20, no. 2, pp. 101–106, 1983.
35. A. D. Frawley, "PHENIX charmonia: what have we learned from d+Au collisions?" Nuclear Physics A, vol. 932, pp. 105–110, 2014.
36. X. Zhao and R. Rapp, "Charmonium in medium: from correlators to experiment," Physical Review C, vol. 82, Article ID 064905, 2010.
37. Y. Liu, Z. Qu, N. Xu, and P. Zhuang, "J/ψ transverse momentum distribution in high energy nuclear

collisions," Physics Letters B, vol. 678, no. 1, pp. 72–76, 2009. ·

38. A. Adare, C. Aidala, N. N. Ajitanand, et al., "Nuclear matter effects on J/ψ production in asymmetric Cu+Au collisions at sNN=200 GeV," Physical Review C, vol. 90, Article ID 064908, 2014. ·
39. A. Iordnova, "Nuclear matter effects on J/ψ production in Cu+Au and U+U collisions in PHENIX," in Proceedings of the International Workshop on Deep-Inelastic Scattering and Related Subjects (DIS ‹14), p. 191, 2014.
40. W. Zha, "Recent measurements of quarkonium production in p + p and A + A collisions from the STAR experiment," Nuclear Physics A, vol. 931, pp. 596–600, 2014.
41. A. Adare, C. Aidala, N. N. Ajitanand, et al., "J/ψ suppression at forward rapidity in Au+Au collisions at sNN=39 and 62.4 GeV," Physical Review C, vol. 86, no. 6, Article ID 064901, 10 pages, 2012.
42. B. Abelev, J. Adam, D. Adamová, et al., "Measurement of quarkonium production at forward rapidity in pp collisions at s=7 TeV," The European Physical Journal C, vol. 74, article 2974, 2014. ·
43. B. Abelev, J. Adam, D. Adamová, et al., "J/ψ suppression at forward rapidity in Pb-Pb collisions at sNN=2.76 TeV," Physical Review Letters, vol. 109, Article ID 072301, 2012.
44. B. Abelev, J. Adam, D. Adamová, et al., "J/ψ production and nuclear effects in p-Pb collisions at sNN=5.2TeV," Journal of High Energy Physics, vol. 2014, no. 2, article 73, 2014.
45. B. Abelev, J. Adam, D. Adamová, et al., "J/ψ Production and nuclear effects in p-Pb collisions at sNN=5.02 TeV," Journal of High Energy Physics, vol. 2014, no. 12, article 73, 2014. ·
46. B. Abelev and ALICE Collaboration, "Production of inclusive Y(1S) and Y (2S) in p–Pb collisions at sNN=5.02 TeV," Physics Letters B, vol. 740, pp. 105–117, 2015.
47. CMS Collaboration, "Measurement of the Y(1S), Y(2S), and Y(3S) cross sections in pp collisions at s=7TeV," Physics Letters B, vol. 727, no. 1–3, pp. 101–125, 2013.
48. S. Chatrchyan, V. Khachatryan, A. M. Sirunyan, et al., "Observation of sequential Y suppression in PbPb collisions," Physical Review Letters, vol. 109, no. 22, Article ID 222301, 15 pages, 2012. ·
49. S. Chatrchyan, V. Khachatryan, A. M. Sirunyan, et al., "Indications of suppression of excited Y states in Pb-Pb collisions at sNN=2.76 TeV," Physical Review Letters, vol. 107, Article ID 052302, 2011.
50. R. Aaij, B. Adeva, M. Adinolfi, et al., "Study of J/ψ production and cold nuclear matter effects in p Pb collisions at sNN=5 TeV," Journal of High Energy Physics, vol. 2014, no. 2, article 72, 2014.
51. R. Aaij, B. Adeva, M. Adinolfi, et al., "Study of Y production and cold nuclear matter effects in pPb collisions at sNN=5 TeV," Journal of High Energy Physics, vol. 2014, no. 7, article 094, 2014.
52. B. Abelev, J. Adam, D. Adamova, et al., "J/ψ production as a function of charged particle multiplicity in pp collisions at s=7 Tev," Physics Letters B, vol. 712, no. 3, pp. 165–175, 2012.
53. H. Fujii and K. Watanable, "Heavy quark pair production in high-energy pA collisions: quarkonium," Nuclear Physics A, vol. 915, pp. 1–23, 2013.
54. J. L. Abbacete, N. Armesto, R. Baier, et al., "Predictions for p+Pb collisions at sNN=5 TeV," International Journal of Modern Physics E, vol. 22, Article ID 1330007, 2013.
55. F. Arleo and S. Peigné, "Heavy-quarkonium suppression in p-A collisions from parton energy loss in cold QCD matter," Journal of High Energy Physics, vol. 2013, no. 3, article 122, 2013.
56. F. Arle1o, R. Kolevativ, S. Peigne, and M. Rustamova, "Centrality and p⊥ dependence of J/ψ suppression in proton-nucleus collisions from parton energy loss," Journal of High Energy Physics, vol. 2013, no. 5, article 155, 2013.
57. R. Aaij, B. Adeva, M. Adinolfi, et al., "Study of Y production and cold nuclear matter effects in p Pb collisions at sNN=5 TeV," Journal of High Energy Physics, vol. 2014, no. 7, article 94, 2014.
58. R. Aaij, C. A. Beteta, B. Adeva, et al., "Measurement of Y production in pp collisions at s=7 TeV," The European Physical Journal C, vol. 72, article 2025, 2012.
59. V. Khachatryan, A. M. Sirunyan, A. Tumasyan, et al., "Upsilon production cross section in pp collisions at s=7 TeV," Physical Review D, vol. 83, no. 11, Article ID 112004, 27 pages, 2011. ·
60. A. Emerick, X. Zhao, and R. Rapp, "Bottomonia in the quark-gluon plasma and their production at RHIC and LHC," The European Physical Journal A, vol. 48, article 72, 2012.
61. B. Abelev, J. Adam, D. Adamova, et al., "Suppression of Y(1S) at forward rapidity in Pb–Pb collisions at sNN=2.76 Tev," Physics Letters B, vol. 738, pp. 361–372, 2014.
62. A. Zhou, N. Xu, Z. Xu, and P. Zhuang, "Medium effects on charmonium production at ultrarelativistic energies available at the CERN Large Hadron Collider," Physical Review C, vol. 89, Article ID 054911, 2014.
63. C. Hohne, "Measurement of dileptons with the CBM experiment at FAIR," Nuclear Physics A, vol. 931, pp. 735–739, 2014.
64. A. Sissakian, A. Sorin, I. Meshkov, et al., "Design and construction of Nuclotron-based Ion Collider fAcility (NICA)," Conceptual Design Report, JINR, Dubna, Russia, 2008.
65. F. Fleuret, F. Arleo, E. G. Ferreiro, P.-B. Gossiauxd, and S. Peigné, "Expression of Interest for an experiment to study charm production with proton and heavy ion beams (CHIC: Charm in Heavy Ion Collisions)," CERN-SPSC-2012-031/IPC-EOI-008, 2012.
66. B. Haag, O. Hajkova, A. Hamed, et al., "A fixed-target program for STAR: extending the low energy reach of the RHIC beam energy scan," QM2014, Quark Matter Poster, 2014.

67. R. Hagedorn, "Multiplicities, pT distributions and the expected hadron-quark-gluon phase transition," La Rivista del Nuovo Cimento, vol. 6, pp. 1–50, 1983.

68. M. Ferro-Luzzi, "Luminosity and luminous region shape for pure Gaussian bunches," Tech. Rep. CERN-LHCb-PUB-2012-016, 2012.

69. A. B. Kurepin, N. S. Topilskaya, and M. B. Golubeva, "Charmonium production in fixed-target experiments with SPS and LHC beams at CERN," Physics of Atomic Nuclei, vol. 74, no. 3, pp. 446–452, 2011.

70. M. Kramer, "Quarkonium production at high-energy colliders," Progress in Particle and Nuclear Physics, vol. 47, no. 1, pp. 141–201, 2001.

71. E. Braaten, S. Fleming, and A. K. Leibovich, "Nonrelativistic QCD analysis of bottomonium production at the Fermilab Tevatron," Physical Review D, vol. 63, no. 9, Article ID 094006, 2001.

72. K. Ehret, "Commissioning of the HERA-B internal target: using the HERA proton ring as a B-factory," Nuclear Instruments and Methods in Physics Research A, vol. 446, no. 1-2, pp. 190–198, 2000.

73. V. D. Barger, W. Y. Keung, and R. J. Phillips, "On ψ and Υ production via gluons," Physics Letters B, vol. 91, no. 2, pp. 253–258, 1980.

74. S. J. Brodsky, F. Fleuret, C. Hadjidakis, and J. P. Lansberg, "Physics opportunities of a fixed-target experiment using LHC beams," Physics Reports, vol. 522, no. 4, pp. 239–255, 2013.

CHAPTER 18

UNDERSTANDING THE COMPOSITION OF NUCLEON SPIN WITH THE PHENIX DETECTOR AT RHIC

Abhay Deshpande

QUANTUM CHROMODYNAMICS AND NUCLEON SPIN

Quantum chromodynamics (QCD) has been established over the last four decades as the theory of strong interaction, quantitatively validated with remarkable precision, by a host of experiments at high energy. QCD differs from quantum electrodynamics (QED) most fundamentally in that the interaction carrier, the gluon, is itself a carrier of color charge, unlike the photon in QED, which is chargeless. It is this fundamental difference between the two that gives rise to the fundamental difference and enormous richness that is associated with QCD. Unlike any other many-body system, the individual quarks and gluons that make up nucleons and atomic nuclei cannot be removed from the systems and studied in isolation: this is the phenomenon of confinement, the details of which we do not yet understand. One of the most exciting discoveries in physics over the past few decades has been the understanding that most of the mass of the visible matter//universe—protons, neutrons, and hence the atomic nuclei—arises predominantly from the interaction energy of the gluons amongst themselves and with quarks. This implies that in all hadrons there is a large sea of gluons constantly fluctuating into quark–anti-quark pairs. This modern picture of a hadron made of a large number of highly relativistic and nearly massless spin-1/2 quarks exchanging spin-1 gluons is very different than the electromagnetic interactions binding electrons and nucleons together to form atoms. Investigation of the sea of gluons and quarks has emerged as a new frontier in QCD experimental physics.

Spin is one of the most fundamental concepts in physics, deeply rooted in the structure of the space-time itself. All elementary particles that we know carry spin, including those that participate in strong interactions (QCD): spin-1/21/2 quarks and spin-1 gluons. It is indeed imperative to expect that spin would play a central role in QCD dynamics, and that needs to be understood. It is for such reasons that it is only natural for physicists to explore this aspect of QCD, using all tools available to them. It should be noted that the discovery of the fact that proton has a spatial structure, i.e., it is *not a point particle*, was due to "spin". It was indeed due to the unexpected discovery by Stern et al. [1] in 1939 that it had an *anomalously large magnetic moment.* Today, after decades of intense experimental and theoretical investigations, the central question of how the proton spin-1/21/2 comes out from the quarks and gluons inside the proton as a combination of helicity and angular momenta of the components inside the proton is not understood. It is through such apparently simple, straightforward, but unsolved questions that nuclear scientists have explored and continue to explore the subtle and not-so-subtle aspects of the nucleon structure and hence understand QCD.

Quarks were originally introduced as theoretical constructs [2, 3] based on symmetry considerations that allowed for the first time physicists to see patterns in the then-known zoo of particles. In order to distinguish amongst baryons such as Δ^{++} and Ω-, which are both made of 3 quarks of the same flavor, they had to carry a different quantum number (based on Pauli's exclusion principle). The property of "color" had to be invented to make sense of this situation. Experiments performed at SLAC in the 1960s with electrons scattering off a nucleon, via exchange of virtual photons at high-enough energies, eventually discovered point-like spin-1/21/2 particles inside the nucleon that carried fractional electric charge. These constituents were eventually identified as quarks. The possible existence of gluons was indirectly inferred [4–6] from the 50% missing contribution to the proton's momentum, even after accounting for the momentum carried by the quarks.

Citation: Abhay Deshpande, Understanding the composition of nucleon spin with the PHENIX detector at RHIC, DOI: 10.1093/ptep/ptv019.

This was followed by a search for and discovery of direct evidence of gluons through 3-jet events in electron–positron annihilation [7]. Over the following decades, the studies of nucleon structure became more precise, partly due to increased luminosities and energies of the lepton machines, leading finally to the e–p collider, HERA, at DESY. As hadron colliders were realized, it was realized that the partonic structure of the nucleon seen in deep inelastic scattering (DIS) is universal in the sense that it can also be studied in inelastic proton–proton scattering. Based on which aspect of the nucleon structure is being studied, either electron–proton or proton–proton colliders became the appropriate tools. When our knowledge of nucleon structure improved, it became the tool to search for new physics. The discovery of electroweak bosons was the primary example, followed later by the discovery of the top-quark and most recently the Higgs boson at the LHC.

An important milestone occurred in electron–proton scattering in the mid–late 1960s. Polarized solid-state targets, which could surrogate for proton targets, were built and efforts to polarize them to a high degree were initiated [8]. Simultaneously, an independent effort to polarize electron beams was successfully launched by the production of the first polarized electron source [9] at Yale University for experiments at SLAC. Polarized DIS experiments were launched to explore and measure the quark›s spin contribution to the proton spin. The first experiments were performed at SLAC, which were then followed by CERN›s EMC experiment using much higher-energy (160–200 GeV/c) polarized muon beam. The EMC experiment at CERN discovered, to the surprise of the physics community, that hardly any of the proton›s spin was due to the spin of the quarks: the spin found was carried by the quarks and the anti-quarks, and only ~(14 ± 19)% [10] of the proton's spin was carried by the quarks. The result became famous as the "spin crisis" in the literature and thus began the worldwide effort to understand the internal dynamics of the proton in terms of quarks and gluons. First, it was important for experiments to confirm the spin crisis by reducing the uncertainties. This was accomplished by the follow-up experiments: SMC at CERN, E142 and E143 at SLAC, and later by the HERMES experiment at DESY, and shown in Fig. 1. By the end of the 1990s, it was known beyond doubt that quarks indeed carry only a small fraction ~(25 ± 3)% of the nucleon's spin [11, 12]. The spin crisis was thus confirmed.

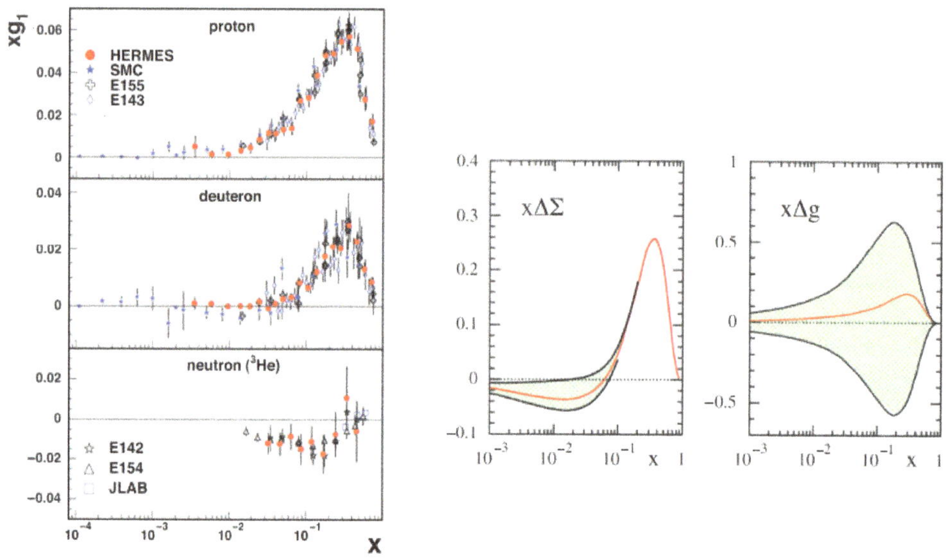

Figure 1: Left: data on spin structure functions xg_1 around the late 1990s [42]. Right: results from an analysis of polarized DIS in terms of spin-dependent nucleon parton densities ΔS and Δg at Q^2=1Q2=1GeV^2. ΔS is the total quark and antiquark helicity contribution and Δg is the total gluon contribution to the nucleon spin. The green shaded bands represent the uncertainty estimates at the time of the compilation of these data. Reproduced with permission from PHENIX Collaboration.

There were two explanations proposed at the time to explain the small contribution from the quarks and anti-quarks. First was that the ΔΣ, the quark + antiquark›s helicity contribution [13], was actually modified due to the axial anomaly, as indicated in the equation below:

$$\Delta\Sigma(Q^2) = \Delta\Sigma'(Q^2) - \frac{n_f}{2\pi}\cdot\alpha_s(Q^2)\cdot\Delta g(Q^2) \tag{1}$$

$\Delta\Sigma'$ is the real contribution that experimenters were seeking, while what was measured directly was ΔΣ. If the term that is subtracted, which has the polarized gluon distribution Δg, is large, then one would get results consistent with the spin crisis. A large Δg would hence result in a small ΔΣ. The next-to-leading-order QCD fits performed under the above definition of relations between ΔΣ and Δg (the Adler–Bardeen scheme) [14] resulted in values that typically fell around 1.0 ± 1.0 in h-bar units. Since the values in such schemes were significantly higher than the spin of the nucleon spin itself, this had very obvious unusual consequences. It was often called the *unnatural solution* of the nucleon spin problem. It implied that the angular momentum of partons would possibly have to be contributing in the *opposite direction* to that of the gluon's helicity contribution, to effectively bring the proton spin down to 1/2 h-bar units. Since the uncertainty in Δg by itself rather large, an experiment to measure the gluon's helicity contribution *directly and precisely* was urgently needed.

By the late 1990s it was also known from the unpolarized gluon distribution measurements from HERA that, at low x (x= the parton momentum fraction of the proton in its infinite momentum frame), the gluon distribution rises significantly, indicating that below ~0.1 the proton is essentially a ball of glue. What better way to measure its helicity contribution (assuming that they constitute a large helicity contribution) than to collide polarized protons at high energy with themselves? Longitudinally polarized proton collisions at high energy, where the gluons dominate the proton's landscape, hence became *the sought-after method* to determine the polarized gluon distribution.

The second possibility why $\Delta\Sigma$ may have been small was related to the interplay of helicity contributions between quarks and anti-quarks. $\Delta\Sigma$ is given in terms of quarks as:

$$\Delta\Sigma = \Delta u + \Delta\bar{u} + \Delta d + \Delta\bar{d} + \Delta s + \Delta\bar{s} \tag{2}$$

The simple suggestion was that, for some reason, anti-quarks and quarks have opposite helicity contributions, thus reducing the $\Delta\Sigma$ in the above equation. This needed an explicit measurement of anti-quark polarizations, which is impossible in the polarized DIS experiment, in which the exchange particle is a virtual photon. However, if the proton collisions were performed at high-enough energy to create $W^{+/-}$, which result from selected quark–anti-quark flavors, then the single spin asymmetry in WW production allows direct access to the quark and anti-quark flavor's helicity. Polarized DIS experiments attempt to do this, by tagging on one of the final-state particles (pions or kaons, including determining their charges) [15], but the lack of knowledge of polarized fragmentation functions at the time these experiments were performed led to speculations of possible large uncertainties in their knowledge, which directly translated to large uncertainties in the polarized anti-quark and quark distributions themselves. The WW-physics method envisioned in a high-energy polarized proton–proton collision went around this limitation of the semi-inclusive deep inelastic scattering (SIDIS) method most naturally, and hence was considered a more reliable measure of anti-quark//quark polarization measurement.

In parallel to the above deep inelastic scattering (DIS) experiments there were many fixed-target polarized proton beam on stationary fixed-target experiments were launched in the US. The expectation for a single spin asymmetry in such experiments based on simple theoretical arguments [16] was small, ~0.001, based on:

$$A_N \sim \frac{\alpha_s \cdot m_q}{p_T} \tag{3}$$

where m_q and p_T are the mass of the quark that fragments into an observable meson and the transverse momenta of the meson measured in the experiment, αsαs is the strong interaction constant, and A_N is the left–right asymmetry measured in the meson production with respect to the transverse polarization of the polarized proton beam. However, unexpectedly large single transverse spin asymmetries were experimentally observed [17–21]. They were not fully understood, i.e., there was no theoretical framework to even begin to understand them, and hence remained an enigma for decades. It was conjectured that they were simply a result of some non-perturbative phenomena not relevant to high-energy physics. However, a new spin physics program at RHIC with polarized beams presented an obvious opportunity to study them systematically. As such, the understanding of the transverse spin phenomenon by exploring it with transverse single and double spin polarized beams at various center-of-mass energies became a third important goal for the RHIC spin program.

THE RELATIVISTIC HEAVY ION COLLIDER (RHIC) WITH POLARIZED PROTON BEAMS

High-polarization proton sources existed before the 1990s, although optically pumped polarized sources were being improved during that decade. Acceleration of polarized protons while preserving the polarization needed development and demonstration of special helical magnets, called "Siberian snake" magnets [22]. These were proposed first by S. Derbenev et al., and experimentally demonstrated at the IUCF in the 1980s and early 1990s, although at low energy. With RIKEN's help, motivated by the need to accelerate protons to 100–250 GeV/cc, these magnets were developed for the RHIC spin program. Part of that development was the establishment of the RIKEN BNL Research Center [23], founded and headed by Prof. T. D. Lee of Columbia University. Demonstration of high-energy polarized protons at successively higher energies from 2 (in the Booster) to 24 (at the AGS at BNL) to 100 GeV at RHIC took a few years, from 1996–2001. A compatible polarimetry technique based on Coulomb nuclear interference (CNI) was also developed during the same time. All these essential and novel accelerator and polarimetry developments are described elsewhere in this volume1313.

The US DOE and the Japanese funding agencies through the RIKEN-BNL Research Center funded the Siberian snakes, the spin rotator magnets for two experiments (PHENIX and STAR) that were critical for the realization of the RHIC spin program at BNL. In this article we will focus briefly on the performance of RHIC as a polarized collider and the PHENIX measurements with polarized beams only. For details of the accelerator achievements of RHIC, see Ref. [14].

Figure 2 shows the polarization achieved and the integrated luminosity delivered by RHIC over the past fourteen years. Two notable effects are: a) that luminosity delivered each year in weeks has improved significantly, from ~45 pb^{-1} in 2006 achieved in 14 weeks with 100 GeV/cc beams with about 55% polarization, to about 470 pb^{-1} delivered in 2013, in the same time, at about 52%. While some of the luminosity improvement is simply the higher boost of 2.5, there were clearly other significant improvements that made this significant achievement possible. Higher beam polarization not only increases the sensitivity of the spin physics measurements, but also allows a better handle on systematic uncertainties in the spin results that come from uncertainties in beam polarization. For more details on the methods and reasons for improved luminosity and beam polarization, see Ref. [14].

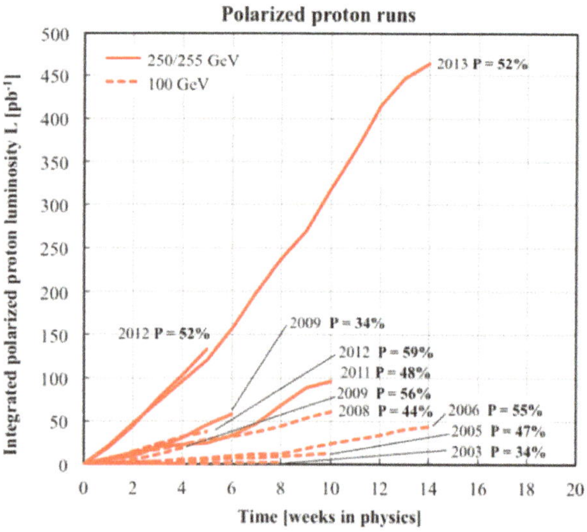

Figure 2: Development of polarization of the proton beam (y-axis) over the fourteen years of operation of RHIC, along with integrated luminosity accumulated in weeks (xx-axis). Dashed lines are for 100 GeV/c proton momentum for the Δg and transverse spin physics programs, while continuous lines are for 250 GeV/c proton beams aimed predominantly at the WW-physics program. The percentage polarization reached in each run is indicated next to the curves. Reproduced with permission from PHENIX Collaboration.

The RHIC comprises two counter-rotating storage rings, colored blue and yellow, each of which operates with as many as 120 polarized proton bunches of 10^{11} protons or more. Protons have been accelerated to 255 GeV/c per proton. Typically, however, the RHIC operated with ~110 bunches. Some of the unfilled bunches are used for background-related diagnostics essential to understanding the beam-related backgrounds in our physics sample.

The stable direction of beam polarization in the RHIC ring is vertical. The polarization direction for each bunch is aligned or anti-aligned with its vertical axis at injection, allowing collisions of multiple-orientation combinations of bunch spins within four consecutive crossings (i.e., 424 ns). Various combinations of spin orientations can be controlled and are called "spin patterns", which are modified within and between runs to understand the various false asymmetry effects that may come from luminosity related to the bunch polarization orientation inherent in the beam from injection. Extensive studies of this effect and correlations have been performed in PHENIX and we have established an upper limit to such effects at the level of $\sim 7\times 10^{-4}$. This is currently the limiting factor for double-longitudinal spin asymmetry in PHENIX.

Beam polarization in RHIC relies on two separate measurements made by two polarimeters. First, the relative polarization is measured several times a fill using a fast high-statistics relative polarimeter in the CNI region of proton–carbon scattering. This measures the relative magnitude of polarization and any variation across the width of the beam [24]. This measurement is further normalized by comparing it with an absolute polarization measurement from a second polarimeter, which is based on scattering of the proton beam from a continuously running polarized hydrogen gas jet target [25].

THE SPIN PHYSICS PROGRAM AT PHENIX

Figure 3 summarizes in a table the longitudinal spin program currently underway at RHIC with the polarized beams. The interactions of p–p and the final-state particles measured are shown in the first column, the partonic interactions responsible for it and the corresponding leading-order (LO) Feynman diagrams are indicated in the second and fourth columns, respectively, and the physical quantities of gluon distribution or anti-quark distributions are shown in the third column. The transverse spin program is not indicated in this table, as the quantities probed are normally a convolution of more than one physical quantity associated with the proton's partonic property, or a correlation of the partonic momentum or spin with the proton's spin or momentum. They will be described separately later.

Reaction	Dom. partonic process	probes	LO Feynman diagram
$\vec{p}\vec{p} \to \pi + X$	$\vec{g}\vec{g} \to gg$ $\vec{q}\vec{g} \to qg$	Δg	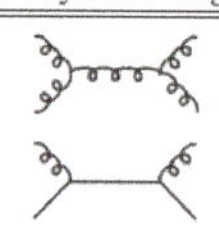
$\vec{p}\vec{p} \to \text{jet(s)} + X$	$\vec{g}\vec{g} \to gg$ $\vec{q}\vec{g} \to qg$	Δg	(as above)
$\vec{p}\vec{p} \to \gamma + X$ $\vec{p}\vec{p} \to \gamma + \text{jet} + X$ $\vec{p}\vec{p} \to \gamma\gamma + X$	$\vec{q}\vec{g} \to \gamma q$ $\vec{q}\vec{g} \to \gamma q$ $\vec{q}\vec{\bar{q}} \to \gamma\gamma$	Δg Δg $\Delta q, \Delta\bar{q}$	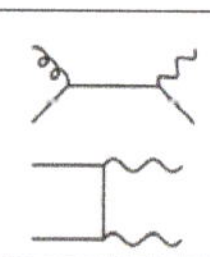
$\vec{p}\vec{p} \to DX, BX$	$\vec{g}\vec{g} \to c\bar{c}, b\bar{b}$	Δg	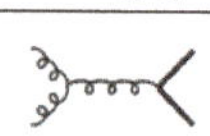
$\vec{p}\vec{p} \to \mu^+\mu^- X$ (Drell-Yan)	$\vec{q}\vec{\bar{q}} \to \gamma^* \to \mu^+\mu^-$	$\Delta q, \Delta\bar{q}$	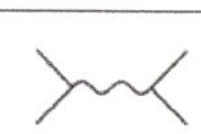
$\vec{p}\vec{p} \to (Z^0, W^\pm)X$ $p\vec{p} \to (Z^0, W^\pm)X$	$\vec{q}\vec{\bar{q}} \to Z^0, \vec{q}'\vec{\bar{q}} \to W^\pm$ $\vec{q}'\bar{q} \to W^\pm, q'\vec{\bar{q}} \to W^\pm$	$\Delta q, \Delta\bar{q}$	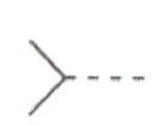

Figure 3: Table summarizing the science of the RHIC spin program with longitudinal spin orientation of the protons indicated in column 1, the partonic interactions of interest shown in column 2, the entity probed indicated in column 3, and the leading-order Feynman diagram indicated in column 4. Reproduced with permission from PHENIX Collaboration.

PHENIX detector system

The PHENIX detector system was designed to perform a broad study of A–A, p–A, and p–p collisions to investigate nuclear matter under extreme conditions. A wide variety of probes sensitive to a range of timescales were expected to be used to study systematic variations in species and energy as well as to measure the spin structure of the nucleon. PHENIX was designed to measure electron and muon pairs, photons, and hadrons with excellent energy and momentum resolution. The detector consists of a large number of subsystems that are shown schematically in Fig. 4 and described below.

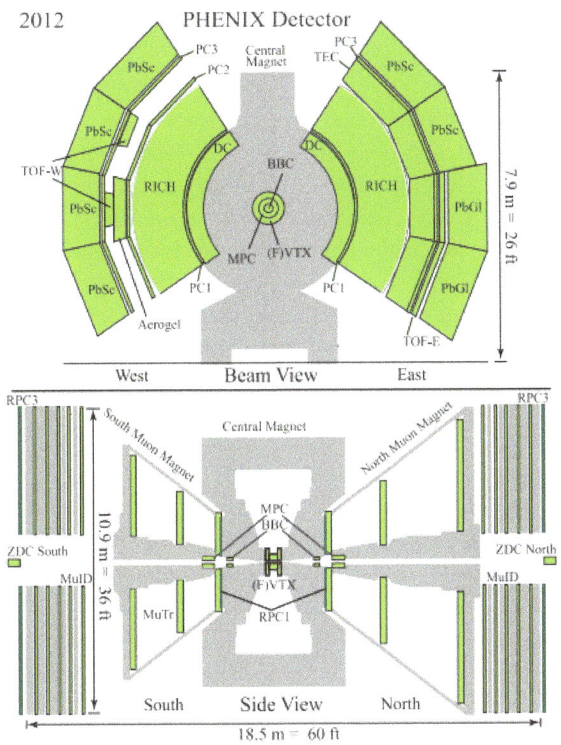

Figure 4: Top: the beam view of PHENIX indicating the radial positions of all detector subsystems as they existed in 2012, close to the present configuration. Bottom: a side view schematic of the PHENIX detector system indicating the z-, h-distributions of various detectors. Reproduced with permission from PHENIX Collaboration.

The PHENIX detector system consists of 4 instrumented spectrometer arms and three global detector systems. The detector consists of number of subsystems. The rapidity (η) and f coverage of each detector system is described in detail in Ref. [26]. The east and west arms visible in Fig. 4(top) are centered around zero rapidity and are excellent spectrometers for the detection of electrons, photons, and charged hadrons. The north and south detectors visible in Fig. 4 (bottom) have full azimuthal coverage but limited η coverage and are instrumented to detecting muons. Each of the four arms has a geometric acceptance of about one steradian. The global detectors that sense the collisions (start time for the rest of the detector), the vertex, and the multiplicity of the interactions comprise two detector systems, a beam–beam counter (BBC) and a zero-degree calorimeter (ZDC). The ZDCs detect neutrons produced in hadronic or nuclear collisions. They are located at $|z|=18.0$m from the interaction point (IP) along the beam axis and cover a region $|\eta|>6$. The BBCs provide a measure of the time-of-flight of forward particles to determine the time of a collision, and provide a trigger as well as providing a measure of the collision position along the beam axis. The VTX tracker in the center is able to measure a more precise position of the event, and is thus able to determine the existence of a secondary vertex distinct from the primary if there is one in an event indicating heavy quark production in the collision. It is made up of 4 concentric barrels of 2 silicon stripixel and pixel layers installed in PHENIX in 2009/2010.

The PHENIX central magnet provides the axial magnetic field for the central spectrometer. The central arms consist of tracking systems for charged particles and electromagnetic calorimetry. The calorimeter is the outermost subsystem of the central arms and is intended for measuring both photons and electrons. A lead-scintillator (PbSc) calorimeter is used for good timing and a lead-glass (PbGl) calorimeter is used for good energy resolution. The tracking system consists of three sets of pad chambers (PC) to provide a 3D space point needed for pattern recognition; the precise projective tracking provided by the drift chamber (DC) provides excellent momentum resolution. A time-of-flight (ToF) and ring-imaging Cherenkov (RICH) provide excellent particle identification. Combining information from the RICH and electromagnetic calorimeter (EMCal), it has been demonstrated that the pion contamination in an electron sample is less than one in 1044 over a wide momentum range.

The forward muon spectrometers give PHENIX an acceptance for J/Ψ decaying to two muons at a rapidity range $-2.25<\eta<-1.15$ for the south arm, and $1.15<\eta<2.441$ for the north arm. Each of these spectrometers includes a muon tracking system based on a radial magnetic field followed by a muon identifier, both with full azimuthal acceptance. The muon trackers consist of three multi-plane drift chambers for precise position resolution. The muon identifier (MuID) system consists of alternating layers of steel absorbers and low-resolution tracking layers of streamer tubes. As an upgrade to this system, directed towards reducing the beam-related backgrounds, more precise tracking was needed and was installed along with upgrades to the MuTR electronics. Overall, the pion contamination in an identified muon sample is less than one in 10^3.

The electronics and triggering in PHENIX has evolved over the years from its beginning and is discussed in more detail in other articles in this volume and in Ref. [27]. Most of the spin physics described in this paper required the central arms to be focused on neutral pions for gluon spin measurement and part of the W physics program, and the muon arms for the anticipated results in the W-physics program in the near future. The transverse spin physics has been achieved with electromagnetic calorimetry installed in the forward muon arms inside the muon piston, also called the muon piston calorimeter (MPC).

The gluon spin contribution to the nucleon spin

The RHIC experiments are connected to Δg through inclusive double-longitudinal spin asymmetry measurements, A_{LL}, defined by:

$$A_{LL} = \frac{\Delta\sigma}{\sigma} = \frac{\sigma_{++} - \sigma_{+-}}{\sigma_{++} + \sigma_{+-}} \tag{4}$$

where σ and Δσ are the polarized and unpolarized cross sections for a given observable, and "++" or "±" indicate the same and opposite helicities in p–p collisions. In perturbative QCD (pQCD), the double spin asymmetry A_{LL} can be written as:

$$A_{LL} = \frac{\sum_{abc} \Delta f_a(x_1, \mu_F^2) \otimes \Delta f_b(x_2, \mu_F^2) \otimes \Delta\sigma^{a+b\to c+X}(x_1, x_2, p_c, \mu_F^2, \mu_R^2, \mu_{FF}^2) \otimes D_c^h(z, \mu_{FF}^2)}{\sum_{abc} f_a(x_1, \mu_F^2) \otimes f_b(x_2, \mu_F^2) \otimes \sigma^{a+b\to c+X}(x_1, x_2, p_c, \mu_F^2, \mu_R^2, \mu_{FF}^2) \otimes D_c^h(z, \mu_{FF}^2)} \quad (5)$$

where $f_{a,b}$ and $\Delta f_{a,b}$ are unpolarized and polarized parton distribution functions (PDFs). D_c^h is the fragmentation function (FF) describing the probability for a parton c with momentum p_c to fragment into a hadron h with momentum p_h, and $z=p_h/p_c$. Δσ and σ are polarized and unpolarized cross sections calculable in pQCD at next-to-leading-order (NLO).

Before the above framework is used for the determination of Δg, it is imperative to demonstrate the applicability of pQCD in the region of measurement by the detector. This is accomplished by measuring the cross sections of the relevant final states—pions, η, photons, etc—and comparing them with the pQCD calculations. This has been demonstrated for 200 GeV collisions of protons for the production of neutral pions [28], ηη-mesons [29], jets [30], and direct photons [31], where the NLO calculations compare very favorably with the measured cross sections. It is only in the regions where these comparisons have been performed that we attempt to extract the polarized gluon distribution information from the double spin asymmetries of that particular final state. A number of different final-state channels can be used to access gluon polarization using Eq. (5), as the pQCD framework seems to agree with the data for the unpolarized cross section. They include jets in the final state and rare probes such as direct photons and heavy flavor. However, so far, neutral pions (predominantly used by PHENIX) and jets (predominantly used by STAR) have been the workhorses of collaborations that have resulted in precise measurements of Δg based on RHIC data. We summarize below the most recently published results from PHENIX at $\sqrt{s}$=200GeV in p–ppcollisions and recently released data, and also for the neutral pions at 510 GeV center of mass. The other final states are not effective yet, but are expected to independently measure Δg in the near future based on data from higher-luminosity data sets.

The results presented below for Δg are based on data taken in 3 years: 2005, 2006, and 2009. The data selection for identifying good neutral events with neutral pions in them has been extensively discussed in a recent publication [32]. The neutral pions are identified using the central arms of PHENIX in |η|<0.35 and have an azimuthal coverage of Δϕ=π/2. The magnetic field configurations in 2005 and 2006 were similar and the central magnet provided a 1.15 Tm field in the central acceptance region. However, in 2009, the field configuration was different because of the existence of the hadron blind detector (HBD) in the run, which required a field-free region near the beam pipe. This had a negligible effect on the π_0 (and η-meson) decay measurements. The details of data selection based on the EMCal response; the EMCal trigger, its biases, and how they were dealt with; and various analysis selection criteria in the tracking detector PC are discussed in Ref. [31]. The main luminosity monitor is the BBC, which comprises two arms (each made of quartz-crystal Cherenkov radiators connected to photomultiplier tubes), is located |z|=144cm from the IP along the beam axis and covers a pseudorapidity range of 3.1<|η|<3.93.1. The second luminosity monitor is the ZDC located 18 m from the IP in the |η|>6 region. They consist of three hadronic calorimeter modules separated by a tungsten absorber. An optical fiber collects the Cherenkov light from the hadron calorimeters. The ZDCs sit behind the beam separator magnets at the end of the interaction region (IR); thus, the charged particles are indeed swept away at the ZDC point, creating a deposit of predominantly neutral particles in the ZDC (mostly neutrons). The ZDC trigger requires a minimum energy deposit of 20 GeV in each of its arms.

For the gluon polarization studies (and for the anti-quark spin studies described in Sect. 3.3) the vertical direction of the proton bunch spin had to be made horizontal and longitudinal. This was accomplished by the spin rotator magnets located in each ring just before and after the interaction region in the blue and yellow rings. A null measurement of a large but accidently measured left–right asymmetry in single spin neutron production in pp–pp scattering at 200 GeV was the benchmark for the local polarimetry, telling us that the spin rotator magnets were doing its job. Typically, in 2009, the polarization fraction along the longitudinal direction was measured and monitored during the run using the ZDCs and was estimated to be larger than (99.4 ± 1.3)% for the blue beam and (97.4 ± 4.3)% for the yellow beam. A typical set of remaining analysis cuts and details of the analysis are described in Ref. [31]. The method to extract a π0π0 sample of events includes understanding of the background, estimating its asymmetry, and then appropriately subtracting that from the signal region. Figure 5 shows a typical plot of this procedure. The red region is identified as the signal region in the reconstructed $M_{\gamma\gamma}$ spectrum. The shaded blue region is the region used to study the background asymmetry in each p_T bin. In this figure, the two-photon invariant mass is plotted for p_T regions $4<p_T<5$GeV/c.

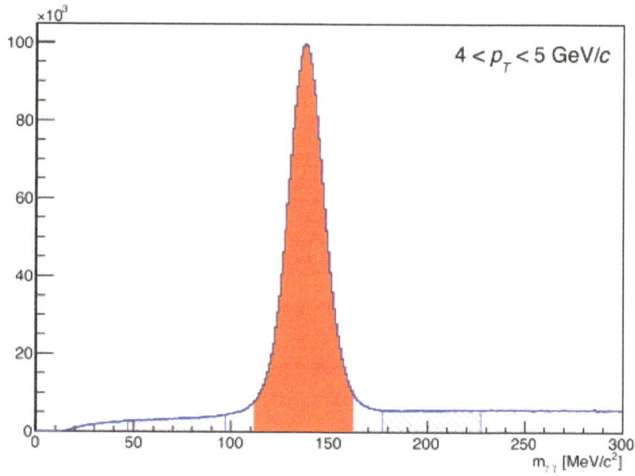

Figure 5: Two-photon invariant mass in the region of π_0 mass for 4<pT<5GeV/c [34]. Also shown are the red and blue shaded regions, which are the π_0 and the background regions used to study and understand the asymmetry of the background, and which are subtracted before the final asymmetry is estimated. Reproduced with permission from PHENIX Collaboration.

Figure 6 shows the remarkable consistency between the cross section for π0π0 measured over a wide p_T range and its comparison with the pQCD calculation performed at next-to-leading-order. The consistency between them over five orders of magnitude over the cross section and a wide pTpTrange gives us confidence in interpreting the double-longitudinal spin asymmetries measured over the same kinematic range in terms of polarized gluon distribution [33]. The cross section measurement, and its excellent agreement with the NLO pQCD calculation, hence becomes the foundation of all our results for polarized gluon distribution. Also shown in the figure is the fact that the charged pion cross section measured by PHENIX, also in the central arm, is consistent within the uncertainty with the neutral pions, when the positive and negative pion cross sections are averaged. While this gives us further motivation to use the charged pions to independently measure the gluon distributions, not having enough of those in the PHENIX trigger so far has prevented us from robustly using them for the determination of the polarized gluon distribution. Should a large-enough triggered sample of them become available in the future, they will be used. In fact, there are clear theoretical expectations that, for a positive value of Δg , the asymmetry in π+π+ is greater than that in π0π0, which in turn is greater than that in π−π−. The pattern reverses itself if Δg is negative. Future high-statistics measurements may explicitly demonstrate these aspects of the measurements.

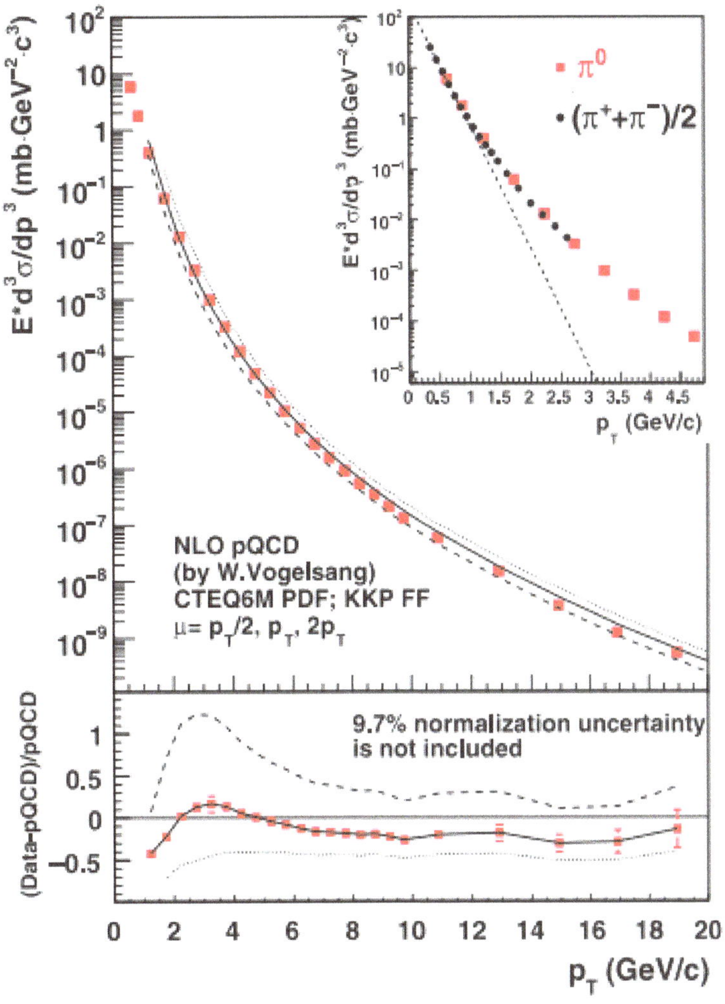

Figure 6: The inclusive π_0 cross section measured with p–ppcollisions at 200 GeV center-of-mass collisions in the central arms of PHENIX, compared with the NLO pQCD calculations and the $\pi^{+/-}$ cross sections measured separately [33]. Reproduced with permission from PHENIX Collaboration.

Based on the above argument, we interpret the double-longitudinal spin asymmetry in terms of the polarization of the glue. Two significant issues need to be addressed in this task: 1) the polarization measurement of the beam as described in Ref. [22], and b) a possible variation of the proton-number correlation in a bunch with its polarization direction, often called the “relative luminosity study”. A very detailed analysis of the systematic uncertainties has been performed in PHENIX of this issue, with the help of the BBC, the ZDC, and the correlations with other fast detectors such as the EMCal in the central arm.

The final results from PHENIX on the double-longitudinal spin asymmetry in the neutral pion have been published in Ref. [34] and are presented in Fig. 7 [34].

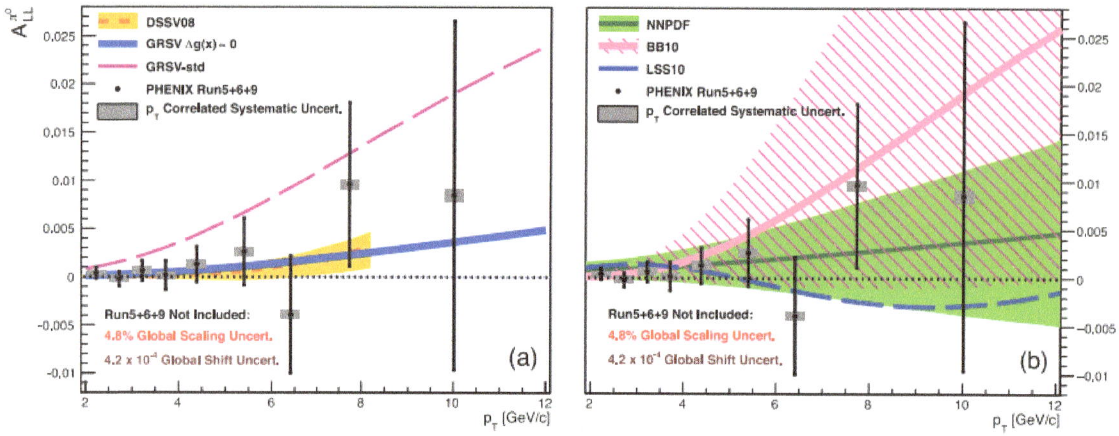

Figure 7: The combined A_{LL} vs. p_T of the π_0 mesons from 2005–2009 with statistical uncertainties [34]. Overlaid on them, in the left figure, are theory curves based on the Theory Group DSSV and GRSV and, in the right figure, the theory curves from NNPDF, BB10, and LSS10, respectively. Reproduced with permission from PHENIX Collaboration.

The final extraction of Δg from the data necessarily requires the data from PHENIX and STAR (not shown here) to be utilized in a global fit at NLO in pQCD. This has been performed by various theoretical collaborations. The most comprehensive in terms of the use of existing data sets (to include all polarized DIS and semi-inclusive data, and the RHIC spin data) is that by de Florian et al [35]. The best-fit results in values of Δg according to these analyses come out to be 0.19 ± 0.06 for $x>0.05$. A slightly older version of the fit by DSSV (the one that was available at the time of writing) is shown in the PHENIX paper (Fig. 8 (left)). The other theoretical collaborations such as NNPDF [36, 37], BB10 [38], and LSS10 [39] have also attempted such analyses; however, they have not included any of the RHIC data. We show them in Fig. 7 (right) only to indicate that the uncertainties in their analysis would be significant reduced should they use PHENIX (and also STAR) data sets in the future. Nevertheless, NNPDF has presented a value and its uncertainty in their extraction of $\Delta g \sim 0.23 \pm 0.07$ in the xx range $0.05<x<0.50$. We note that it is almost exactly the same in magnitude as the one extracted for $\Delta\Sigma$ before the RHIC data became available.

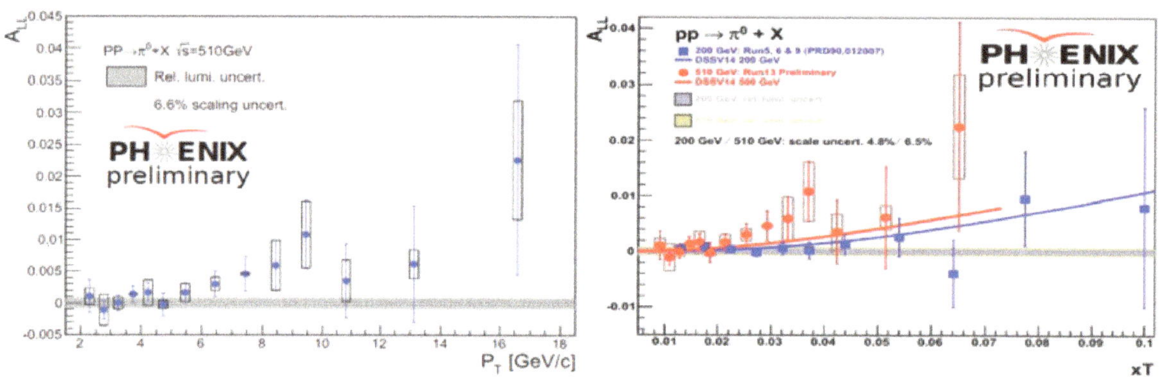

Figure 8: Double spin asymmetry in neutral pions produced at 510 GeV center of mass. Left plot: vs. p_T, right plot: vs. x_T [40]. Reproduced with permission from PHENIX Collaboration.

As this paper was being prepared, PHENIX released a preliminary version of the double-longitudinal spin asymmetry measured in neutral pions at 510 GeV center of mass [40]. A large asymmetry is observed, as shown in Fig. 8. It is also confirmed that, at higher center of mass, the gluon distribution for the same x_T should increase. The observed increase is consistent with the theoretical expectation based on evolution.

Other probes of Δg from PHENIX include possible future extraction of this quantity by including η and direct photons in the global fits. However, these are currently limited by the statistical uncertainties on the measurements. A substantial increase in the luminosity in the near future is not expected at 200 GeV center of mass, but if RHIC were to run for much longer at 500 GeV, then there is a possibility that these final states would become competitive and would give access to a somewhat lower −x−xregion in the proton [41]. We note that, while the direct photon was called the "golden channel" for extraction of the gluon polarization, since it does not involve any quark fragmentation function in the final state (unlike $\pi_0$and η), the low rate of production in this channel is the only hindrance so far. In theory, indeed, it is one of the cleanest channels to access the gluon's contribution to nucleon spin from polarized p–pp collisions. If the statistical uncertainties in the data sets presented so far can be reduced in future by about a factor of 5 or so (i.e., increase the data sample by about 25 times), it is possible that the direct-photon events would start having an effect on the global fits. There have also been speculations about the possible use of heavy quarks $(c\text{–}\bar{c},\, d\text{–}\bar{d})$ to access the polarized gluon distribution. However, the double-longitudinal spin asymmetries associated with the currently expected polarized gluon distributions are small, in fact significantly smaller, than those expected before the RHIC spin data on π_0 (from PHENIX, presented above) and from STAR jets became available. So, unless there are novel QCD effects associated with fragmentation functions that have not been observed so far, we do not anticipate significant results from heavy quarks related to the gluon spin contribution based on future measurements in this area.

The anti-quark polarization measurements

One reason for investigating the anti-quark sea at RHIC was mentioned earlier in the introduction to this paper: to understand the small value of ΔS. Could it be because the sea quarks (anti-quarks included) have their spins aligned opposite to the valence quarks? There is no way to directly distinguish quarks from anti-quarks using SIDIS (without using fragmentation functions, which have large uncertainties). An alternate method is needed, and polarized proton collisions at RHIC provide just such a method. Another unexplained and surprising discovery was made regarding the anti-quark densities by the E866 experiment at FNAL [42]. It was shown that the ratio of anti-d/anti-u distribution as a function of x is not flat, but rises from low to high xx up to about x=0.15 and then decreases at high x. This is a very clear indication that the anti-d sea quark is clearly more abundant than anti-u. This poses a natural question for the quark spin contributions: are the anti-d and anti-u spin contributions also asymmetric? Some configuration selected by nature for them could be the reason for the low value of ΔΣ. It was important that this be measured and understood; see Ref. [43] for further discussion on this topic. Semi-inclusive DIS, in principle, does shed light on this question; however, the experimental requirement of flavor tagging and its understanding in terms of pQCD requires use of fragmentation functions, which were until recently only parametrically known, and required assignment of arbitrarily large uncertainty. At RHIC the polarization of u, d, anti-u, and anti-d quarks in the proton can be measured directly and precisely using maximally parity-violating production of u–anti-$d \rightarrow W+$ and d–anti-$u \rightarrow W$-[44–52].

Within the Standard Model (SM), the WW boson is produced via a pure V–A interaction, which fixes the helicities of the quarks and anti-quark that participate in its production. Further, the WW boson couples directly to a weak charge that correlates to flavors (per generation). While the production of WW at RHIC in pp–pp collisions is predominantly through u, d, anti-u, and anti-d quarks, a small mixture of heavier quark flavor contamination also exists, through quark mixing in the Cabibbo–Kobayashi–Maskawa (CKM) matrix. As such, WW production is an ideal tool for the study of the flavor structure of the proton.

The leading-order production of WW in polarized protons with positive and negative helicity is illustrated in Fig. 9. At RHIC, trains of bunches are alternately right- (+) and left- (−) handed. The parity-violating single spin asymmetry is the difference between left- and right-handed production of WW divided by the sum and normalized by the degree of beam polarization:

$$A_L^W = \frac{1}{P} \cdot \frac{N_-(W) - N_+(W)}{N_-(W) + N_+(W)} \tag{6}$$

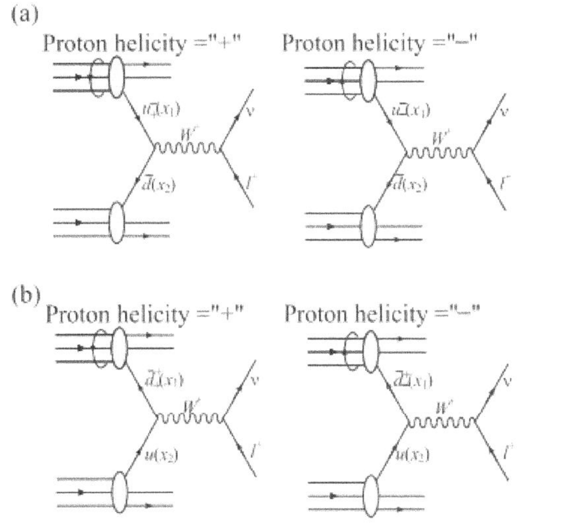

Figure 9: Production of W⁺ in a single-longitudinal p–ppcollision at the lowest order. (a) Δu is proved in the polarized proton and (b) Δanti-d is probed. Reproduced with permission from PHENIX Collaboration.

We construct this asymmetry from either beam by summing over the polarizations of the other beam, thus naturally doubling the statistical sample of events. Because of the maximal parity-violating production of the left-handed WW boson, if the W boson is produced as in Fig. 9(a) or (b), then the asymmetry would be equal to the longitudinal polarization asymmetry of the u or d-bar quark in the proton (high-x):

$$A_L^{W+} = \frac{\Delta u(x_1)}{u(x_1)} \quad A_L^{W+} = \frac{\Delta u(x_1)}{u(x_1)} \tag{7}$$

In general, the asymmetry is a superposition of these two. Interchangeofuanddintheaboveequationyieldstheexpressions for the WW-boson asymmetries. At RHIC, at 500 GeV, the quark is predominantly a valence quark. By identifying the rapidity of W, y_W, relative to the polarized proton we can get to the flavor-separated quark and anti-quark polarizations, particularly at high $y_W \gg 0$ (for Δu/u) and $y_W \ll 0$ (for −Δ(anti-d)/(anti-d)). Higher-order corrections have been estimated and are expected to be small. In measurements of these asymmetries in which one cannot make an approximation of very high values of $|y_W| \gg 0$, the experimental separation amongst the quark and anti-quark flavors becomes more complicated and has to rely on a global analysis in which these are fitted to get the most optimal functions. The measurements of W in the central arm of PHENIX, presented below, will be used in this way. Future measurements of W in the muon arms of PHENIX will be more direct, although obviously they can be utilized in the global fits as well.

PHENIX has pursued two methods to get to the flavor-separated anti-quark polarization: 1) through the W→e decay in the central arms, where we have excellent calorimeter and tracking systems, and 2) with W→μ decay in the forward and backward muon arms. To demonstrate that the detector acceptance and the analysis of W→l are well understood, as a first step, we measured the cross section of W→e and compared it with the Standard Model predictions. Only after we confirm a good match between our measurements and the SM prediction do we go forward to evaluating and interpreting the asymmetries. Figure 10 shows a comparison of theory with the experimental measurements from PHENIX [53] and the STAR [54] detector collaboration.

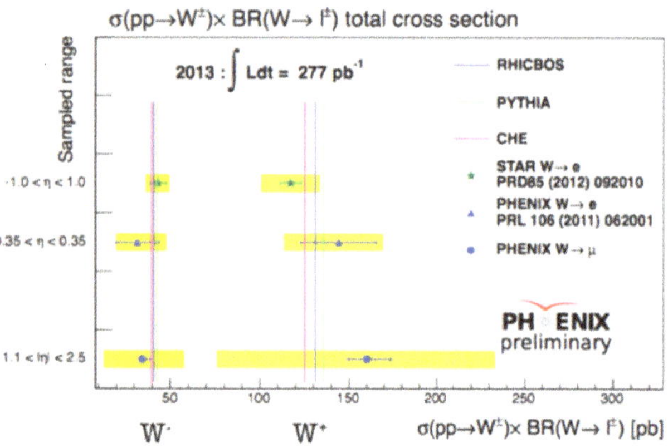

Figure 10: The measured cross sections for W^+ and W^- in the W→e and W→m decays in PHENIX acceptances |η|<0.35 and 1.1<|η|2.5, respectively. Also shown are the STAR published result and calculations based on RHICBOS, PYTHIA, and CHE using the SM parameters [53, 54]. Reproduced with permission from PHENIX Collaboration.

Figure 11 shows the results from PHENIX at different rapidities, and hence with different final states (e or mm) for data taken in 2012 and 2013. These are our most recent results. The 2012 and 2013 runs corresponded to a total sampled luminosity by PHENIX of 53 and 277 pb^{-1}, respectively. The preliminary results are consistent with the expectations from the existing parameterizations of the quark and anti-quark polarizations calculated in the corresponding kinematic regions based on various theoretical models.

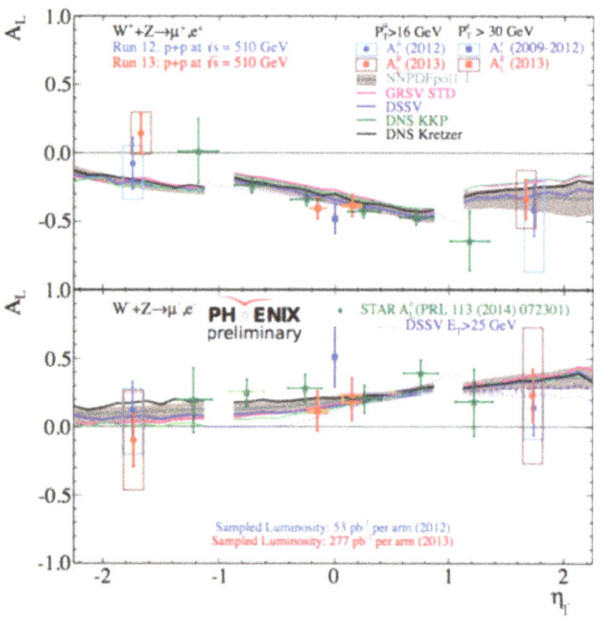

Figure 11: Single-longitudinal spin asymmetries for the various W→e and W→μ decays measured at RHIC. The central arm results shown for PHENIX (blue and red) represent Runs 12 and 13 for the electron in the final state. Forward and backward arm results for muons in the final state are shown for high-η regions. Theoretical expectations for polarized quark and anti-quark polarizations at these rapidities are also shown [53, 54]. Reproduced with permission from PHENIX Collaboration.

It is expected that the statistical and systematic uncertainties in the PHENIX data will reduce in the near future. This is especially true of the systematic uncertainties in the high-η muon results as the understanding of the various backgrounds from hadron and meson decays coming from secondary sources is improved. The primary message, that the asymmetries observed are consistent with the theory calculations, which are fitted to the SIDIS, is unlikely to be modified.

Exploring the transverse spin structure of the proton

The third and last major thrust of the nucleon spin studies at RHIC using the large detector systems is the transverse spin structure of the nucleon. Having tried to explain the 1/21/2 h-bar units of spin purely based on the quark and gluon helicity, and having found that they may not explain all the spin, one is compelled to look elsewhere. The transverse spin structure of the proton is expected to be connected to quantities such as the *confined transverse momentum* of the partons. It was understood early on that, having seen about 400 hundred times larger spin asymmetries in proton–proton single spin scattering, clearly and overwhelmingly disagreeing with the KPL1978 argument of small asymmetries, there was much more to the transverse spin structure of the proton than was earlier envisioned. If the partons do indeed have transverse motion k_T and are still confined within the proton, the purely physical picture of this requires the connection of this to the orbital angular momentum of the partons, which may contribute to the remaining spin of the proton (beyond the quark and gluon helicity contributions). Further, when a proton is transversely polarized with respect to its momentum or the collisions axis, a novel helicity-flip chiral odd twist-2 quark distribution, known as the transversity distribution, δq(x), appears. The transversity distribution is as fundamental as the unpolarized and polarized quark distributions in QCD and has its unique factorization scale dependence and a spin sum rule:

$$\frac{1}{2} = \frac{1}{2}\sum_{\alpha=q\cdot\bar{q}}\int dx\delta q_a(x\cdot Q^2) + \sum_{\alpha=q\cdot\bar{q}\cdot g}(L_{S_T})_a(Q^2) \tag{8}$$

where LS is the component of orbital angular momentum along the transverse spin s_T of the nucleon. $\delta_q(x)$ and $\Delta_q(x)$ are not identical because boosts and rotations are not commutative in relativistic theory. The difference between them hence gives us an insight into the non-perturbative structure of the proton structure, which is indeed of great interest. One of the most coveted aims of scientists in this field is to measure $\delta_q(x)$ over a large range in x so that we can extract the contribution to the nucleon's tensor charge $\int_0^1 dx(\delta q(x) - \delta\bar{q}(x))$, which may be compared directly to the evaluations in lattice QCD.

Transversity can be measured in measurements with transverse spin scattering in two ways: 1) observables dominated by quark-initiated partonic process like A_{TT} of Drell–Yan physics or 2) single transverse spin asymmetries (SSA), where $A_N{\sim}\delta_q(x)$. The latter requires knowledge of another chiral odd unpolarized non-perturbative function known as the Collins function [55], which contains information about the influence of the transverse spin component of the parton in a transversely polarized proton in imparting a left light final-state momentum bias to the fragmented hadron measured in a single spin asymmetry.

Another source of single transverse spin asymmetry, now known as the Sivers function [56], has been suggested by D. Sivers; this relates the influence of the initial-state parton transverse momentum to the left–right single spin asymmetry in the production of the final-state quark/meson. This is by its physical origin easily connected to the orbital motion of the parton and hence to the orbital angular motion of the partons in the initial state.

In proton–proton collisions, clearly the Collins and Sivers functions, if they are both non-zero, will play a role in determining the observed asymmetries, and by themselves cannot be without additional input from experiments performed with e+e-collisions, which measure polarized and unpolarized fragmentation functions for meson production (pions in our case), thus allowing us to unfold the Collins asymmetries and hence transversity, and e–pp collisions, which impart direct input to the initial-state k_T of partons. There are some fundamental rules regarding the color flow that are of great interest to check if the Sivers asymmetries in DIS and in Drell-Yan measurable at RHIC have a reversed analyzing power, which may lead to further fundamental understanding about the factorizability of the transverse-momentum-dependent parton distributions.

All of these constitute a complex and multifaceted program for the PHENIX detector to be carried out in the remaining years of the RHIC operations. A slew of transverse spin measurements have been pursued that have gone in to our understanding of the rich dynamics of the parton and their correlations described so far. Typically, the transverse spin measurements are based on π_0, η, or "energy cluster" measurements in the central arms of PHENIX in the |η|<0.35 or forward direction (3.0<|h|<3.8) using the muon piston calorimeter (MPC) [57].

Figure 12 shows the single spin asymmetries measured in neutral pion and η production at 200 GeV center-of-mass collisions of transversely polarized protons at RHIC measured in the PHENIX detector's central arms. Evidently, the A_N in both cases is consistent with zero.

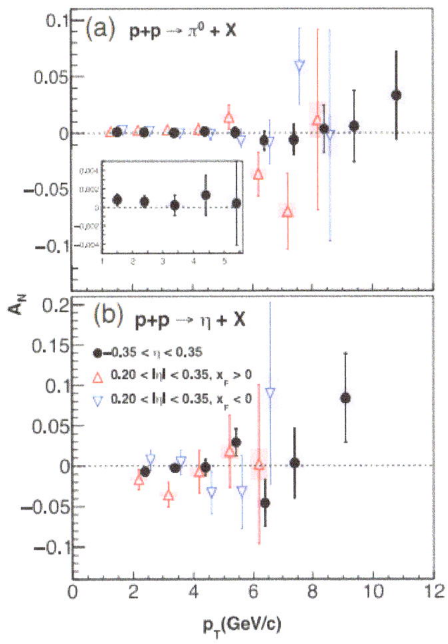

Figure 12: Single spin asymmetry in $\pi_0$and η produced in the central arm of PHENIX [57]. Reproduced with permission from PHENIX Collaboration.

Figure 13 shows a comparison of the PHENIX ANAN measurement at 62.4 GeV center of mass with E704 (fixed-target single spin physics measurements at FNAL) and with the STAR detector at 200 GeV center of mass. All measurements are mostly consistent with each other within the experimental uncertainties and who now center of mass or η dependence.

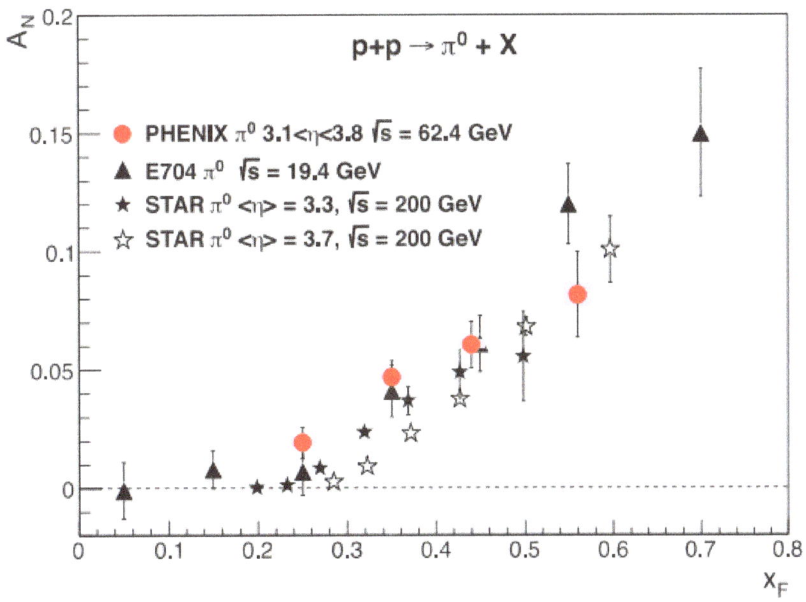

Figure 13: Comparison of A_N of π_0 in PHENIX measured at high h at 62.4 GeV center of mass compared with other measurements from E704 and STAR [57]. Reproduced with permission from PHENIX Collaboration.

Figure 14 (left) shows the PHENIX π^+ transverse single spin asymmetry compared to the measurements by the BRAHMS detector collaboration in the early 2010s. The origin of this A_N is inconsistent with the u and d quark Sivers functions (Fig. 15, right) extracted from SIDIS experiments.

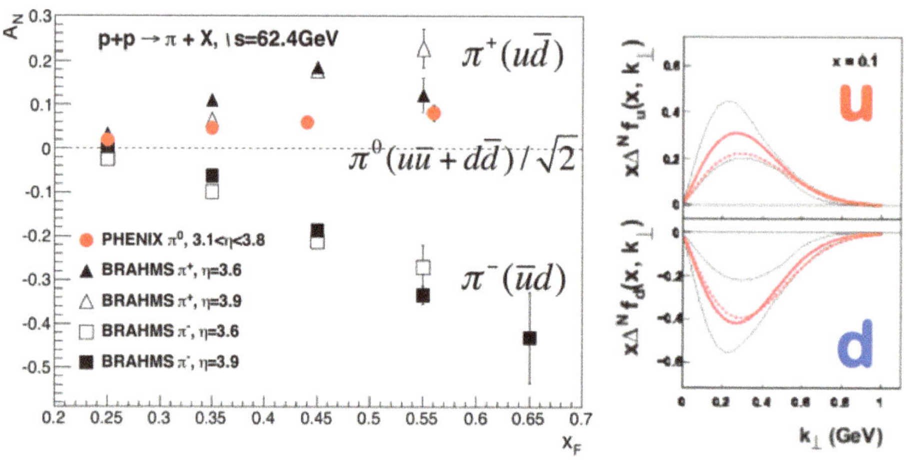

Figure 14: Left: Comparison of charged pion production asymmetries in PHENIX measured at 62.4 GeV center-of-mass operation of RHIC [57]. Right: the u and d quark Sivers functions extracted from SIDIS experiments. Reproduced with permission from PHENIX Collaboration.

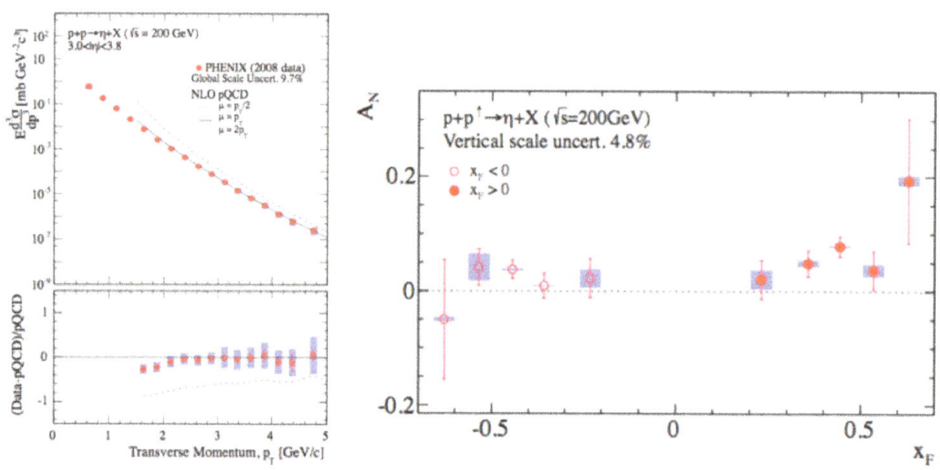

Figure 15: Left: Absolute cross section of the h meson measured at 200 GeV center of mass at RHIC with the PHENIX detector, compared with the NLO calculation in pQCD [57]. Right: Single transverse spin asymmetry in h production at 200 GeV measured at positive and negative x_F[57]. Reproduced with permission from PHENIX Collaboration.

Figure 15 (left) shows a comparison of the measured cross section of hhproduction at 200 GeV center of mass and a next-to-leading-order perturbative QCD calculation [57]. A very impressive consistency is demonstrated, indicating the validity of pQCD effects in the measured high-h range indicated in the figure. The right-hand figure shows the single spin asymmetry measured in the same kinematic region. As expected, the negative x_F asymmetries are zero, while the positive x_F asymmetries are indeed positive. Although it is not shown, they are consistent with the π^0 asymmetries measured by PHENIX in the same region, shown in Fig. 13, and by the STAR and E704 experiments.

While no concrete conclusions can be drawn from the above measurements regarding the transverse dynamics of partons in the proton, it is clearly a topic of great experimental and theoretical interest. Short- and long-term future investigations will allow many aspects of this puzzle to be uncovered. However, a complete picture of the transverse momentum and position distribution of the partons needs more measurements from RHIC, complementary to those currently planned at the Jefferson Lab and with the COMPASS detector at CERN, as well as the future electron ion collider (EIC).

SUMMARY AND OUTLOOK

The realization of RHIC with polarized beams has been a major advance in the field of nucleon spin. It has addressed some of the most fundamental open questions that have faced the nucleon spin community since the 1990s. It has decisively measured the contribution of polarized gluons to the nucleon spin in x>0.05. We *now know* that $\Delta g \sim 25\%$ of the proton's spin 1/2, far less than what was envisioned before RHIC came about. The current best estimates of the quark and gluon's contribution to the nucleon spin are about 20–25% each. This leaves room for a substantial contribution to the nucleon spin from the orbital angular momentum from quarks and gluons.

RHIC has also initiated a program to directly measure the sea quark's contribution to the nucleon's spin. Unambiguous separation of the quark vs. anti-quark's contribution to the nucleon spin was not possible in the past. It had to depend on SIDIS data from fixed-target experiments, which relied heavily on insufficient and model-dependent polarized and unpolarized fragmentation functions. RHIC has worked around this problem very naturally by invoking the parity-violating generation and decay of WW bosons. Early results summarized in this paper indicate that the expected quark and anti-quark polarized distributions are indeed what the pre-RHIC models had predicted, with at most minor modifications. The data collected recently in 2014 may solidify this observation further over a wide kinematic region, as results from PHENIX and STAR get published in the coming year.

The transverse spin collisions have been pursued intermittently at RHIC and have provided surprising and exciting new results, which promise to help shed light on the transverse dynamics of partons inside the proton when combined with the transverse spin physics results from the DIS and SIDIS fixed-target experiments at COMPASS and the Jefferson Lab in the near future. The complementary but essential nature of these methods has been highlighted by the global analyses now underway with the participation of both theorists and experimentalists.

The need for even higher precision in understanding the longitudinal spin structure of the proton, and many tests of QCD possible only through transverse spin collisions in p–p and e–p, has become apparent over the years. It has also been acknowledged that some of these DIS and SIDIS data that may be needed for these physics goals would have to be at very high energy. As a result, a high-luminosity high-energy polarized electron ion collider (EIC) is currently under consideration by the US nuclear science community as their next major research facility in the US after the construction of the FRIB (the radioactive beam facility at Michigan State University). The science of the EIC has been laid out in a recent White Paper [58], prepared and released recently for the Nuclear Science Advisory Committee's (NSAC) deliberations. This White Paper presents measurements that will help us fully elucidate and understand the nucleon's spin structure in terms of the helicity of quarks and gluons, and construct a very detailed picture of the quark and gluon's positions and momentum distributions in the transverse direction to its motion, akin to realizing tomographic images of the proton in its polarized state. A natural consequence of the precision enabled by the EIC will be to allow us to access and possibly also measure the orbital angular momentum of the partons. Progress in lattice QCD also promises to evaluate various parton distribution functions [59–62] in polarized protons by the mid-2020s. The EIC thus provides an exciting possibility to enable a direct comparison of these quantities from experimental data and lattice QCD. High-energy nuclei colliding with the electron beam is another major thrust of the EIC's physics program. Precise measurements leading to unprecedented understanding of the role of gluons in QCD at the highest energy (partonic density) would be possible. We refer readers to the White Paper[29] for further details of the science case.

Currently, there are two designs for the US EIC. One uses one of the beams of RHIC and collides with a 5–20 GeV/c (upgradable to 30 GeV/cc at a lower luminosity) electron beam facility to be constructed within the RHIC tunnel. The other proposal uses the recently upgraded 12 GeV CEBAF facility at the Jefferson Lab and collides it with a new hadron beam facility to be built near it. The physics program could start with the lower-energy electron beam and then be upgraded to the full 20 (30) GeV/c at a later date. The luminosity aimed at for the physics is 10^{33-34} cm^{-2} sec^{-1} (about 100–1000 times the luminosity achieved at the HERA collider at DESY). The US EIC would be the first polarized electron–proton collider, as well as the first electron–nucleus collider.

Should the site for EIC be RHIC, i.e., eRHIC, then the existing IRs (currently occupied by PHENIX and STAR) would be natural places to consider locating the EIC detectors. Recently, the PHENIX collaboration and Brookhaven National Laboratory submitted a proposal for a major upgrade of the PHENIX detector [63], called the sPHENIX detector, and a letter of intent to further enhance this detector to be optimized as a first-stage eRHIC detector [64]. The sPHENIX consists of new large-acceptance electromagnetic and hadronic calorimetry built around the superconducting solenoid acquired from the decommissioned BaBar experiment at SLAC. sPHENIX proposes to make key new measurements of probes of strongly coupled quark–gluon plasma (sQGP) and study its internal dynamics. This program is being considered at RHIC before the transition to eRHIC. It is envisioned that the sPHENIX detector would be the central detector for the EIC detector. Figure 16 shows engineering drawings of the two concepts mentioned, indicating how sPHENIX could be part of the EIC detector under consideration in the long term.

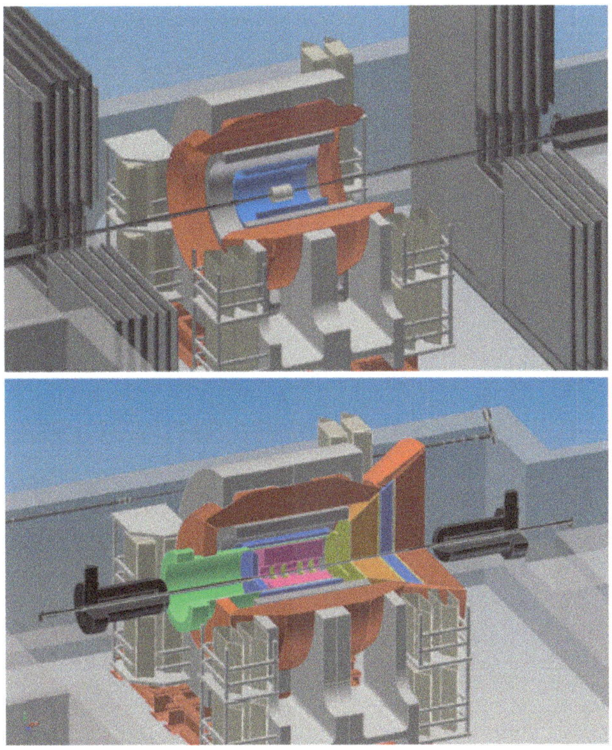

Figure 16: Top: A schematic engineering drawing with a cutout of the sPHENIX detector. Bottom: The future EIC detector that uses the sPHENIX as its central barrel, and augments a forward detector for the hadrons and nuclear fragments on one side (red/blue), and the electromagnetic calorimetry and appropriate tracking (green/blue) on the left, in the backward direction for the scattered electrons, in an electron–ion collision. Reproduced with permission from PHENIX Collaboration.

It is possible that, after the sPHENIX detector is constructed, the forward (hadron-going side; red detectors on the right in

Fig. 16 (bottom)) and backward (electron-going side; green detectors on the left in Fig. 16(bottom)) detectors of the eRHIC detector may be realized sequentially in time. If the forward detectors (hadron-going side) can be realized earlier than the transition of RHIC to eRHIC (in which case, according to the current plans of the BNL management, the yellow RHIC ring will be dismantled), then sPHENIX with its forward detector upgrade could become a detector for transverse spin physics and a p–A physics program at RHIC before eRHIC comes online. Interested scientists in the PHENIX collaboration are considering this stage beyond the sPHENIX upgrade, as yet another upgrade: forward-sPHENIX or fsPHENIX. Arguments for the scientific case for such an interim program, based on measurements of transverse spin phenomena extending to the Drell–Yan physics, direct-photon measurements in single spin polarized collisions, and also exploration of the nuclear-size dependence of the saturation scale in polarized p–A collisions, are currently being developed [65].

ACKNOWLEDGEMENTS

The author is immensely grateful to the PHENIX Collaboration for the fruitful scientific collaboration. The author gratefully acknowledges the support of the RIKEN-BNL Research Center and RIKEN/Japan for the spin physics program at RHIC in collaboration with the US DoE, Office of Nuclear Science. Last but not the least, the author is enormously grateful for the patience of the editors of PTEP for allowing an overly generous extension of the submission of this write-up.

REFERENCES

1. I. Estermann et al., Nature 132, 169 (1933).
2. M. Gell-Mann, Phys. Lett. 8, 214 (1964).
3. G. Zweig, CERN-TH-412.
4. R. P. Feynman, Phys. Rev. Lett. 23, 1415 (1969).
5. J. D. Bjorken, Phys. Rev. 179, 1547 (1969).
6. J. D. Bjorken and E. Paschos, Phys. Rev. 185, 1975 (1969).
7. R. Brandelik et al. [Tasso Collaboration], Phys. Lett. B 86, 243 (1979).
8. D. G. Crabb and W. Meyer, Ann. Rev. Nucl. Part. Sci. 47, 67 (1997).
9. M. S. Lubell, AIP Conf. Proc. 35, 418 (1976).
10. J. Ashman et al. [European Muon Collaboration], Phys. Lett. B 206, 364 (1988).
11. B. Adeva et al. [Spin Muon Collaboration], Phys. Rev. D 58, 112001 (1998).
12. B. Adeva et al. [Spin Muon Collaboration], Phys. Rev. D 58, 112002 (1998).
13. H. Itoyama and A. H. Mueller, Nucl. Phys. B 218, 349 (1983).
14. G. Altarelli, R. Ball, S. Forte, and G. Ridolfi, Nucl. Phys. B 496, 337 (1997).
15. W. Lorenzon et al. [HERMES Collaboration], http://inspirehep.net/record/668697?ln=en.
16. G. L. Kane, J. Pumplin and W. Repko, Phys. Rev. Lett. 41, 1689 (1978).
17. D. Adams et al. [FNAL E704 Collaboration], Phys. Lett. B 261, 201 (1991).
18. K. Krueger et al., Phys. Lett. B 459, 412 (1999).
19. G. Bunce et al., Phys. Rev. Lett. 36, 1113 (1976).
20. K. Heller et al., Phys. Lett. B 68, 480 (1977).
21. S. A. Gourlay et al. Phys. Rev. Lett. 56, 2244 (1986).
22. S. Ozaki and T. Roser, Prog. Theor. Exp. Phys. 2015, 03A102 (2015).
23. RIKEN BNL Research Center. Available at: http://www.bnl.gov/riken/
24. I. Nakagawa et al., Eur. Phys. J. ST 162, 259 (2008).
25. A. Zelenski et al., Nucl. Instrum. Meth. A 536, 248 (2005).
26. K. Adcox et al., Nucl. Instrum. Meth. A 499, 469 (2003).
27. M. Harrison, T. Ludlam and S. Ozaki, Nucl. Instrum. Meth. A 499, 235 (2003).
28. A. Adare et al. [PHENIX Collaboration], Phys. Rev. D 76, 051106 (2007).
29. A. Adare et al. [PHENIX Collaboration], Phys. Rev. D 83, 032001 (2011).
30. L. Adamczyk et al. [STAR Collaboration], Phys. Rev. D 86, 032006 (2012).
31. S. S. Adler et al. [PHENIX Collaboration], Phys. Rev. Lett. 98, 012002 (2007).
32. A. Adare et al. [PHENIX Collaboration], Phys. Rev. D 90, 012007 (2014).
33. A. Adare et al., Phys. Rev. D 76, 051106(R) (2007).
34. A. Adare et al., Phys. Rev. D 90, 012007 (2014).
35. D. de Florian et al., Phys. Rev. Lett. 113, 012001 (2013).
36. R. D. Ball et al. [NNPDF Collaboration], Nucl. Phys. B 874, 36 (2013).
37. R. D. Ball et al. [NNPDF Collaboration], Nucl. Phys. B 867, 244 (2013).
38. J. Blümline and H. Böttcher, Nucl. Phys. B 841, 205 (2010).
39. E. Leader et al., Phys. Rev. D 82, 114018 (2010).
40. Y. Insoek for the PHENIX Collaboration, SPIN 2014, Beijing, China.
41. E. Aschenauer et al., RHIC spin program. Available at: http://www.bnl.gov/npp/docs/RHIC-Spin-WriteUp-121105.pdf.
42. R. S. Towell et al., Phys. Rev. D 64, 052002 (2001).
43. C. Aidala et al., Research Plan for Spin Physics at RHIC, February 11, 2005, BNL Report: BNL-73798-2005.
44. P. Ratcliffe, Nucl. Phys. B 223, 45 (1983).
45. A. Weber, Nucl. Phys. B 382, 63 (1992).
46. B. Kamal, Phys. Rev. D 57, 6663 (1998).
47. T. Gehrmann, Nucl. Phys. B 534, 21 (1998).
48. N. Craigie et al. Phys. Rep. 99, 69 (1983).
49. C. Bourrely, J. Soffer, and E. Leader, Phys. Rep. 59, 95 (1980).
50. C. Bourrely and J. Soffer, Phys. Lett. B 314, 132 (1993).
51. C. Bourrely and J. Soffer, Nucl. Phys. B 432, 329 (1994).

52. P. Chiappetta and J. Soffer, Phys. Lett. B 152, 126 (1985).
53. A. Adare et al. [PHENIX Collaboration], Phys. Rev. Lett. 106, 062001 (2011).
54. L. Adamczyk et al. [STAR Collaboration], Phys. Rev. D 85, 092010 (2012).
55. J. C. Collins, Nucl. Phys. B 396, 161 (1993).
56. D. Sivers, Phys. Rev. D 41, 83 (1990).
57. A. Adare et al., [PHENIX Collaboration], Phys. Rev. D 90, 012006 (2014).
58. A. Accardi et al., [arXiv:1212.1701] [Search inSPIRE].
59. Y. Hatta et al., Phys. Rev. D 89, 085030 (2014) [arXiv:1310.4263] [Search inSPIRE]
60. X. Xiong et al., Phys. Rev. D 90, 014051 (2014).
61. H.-W. Lin et al., [arXiv:1402.1462] [Search inSPIRE].
62. Y.-Q. Ma and J.-W. Qiu, [arXiv:1404.6860] [Search inSPIRE].
63. A. Adare et al., [arXiv:1207.6378] .
64. A. Adare et al., [arXiv:1402.1209] [Search inSPIRE].
65. A. Adare et al., Future opportunities in p + p and p + A collisions at RHIC with Forward sPHENIX detector. Available at: http://www.phenix.bnl.gov/phenix/WWW/publish/publish/dave/sPHENIX/ pp_pA_whitepaper.pdf.

CHAPTER 19 DARK SECTOR SPECTROSCOPY AT THE ILC

Jeppe R. Andersen, Michael Rauch, Michael Spannowsky

INTRODUCTION

Astronomical observations strongly indicate the existence of dark matter [1–5]. Many extensions of the Standard Model take this into account by incorporating a so-called dark sector: a sector of particles that do not carry electric or color charge. The interactions of the dark sector can be protected by global symmetries and the particles can have a long lifetime. Often the dark sector is not completely decoupled, but it interacts with the Standard Model particles by the exchange of a Z boson, or a mediator of a yet unknown force.

In recent years several experiments, i.e. PAMELA [6, 7], Fermi LAT [8] and most recently AMS [9], have observed an excess in the positron fraction in the electron–positron energy spectrum at energies above ~10 GeV. A possible explanation for this excess could be the decay of an invisible particle into an e^+e^- pair. Leptophilic dark sectors have been identified as a possible explanation of the observed excess [10–15].

Unfortunately, due to the large uncertainties in evaluating the astrophysical backgrounds, indirect detection experiments face challenges in claiming the discovery of a potential dark matter candidate. Direct detection experiments try to measure the momentum transfer between the weakly interacting massive particle (WIMP) and the detector. If the WIMP is light, the sensitivity of the experiments is strongly reduced [16, 17]. Interpretations of searches for WIMPs at direct detection experiments usually assume that the dark sector is minimal, i.e. consists of only one particle, and the WIMP accounts for the total dark matter abundance in the universe. We will not make those strong assumptions. Indeed, if the dark sector is not minimal, constraints from direct detection experiments are strongly relaxed.

Another possibility to evade the bounds from detection experiments is that the dark sector particle (DSP) is not even stable on cosmological time scales, and might even be unconnected to the dark matter solution. For high-energy collider experiments, it is sufficient that the DSP has a life time of the order of microseconds to give a signature of missing transverse momentum only, or even that it does not decay back into detectable Standard Model particles with a large enough rate.

Recently it has been pointed out that in case the DSP couples with reasonable strength to quarks, mono-jet searches at the Tevatron and LHC can be a superior way in discovering them [18–26]. If the DSP couples predominantly to leptons, constraints can be derived from LEP in mono-photon searches [27] or a future electron–positron collider [28].

Unfortunately, both direct detection experiments and mono-jet/mono-photon searches are very limited tools in unravelling the detailed structure of the dark sector, e.g. the spin of the force mediator in combination with its mass and the mass of the DSP(s). In mono-photon searches the DSPs recoil against a high-pT photon. Therefore, only the total amount of missing transverse energy in the event can be measured, i.e. the differential cross section for one observable.

The observables discussed in this article allow an unbiased view into the dark sector, as long as a mediator couples the dark sector to electrons. Because of the lack of an existing electron–positron collider we will discuss these observables in the context of the International Linear Collider (ILC, 500 GeV beam energy). So far, the ILC's great potential in studying the structure of dark sectors has not been completely appreciated [29, 30]. We assume that the particles in the dark sector are stable on collider time scales and escape detection at the LHC. Therefore, our signal will consist of electrons and missing energy, i.e. our process is $e^+e^- \to e^+e^- + \not{E}_{T,\text{miss}}$. This signature is relatively rare in the Standard Model,

Citation: Jeppe R. Andersen, Michael Rauch, Michael Spannowsky, Dark sector spectroscopy at the ILC, DOI: 10.1140/epjc/s10052-014-2908-1.

predominantly generated in the production of Z bosons and photons with subsequent decay/splitting to an e^+e^- pair and neutrinos. Some example Feynman diagrams for both signal and background processes are depicted in Figure. 1.

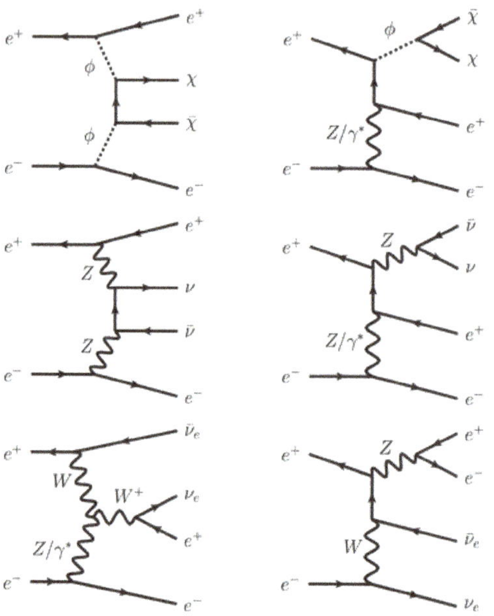

Figure 1: Representative Feynman diagrams for the process considered in this article. The *top row*shows the signal process with mediator ϕ and DSP χ, while the *middle and bottom row*depict Standard Model backgrounds, where ν *without index* denotes a neutrino of any generation.

An important topology for the study of the structure of the dark sector is the so-called vector boson fusion (VBF) topology (i.e., two possibly forward well-separated electrons), even though the DSPs are not necessarily produced by exchanging a vector boson. Tight cuts on the electron–positron system can reduce the Standard Model backgrounds and increase the average energy flowing through the DSP–e^+e^- coupling. A similar strategy is used when studying the coupling structure of the Higgs boson to quarks: Producing a Higgs boson in the VBF channel, angular correlations of the jets can be used to distinguish a CP-even from a CP-odd Higgs boson [31–33] and an invisibly decaying Higgs boson can be disentangled from the backgrounds [34]. Similar kinematic conFigureurations can be used to study dark sectors with DSP–quark couplings at the LHC. However, we find that at the LHC the DSP–quark coupling has to be of the order of the electroweak coupling to give a significant event shape contribution, reflected in the invariant mass $m_{jj} = \sqrt{(p_{j_1} + p_{j_2})^2}$ or the rapidity difference $\Delta y_{jj} = |y_{j_1} - y_{j_2}|$ of the two so-called tagging jets, the two jets with the largest transverse momentum. Furthermore, at the LHC large systematic uncertainties in final states with missing transverse energy (MET) and jets render a dark sector spectroscopy a difficult task [35].

This article is organised as follows: In Sect. 2 we discuss our benchmark models and assumptions which specify the DSPs–electron–positron interactions. Kinematic observables are identified and interpreted in the context of Regge theory in Sect. 3. We further evaluate how well the different benchmark models can be discriminated. In Sect. 4 we present our conclusions.

BENCHMARK MODELS AND EXPERIMENTAL CONSTRAINTS

For simplicity we assume that the DSP is a Dirac fermion. We have also investigated scenarios where the DSP is a scalar instead, but we find no significant differences compared to the Dirac case. As we will see later in Sect. 3, such a behaviour is expected, as we are predominantly sensitive to the mass of the DSP only. The mediator can be either a scalar particle or a vector particle, which couples only electrons from the Standard Model to the dark sector. Extending this coupling to all SM leptons would be straightforward. The only place where this matters is the width of the mediator particle, which gets increased by additional couplings to muons and taus. The width is calculated using the program BRIDGE [36]. As the width turns out to be small for the coupling values considered in the following, taking for example a generation-blind scenario instead would have no relevant effect on our results. We also assume that the coupling between mediator and neutrino vanishes. If such a coupling was of the same strength as those to electrons, this would yield an additional contribution to the mono-photon searches, as it has the same signature, and is therefore significantly constrained.

The interaction terms appearing in the Lagrangian are denoted in Table 1.

Table 1: Terms in the Lagrangian describing interactions between the mediator and the electron or the DSP particle

	Scalar	Vector
e	$g_{ee\phi.S}\,\bar{e}e\,\phi_S$	$g_{ee\phi.V}\,\bar{e}\gamma_\mu e\,\phi_V^\mu$
χ	$g_{\chi\chi\phi.S}\,\bar{\chi}\chi\,\phi_S$	$g_{\chi\chi\phi.V}\,\bar{\chi}\gamma_\mu\chi\,\phi_V^\mu$

Our analysis uses light mediator particles. Therefore, the momentum dependence in the propagator of the mediator plays an important role and cannot be neglected. Hence, it is not possible to formulate the results in terms of an effective dimension-six operator of the form $\bar{e}e\bar{\chi}\chi$ or $\bar{e}\gamma_\mu e\bar{\chi}\gamma^\mu\chi$ for scalar or vector mediator, respectively, where the mediator is integrated out. Nevertheless, one can still define an effective mass M_* as

$$M_* = \frac{M_\phi}{\sqrt{g_{ee\phi} g_{\chi\chi\phi}}} \tag{1}$$

as in Reference [25]. In the limit of a heavy mediator, the term $1/M_*^2$ becomes the prefactor of the dimension-six operator.

In the following, we define eight different model scenarios. They are characterised by three different options, namely the spin and mass of the mediator particle and the mass of the DSP. For the spin of the mediator particle we investigate the two possibilities already mentioned, namely a scalar and a vector particle. In many scenarios the dark sector is linked to the Standard Model via kinetic mixing of a dark photon of a hidden U(1) with the U(1)hypercharge of the Standard-Model [37–40]. In the scalar case we assume that the mediator couples chirally and exclusively to electrons and DSPs. By measuring the spin of the mediator such models can be either confirmed or disfavoured.

For the two masses we define a light and a heavy scenario each, with values of 8 and 200 GeV for the mediator particle and 5 and 120 GeV for the DSP mass.

An overview is shown in Table 2 together with the effective mass M_* used for each scenario. The exact choice of masses is somewhat arbitrary, but it has been guided by the following considerations. The mass of the light DSP is chosen such that it is below the typical reach of direct detection experiments, while the heavy scenario has a mass which is beyond the reach of direct searches at LEP. The two choices for the mediator mass have then been chosen such that in the light–light and heavy–heavy models the on-shell decay of a mediator particle into two DSPs is kinematically forbidden.

Table 2: Overview of the different model scenarios used in our analysis. The first and third letter of the model name denote the mass (light or heavy) of the mediator and DSP, respectively, while the middle one describes the spin nature of the mediator (scalar or vector)

Model	Mediator mass (GeV)	Mediator spin	DSP mass (GeV)	M_* (GeV)
LSL	8	0 (scalar)	5	30
LVL	8	1 (vector)	5	30
LSH	8	0 (scalar)	120	27.4
LVH	8	1 (vector)	120	21
HSL	200	0 (scalar)	5	1250
HVL	200	1 (vector)	5	1250
HSH	200	0 (scalar)	120	332.4
HVH	200	1 (vector)	120	511.8

The coupling parameters for the light DSP models are already constrained by direct searches at LEP. Therefore, we choose our effective mass such that they are at the 90 % CL exclusion boundary given in Reference [25]. For the 200 GeV mediator the effective mass parameter can be taken directly from there, while for the 8 GeV mediator we have instead used the given value for the 10 GeV curve. This is slightly more restrictive than the true 8 GeV value, so with this choice we are erring on the safe side. The heavy DSP scenarios with a DSP mass of 120 GeV are beyond the reach of LEP. As the mediator couples only to electrons, also searches at hadron colliders cannot significantly constrain the coupling parameters. Only a direct interpretation of the DSP as dark matter candidate would immediately yield strong constraints by direct detection experiments [16, 17], and in fact reduce the signal-to-background ratio to a value too low for realistic studies. Therefore, we set the couplings in the heavy DSP cases to a value that gives similar cross sections as the corresponding light DSP scenario. This choice also simplifies comparisons between the two options.

DISCRIMINATIVE OBSERVABLES

The spin of the mediator can be assessed by appealing to the analytic behaviour of the scattering amplitude dictated by Regge theory [41, 42] in the limit of large invariant mass between each produced particle compared to the propagating momentum (and any mass of fields), $s_{ij} \gg |t_i|$. In this *multi-Regge kinematic* limit, which is attained within the VBF cuts, the analytic behaviour of a 2→n scattering is determined by

$$\mathcal{M}^{p_a p_b \to p_1 p_2 p_3 p_4} \to s_{12}^{\alpha_1(t_1)} s_{23}^{\alpha_2(t_2)} s_{34}^{\alpha_3(t_3)} \gamma, \tag{2}$$

$p_1,\ldots,p_4$ are the final-state momenta ordered in rapidity, and γ depends on the couplings, the t-channel momenta t_i and ratios of $s_{ab}/(\prod s_{ij})$ only. The powers α_i determining the scaling behaviour with s_{ij} depend on the spin of the particle exchanged in the t-channel, $\alpha_i = J_i$ up to radiative corrections. In cases where the mass of the exchanged particles is negligible, the spin of the exchanged particle can therefore be inferred by studying the scaling of the cross section with the invariant mass between the electron/positron pair, see Figure. 2. Since $s_{ab} = 2p_{a\perp} p_{b\perp}(\cosh(y_a - y_b) - \cos(\phi_a - \phi_b))$, the constraint on the analytic behaviour of the scattering amplitude means that the spin of the exchanged messenger particle can be directly probed by investigating the scaling of the cross section with either the rapidity difference or the invariant mass between the electron–positron pair.

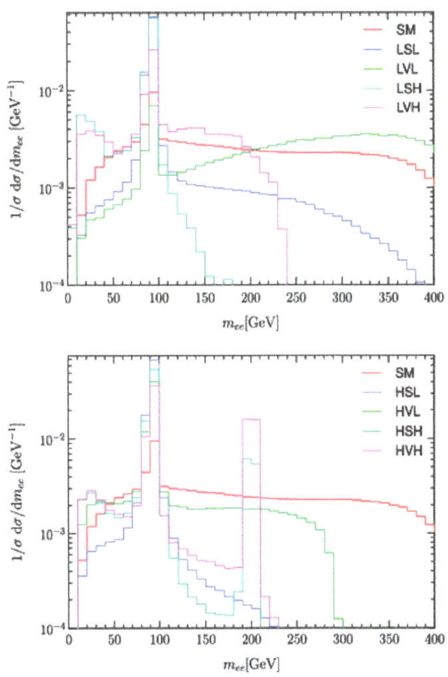

Figure 2: $1/\sigma\, d\sigma/dm_{ee}$ for the Standard Model, and the models with light or heavy dark matter candidates and (*top*) a light mediator, (*bottom*) a heavy mediator. See text for more details.

However, if the exchanged particle has a mass which is large compared to the other scales of the process, then there will be modifications to this simple picture. Other distributions would then be consulted to differentiate the spin and the mass simultaneously.

Figures 2 and 3 illustrate the different scaling of the cross section with the invariant mass of the electron-positron pair, m_{ee}, and the rapidity difference between the electron and the positron, Δy_{ee}, respectively, within our models. All shown differential cross sections are obtained from an evaluation of the full matrix elements. The predictions from the Regge analysis are observed: for a fixed setup of masses, the scalar exchange is suppressed at large m_{ee} and Δy_{ee} compared to the models with a vector mediator. While the impact of the heavy mediator mass is significant for a fixed mass of the dark matter candidate, the dominance at large Δy_{ee} (Figure. 3) of vector exchanges over scalar exchanges still holds as predicted by the Regge analysis. The same is true for $1/\sigma\ d\sigma/dm_{ee}$ (Figure. 2). For this distribution, two peaks in the spectrum are visible. In all cases, there is one at M_Z. This originates from diagrams which can be viewed as $Z\phi$ production, where the mediator ϕ is possibly off-shell and decays into $\chi\bar{\chi}$, and the Z subsequently decays into e^+e^-. The second peak, which is only visible for the combination of heavy mediator and heavy DSP, has its origin in diagrams where the mediator is produced on-shell together with $\chi\bar{\chi}$ and decays into e^+e^-. Such a process is in principle also possible for the other combinations, but for light DSP the heavy mediator predominantly decays into DSPs, leaving only a tiny branching fraction, and for light mediators the final-state cuts on electron and positron remove this contribution. For polarised beams, namely + polarisation of the electron beam and both possibilities for the positron beam, and vector mediators, the ratio S/B can reach 27 % (for the model LVL) at $\Delta y \geq 4$, which is where the vector like signal processes will peak.

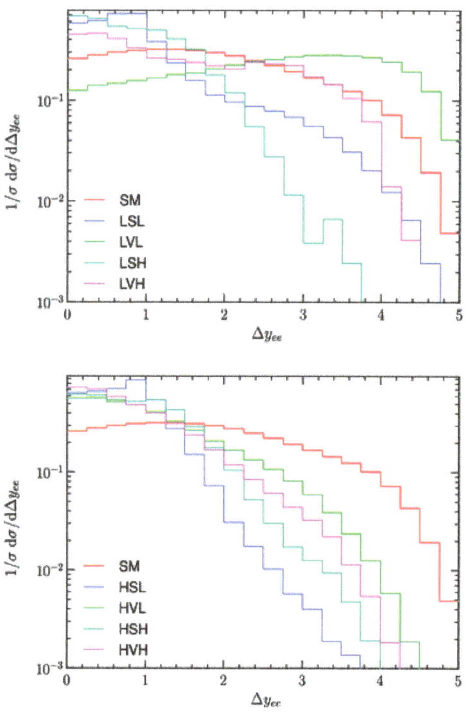

Figure 3: $1/\sigma\ d\sigma/d\Delta y_{ee}$ for the Standard Model, and the models with light or heavy dark matter candidates and (*top*) a light mediator, (*bottom*) a heavy mediator. See text for more details.

Conversely, scalar exchange models can get S/B enhanced by studying only regions of small Δy_{ee}.

In Figure. 4 we plot the normalised differential spectrum of the invariant mass of the invisible 4-momentum in the event. This is obviously bounded from below by the sum of the masses of the two dark matter particles. If this bound is below the mass of the mediator, then the spectrum has a pronounced peak at this mass for both scalar and vector mediators. The shape of the spectrum clearly identifies the mass-hierarchy of the mediator and DM particles: When the bound from the DM particles is above the mass of the mediator, the spectrum is very broad, as contrasted with the pronounced peak at the mediator mass.

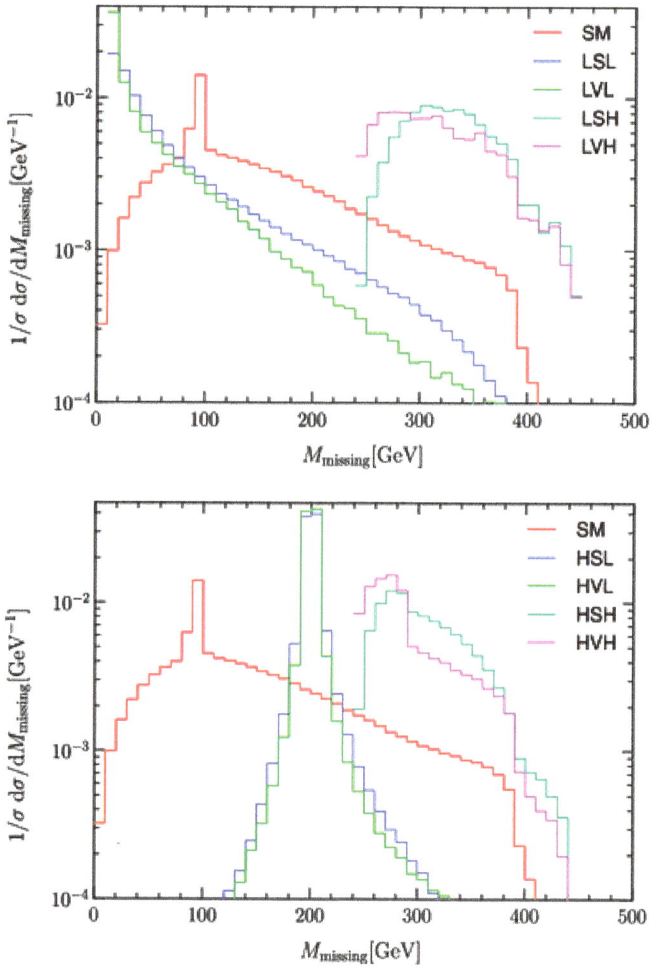

Figure 4: $1/\sigma\ d\sigma/dM_{missing}$ for the Standard Model, and the models with light or heavy dark matter candidates and (*top*) a light mediator, (*bottom*) a heavy mediator. See text for more details.

Let us now consider shortly the case of scalar DSPs. Only those models with scalar mediator exchange give viable scenarios, where there is sufficient cross section for our signal process after the constraints from mono-photon searches have been applied. Therefore we will consider only those. We use the same setup for masses and coupling strength as in Table 2 and denote the models with an additional "S" at the beginning. Corresponding distributions are depicted in Figure. 5. The upper part shows the rapidity difference between the final-state electron and positron and corresponds to Figure. 3 for fermionic DSPs. We see that there are only small differences between the CP-even scalar and fermion DSP case. This is also expected from Regge theory, as the spin of the DSP does not enter there. In the lower part of Figure. 5 we present the invariant mass of the invisible 4-momentum, corresponding to Figure. 4 for the fermionic case. Here we do observe small differences. Probably most striking is the SHSL model, which is, however, simply an effect of the different width of the mediator particle, which is predominantly produced on-shell. With large enough statistics, these differences could eventually be exploited for a distinction of scalar and fermionic dark matter. In the following, we will, however, concentrate on the much larger effects induced by the spin of the mediator and the different masses.

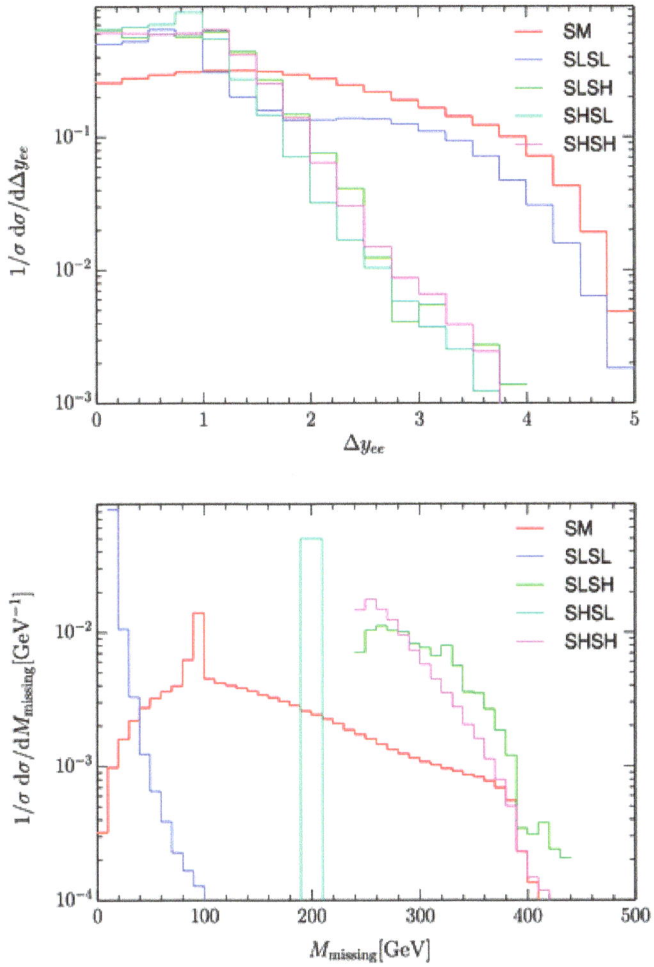

Figure 5: $1/\sigma\ d\sigma/d\Delta y_{ee}$ (*top*) and $1/\sigma\ d\sigma/dM_{missing}$ for scalar CP-even DSPs, scalar mediator particle and different mass scenarios of the two particles. See text for more details.

In case we studied scalar CP-odd DSPs, $\Delta\phi_{e^-e^+}$ can be helpful for their discrimination. For the eight models with fermionic DSPs we study here, $\Delta\phi_{e^-e^+}$is of minor importance.

In conclusion, the distribution of the invariant mass of the invisible momentum, $M_{missing}$ can uniquely determine the mass scale and hierarchies of the mediator and dark matter particle, but does not discriminate between scalar and vector mediators. However, once the mass scales are determined, the spin of the mediator can be determined from the shape of the cross section with respect to Δy_{ee} or equivalently m_{ee}, due to the spin-dependence dictated by the Regge analysis. Other distributions on e.g. the transverse momentum of the hardest lepton can then be used to check for consistency.

DISCRIMINATING MODELS

As discussed in Sect. 3, we expect the observables $m_{e^+e^-}$ and the invariant mass of the DSPs, precisely measured as recoil system of e^+e^-, to be the strongest discriminators for the benchmark models. At the ILC the lepton's energy is precisely determined and the beams can be partly polarised. We assume conservatively that the degree of polarisation is 80 %80 % for the electrons and 30 % for the positrons [43]. In Table 3 we show the cross section for three different beam polarisations: unpolarised, ++ and +−. The first (second) index refers to the electron (positron) beam. Only fairly inclusive cuts have been placed on the final-state particles, namely $p_{T,e^+,e^-} > 20$ GeV, $|\eta_{e^+,e^-} < 2.5|$, $R_{e^+e^-} > 0.4$

We find that the ratio between signal and background cross section can be improved for all models by using polarised beams.

Table 3: Cross sections in femtobarn for the different models imposing the constraints as outlined in the text. The three columns refer to the three different beam polarisations: averaged, ++ and +−. The first index refers to the electron beam which we assume to be able to polarise 80 % (always +). The second index represents the positron beam which we assume to have a polarisation degree of 30 %. The cross sections vary between (0.7–13.0, 1.3–26.1, 2.4–46.8 %) of the Standard Model background for three polarisations (all, ++, +−), respectively

Model	σ_{unpol}	σ_{++}	σ_{+-}
SM	115.8	49.1	36.4
LSL	1.60	1.79	1.40
LVL	15.07	12.80	17.02
LSH	1.45	1.80	1.10
LVH	9.99	7.64	12.33
HSL	1.17	1.43	0.92
HVL	0.85	0.71	0.89
HSH	1.18	1.45	0.90
HVH	0.85	0.64	0.98

Note, the signal models' cross sections respond differently to a change in the polarisation of the beams. Hence the inclusive production cross section can be used to discriminate between the models as well, i.e. for scalar mediators $\sigma_{++} > \sigma_{\mathrm{unpol}}$ while for vector mediators $\sigma_{+-} > \sigma_{\mathrm{unpol}}$. . However, in this work we will focus on the observables identified in Sect. 3 only and because of the recent interest in vector mediators [37–40] we will choose the +− beam polarisation in the following.

We perform a binned log-likelihood hypothesis test [44–47] using the CL_S method [48, 49] to evaluate how well the eight benchmark models can be discriminated from each other and from the Standard Model. For the graphs in Figure. 6 we assume only the Standard Model is realised in nature. Hence, the Standard Model is used as a background hypothesis in the CL_S analysis. Due to the large allowed cross section of the LVL and LVH models they can be excluded for the given coupling strength with less than 10 fb^{-1}. However, with an integrated luminosity of roughly 300 fb^{-1} all models can be disfavoured at the 5σ confidence level.

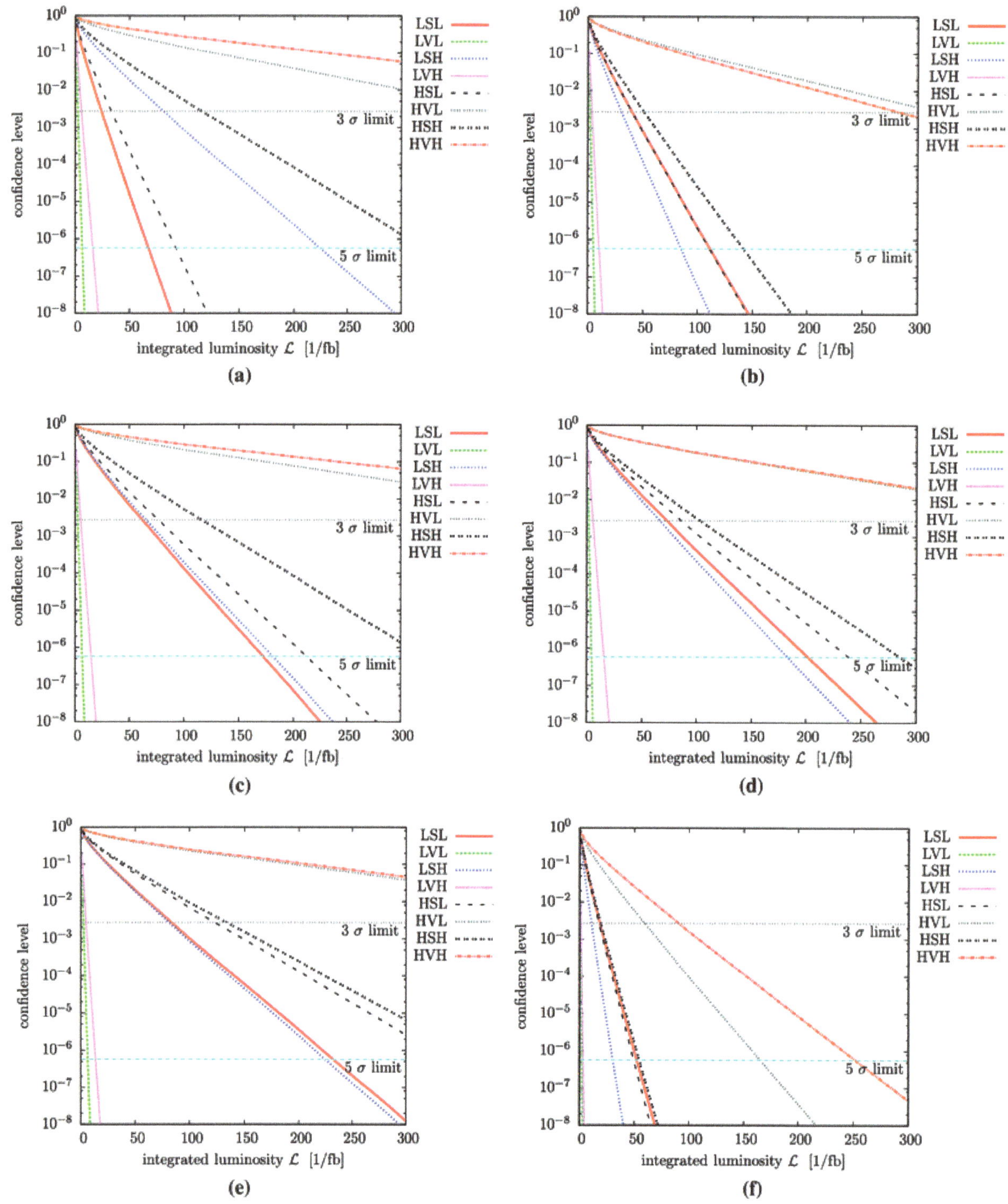

Figure 6: Assuming realisation of the Standard Model only, these plots show the confidence level to which the respective Dark Sector model can be disfavoured using just one observable. The *curves* take into account not only the shapes but also the normalisation of the model's cross section. **a** Missing transverse energy, **b** invariant mass of e^+e^- system, **c** $\Delta\phi_{e^+e^-}$, **d** $\Delta y_{e^+e^-}$, **e** p_T of hardest lepton, **f** invariant mass of DSP system, $M_{missing}$.

For the results shown in Figure. 7 we assume, respectively, that one of the models is realised in nature, as indicated by the caption of each sub-Figure. Thus, the distributions used for the CL_S method consist of the sum of the Standard Model and the respective benchmark model distribution, i.e. we take the statistical uncertainty of the Standard Model distributions into account for the hypothesis test, while we neglect interference effects between the Standard Model and the new physics scenarios. The coupling strength in each model is chosen such that it is respecting present bounds. However, as discussed in Sect. 2, bounds from direct and indirect detection experiments can be avoided in case the DSPs are not stable on cosmological time scales. Only bounds from direct searches at LEP pose a stringent constraint. To evaluate how well the different quantum numbers of the benchmark models can be discriminated we assume all benchmark models have a cross section of 2.5 % of the Standard Model cross section. All CL_S results are based on the distributions of Figures. 2, 3, 4 and 5.

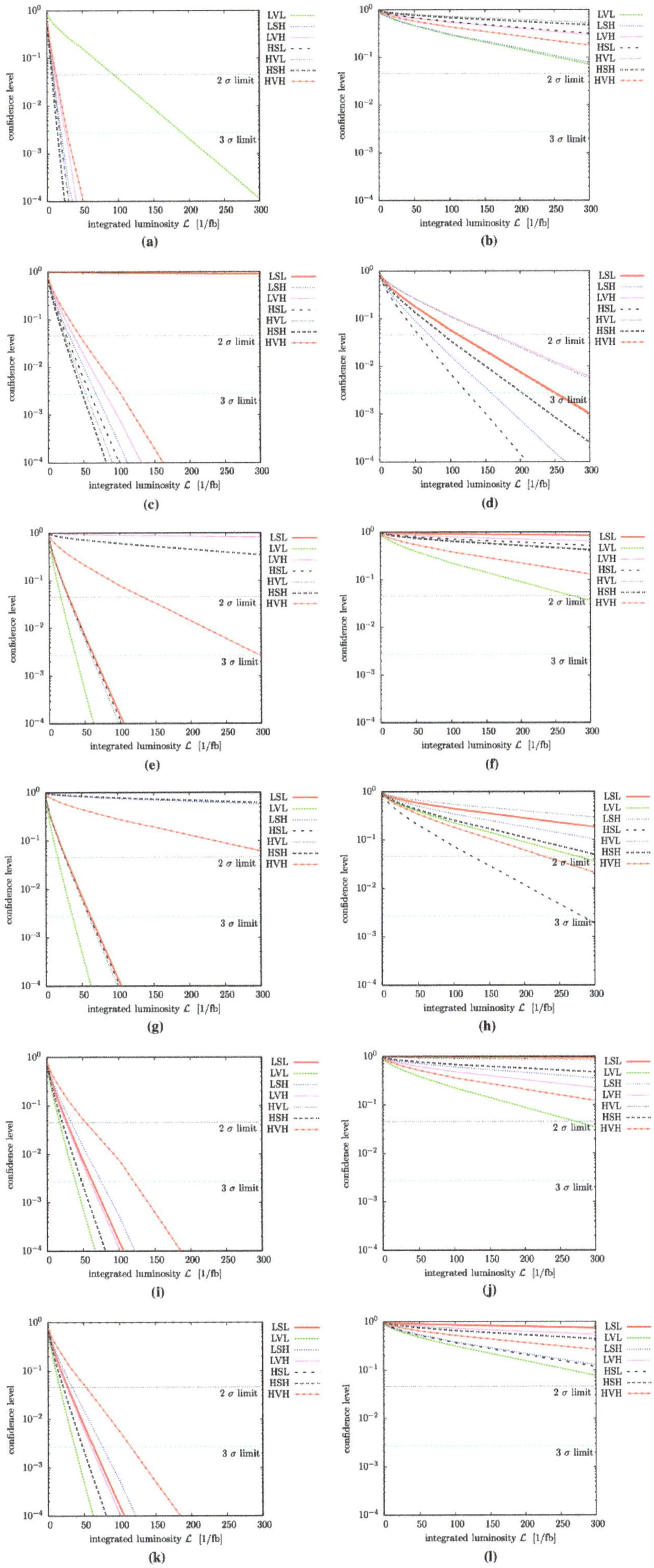
confidence level
integrated luminosity $\mathcal{L}$ [1/fb]
2 σ limit
3 σ limit
LVL
LSH
LVH
HSL
HVL
HSH
HVH
(a)
(b)
LSL
LSH
LVH
HSL
HVL
HSH
HVH
(c)
(d)
LSL
LVL
LVH
HSL
HVL
HSH
HVH
(e)
(f)
LSL
LVL
LSH
HSL
HVL
HSH
HVH
(g)
(h)
LSL
LVL
LSH
LVH
HVL
HSH
HVH
(i)
(j)
LSL
LVL
LSH
LVH
HSL
HSH
HVH
(k)
(l)

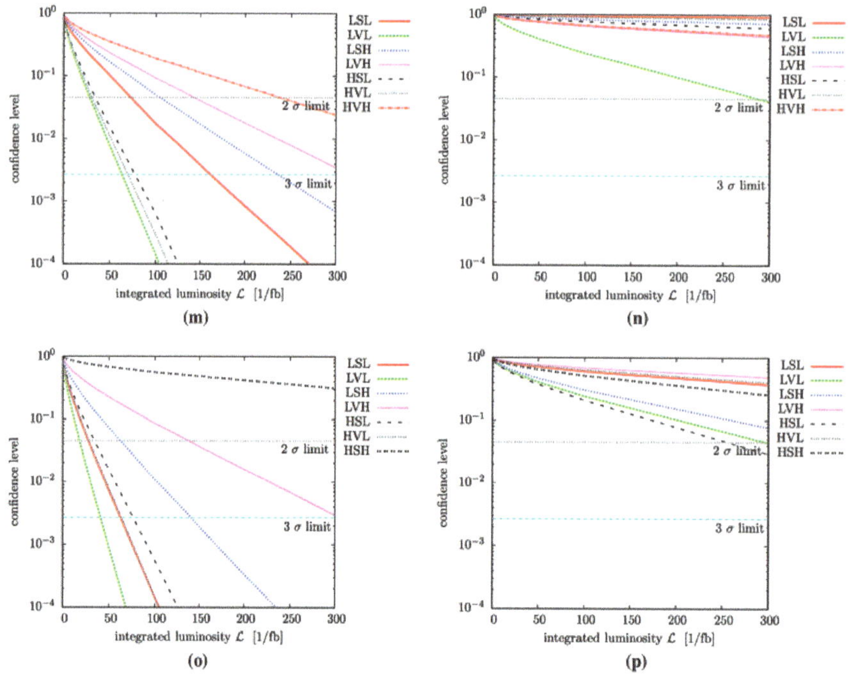

Figure 7: The curves illustrate the level to which the $M_{missing}$ and m_{e+e-} can discriminate between the different models, assuming the realisation of LSL (a–b), LVL (c–d), LSH (e–f), LVH (g–h), HSL (i–j), HVL (k–l), HSH (m–n) and HVH (o–p). All cross sections are assumed to be 2.5 % of the Standard Model background cross section. a $M_{missing}$ for LSL, b invariant mass of e^+e^- system for LSL, c $M_{missing}$ for LVL, d invariant mass of e^+e^- system for LVL, e $M_{missing}$ for LSH, f invariant mass of e^+e^- system for LSH, g $M_{missing}$ for LVH, h invariant mass of e^+e^- system for LVH,i $M_{missing}$ for HSL, j invariant mass of e^+e^- system for HSL,k Mmissing for HVL, l invariant mass of e^+e^- system for HVL, m $M_{missing}$ for HSH, n invariant mass of e^+e^- system for HSH,o $M_{missing}$ for HVH and p invariant mass of e^+e^- system for HVH.

We have studied the impact of all the observables shown in Figure. 6, namely missing transverse energy, $m_{e^+e^-}$, $\Delta\phi_{e^+e^-}$, $\Delta y_{e^+e^-}$, p_T of the hardest lepton and M_{missing}. We find that m_{e+e-} and $M_{missing}$ are sufficient to discriminate the quantum numbers of our candidate models confidently. $M_{missing}$ gives a handle on the mass scales, while m_{e+e-} discriminates the spin of the mediators. This is clearly demonstrated by comparing Figure. 7c and 7d, where LVL and LSL are strongly discriminated by including m_{e+e-} whereas $M_{missing}$ has almost no discriminating power. The situation is reversed for the discrimination of LSL and LSH, as shown in Figure. 7e and 7f. Here both, LSL and LSH have a pronounced peak in m_{e+e-} at the Z boson mass (thus m_{e+e-} is not discriminating the two models), but the large difference of the DSP mass is reflected in $M_{missing} \geq 2M_{DSP}$. However, if one wants to discriminate quantum numbers beyond mass and spin (e.g. CP structure) other observables should be included, e.g. $\Delta\phi_{e+e-}$. A combination of the variables discussed will improve the statistical significance in discriminating the models' quantum numbers.

In general and in particularly for a leptophilic scenario, studying a dark sector using *t*-channel mediated forces is a challenging task at the LHC. This study has demonstrated that even with conservative assumptions on the level of polarisation, the ILC can conclusively explore the quantum numbers of a dark sector by using a combination of $M_{missing}$ and m_{e+e-}.

ACKNOWLEDGMENTS

MR acknowledges partial support by the Deutsche Forschungsgemeinschaft via the Sonderforschungsbereich/Transregio SFB/TR-9 "Computational Particle Physics" and the Initiative and Networking Fund of the Helmholtz Association, contract HA-101 ("Physics at the Terascale").

REFERENCES

1. G. Bertone, D. Hooper, J. Silk, Phys. Rep. 405, 279 (2005). (hep-ph/0404175)
2. E. Komatsu et al., WMAP collaboration. Astrophys. J. Suppl. 192, 18 (2011) (arXiv:1001.4538 [astro-ph.CO])
3. L. Bergstrom, Ann. Phys. 524, 479 (2012). (arXiv:1205.4882 [astro-ph.HE])
4. P.A.R. Ade et al., Planck collaboration (arXiv:1303.5062 [astro-ph.CO])
5. P.A.R. Ade et al., Planck collaboration (arXiv:1303.5076 [astro-ph.CO])
6. P. Picozza, A.M. Galper, G. Castellini, O. Adriani, F. Altamura, M. Ambriola, G.C. Barbarino, A. Basili et al., Astropart. Phys. 27, 296 (2007)
7. O. Adriani et al., PAMELA collaboration. Nature 458, 607 (2009)
8. M. Ackermann et al., Fermi LAT collaboration. Phys. Rev. Lett. 108, 011103 (2012) (arXiv:1109.0521 [astro-ph.HE])
9. M. Aguilar et al., AMS collaboration. Phys. Rev. Lett. 110(14), 141102 (2013) (main results at http://www.ams02.org/)
10. P.J. Fox, E. Poppitz, Phys. Rev. D 79, 083528 (2009). (arXiv:0811.0399)
11. M. Cirelli, M. Kadastik, M. Raidal, A. Strumia, Nucl. Phys. B813, 1 (2009) (addendum-ibid. B 873, 530, 2013) (arXiv:0809.2409)
12. C.-R. Chen, F. Takahashi, JCAP 0902, 004 (2009). (arXiv:0810.4110)

13. M.J. Strassler, K.M. Zurek, Phys. Lett. B 651, 374 (2007). (hep-ph/0604261)
14. D.E. Kaplan, M.A. Luty, K.M. Zurek, Phys. Rev. D 79, 115016 (2009). (arXiv:0901.4117)
15. T. Cohen, K.M. Zurek, Phys. Rev. Lett. 104, 101301 (2010). (arXiv:0909.2035)
16. C. Savage, G. Gelmini, P. Gondolo, K. Freese, Phys. Rev. D 83, 055002 (2011)
17. J.M. Cline, Z. Liu, W. Xue, Phys. Rev. D 87, 015001 (2013)
18. A. Birkedal, K. Matchev, M. Perelstein, Phys. Rev. D70, 077701 (2004). (hep-ph/0403004)
19. J.L. Feng, S. Su, F. Takayama, Phys. Rev. Lett. 96, 151802 (2006). (hep-ph/0503117)
20. M. Beltran, D. Hooper, E.W. Kolb, Z.C. Krusberg, Phys. Rev. D 80, 043509 (2009). (arXiv:0808.3384)
21. Q.-H. Cao, C.-R. Chen, C.S. Li, H. Zhang, JHEP 1108, 018 (2011). (arXiv:0912.4511)
22. M. Beltran, D. Hooper, E.W. Kolb, Z.A.C. Krusberg, T.M.P. Tait, JHEP 1009, 037 (2010). (arXiv:1002.4137)
23. J. Goodman, M. Ibe, A. Rajaraman, W. Shepherd, T.M.P. Tait, H.-B. Yu, Phys. Lett. B695, 185–188 (2011). (arXiv:1005.1286)
24. Y. Bai, P.J. Fox, R. Harnik, JHEP 1012, 048 (2010). (arXiv:1005.3797)
25. P.J. Fox, R. Harnik, J. Kopp, Y. Tsai, Phys. Rev. D 84, 014028 (2011). (arXiv:1103.0240)
26. A. Rajaraman, W. Shepherd, T.M.P. Tait, A.M. Wijangco, Phys. Rev. D 84, 095013 (2011). (arXiv:1108.1196)
27. P.J. Fox, R. Harnik, J. Kopp, Y. Tsai, Phys. Rev. D 85, 056011 (2012)
28. H. Dreiner, M. Huck, M. Kramer, D. Schmeier, J. Tattersall, Phys. Rev. D 87, 075015 (2013). (arXiv:1211.2254)
29. ILC Technical Design Report, ILC collaboration. http://www.linearcollider.org/ILC/Publications/Technical-Design-Report. (arXiv:1306.6352)
30. G. Aarons et al., ILC collaboration (arXiv:0709.1893)
31. T. Plehn, D.L. Rainwater, D. Zeppenfeld, Phys. Rev. Lett. 88, 051801 (2002). (hep-ph/0105325)
32. G. Klamke, D. Zeppenfeld, JHEP 0704, 052 (2007). (hep-ph/0703202)
33. J.R. Andersen, K. Arnold, D. Zeppenfeld, JHEP 1006, 091 (2010). (arXiv:1001.3822)
34. O.J.P. Eboli, D. Zeppenfeld, Phys. Lett. B495, 147–154 (2000). (hep-ph/0009158)
35. H. An, L.-T. Wang, H. Zhang (arXiv:1308.0592)
36. P. Meade, M. Reece (hep-ph/0703031)
37. B. Holdom, Phys. Lett. B 166, 196 (1986)
38. R. Foot, X.-G. He, Phys. Lett. B 267, 509 (1991)
39. S.A. Abel, M.D. Goodsell, J. Jaeckel, V.V. Khoze, A. Ringwald, JHEP 0807, 124 (2008)
40. N. Arkani-Hamed, N. Weiner, JHEP 0812, 104 (2008)
41. T. Regge, Nuovo Cim. 14, 951 (1959)
42. R.C. Brower, C.E. DeTar, J.H. Weis, Phys. Rep. 14, 257 (1974)
43. M.E. Peskin (ed.) et al., Physics at the International Linear Collider, ILC Detailed Baseline Design Report (2013)
44. T. Junk, Nucl. Instrum. Method A 434, 435 (1999)
45. T. Junk, CDF Note 8128 (cdf/doc/statistics/public/8128)
46. T. Junk, CDF Note 7904 (cdf/doc/statistics/public/7904)
47. H. Hu, J. Nielsen, in 1st Workshop on Confidence Limits, CERN 2000–005 (2000)
48. A.L. Read, CERN-OPEN-2000-205
49. A.L. Read, J. Phys. G G28, 2693–2704 (2002)

INDEX